UNITEXT

# La Matematica per il 3+2

Volume 159

The **UNITEXT - La Matematica per il 3+2** series is designed for undergraduate and graduate academic courses, and also includes books addressed to PhD students in mathematics, presented at a sufficiently general and advanced level so that the student or scholar interested in a more specific theme would get the necessary background to explore it.

Originally released in Italian, the series now publishes textbooks in English addressed to students in mathematics worldwide.

Some of the most successful books in the series have evolved through several editions, adapting to the evolution of teaching curricula.

Submissions must include at least 3 sample chapters, a table of contents, and a preface outlining the aims and scope of the book, how the book fits in with the current literature, and which courses the book is suitable for.

For any further information, please contact the Editor at Springer: francesca.bonadei@springer.com

**THE SERIES IS INDEXED IN SCOPUS**

***

UNITEXT is glad to announce a new series of free webinars and interviews handled by the Board members, who rotate in order to interview top experts in their field.

Access this link to subscribe to the events: https://cassyni.com/s/springer-unitext

Andrea Bandini • Patrizia Gianni • Enrico Sbarra

# Commutative Algebra through Exercises

Springer

Andrea Bandini
Department of Mathematics
University of Pisa
Pisa, Italy

Patrizia Gianni
Department of Mathematics
University of Pisa
Pisa, Italy

Enrico Sbarra
Department of Mathematics
University of Pisa
Pisa, Italy

ISSN 2038-5714 ISSN 2532-3318 (electronic)
UNITEXT
ISSN 2038-5722 ISSN 2038-5757 (electronic)
La Matematica per il 3+2
ISBN 978-3-031-56909-8 ISBN 978-3-031-56910-4 (eBook)
https://doi.org/10.1007/978-3-031-56910-4

Translation from the Italian language edition: "Esercizi di algebra commutativa" by Andrea Bandini et al., © Pisa University Press 2022. Published by Pisa University Press. All Rights Reserved.



This Springer imprint is published by the registered company Springer Nature Switzerland AG
The registered company address is: Gewerbestrasse 11, 6330 Cham, Switzerland

# Preface

This book contains a carefully selected collection of exercises on commutative algebra, based on the course of Algebra 2 at the University of Pisa that we have been teaching for the last ten years. This course is intended for students in the third year of the bachelor's degree or the first year of the master's degree in Mathematics, and provides an introduction to the fundamental concepts of commutative algebra, including rings, ideals, modules, tensor product, localization, Noetherian rings, and primary decomposition. A few prerequisites are needed, namely a basic knowledge of elementary number theory, Abelian groups, and linear algebra.

We have included more than 400 exercises of varying levels of difficulty to introduce basic concepts in an intuitive and engaging manner and to facilitate gradual and motivated learning of topics that are often abstract and challenging. Some of these exercises are exam questions, some were created to supplement lectures, and others are derived from commonly used textbooks, which are cited in the bibliography. We believe that exercises are crucial for a genuine understanding of the theory, and as such, we have included both theoretical and computational ones. These are meant to help the reader apply abstract concepts to concrete situations, construct examples, and clarify the concepts introduced in the book. Detailed solutions accompany both theoretical and computational exercises.

Most books on commutative algebra tend to favor a theoretical approach, whereas our approach is practical and constructive. We aim to illustrate fundamental theoretical concepts by applying them to the study of multivariate polynomials rings and modules over a principal ideal domain (PID), whenever possible. Algorithms for the manipulation of these objects support the construction of many examples; thus, we also present some computational methods. The structure of finitely generated modules over a PID is described using the Smith canonical form of matrices. The ability to compute this form makes the structure of a finitely generated module evident, emphasizing the significance of canonical forms in a concrete application. Equally meaningful are the examples that can be constructed with multivariate polynomials. In this context, we present Buchberger's Algorithm for computing Gröbner bases, which have natural applications in the study of ideals, varieties, and in solving systems of polynomial equations. Nowadays, Gröbner bases are widely applied in diverse areas, such as cryptography, coding theory, computer vision, signal and image processing, and robotics, to name a few. We believe knowledge of these methods should be an integral part of the cultural background of a mathematics graduate.

The book is organized into three parts. The first part includes key definitions, properties, and results that are essential for understanding and completing the exercises. It has been designed as a reference text, and to make it quick and easy to use, we have only included the most instructive proofs that present essential techniques and methods. All remaining proofs have been included in a separate chapter at the end of the book. Our aim is to stimulate curiosity and more thoughtful and active learning, and we encourage the readers to find proofs on their own, providing at the same time detailed solutions to support them.

The second part of the book includes exercises organized by chapter corresponding to the theory. Additionally, there is a "True or False?" section without any topic categorization, which serves as a tool for testing the level of learning and comprehension achieved. Finally, there is a section of review exercises.

The third and final part of the book contains the proofs of the theoretical results we have chosen to postpone, and the solutions to all the exercises. We have invested a lot of time and effort in preparing the solutions with the appropriate level of detail in order to guide the student gradually in the challenging task of writing formal, correct, and complete proofs in mathematics. However, we would like to emphasize that the proposed solutions are not the only possible ones, and exploring alternatives can be an additional source of learning.

Overall, the book aims at inspiring the curiosity of the readers and encouraging them to find their own proofs while providing detailed solutions to support their learning. It also provides students with the necessary tools to pursue more advanced studies in commutative algebra and related subjects.

Pisa, February 1st, 2024

Andrea Bandini
Patrizia Gianni
Enrico Sbarra

# Notation

In this book the sets of natural, integer, rational, real, and complex numbers are denoted by $\mathbb{N}$, $\mathbb{Z}$, $\mathbb{Q}$, $\mathbb{R}$, and $\mathbb{C}$, respectively. We use the notation $\mathbb{N}_+$ for $\mathbb{N} \setminus \{0\}$. The set of cosets of $\mathbb{Z}$ modulo $n$ is denoted by $\mathbb{Z}/(n)$.

The symbols $\bigcap$, $\bigcup$, and $\bigsqcup$ denote - as usual - the intersection, the union, and the disjoint union, respectively. The symbols $\subseteq$ and $\subset$ are both used to denote a containment, while $\subsetneq$ is used for the strict containment. The symbols $\equiv$ and $\simeq$ denote congruences and isomorphisms, respectively.

For any ring $A$ and any positive integers $r$ and $s$, we denote by $M_r(A)$, respectively by $M_{rs}(A)$, the $r \times r$ matrices, respectively $r \times s$ matrices, with coefficients in $A$. Finally, for any matrix $B$, we denote by $B^{\mathrm{t}}$ its transpose.

For any ring $A$, the ring of polynomials and the ring of formal power series in the variable, or indeterminate, $x$ with coefficients in $A$ are denoted by $A[x]$ and $A[[x]]$, respectively. Similarly $A[x_1, \ldots, x_n]$ denotes the ring of polynomials in the variables $x_1, \ldots, x_n$ with coefficients in $A$. If $A$ is a domain, $Q(A)$ is the total quotient ring of $A$. The letter $K$ always denotes a field and $\overline{K}$ its algebraic closure. For any field $K$, we denote by $K(x_1, \ldots, x_n)$ the smallest field containing $K$ and $x_1, \ldots, x_n$.

**Navigating Through This Book**

The initial section of this book serves as a comprehensive reference. We have chosen not to make distinctions between Theorems, Propositions, and similar classifications. All these results are uniformly labeled as **T** for simplicity. In cases where the proof is deferred, readers are encouraged to treat it as an exercise, fostering active engagement and problem-solving.

The second part of the book focuses on exercises, which are marked by the symbol **E**. This section offers a diverse range of problem-solving opportunities aimed at solidifying the covered concepts. Exercises presented in a True or False format are specifically labeled as **ToF**, creating a distinct category within the exercise set.

# Contents

## Part I Theory

# Part I
# Theory

# Chapter 1
# Rings

In this chapter, we provide an introduction to the general theory of commutative rings with identity. We begin by reviewing fundamental concepts and properties related to such rings and their elements. We then proceed to discuss ring homomorphisms and homomorphism theorems, ideals and various operations commonly associated with them. It is worth noting that many of the definitions and properties of commutative rings were initially studied and developed for the ring of integers $\mathbb{Z}$ and the ring of polynomials $K[x]$ with coefficients in a field $K$.

## 1.1 Rings and Ideals

A *ring* $(A, +, \cdot)$ is an Abelian group $(A, +)$ with a product operation

$$A \times A \stackrel{\cdot}{\longrightarrow} A, \quad (a, b) \mapsto ab,$$

such that for each $a$, $b$, $c \in A$:

i) $(ab)c = a(bc)$;
ii) $(a + b)c = ac + bc$;
iii) $a(b + c) = ab + ac$.

A ring is *commutative* if its product is commutative. A ring is called *unitary* or *with identity* if there exists an identity element for the product, denoted by 1 or $1_A$. If such an element exists, then it is easy to verify that it is unique. We remark that it is possible to have $1 = 0$, but in that case $A = 0$, because $a = 1 \cdot a = 0 \cdot a = 0$ for every $a \in A$.

For example, let $n$ be a positive integer, and let $K$ be a field, *e.g.*, $\mathbb{Q}$, $\mathbb{R}$ or $\mathbb{C}$. The sets $\mathbb{Z}$, $\mathbb{Z}/(n)$, $K[x]$, $K[x_1, \ldots, x_n]$, and $K[x_n : n \in \mathbb{N}]$ with the usual sum and product operations, are commutative rings with identity.



A. Bandini et al., *Commutative Algebra through Exercises*, UNITEXT 159,
https://doi.org/10.1007/978-3-031-56910-4_1

When $n > 1$, the set $M_n(K)$ of square matrices of dimension $n$ with coefficients in $K$ is a ring with identity that is not commutative, and the group $n\mathbb{Z}$ is a commutative ring without identity.

Throughout this book all rings are assumed to be commutative with identity and different from 0, unless otherwise specified.

A subset $B \subseteq A$ is a *subring* of $A$ if $B$ is an additive subgroup of $A$, closed with respect to the product of $A$, and with identity element $1_B = 1_A$.
It is easy to verify that the intersection of any family $\{B_h\}_{h\in H}$ of subrings of $A$ is a subring of $A$.

**Special Elements of a Ring *A***

Let $a \in A$.

$a$ is **invertible** or a *unit* if there exists $b \in A$ such that $ab = 1$.
The set of invertible elements of $A$ is denoted by $A^*$.

$a$ is a **zero-divisor** if there exists $b \in A$, $b \neq 0$, such that $ab = 0$.
The set of zero-divisors of $A$ is denoted by $\mathcal{D}(A)$.

$a$ is **nilpotent** if there exists $n \in \mathbb{N}$ such that $a^n = 0$.
The set of nilpotent elements of $A$ is denoted by $\mathcal{N}(A)$.

$a$ is **idempotent** if $a^2 = a$.

From the definitions, it immediately follows that $\mathcal{N}(A) \subseteq \mathcal{D}(A)$.
A ring $A$ in which all non-zero elements are invertible, *i.e.*, $A^* = A \setminus \{0\}$, is a *field*. If $\mathcal{D}(A) = \{0\}$, then the ring $A$ is called an *integral domain* or simply a *domain*.
The set $\mathcal{N}(A)$ is the *nilradical* of $A$. A ring $A$ such that $\mathcal{N}(A) = \{0\}$ is said to be *reduced*.
A ring $A$ is *Boolean* if all its elements are idempotent.
Finally, recall that if $A \neq 0$, then $\mathcal{D}(A) \cap A^* = \emptyset$.

As an example, consider the ring $A = \mathbb{Z}/(12) = \{\overline{0}, \dots, \overline{11}\}$.
In this case, we have $A^* = \{\overline{1}, \overline{5}, \overline{7}, \overline{11}\}$, $\mathcal{D}(A) = \{\overline{0}, \overline{2}, \overline{3}, \overline{4}, \overline{6}, \overline{8}, \overline{9}, \overline{10}\}$, and $\mathcal{N}(A) = \{\overline{0}, \overline{6}\}$. Additionally, the idempotents are $\{\overline{0}, \overline{1}, \overline{4}, \overline{9}\}$.
Note that $\mathcal{D}(A) \cap A^* = \emptyset$, $A = A^* \sqcup \mathcal{D}(A)$, and $\mathcal{N}(A) \subsetneq \mathcal{D}(A)$.

**T. 1.1.** ($\rightarrow$ p. 207) Let $A$ be a finite ring. Then, $A = A^* \sqcup \mathcal{D}(A)$.

In particular, a finite domain is a field.

A subset $I \subseteq A$ is an *ideal* of $A$ if:

i) $(I, +)$ is a subgroup of $A$, *i.e.*, $0 \in I$ and if $a$, $b \in I$, then $a - b \in I$;
ii) for every $a \in A$ and $i \in I$ we have $ai \in I$.

An ideal is *proper* if it is a proper subset of $A$. It follows directly from the definition that an ideal $I$ is proper if and only if $1 \notin I$, *i.e.*, if and only if $A^* \cap I = \emptyset$. A proper ideal of $A$ is *maximal* if it is not strictly contained in any other proper ideal of $A$.

Let $A$ be a ring, and let $S \subset A$ be a non-empty subset. The set

$$(S) = \left\{ \sum_{i=1}^{n} a_i s_i \colon a_i \in A,\ s_i \in S \ \text{ for some } n \in \mathbb{N} \right\}$$

consisting of all finite linear combinations of elements of $S$ with coefficients in $A$ is an ideal, which we refer to as *the ideal generated* by $S$. It is the smallest ideal of $A$ containing $S$. The elements of $S$ are called *generators* of $(S)$. If an ideal $I$ is generated by a finite set $S = \{s_1, \ldots, s_n\}$, then $I$ is said to be *finitely generated*, and we simply write $I = (s_1, \ldots, s_n)$. In the special case where $S = \{s\}$, the ideal $(s)$ is called *principal.*
A ring $A$ is referred to as *Principal Ideal Ring* (PIR) if all of its ideals are principal. If $A$ is also a domain, we call $A$ a *Principal Ideal Domain* (PID).

A ring $A$ always contains the ideals $(0)$ and $(1) = A$. If $A$ is not a field, then $A$ also contains at least another ideal, as proved in **T.1.2** below. This is a consequence of Zorn's Lemma, which we introduce next.
A *poset* $(\Sigma, \leq)$ is a non-empty partially ordered set with an order relation $\leq$ which is reflexive, transitive, and antisymmetric. A *chain* of $\Sigma$ is a totally ordered subset of $\Sigma$.

**Zorn's Lemma**

Let $(\Sigma, \leq)$ be a poset such that every chain has an upper bound in $\Sigma$. Then, $\Sigma$ contains maximal elements with respect to $\leq$.

**T. 1.2.** Let $A \neq 0$ be a ring. Then:

1. $A$ has at least one maximal ideal $\mathfrak{m}$;
2. for any proper ideal $I$ of $A$ there exists a maximal ideal $\mathfrak{m} \supseteq I$;
3. every non-invertible element of $A$ is contained in a maximal ideal.

**Proof T. 1.2.** 1. Consider the set of all proper ideals of $A$

$$\Sigma = \{I \subsetneq A \colon I \text{ ideal of } A\},$$

partially ordered by $\subseteq$. Since $(0) \in \Sigma$, we have $\Sigma \neq \emptyset$. Consider a chain $\mathcal{C} = \{I_h \colon h \in H\}$ of elements of $\Sigma$. We claim that $I = \bigcup_{h \in H} I_h$ is an upper bound in $\Sigma$ for $\mathcal{C}$.
To prove that $I$ is an ideal of $A$, we take $a, b \in I$. There exist $h, k \in H$ such that $a \in I_h$ and $b \in I_k$. Since $\mathcal{C}$ is totally ordered, we can assume without loss of generality, that $I_h \subseteq I_k$, and hence, $a, b \in I_k$. This implies that $a - b \in I_k \subseteq I$.
Furthermore, if $c \in A$ and $a \in I$, then $a \in I_k$ for some $k$ and $ca \in I_k \subseteq I$. Finally, $I \subsetneq A$ since $1 \notin I_h$ for each $h$. This shows that every chain has an upper bound in $\Sigma$, and hence, by Zorn's Lemma, there exists at least one maximal element in $\Sigma$, *i.e.*, a maximal ideal of $A$.

2. It is sufficient to modify the previous proof to use the set

$$\Sigma_I = \{J \subsetneq A \colon J \text{ ideal of } A,\ I \subseteq J\}.$$

3. If $a \in A$ is not invertible, then $(a) \neq (1)$ is a proper ideal. The conclusion follows from part 2. □

**Ideal Operations**

**Intersection.** For any family $\{I_h\}_{h\in H}$ of ideals of $A$, the intersection $\bigcap_{h\in H} I_h$ is an ideal of $A$.

**Sum.** For any family $\{I_h\}_{h\in H}$ of ideals of $A$, the ideal generated by the set $\bigcup_{h\in H} I_h$ is called the *sum* of the ideals $I_h$. It is denoted by $\sum_{h\in H} I_h$. In particular, for any $I$, $J$ ideals of $A$

$$I + J = \{i + j \colon i \in I,\ j \in J\}.$$

**Product.** For any <u>finite</u> family $I_1, \ldots, I_k$ of ideals of $A$, the ideal generated by all the products $i_1 \cdots i_k$, with $i_j \in I_j$ for all $j$, is called the *product* of the ideals $I_1, \ldots, I_k$. It is denoted by $\prod_{i=1}^{k} I_i$.
In particular, when $I_i = I$ for all $i$, we can consider the powers $I^k$ of $I$, for any $k \in \mathbb{N}$, by letting $I^0 = A$.

**Quotient and Annihilator.** For any $I$ and $J$ ideals of $A$, we define the *quotient* or *colon* of $I$ by $J$ as the set

$$I : J = \{a \in A \colon aJ \subseteq I\}.$$

In particular, when $I = (0)$, the quotient $(0) : J$ is called the *annihilator* of $J$, and is also denoted by $\operatorname{Ann}_A J$ or simply by $\operatorname{Ann} J$. When $J = (a)$, we simply denote it by $\operatorname{Ann} a$.

**Radical.** For any ideal $I$ of $A$, we define the *radical* of $I$, denoted by $\sqrt{I}$, as the set $\{a \in A \colon a^n \in I \text{ for some } n \in \mathbb{N}\}$.

Note that for any ideal $I$, we have $I \supseteq I^2 \supseteq I^3 \supseteq \ldots \supseteq I^h \supseteq \ldots$.
It is possible to have $I = I^2$, for example $(\overline{2}) = (\overline{4})$ in $\mathbb{Z}/(6)$.

**T. 1.3.** ($\rightarrow$ p. 207) Let $I$, $J$ be ideals of a ring $A$. Then, the quotient $I : J$ and the radical $\sqrt{I}$ are ideals.

In particular, the set of nilpotent elements of a ring $\mathcal{N}(A) = \sqrt{(0)}$ is an ideal. In general, the set $\mathcal{D}(A)$ is not an ideal.

For example consider $\overline{2}, \overline{3} \in \mathbb{Z}/(6)$. Both elements belong to $\mathcal{D}(\mathbb{Z}/(6))$, but their sum is in $(\mathbb{Z}/(6))^*$.

The definitions immediately yield that if $I$ and $J$ are ideals of a ring $A$, then $IJ \subseteq I \cap J$. In general, this inclusion is strict, however it holds

$$I + J = (1) \implies IJ = I \cap J.$$

Indeed, in this case, there exist $i \in I$ and $j \in J$ such that $i + j = 1$. Hence, for each element $a \in I \cap J$, we have

$$a = 1 \cdot a = (i + j)a = ia + ja \in IJ.$$

If two ideals $I$ and $J$ satisfy the condition $I + J = (1)$, we say that $I$ and $J$ are *comaximal.*

**T. 1.4.** (→ p. 207) Let $I_1, \ldots, I_n$ be pairwise comaximal ideals of $A$. Then, $\bigcap_{i=1}^n I_i = \prod_{i=1}^n I_i$.

**T. 1.5.** (→ p. 208) Let $I$, $J$, and $H$ be ideals of a ring $A$. Then:

1. $(I + J)H = IH + JH$;
2. $(I + J)(I \cap J) \subseteq IJ$;
3. $I \cap (J + H) \supseteq (I \cap J) + (I \cap H)$;
4. [**Modular Law**] if $I \supseteq J$ or $I \supseteq H$, then
$$I \cap (J + H) = (I \cap J) + (I \cap H);$$
5. $I + JH \subseteq (I + J) \cap (I + H)$.

The equality in part 5 does not hold in general. For instance, take $A = \mathbb{Z}$, $I = (2^2)$ and $J = H = (2)$.
Another counterexample can be obtained by taking $A = K[x, y]$, where $K$ is a field, $I = (x + y)$, $J = (x)$, and $H = (y)$. This also provides a counterexample to the equality in part 3, since $I \cap (J + H) = I \cap (x, y) = I$ whereas $(I \cap J) + (I \cap H) = (x^2 + xy) + (xy + y^2) \neq I$.
Is it possible to find a counterexample to the equality in part 3 with $A = \mathbb{Z}$?

**Properties of the Radical**

**T. 1.6.** (→ p. 208) Let $A$ be a ring, and let $I$, $J$, and $H$ be ideals of $A$. Then:

1. if $I \subseteq J$, then $\sqrt{I} \subseteq \sqrt{J}$;
2. $I \subseteq \sqrt{I}$ and $\sqrt{\sqrt{I}} = \sqrt{I}$;
3. $\sqrt{IJ} = \sqrt{I \cap J} = \sqrt{I} \cap \sqrt{J}$;
4. $\sqrt{I^n} = \sqrt{I}$ for any positive integer $n$;
5. $\sqrt{I} = (1)$ if and only if $I = (1)$;
6. $\sqrt{I + J} = \sqrt{\sqrt{I} + \sqrt{J}}$;
7. $\sqrt{I + JH} = \sqrt{I + J} \cap \sqrt{I + H}$.

As we have seen in **T.1.2**, all rings $A \neq 0$ have maximal ideals. We will now define other types of ideals which will be important in the following.

**Special Ideals 1**

Let $A$ be a ring, and let $I$, $I_1$, $I_2 \subset A$ be proper ideals. Then, $I$ is

**prime**, if $ab \in I$ implies $a \in I$ or $b \in I$;
**radical**, if $I = \sqrt{I}$;
**primary**, if $ab \in I$ implies $a \in I$ or $b \in \sqrt{I}$;
**irreducible**, if $I = I_1 \cap I_2$ implies $I = I_1$ or $I = I_2$.

The set of prime, respectively maximal, ideals of a ring $A$ is denoted by $\operatorname{Spec} A$, respectively $\operatorname{Max} A$.

**T. 1.7.** ($\rightarrow$ p. 209) Let $I$ be a proper ideal of $A$.

1. If $I$ is primary, then $\sqrt{I}$ is prime.
2. If $\sqrt{I} = \mathfrak{m}$ is maximal, then $I$ is primary.

**Krull Dimension**

Let $A$ be a ring. The *Krull dimension* of $A$ is defined as

$$\sup\{k\colon \text{ there exists } \mathfrak{p}_0 \subsetneq \mathfrak{p}_1 \subsetneq \ldots \subsetneq \mathfrak{p}_k, \text{ with } \mathfrak{p}_i \in \operatorname{Spec} A \text{ for all } i\}.$$

We denote the Krull dimension of $A$ by $\dim A$.

The definition immediately yields that $\dim A = 0$, when $A$ is a field, and that $\dim A = 1$, when $A$ is a PID but not a field, see **T.1.22**.5, **T.1.23**, and **E.8.57**.2. Moreover, if $A = K[x_1, \ldots, x_n]$, then we can consider the chain $(0) \subsetneq (x_1) \subsetneq (x_1, x_2) \subsetneq \ldots \subsetneq (x_1, \ldots, x_n)$. Since all ideals $(x_1, \ldots, x_i)$ are prime, see **T.1.11**.3, we obtain $\dim A \geq n$. Proving that $\dim A = n$ is significantly more challenging. We do not include a proof in this text, since we do not cover Dimension Theory.

## 1.2 Homomorphisms and Quotient Rings

Let $A$ be a ring. Every ideal $I$ of $A$ is a normal subgroup of $A$, thus we can consider the quotient group $A/I$. For any element $a \in A$, we denote by $\overline{a}$ the coset, or equivalence class, of $a$ modulo $I$, that is, $a + I$. Hence $\overline{a} = \overline{b}$ if and only if $a - b \in I$, *i.e.*, if and only if $a \equiv b \mod I$. We define a ring structure on $A/I$ by setting $\overline{a}+\overline{b}=\overline{a+b}$ and $\overline{a}\overline{b}=\overline{ab}$. The ring $A/I$ is called the *quotient ring* of $A$ modulo $I$.

**T. 1.8.** ($\rightarrow$ p. 209) The product in $A/I$ described above is well-defined and satisfies the required properties.

Let $A$ and $B$ be rings. A map $f\colon A \longrightarrow B$ is a *ring homomorphism* if for all $a,\ b \in A$:

i) $f(a+b)=f(a)+f(b)$;
ii) $f(ab)=f(a)f(b)$;
iii) $f(1_A)=1_B$.

It follows from the definition that $f=0$ is not a ring homomorphism, unless $B=0$.
For any ring homomorphism $f$, the *kernel* of $f$ is

$$\operatorname{Ker} f=\{a \in A\colon f(a)=0\}=f^{-1}((0)),$$

and the *image* of $f$ is

$$\operatorname{Im} f=\{b \in B\colon b=f(a) \text{ for some } a \in A\}=f(A).$$

The definition of ring homomorphism immediately implies that:

i) $\operatorname{Ker} f$ is an ideal of $A$;
ii) $f$ is injective if and only if $\operatorname{Ker} f=(0)$;
iii) $\operatorname{Im} f$ is a subring of $B$,

see **E.8.8**. More generally, given a ring homomorphism $f\colon A \longrightarrow B$ and ideals $I \subset A,\ J \subset B$, we have:

i) $f^{-1}(J)=\{a \in A\colon f(a) \in J\}$ is an ideal of $A$;
ii) if $f$ is surjective, then $f(I)$ is an ideal of $B$,

see Section 1.4.

A bijective ring homomorphism is called *ring isomorphism*. If there exists an isomorphism $f\colon A \longrightarrow B$, then $A$ and $B$ are *isomorphic* and we write $A \simeq B$. The composition of homomorphisms is a homomorphism and the inverse of an isomorphism is an isomorphism, therefore $\simeq$ defines an equivalence relation on the set of rings.

Let $f\colon A \longrightarrow B$ be a ring homomorphism and $I \subset A$ an ideal of $A$.
If $I \subseteq \operatorname{Ker} f$, then $f$ induces a unique homomorphism $\overline{f}\colon A/I \longrightarrow B$ such that $f=\tilde{f} \circ \pi$, *i.e.*, the following diagram commutes:

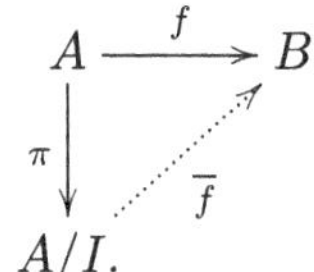

In the above diagram, the map $\pi\colon A \longrightarrow A/I$ denotes the homomorphism defined by $a \mapsto \overline{a}$, *i.e.*, the *canonical projection* of $A$ onto $A/I$, which is clearly surjective.
In order for the diagram to be commutative, we set $\overline{f}(\overline{a}) = f(a)$.
In this way, $\overline{f}$ is well-defined. Indeed, if $\overline{a} = \overline{b}$, then $a - b \in I \subseteq \operatorname{Ker} f$. Thus, $0 = f(a-b) = f(a) - f(b)$, which implies $\overline{f}(\overline{a}) = \overline{f}(\overline{b})$.
Furthermore, since $f$ is a homomorphism, $\overline{f}$ is also a homomorphism.

**Ring Homomorphism Theorems**

**T. 1.9.** (→ p. 209) Let $A$ and $B$ be rings.

1. A homomorphism $f\colon A \longrightarrow B$ induces an isomorphism
$$A/\operatorname{Ker} f \simeq \operatorname{Im} f.$$
2. Let $I$, $J \subset A$ be ideals with $I \subset J$. Then, $J/I$ is an ideal of $A/I$ and
$$(A/I)/(J/I) \simeq A/J.$$
3. Let $I \subset A$ be an ideal, and let $B \subset A$ be a subring. Then:
   a. $B + I = \{b + i\colon b \in B,\, i \in I\} \subset A$ is a subring;
   b. $I$ is an ideal of $B + I$;
   c. $B \cap I$ is an ideal of $B$.

   Moreover,
$$(B+I)/I \simeq B/(B \cap I).$$

**T. 1.10.** (→ p. 210) There is a one-to-one correspondence between ideals of $A$ containing $I$ and ideals of $A/I$. Moreover, via this bijection, prime ideals correspond to prime ideals and maximal ideals to maximal ideals.

In particular, if all primes containing an ideal $I$ are maximal, then $\dim A/I = 0$. In this case we say that $I$ is 0-*dimensional.*

Moreover, many of the special ideals previously defined can be characterized in terms of their quotient rings.

**Special Ideals 2**

**T. 1.11.** (→ p. 210) Let $I$ be an ideal of a ring $A$. Then:

1. $I$ is proper if and only if $A/I \neq 0$;
2. $I$ is maximal if and only if $A/I$ is a field;
3. $I$ is prime if and only if $A/I$ is a domain;

4. $I$ is radical if and only if $A/I$ is reduced;
5. $I$ is primary if and only if $\mathcal{N}(A/I) = \mathcal{D}(A/I)$.

Moreover,

6. $I$ maximal $\Longrightarrow I$ prime;
7. $I$ prime $\Longrightarrow I$ radical;
8. $I$ prime $\Longrightarrow I$ primary.

Intersections and unions of prime ideals have the following properties, although the union of ideals is not necessarily an ideal.

**T. 1.12.** 1. [**Prime Avoidance Lemma**] Let $\mathfrak{p}_1, \dots, \mathfrak{p}_n$ be prime ideals, and let $I$ be an ideal such that $I \subseteq \bigcup_{j=1}^n \mathfrak{p}_j$. Then, there exists $j_0$ such that $I \subseteq \mathfrak{p}_{j_0}$.

2. Let $I_1, \dots, I_n$ be ideals, and let $\mathfrak{p}$ be a prime such that $\bigcap_{j=1}^n I_j \subseteq \mathfrak{p}$. Then, there exists $j_0$ such that $I_{j_0} \subseteq \mathfrak{p}$.
Furthermore, if $\bigcap_{j=1}^n I_j = \mathfrak{p}$, then $I_{j_0} = \mathfrak{p}$.

**Proof T. 1.12.** 1. We will prove, by induction on $n$, that if $I \nsubseteq \mathfrak{p}_j$ for every $1 \leq j \leq n$, then $I \nsubseteq \bigcup_{j=1}^n \mathfrak{p}_j$. This means that if $I$ "avoids" every $\mathfrak{p}_j$, then it also "avoids" their union.
The statement is certainly true for $n = 1$.
Let $n > 1$, and assume the statement holds for $n-1$. Consider all possible subsets of $\{\mathfrak{p}_1, \dots, \mathfrak{p}_n\}$ consisting of $n-1$ elements.
Then, for every $i$, there exists an $a_i \in I$ such that $a_i \notin \bigcup_{j=1,\, j\neq i}^{n} \mathfrak{p}_j$.
If $a_i \notin \mathfrak{p}_i$ for some $i$, then $I \nsubseteq \bigcup_{j=1}^n \mathfrak{p}_j$.
Otherwise, $a_i \in \mathfrak{p}_i$ for each $i$. It follows that the element $b = \sum_{i=1}^{n} \prod_{j=1,\, j\neq i}^{n} a_j \in I$, but $b \notin \mathfrak{p}_i$ for each $i$, because all except one of its terms belong to $\mathfrak{p}_i$.
As a result, $I \nsubseteq \bigcup_{j=1}^n \mathfrak{p}_j$.
2. Assume, by contradiction, that $\mathfrak{p} \nsupseteq I_j$ for all $j$. In this case, for every $j$, there exists at least one element $a_j \in I_j \setminus \mathfrak{p}$.
Thus,

$$\prod_{j=1}^{n} a_j \in \bigcap_{j=1}^{n} I_j \quad \text{and} \quad \prod_{j=1}^{n} a_j \notin \mathfrak{p},$$

because $\mathfrak{p}$ is a prime ideal. It follows that $\mathfrak{p} \nsupseteq \bigcap_{j=1}^n I_j$.
Finally, if $\bigcap_{j=1}^n I_j = \mathfrak{p}$, then there exists $j_0$ such that $\mathfrak{p} = \bigcap_{j=1}^n I_j \subseteq I_{j_0} \subseteq \mathfrak{p}$. Therefore, it follows that $I_{j_0} = \mathfrak{p}$. □

As a consequence, we obtain the following result.

**T. 1.13.** ($\rightarrow$ p. 211) Every prime ideal is irreducible.

## 1.3 Nilradical, Jacobson Radical and Local Rings

By construction, the ring $A/\mathcal{N}(A) = A/\sqrt{(0)}$ is reduced. Indeed, $\overline{a}^k = \overline{0}$ means that $a^k$ and, consequently, $a$ are nilpotent. Hence, $\overline{a} = \overline{0}$.

**T. 1.14.** **[Characterization of the Radical of an Ideal]**

1. For every ring $A$,
$$\mathcal{N}(A) = \bigcap_{\mathfrak{p} \in \operatorname{Spec} A} \mathfrak{p}.$$
2. Let $I \subset A$ be an ideal. Then,
$$\sqrt{I} = \bigcap_{\substack{\mathfrak{p} \in \operatorname{Spec} A \\ \mathfrak{p} \supseteq I}} \mathfrak{p}.$$

**Proof T. 1.14.** 1. We first prove that a nilpotent element $a$ belongs to every prime ideal of $A$. If $a^n = 0$ for some $n \in \mathbb{N}$, then $a^n \in \mathfrak{p}$ for each prime $\mathfrak{p}$, hence $a \in \mathfrak{p}$ for all $\mathfrak{p}$.

To prove the opposite inclusion, we apply Zorn's Lemma. We prove that if $a \notin \mathcal{N}(A)$, then there exists a prime $\mathfrak{p}$ such that $a \notin \mathfrak{p}$.
Consider the family
$$\Sigma = \{J \subset A\colon a^n \notin J \text{ for every } n \in \mathbb{N}\},$$
partially ordered by set inclusion $\subseteq$.
Since, by hypothesis, $a$ is not nilpotent, $(0) \in \Sigma$, and $\Sigma$ is not empty.
It is easy to verify that the union of the ideals in a chain of $\Sigma$ is still an element of $\Sigma$. By Zorn's Lemma, there exists a maximal element $I$ of $\Sigma$. Since $I$ does not contain $a$, it suffices to prove that $I$ is prime.
Assume that there are elements $b$, $c \notin I$ such that $bc \in I$. Then, $I : (b)$ properly contains $I$, therefore, since $I$ is maximal in $\Sigma$, the ideal $I : (b)$ does not belong to $\Sigma$. Hence, there exists $n \in \mathbb{N}_+$ such that $a^n \in I : (b)$. Thus, $I \subsetneq (I, b) \subseteq I : (a^n) \notin \Sigma$ and there exists $m \in \mathbb{N}_+$ such that $a^m \in I : (a^n)$. Therefore, $a^{n+m} \in I$, which is the desired contradiction.

2. Consider the ring $A/I$, and note that $\mathcal{N}(A/I) = \sqrt{I}/I$ can be decomposed as the intersection of the primes of $A/I$ by part 1. The conclusion follows from the one-to-one correspondence between primes of $A/I$ and primes of $A$ containing $I$, see **T.1.10**. □

The previous result characterizes the nilradical of a ring $A$ as the intersection of all its prime ideals. The *Jacobson radical* of a ring $A$ is instead defined as the intersection of all maximal ideals of $A$. We denote it by $\mathcal{J}(A)$.
The definition immediately yields

$$\mathcal{N}(A) = \bigcap_{\mathfrak{p} \in \operatorname{Spec} A} \mathfrak{p} \subseteq \bigcap_{\mathfrak{p} \in \operatorname{Max} A} \mathfrak{p} = \mathcal{J}(A).$$

The elements of the Jacobson radical are characterized by the following property.

**T. 1.15.** (→ p. 211) [**Characterization of the Jacobson Radical**] Let $A$ be a ring. Then,

$$a \in \mathcal{J}(A) \ \text{ if and only if } \ 1 - ab \in A^* \ \text{ for every } \ b \in A.$$

A ring $A$ is said to be *local* if it has a unique maximal ideal $\mathfrak{m}$. In this case, the field $K = A/\mathfrak{m}$ is *the residue field* of $A$, and we use the notation $(A, \mathfrak{m})$ or $(A, \mathfrak{m}, K)$. A ring with a finite number of maximal ideals is said to be *semilocal.*
There are several examples of local rings, such as fields, $\mathbb{Z}/(p^n)$ with $p \in \mathbb{Z}$ a prime, and the ring of formal power series in one variable over a field $K$, see **E.8.73**. Their defining property makes them easier to study, although it does not provide a description of the set of prime ideals of the ring.

**T. 1.16.** (→ p. 211) [**Characterization of Local Rings**] Let $A$ be a ring.

1. If there exists an ideal $\mathfrak{m} \subset A$ such that $A \setminus \mathfrak{m} \subseteq A^*$, then $A$ is local with maximal ideal $\mathfrak{m}$.
2. If $\mathfrak{m} \subset A$ is a maximal ideal such that $1 + a \in A^*$ for every $a \in \mathfrak{m}$, then $A$ is local with maximal ideal $\mathfrak{m}$.

## 1.4 Extension and Contraction of Ideals

We have already seen the correspondence induced by the projection homomorphism $\pi : A \longrightarrow A/I$, between ideals of $A$ containing $I$ and ideals of the quotient $A/I$. More generally, given two rings $A$ and $B$, and a ring homomorphism $f\colon A \longrightarrow B$, we want to study the correspondence determined by the homomorphism $f$ between ideals of $A$ and those of $B$.

If $I \subset A$ is an ideal, in general, the image $f(I)$ is not an ideal of $B$. We define the *extension* of $I$ with respect to $f$ to be the ideal generated by the image $f(I)$, denoted by $I^e$.

On the other hand, if $J \subset B$ is an ideal, then $f^{-1}(J) = \{a \in A\colon f(a) \in J\}$ is always an ideal of $A$ and it contains $\operatorname{Ker} f$. We call this ideal the *contraction* of $J$ with respect to $f$, denoted by $J^c$.

**Extension and Contraction of Ideals**

**T. 1.17.** (→ p. 211) Let $f \colon A \longrightarrow B$ be a ring homomorphism, and let $I,\ I_1,\ I_2 \subseteq A$ and $J,\ J_1,\ J_2 \subseteq B$ be ideals. Then:

1. $I_1 \subseteq I_2 \Longrightarrow I_1^e \subseteq I_2^e$;
2. $J_1 \subseteq J_2 \Longrightarrow J_1^c \subseteq J_2^c$;
3. $I \subseteq I^{ec}$, but the containment may be strict;
4. $J^{ce} \subseteq J$, but the containment may be strict;
5. $I^{ece} = I^e$;
6. $J^{cec} = J^c$;
7. $J$ prime $\Longrightarrow J^c$ prime;
8. $J$ primary $\Longrightarrow J^c$ primary;
9. $J$ radical $\Longrightarrow J^c$ radical.

In general, 7, 8 and 9 do not hold for extension of ideals.

In the case of a surjective homomorphism, *e.g.*, the projection $\pi : A \longrightarrow A/I$, this correspondence also satisfies the following properties.

**T. 1.18.** (→ p. 212) Let $A$ and $B$ be rings, and let $f \colon A \longrightarrow B$ be a surjective homomorphism. If $\operatorname{Ker} f \subseteq I \subset A$ and $J \subset B$ are ideals, then:

1. the image $f(I)$ is an ideal of $B$, therefore $I^e = f(I)$;
2. $I = I^{ec}$ and $J = J^{ce}$, thus the operations of extension and contraction establish a one-to-one correspondence between ideals of $A$ containing $\operatorname{Ker} f$ and ideals of $B$;
3. with respect to this correspondence, maximal, prime, primary, and radical ideals correspond to maximal, prime, primary, and radical ideals, respectively.

## 1.5 The Chinese Remainder Theorem

Let $A_1, \ldots, A_n$ be rings. A ring structure can be defined on the Cartesian product $A = \prod_{i=1}^{n} A_i = A_1 \times A_2 \times \cdots \times A_n$ by setting

$$a + b = (a_1 + b_1, \ldots, a_n + b_n) \quad \text{and} \quad ab = (a_1 b_1, \ldots, a_n b_n),$$

for any $a = (a_1, \ldots, a_n),\ b = (b_1, \ldots, b_n) \in A$.
In other words, sum and product are defined componentwise using the sum and product defined on each $A_i$.
Obviously, $0_A = (0_{A_1}, \ldots, 0_{A_n})$ and $1_A = (1_{A_1}, \ldots, 1_{A_n})$.

We call this ring the *direct sum* of $A_1, \ldots, A_n$. Such a ring can never be a domain if $n > 1$. Moreover, $A$ always contains non-trivial idempotents.

**T. 1.19.** A ring $A$ is isomorphic to a direct sum $A_1 \times \cdots \times A_n$ if and only if there exist $n$ idempotent elements $e_1, \ldots, e_n \in A$ which are *orthogonal, i.e.*, such that $e_i e_j = 0$ for every $i \neq j$, and verify $\sum_i e_i = 1_A$.

**Proof T. 1.19.** Let $f\colon A \longrightarrow \prod_{i=1}^{n} A_i$ be a ring isomorphism.
We claim that the elements $e_i = f^{-1}(0, \ldots, 0, 1_{A_i}, 0, \ldots, 0)$ are orthogonal idempotents such that $\sum_i e_i = 1_A$.
Indeed, since $f$ is a ring homomorphism, $f(e_i e_j) = f(e_i)f(e_j) = 0$ for all $i \neq j$, and $f(e_i^2 - e_i) = f(e_i)^2 - f(e_i) = 0$ for all $i$. Therefore, $e_i e_j = 0$ for each $i \neq j$ and $e_i^2 = e_i$ for every $i$, because $f$ is injective.
Finally,

$$f(\sum_i e_i) = \sum_i f(e_i) = (1_{A_1}, \ldots, 1_{A_n})$$

yields $\sum_i e_i = 1_A$.

Conversely, let $A_i = e_i A$ be commutative rings whose identity is $e_i$, and define $f\colon A \longrightarrow \prod_{i=1}^{n} A_i$ by setting $f(a) = (e_1 a, \ldots, e_n a)$.
It is easy to verify that $f$ is a ring homomorphism. We prove that $f$ is an isomorphism.
It is injective, because if $f(a) = f(b)$, then for every $i$ we have $e_i a = e_i b$, which implies

$$a = a \cdot 1 = a \sum_i e_i = \sum_i e_i a = \sum_i e_i b = b \sum_i e_i = b.$$

It is also surjective, since for every $(e_1 a_1, \ldots, e_n a_n) \in \prod_{i=1}^{n} A_i$ the element $a = e_1 a_1 + \ldots + e_n a_n \in A$ verifies $f(a) = (e_1 a_1, \ldots, e_n a_n)$, because the elements $e_i$ are orthogonal. □

**The Chinese Remainder Theorem**

**T. 1.20.** Let $I_1, \ldots, I_n \subset A$ be ideals such that $I_i + I_j = (1)$ when $i \neq j$, and let $a_1, \ldots, a_n \in A$. Then, there exists $a \in A$ such that $a \equiv a_i \mod I_i$ for each $i$.

**Proof T. 1.20.** By hypothesis, for every $i \neq j$ there exist elements $\gamma_i^{(j)} \in I_i$ such that $\gamma_i^{(j)} + \gamma_j^{(i)} = 1$. For each $i$, define $L_i = \prod_{j \neq i} \gamma_j^{(i)}$.
Thus,

$$L_i \equiv 0 \mod I_j \text{ when } j \neq i, \text{ and } L_i = \prod_{i \neq j}(1 - \gamma_i^{(j)}) \equiv 1 \mod I_i.$$

The element $a = \sum_i a_i L_i \in A$ satisfies the required conditions. □

**T. 1.21.** (→ p. 212) Let $I_1, \ldots, I_n \subset A$ be ideals of $A$, and define a homomorphism $f\colon A \longrightarrow \prod_{i=1}^{n} A/I_i$ by $f(a) = (\overline{a}_1, \ldots, \overline{a}_n)$, where $a_i \equiv a \mod I_i$. Then, the following hold:

1. $f$ is surjective if and only if $I_i + I_j = (1)$ for each $i \neq j$;
2. $f$ is injective if and only if $\bigcap_{i=1}^{n} I_i = 0$.

In conclusion, when $A$ contains pairwise comaximal ideals $I_i$, with $i = 1, \ldots, n$, such that $\prod_{i=1}^{n} I_i = 0$, then $f\colon A \longrightarrow \prod_{i=1}^{n} A/I_i$ is an isomorphism by **T.1.4** and the previous result.

## 1.6 Factorization in Integral Domains: PID and UFD

In this section we assume that $A$ is a domain, and thus the cancellation law holds, see **E.8.2**. We analyze the problems of division and factorization in $A$.

For any $a,\ b \in A$, we say that $a$ *divides* $b$, and we write $a \mid b$, if there exists $c \in A$ such that $b = ac$. In this case, we say that $a$ is a *divisor* of $b$ and that $b$ is a *multiple* of $a$. The divisors of 1 are exactly the elements of $A^*$. If $a$ is a divisor of $b$ and $b$ is a divisor of $c$, then $a$ is a divisor of $c$. Note that, if $b \neq 0$ and $b = ac$, then $c$ is uniquely determined by $a$ and $b$. If $a,\ c \notin A^*$, we say that $a$ is a *proper divisor* of $b$.

Two elements $a,\ b \in A$ are *associate* when there exists $u \in A^*$ such that $b = ua$. It is easy to verify that the property of being associate defines an equivalence relation $\mathcal{R}$ on $A$.
An element $a \notin A^*$ is called *prime* if for every $b,\ c \in A$, we have $a \mid bc$ if and only if either $a \mid b$ or $a \mid c$.
An element $a \notin A^*$ is called *irreducible* if $a = bc$ implies either $b \in A^*$ or $c \in A^*$. In other words, every divisor of $a$ is either an associate of $a$ or a unit, which means that $a$ has no proper divisors.
These concepts can be easily expressed in terms of properties of principal ideals.

**T. 1.22.** (→ p. 212) Let $a,\ b \in A$. Then:

1. $a$ is invertible if and only if $(a) = (1)$;
2. $a$ divides $b$ if and only if $(b) \subseteq (a)$;
3. $a$ and $b$ are associate if and only if $(a) = (b)$;
4. $a$ is a proper divisor of $b \neq 0$ if and only if $(b) \subsetneq (a) \subsetneq (1)$;
5. $a$ is prime if and only if $(a)$ is a prime ideal.

**T. 1.23.** ($\rightarrow$ p. 213) Let $A$ be a domain. Then:

1. if $a \in A \setminus \{0\}$ is prime, then $a$ is irreducible;
2. if $A$ is a PID and $a \in A$ is irreducible, then $a$ is prime.

For example, $\mathbb{Z}$ is a PID, and its non-zero prime ideals, which are also maximal, are precisely the ideals $(p)$, with $p$ a prime number. In the ring $\mathbb{Z}[x]$ the element $x$ is irreducible and prime, but the ideal $(x)$ is prime and not maximal. Finally, in the ring $\mathbb{Z}[\sqrt{-5}]$ the element 2 is irreducible but not prime.

**T. 1.24.** ($\rightarrow$ p. 213) Let $A$ be a PID. Then, every ascending chain $\mathcal{C} = \{I_h\}_{h \in H}$ of ideals of $A$ is stationary, *i.e.*, there exists $h_0 \in H$ such that $\bigcup_{h \in H} I_h = I_{h_0}$.

A domain $A$ is said to be a *unique factorization domain* (UFD), if it satisfies the following properties:

(UFD1) *existence of a factorization into irreducible elements*: for each element $a \in A \setminus \{A^* \cup \{0\}\}$ there exist irreducible elements $b_1, \ldots, b_k \in A$ and $u \in A^*$ such that $a = ub_1 \cdots b_k$;

(UFD2) *uniqueness of the factorization*: if $a = ub_1 \cdots b_k = vc_1 \cdots c_h$ are two factorizations of $a$ into irreducible elements, with $u, v \in A^*$, then $k = h$ and, up to reordering of the terms, the elements $b_i$ and $c_i$ are associate for each $i$.

**T. 1.25.** Let $A$ be a domain that satisfies (UFD1). Then, $A$ satisfies (UFD2) if and only if the property

(UFD3) *every irreducible element is prime*

holds in $A$.

**Proof T. 1.25.** We show that if $A$ is a unique factorization domain, then (UFD3) holds in $A$. Assume that $a$ is an irreducible element, and let $b, c \in A$ be such that $a \mid bc$. We aim to prove that $a$ divides either $b$ or $c$.
If either $b$ or $c$ is equal to 0, then the statement is trivially true.
Moreover, we can also assume that neither $b$ nor $c$ are invertible, that is, $b, c \in A \setminus \{A^* \cup \{0\}\}$. Then, there exists a non-zero element $d \in A$ such that $da = bc$. By (UFD1), there exist irreducible elements $b_i$, $c_j$, and units $u, v \in A^*$, such that $b = u \prod_i b_i$ and $c = v \prod_j c_j$. By (UFD2), there exists an irreducible factor $e$ of $b$ or $c$ such that $a$ and $e$ are associate, since we assumed that $a$ is irreducible. Hence, $a \mid b$ or $a \mid c$.

Conversely, we assume (UFD3) and prove that (UFD2) holds.
Consider $a \in A \setminus \{A^* \cup \{0\}\}$ and two factorizations of $a$ into irreducible elements

$$a = u \prod_{i=1}^{m} a_i = v \prod_{j=1}^{n} b_j,$$

where $u$, $v \in A^*$. Without loss of generality, we can assume that $m \leq n$.
We note that $\prod_{j=1}^{n} b_j \in (a_1)$ which is a prime ideal, since by hypothesis (UFD3) holds. Hence, there exists $j_1 \in \{1, \ldots, n\}$ such that $b_{j_1} \in (a_1)$.
Reordering the indices, we can assume that $j_1 = 1$, and $b_1 = a_1 c_1$. Since $b_1$ is irreducible and $a_1 \notin A^*$, we have $c_1 \in A^*$. Therefore, $a_1$ and $b_1$ are associate. Cancelling $a_1$, we obtain

$$u \prod_{i=2}^{m} a_i = v c_1 \prod_{j=2}^{n} b_j.$$

Reasoning in the same way, we find that $b_2 = a_2 c_2$ with $c_2 \in A^*$ and so on. After $m$ steps, we have that $a_i$ and $b_i$ are associate for all $i = 1, \ldots, m$, and

$$v \prod_{i=1}^{m} c_i \prod_{j=m+1}^{n} b_j = u \in A^*.$$

Thus, we have proved that two factorizations of $a$ are identical, up to a change of order and unit factors. □

**T. 1.26.** (→ p. 213) If $A$ is a PID, then $A$ is a UFD.

An example of a unique factorization domain which is not a PID is $\mathbb{Z}[x]$, where $(3, x)$ is not principal.
Another example is a polynomial ring $A$ with coefficients in a field and at least two variables $x_1$ and $x_2$. Then, Gauss' Lemma **E.8.13** implies that $A$ is a UFD.
However, it is not a PID. Let $I$ be the ideal $(x_1, x_2)$. Then, $I$ is a proper and non-trivial ideal of $A$ such that $I \neq (x_1)$ and $I \neq (x_2)$.
If there exists $d \in A$ such that $(d) = (x_1, x_2)$, then $d \neq 0$ and $d \notin A^*$. Since $A$ is a domain, $x_i = c_i d$ implies $\deg d = 1$ and $c_i \in K^*$ because $x_i$ has degree 1. This means that $x_i$ and $d$ are associate. Thus, we would have $(x_1) = (d) = (x_2)$, which is impossible.

Let $A$ be a UFD, and let $a$, $b \in A$, not both zero. An element $d \in A$ such that:

i) $d \mid a$ and $d \mid b$ (*common divisor*);
ii) for each $c \in A$ such that $c \mid a$ and $c \mid b$, we have $c \mid d$ (*greatest divisor*);

is called *greatest common divisor* of $a$ and $b$, and is denoted by $\gcd(a, b)$.
The definition immediately yields that if $d_1$ and $d_2$ are both greatest common divisors of $a$ and $b$, then they are associate. With this in mind, with a slight abuse of notation, we call an element $d$ satisfying conditions i) and ii) of the above definition *the greatest common divisor* of $a$ and $b$, meaning that it is unique up to associates.

Similarly, we introduce the definition of *least common multiple*. This is an element $m \in A$ such that:

i) $a \mid m$ and $b \mid m$ (*common multiple*);

ii) for each $c \in A$ such that $a \mid c$ and $b \mid c$, we have $m \mid c$ (*least multiple*);

we denote it by $\operatorname{lcm}(a, b)$.
Note that for any non-zero $a$, we have $\gcd(a, 0) = a$ and $\operatorname{lcm}(a, 0) = 0$.

Let $\mathcal{R}$ be the equivalence relation defined by the property of being associate. Consider the set $\{p \neq 0 \colon p \text{ prime }\}/\mathcal{R}$ of non-zero prime elements of $A$, or, equivalently, by **T.1.23** and **T.1.25**, the set of irreducible elements, modulo the equivalence relation $\mathcal{R}$, and fix a set of representatives $\operatorname{Irr} A$.
Each pair of elements $a$, $b \in A \setminus \{A^* \cup \{0\}\}$ can be uniquely written as

$$a = u \prod_{p \in \operatorname{Irr} A} p^{a_p} \quad \text{and} \quad b = v \prod_{p \in \operatorname{Irr} A} p^{b_p},$$

where $u$, $v \in A^*$ and $a_p$, $b_p$ almost all zero.
Then,

$$\gcd(a, b) = \prod_{p \in \operatorname{Irr} A} p^{\min\{a_p, b_p\}} \quad \text{and} \quad \operatorname{lcm}(a, b) = \prod_{p \in \operatorname{Irr} A} p^{\max\{a_p, b_p\}}.$$

An integral domain $A$ is an *Euclidean domain* if there exists a function $\delta \colon A \setminus \{0\} \longrightarrow \mathbb{N}$, called *degree function*, such that:

i) for any $a$, $b \in A \setminus \{0\}$, we have $\delta(a) \leq \delta(ab)$;

ii) for any $a \in A$ and $b \in A \setminus \{0\}$, there exist $q$, $r \in A$ such that $a = qb + r$, and either $r = 0$ or $\delta(r) < \delta(b)$.

The elements $q$ and $r$ are called, respectively, a *quotient* and a *remainder* of the division of $a$ by $b$.

**T. 1.27.** ($\rightarrow$ p. 213) A Euclidean domain is a PID, therefore it is also a UFD.

For example, the rings $\mathbb{Z}$, $K[x]$, with $K$ a field, and $\mathbb{Z}[i]$ are Euclidean rings, with degree functions $\delta(n) = |n|$, $\delta(f) = \deg(f)$, and $\delta(a+ib) = a^2+b^2$, respectively. Therefore, they are also PID and UFD.

**T. 1.28.** ($\rightarrow$ p. 214) Let $A$ be a UFD, and let $a \in A$ be a non-invertible and non-zero element. If $a$ is irreducible, then the ideal $(a)$ is irreducible.

We remark that if $A$ is not a UFD, then the previous statement does not hold in general, see **E.8.67**. Furthermore, even if $A$ is a PID, it is not necessarily true that irreducible ideals are generated by irreducible elements.

For example, consider the ideal $(9) \subset \mathbb{Z}$. This ideal is irreducible, because if $(9) = (a) \cap (b) = (\operatorname{lcm}(a, b))$, then either $(a) = (9)$ or $(b) = (9)$, but 9 is not an irreducible element. In addition, an irreducible ideal is not necessarily prime, whereas every prime ideal is always irreducible, as shown in **T.1.13**.

# Chapter 2
# The Ring $K[x_1, \ldots, x_n]$

In this chapter we study the properties of the ring $A = K[x_1, \ldots, x_n]$ of polynomials in $n$ indeterminates, or variables, with coefficients in a field $K$. To simplify notation, we use $X = x_1, \ldots, x_n$ to denote the set of indeterminates and write $A$ as $K[X]$. Throughout the chapter, we define monomial ideals and investigate their properties as well as introduce the fundamentals of Gröbner bases, an essential tool for solving many problems related to the study of ideals and their operations.

## 2.1 Monomial Ideals and $\mathcal{E}$-subsets

Let $\mathbf{a} = (a_1, \ldots, a_n) \in \mathbb{N}^n$. We define a *monomial* of $A$ with *exponent* $\mathbf{a}$ as the element $x_1^{a_1} \cdots x_n^{a_n} \in A$, which we denote as $X^{\mathbf{a}}$. The set $\operatorname{Mon} A$ of the monomials of $A$ is a basis of the $K$-vector space $A$. Moreover, there exists a natural one-to-one correspondence between the sets $\operatorname{Mon} A$ and $\mathbb{N}^n$. Under this bijection the monomial 1 corresponds to the exponent $\mathbf{0} \in \mathbb{N}^n$.
If $\mathbf{a}, \mathbf{b} \in \mathbb{N}^n$, then $X^{\mathbf{a}}$ divides $X^{\mathbf{b}}$ if and only if $\mathbf{b} - \mathbf{a} \in \mathbb{N}^n$.

A *polynomial* is an element $f \in A$ of the form

$$f = \sum_{\mathbf{a}} c_{\mathbf{a}} X^{\mathbf{a}} \quad \text{with} \quad \mathbf{a} \in \mathbb{N}^n \quad \text{and} \quad c_{\mathbf{a}} \neq 0 \quad \text{for finitely many} \quad \mathbf{a} \in \mathbb{N}^n,$$

*i.e.*, a $K$-linear combination of monomials.
Any non-zero addend $c_{\mathbf{a}} X^{\mathbf{a}}$ of $f$ is called *term* of $f$.

In order to study the ideals of $A$, it is convenient to define the notion of monomial ideal.

An ideal $I \subset A$ is a *monomial ideal* if it has a set of generators consisting of monomials.



A. Bandini et al., *Commutative Algebra through Exercises*, UNITEXT 159,
https://doi.org/10.1007/978-3-031-56910-4_2

The following result provides a criterion to decide whether a polynomial belongs to a monomial ideal.

**T. 2.1.** (→ p. 215) Let $I$ be a monomial ideal and let $f = \sum_{\mathbf{a}} c_{\mathbf{a}} X^{\mathbf{a}} \in A$. Then,

$$f \in I \iff X^{\mathbf{a}} \in I \text{ for all } \mathbf{a} \text{ such that } c_{\mathbf{a}} \neq 0.$$

The previous property characterizes monomial ideals. If $I$ is an ideal which satisfies the condition in **T.2.1**, then $I$ is generated by the monomials of the polynomials contained in $I$. Consequently, $I$ is a monomial ideal.

Consider $E \subseteq \mathbb{N}^n$, and let $I = (X^{\mathbf{a}}\colon \mathbf{a} \in E)$ be a monomial ideal. A monomial $X^{\mathbf{b}} \in I$ if and only if there exists $\mathbf{a} \in E$ such that $\mathbf{b} - \mathbf{a} \in \mathbb{N}^n$. The set of exponents $E$, and therefore, a set of generators of $I$, may not be finite. However, we will prove that there always exists a finite subset $E' \subseteq E$ such that $I = (X^{\mathbf{a}}\colon \mathbf{a} \in E) = (X^{\mathbf{b}}\colon \mathbf{b} \in E')$.

The bijection between monomials and exponents shows that studying monomial ideals is equivalent to studying special subsets of $\mathbb{N}^n$. For this reason, we introduce the following definitions, which correspond to those of monomial ideals and monomial set of generators.

A non-empty subset $E$ of $\mathbb{N}^n$ is an *$\mathcal{E}$-subset* if

$$\mathbf{a} \in E \Longrightarrow \mathbf{a} + \mathbf{b} \in E \quad \text{for all } \mathbf{b} \in \mathbb{N}^n.$$

A subset $F \neq \emptyset$ of an $\mathcal{E}$-subset $E$ is called *boundary of $E$* if, for every $\mathbf{a} \in E$, there exist $\mathbf{b} \in F$ and $\mathbf{c} \in \mathbb{N}^n$ such that $\mathbf{a} = \mathbf{b} + \mathbf{c}$.

**T. 2.2.** (→ p. 215) Let $I \neq 0$ be a monomial ideal. Then, there exists an $\mathcal{E}$-subset $E$ such that $I = (X^{\mathbf{a}}\colon \mathbf{a} \in E)$. Moreover, if $F$ is a boundary of $E$, then

$$I = (X^{\mathbf{a}}\colon \mathbf{a} \in F).$$

The following result proves that every $\mathcal{E}$-subset has a finite boundary.

**Dickson's Lemma**

**T. 2.3.** Every $\mathcal{E}$-subset has a finite boundary.
Therefore, every monomial ideal of $A$ is finitely generated.

**Proof T. 2.3.** The proof proceeds by induction on the number of variables $n$. If $n = 1$, then $E \subset \mathbb{N}$ and $F = \{\min E\}$ is a finite boundary of $E$.

Now we assume that the statement holds for $n$ and prove it for $n + 1$. Define the projection $\pi\colon \mathbb{N}^{n+1} \longrightarrow \mathbb{N}^n$, $\pi(a_0, \ldots, a_n) = (a_1, \ldots, a_n)$.

Clearly, $\pi(E) \neq \emptyset$. Moreover, $\pi(E)$ is an $\mathcal{E}$-subset of $\mathbb{N}^n$, since for every $\mathbf{a} \in E$ and $\mathbf{b} \in \mathbb{N}^n$ the element $\mathbf{a} + (0, \mathbf{b})$ belongs to $E$, and thus,

$$\pi(\mathbf{a}) + \mathbf{b} = \pi(\mathbf{a} + (0, \mathbf{b})) \in \pi(E).$$

Therefore, by the inductive hypothesis and the surjectivity of $\pi$, there exists $F' = \{\mathbf{a}_1, \ldots, \mathbf{a}_k\} \subset E$ such that $\pi(F')$ is a finite boundary of $\pi(E)$.
In order to complete the construction of a finite boundary of $E$, we write $\mathbf{a}_i = (a_0^{(i)}, \ldots, a_n^{(i)})$ for every $i$, and let $\overline{a} = \max_i\{a_0^{(i)}\}$.
For each $0 \leq a < \overline{a}$, let

$$E_a = E \cap (\{a\} \times \mathbb{N}^n)$$

be the subset of all $(n+1)$-tuples in $E$ whose first coordinate is $a$.
If $E_a$ is non-empty, then $\pi(E_a) \subset \mathbb{N}^n$ is an $\mathcal{E}$-subset of $\mathbb{N}^n$. Hence, by the inductive hypothesis, there exists a finite set $F_a \subset E_a$ such that $\pi(F_a)$ is a finite boundary of $\pi(E_a)$.
If $E_a = \emptyset$, then we define $F_a = \emptyset$.
We claim that

$$F = \left( \bigcup_{0 \leq a < \overline{a}} F_a \right) \cup F'$$

is a finite boundary of $E$.
Let $\mathbf{a} = (a_0, \ldots, a_{n+1}) \in E$. We have to show that there exists $\mathbf{b} \in F$ such that $\mathbf{a} - \mathbf{b} \in \mathbb{N}^{n+1}$. If $a_0 < \overline{a}$, then $\mathbf{a} \in E_{a_0}$, and there exists $\mathbf{b} \in F_{a_0}$ such that $\mathbf{a} - \mathbf{b} \in \mathbb{N}^{n+1}$. Otherwise, since $\pi(\mathbf{a}) \in \pi(E)$, there exists $\mathbf{a}_i \in F'$ such that $\pi(\mathbf{a} - \mathbf{a}_i) \in \mathbb{N}^n$, and thus $\mathbf{a} - \mathbf{a}_i \in \mathbb{N}^{n+1}$, since $a_0 \geq \overline{a} \geq a_0^{(i)}$. □

Let $F$ be a boundary of an $\mathcal{E}$-subset. By removing the elements $\mathbf{b} \in F$ of the form $\mathbf{a} + \mathbf{c}$, for some $\mathbf{a} \in F$ and $\mathbf{c} \in \mathbb{N}^n \setminus \{\mathbf{0}\}$ we obtain what is called a *minimal* boundary.

**T. 2.4.** (→ p. 215) Let $E$ be an $\mathcal{E}$-subset of $\mathbb{N}^n$. Then, every minimal boundary of $E$ is finite.

**T. 2.5.** (→ p. 215) Every $\mathcal{E}$-subset has a unique minimal boundary. Therefore, every monomial ideal $I$ has a unique minimal set of monomial generators.

The minimal boundary of an $\mathcal{E}$-subset $E$ is called the *staircase* of $E$.

In summary, a monomial ideal $I = (X^{\mathbf{a}} \colon X^{\mathbf{a}} \in M)$, generated by a set of monomials $M$, corresponds to the $\mathcal{E}$-subset $E = \{\mathbf{a} + \mathbb{N}^n \colon X^{\mathbf{a}} \in M\}$. By definition of $\mathcal{E}$-subset, each boundary of $E$ corresponds to a set of generators of $I$. In this way Dickson's Lemma, which asserts the existence of a finite boundary $E'$ of $E$, proves that any monomial ideal has a finite set of generators. Furthermore, if $F = \{\mathbf{a}_1, \ldots, \mathbf{a}_k\}$ is the minimal boundary of $E$, then the set $M' = \{X^{\mathbf{a}_1}, \ldots, X^{\mathbf{a}_k}\} \subseteq M$ of the corresponding monomials is a minimal set of generators of $I$. In particular, $M'$ is the set of the monomials of $I$ which are minimal with respect to divisibility.

In conclusion, every monomial ideal $I$ has a unique minimal set of monomial generators, and we denote it by $G(I)$.

Performing ideal operations on monomial ideals and computing a monomial generating set for the resulting ideal is straightforward.
The simplest operation is the sum. If $I$ and $J$ are monomial ideals, then $I+J$ is generated by $G(I)\cup G(J)\supseteq G(I+J)$.

**Operations on Monomial Ideals**

**T. 2.6.** Let $I=(m_1,\ldots,m_s)$ and $J=(n_1,\ldots,n_t)$ be monomial ideals of $A$, with $m_i$, $n_j\in\operatorname{Mon}A$, for $i=1,\ldots,s$ and $j=1,\ldots,t$.

1. **Decomposition.** Let $m$, $u\in A$ be relatively prime monomials. Then,
$$(I,mu)=(I,m)\cap(I,u).$$
2. **Intersection.** The ideal $I\cap J$ is generated by the monomials $\operatorname{lcm}(m_i,n_j)$, for $i=1,\ldots,s$ and $j=1,\ldots,t$.
3. **Quotient.** Let $m$ be a monomial of $A$. Then, the ideal $I:(m)$ is generated by the monomials $\dfrac{m_i}{\gcd(m_i,m)}$, for $i=1,\ldots,s$.
4. **Radical.** For any monomial $m$, let $\sqrt{m}=\prod_{x_h|m}x_h$ denote its *squarefree part*. Then, the ideal $\sqrt{I}$ is generated by the monomials $\sqrt{m_i}$, for $i=1,\ldots,s$.

**Proof T. 2.6.** At first, we observe that the intersection of monomial ideals is a monomial ideal. Indeed, if $f\in I\cap J$, then all of its monomials belong to both $I$ and $J$, *i.e.*, to $I\cap J$.
Therefore, $I\cap J$ is monomial by the remark after **T.2.1**.

1. It is evident that $(I,m)\cap(I,u)\supseteq(I,mu)$.
To prove the opposite inclusion, write $f=\sum_{\mathbf{a}}c_{\mathbf{a}}X^{\mathbf{a}}\in(I,m)\cap(I,u)$ as a sum of terms. Since $(I,m)$ and $(I,u)$ are monomial ideals, every monomial $X^{\mathbf{a}}\in(I,m)\cap(I,u)$. Thus, we can assume that $f$ is a monomial.
If $f\in I$, then $f\in(I,mu)$.
If $f\notin I$, then $f=am=bu$, hence $m\mid b$, since $m$ and $u$ are relatively prime. Therefore, $f=cmu$ for some $c\in\operatorname{Mon}A$.

2. It is clear that the elements $d_{ij}=\operatorname{lcm}(m_i,n_j)$ generate an ideal contained in $I\cap J$.

To prove the opposite inclusion, we can again assume that $f \in I \cap J$ is a monomial. Since $f \in J$, we have $f = X^{\mathbf{a}} n_j$ for some $j$. Since $X^{\mathbf{a}} n_j$ is also in $I$, there exists $m_i$ such that $m_i \mid X^{\mathbf{a}} n_j$. Therefore, $d_{ij} \mid X^{\mathbf{a}} n_j$.

3. If $d_i = \gcd(m_i, m)$ for $i = 1, \ldots, s$, then there exist $a_i$, $b_i \in \operatorname{Mon} A$ with $\gcd(a_i, b_i) = 1$, such that $m_i = a_i d_i$ and $m = b_i d_i$.
Moreover, $a_i m = a_i b_i d_i = b_i m_i \in I$, and hence,

$$(a_1, \ldots, a_s) \subset I : (m) = \{f \in A\colon fm \in I\}.$$

To prove the opposite inclusion, write $f = \sum_i c_i n_i$ as a sum of terms, with $c_i \in K^*$ and $n_i \in \operatorname{Mon} A$. Then, from $fm = \sum_i c_i n_i m \in I$, it follows that, for every $i$, there exist monomials $m_{k_i} \in I$ and $e_i \in \operatorname{Mon} A$ such that

$$n_i m = e_i m_{k_i} = e_i a_{k_i} d_{k_i}.$$

Moreover, by the definition of $d_{k_i}$, we can write $n_i m = n_i b_{k_i} d_{k_i}$, and thus, $e_i a_{k_i} d_{k_i} = n_i b_{k_i} d_{k_i}$.
This implies that $a_{k_i} \mid n_i$ for each $i$, and therefore, $f \in (a_1, \ldots, a_s)$.

4. If $H = (\sqrt{m_1}, \ldots, \sqrt{m_s})$, then

$$I \subseteq H \subseteq \sqrt{I}.$$

By **T.1.6**.1 and 2, we have $\sqrt{I} = \sqrt{H}$. Thus, it is sufficient to prove that $H$ is radical.
Since for each $i$, $\sqrt{m_i}$ is a product of variables, by repeatedly applying the decomposition from part 1, we can express the ideal $H$ as an intersection of ideals generated by variables. These ideals are prime by **T.1.11**.3, thus $H$ is radical by **T.1.6**.3 and **T.1.11**.7. □

We remark that, by **T.2.6**.3 and **E.8.21**.5, for every pair of monomial ideals $I = (m_1, \ldots, m_s)$ and $J = (n_1, \ldots, n_t)$, we have

$$I : J = \bigcap_{j=1}^{t} I : (n_j) = \bigcap_{j=1}^{t} \left( \frac{m_1}{\gcd(m_1, n_j)}, \ldots, \frac{m_s}{\gcd(m_s, n_j)} \right).$$

**Special Monomial Ideals**

**T. 2.7.** Let $I$ be a monomial ideal of $A$.

1. **Primality Test.** $I$ is prime if and only if for every $m \in G(I)$ there exists $j \in \{1, \ldots, n\}$ such that $m = x_j$, *i.e.*, if and only if $G(I)$ is a set of variables.
2. **Radical Test.** $I$ is radical if and only if $m = \sqrt{m}$ for every $m \in G(I)$, *i.e.*, if and only if each element of $G(I)$ is squarefree.

3. **Irreducibility Test.** $I$ is irreducible if and only if for every $m \in G(I)$ there exist $j \in \{1, \ldots, n\}$ and $b > 0$ such that $m = x_j^b$, *i.e.*, if and only if $G(I)$ is a set of pure powers of variables.
4. **Primary Test.** $I$ is primary if and only if for every $m = X^{\mathbf{a}} \in G(I)$, with $\mathbf{a} = (a_1, \ldots, a_n)$, it holds:
$$\text{if } a_j \neq 0, \text{ then there exists } b > 0 \text{ such that } x_j^b \in G(I),$$
*i.e.*, if and only if every variable appearing in an element of $G(I)$ has some pure power in $G(I)$.

As a corollary, we obtain that if $I$ is an irreducible monomial ideal, then it is primary.

**Proof T. 2.7.** 1. Let $I = (x_{i_1}, \ldots, x_{i_k})$. Then, $A/I$ is a domain, *i.e.*, $I$ is prime.

Conversely, for every $m \in G(I)$ there exists $j \in \{1, \ldots, n\}$ such that $x_j \mid m$, that is, $m = x_j u$ for some $u \in \operatorname{Mon} A$. Since $G(I)$ is minimal, $u \notin I$, and hence, $x_j \in I$, because $I$ is prime.
This implies $m = x_j$, by the minimality of $G(I)$.

2. It is a straightforward application of **T.2.6**.4.

3. Let $I$ be irreducible. If there exists $m \in G(I)$ with $m = uv$ and $\gcd(u, v) = 1$, then $I \subsetneq (I, u)$ and $I \subsetneq (I, v)$, since $G(I)$ is minimal.
Moreover, $(I, u) \cap (I, v) = I$ by **T.2.6**.1 and 2, which is a contradiction.

Conversely, reordering the variables and the monomials of $G(I)$, if necessary, we can write $G(I) = \{x_1^{a_1}, \ldots, x_k^{a_k}\}$ for some $k \leq n$.
Let $Y = x_1, \ldots, x_k$ and $Z = x_{k+1}, \ldots, x_n$. Moreover, let $J$ and $L$ be ideals of $A = K[Y, Z]$, not necessarily monomial, such that $I = J \cap L$, with $I \subsetneq J$ and $I \subsetneq L$. We will prove that there exists $p(Z) \in K[Z]$ such that
$$p(Z) x_1^{a_1 - 1} \cdots x_k^{a_k - 1} \in I,$$
which is impossible, since $I = (x_1^{a_1}, \ldots, x_k^{a_k})$.

Take $f \in J \setminus I$ and $g \in L \setminus I$. Then, $fg \in I$, and we can assume that the monomials of $f$ and of $g$ do not belong to $I$.
Write $f$ as
$$f = \sum_{\mathbf{a}} c_{\mathbf{a}}(Z) Y^{\mathbf{a}} \in K[Z][Y],$$
and let $Y^{\boldsymbol{\delta}}$ be a monomial of $f$ of minimal total degree $|\boldsymbol{\delta}| = \delta_1 + \ldots + \delta_k$. Since $Y^{\boldsymbol{\delta}} \notin I$, for every $i$ we have $\delta_i < a_i$, and, for every other monomial $u$ of $f$, there exists an index $i$ such that $\deg_{x_i} u > \delta_i$.
Define $\boldsymbol{\gamma} = (\gamma_1, \ldots, \gamma_k)$ with $\gamma_i = a_i - \delta_i - 1 \geq 0$.
By construction,
$$Y^{\boldsymbol{\gamma} + \boldsymbol{\delta}} = x_1^{a_1 - 1} \cdots x_k^{a_k - 1} \notin I,$$
and $Y^{\boldsymbol{\gamma}} u \in I$ for each monomial $u$ of $f$ different from $Y^{\boldsymbol{\delta}}$. Thus,

$$Y^{\gamma}(f - c_{\boldsymbol{\delta}}(Z)Y^{\boldsymbol{\delta}}) \in I \subset J.$$

Hence,

$$c_{\boldsymbol{\delta}}(Z)x_1^{a_1-1}\cdots x_k^{a_k-1} = Y^{\gamma}f - Y^{\gamma}(f - c_{\boldsymbol{\delta}}(Z)Y^{\boldsymbol{\delta}}) \in J.$$

Applying the same argument to the polynomial $g = \sum_{\mathbf{a}} d_{\mathbf{a}}(Z)Y^{\mathbf{a}} \in L \setminus I$, we find $\boldsymbol{\varepsilon}$, $\boldsymbol{\eta}$, and $d_{\boldsymbol{\varepsilon}}(Z)$ such that

$$d_{\boldsymbol{\varepsilon}}(Z)x_1^{a_1-1}\cdots x_k^{a_k-1} = Y^{\boldsymbol{\eta}}g - Y^{\boldsymbol{\eta}}(g - d_{\boldsymbol{\varepsilon}}(Z)Y^{\boldsymbol{\varepsilon}}) \in L.$$

The polynomial $p(Z) = c_{\boldsymbol{\delta}}(Z)d_{\boldsymbol{\varepsilon}}(Z)$ has the required property, since

$$p(Z)x_1^{a_1-1}\cdots x_k^{a_k-1} \in J \cap L = I.$$

4. Suppose $I$ is primary. If $m = x_k u \in G(I)$ for some $u \in \operatorname{Mon} A$, then, since $G(I)$ is minimal, $u \notin I$, hence $x_k^a \in I$ for some $a \in \mathbb{N}$.

Conversely, after reordering the variables, we may assume that all the monomials of $G(I)$ belong to $K[x_1, \dots, x_r]$ with $r \leq n$. Then, from the hypothesis it follows that $\sqrt{I} = (x_1, \dots, x_r)$ by **T.2.6**.4.
If

$$\varphi \colon K[x_1, \dots, x_n] \longrightarrow K(x_{r+1}, \dots, x_n)[x_1, \dots, x_r]$$

is the inclusion homomorphism, then $\sqrt{(\varphi(I))} = (x_1, \dots, x_r)$.
This ideal is maximal in $K(x_{r+1}, \dots, x_n)[x_1, \dots, x_r]$, hence $(\varphi(I))$ is primary by **T.1.7**.2. Since $(\varphi(I))^c = I$, because the ideals share the same set of generators, $I$ is primary by **T.1.17**.8. □

Applying **T.2.6**.1, every monomial ideal can be expressed as the intersection of irreducible monomial ideals.
For example, let

$$I = (x_1^3x_2,\ x_1^4x_2x_3^3,\ x_1^7,\ x_1^2x_3^3).$$

Since $x_1^2x_3^3 \mid x_1^4x_2x_3^4$, we first reduce the set of generators, and we can write $I = (x_1^3x_2,\ x_1^7,\ x_1^2x_3^3)$.
By **T.2.6**.1, using the monomial $x_1^2x_3^3$, we decompose the ideal as

$$I = (I, x_1^2) \cap (I, x_3^3) = (x_1^2) \cap (x_1^3x_2,\ x_1^7,\ x_3^3).$$

Repeating this procedure using the monomial $x_1^3x_2$, and reducing the generators, we obtain

$$I = (x_1^2) \cap (x_1^3,\ x_3^3) \cap (x_1^7,\ x_2,\ x_3^3).$$

This is a decomposition of $I$ as an intersection of irreducible, hence primary, monomial ideals.

## 2.2 Monomial Orderings

In order to define a division for multivariate polynomials, we need to extend the concepts of degree and leading coefficient to polynomials in several variables. For this purpose, we define an ordering on the set Mon $A$. This is equivalent to defining an ordering on $\mathbb{N}^n$.

We consider orderings having the following properties, and we will call them *monomial orderings.*

**Monomial Orderings**

A monomial ordering is an order relation $>$ on $\mathbb{N}^n$ or, equivalently, on Mon $A$ which satisfies the following properties:

i) $>$ is a total order;

ii) $>$ is a well-order, *i.e.*, every non-empty subset of $\mathbb{N}^n$ has a minimal element with respect to $>$, or, equivalently, every descending chain in $\mathbb{N}^n$ is stationary, see **E.9.1** or **T.7.1**;

iii) let $\mathbf{a}$, $\mathbf{b}$, $\mathbf{c} \in \mathbb{N}^n$, if $\mathbf{a} > \mathbf{b}$, then $\mathbf{a}+\mathbf{c} > \mathbf{b}+\mathbf{c}$.

Some of the most commonly used monomial orderings are the following, see **E.9.2** for a proof.

Let $\mathbf{a} = (a_1, \ldots, a_n)$, $\mathbf{b} = (b_1, \ldots, b_n) \in \mathbb{N}^n$, and set $|\,\mathbf{a}\,| = \sum_{i=1}^n a_i$.

Lex or *lexicographic order*
$\mathbf{a} >_{\text{lex}} \mathbf{b}$ if and only if the first non-zero entry of $\mathbf{a} - \mathbf{b}$ is positive.

Deglex or *degree lexicographic order*
$\mathbf{a} >_{\text{deglex}} \mathbf{b}$ if and only if $|\,\mathbf{a}\,| > |\,\mathbf{b}\,|$ or if $|\,\mathbf{a}\,| = |\,\mathbf{b}\,|$ and $\mathbf{a} >_{\text{lex}} \mathbf{b}$.

Degrevlex or *degree reverse lexicographic order*
$\mathbf{a} >_{\text{degrevlex}} \mathbf{b}$ if and only if $|\,\mathbf{a}\,| > |\,\mathbf{b}\,|$ or if $|\,\mathbf{a}\,| = |\,\mathbf{b}\,|$ and the last non-zero component of $\mathbf{a} - \mathbf{b}$ is negative.

We remark that, in the correspondence

$$X^{\mathbf{a}} = x_1^{a_1} x_2^{a_2} \cdots x_n^{a_n} \longleftrightarrow (a_1, a_2, \ldots, a_n),$$

the lex order sorts the variables as $x_1 > x_2 > \ldots > x_n$. We say that this is the lex order with $x_1 > x_2 > \ldots > x_n$.

For example, let $m_1 = x^2y$ and $m_2 = xy^3$ in $K[x, y]$. If $>_1$ is the lex order with $x > y$, then $m_1 = x^2y >_1 xy^3 = m_2$. If $>_2$ is the lex order with $y > x$, then $m_2 = y^3x >_2 yx^2 = m_1$.

Assume a monomial ordering $>$ is fixed.
For every polynomial $0 \neq f = \sum_{\mathbf{a}} c_{\mathbf{a}} X^{\mathbf{a}} \in A$ with $c_{\mathbf{a}} \in K$, we introduce the following definitions:

$\mathrm{Deg}(f) = \max_{>}\{\mathbf{a} \in \mathbb{N}^n : c_{\mathbf{a}} \neq 0\}$ is the *multidegree* of $f$;

$\mathrm{lc}(f) = c_{\mathrm{Deg}(f)}$ is the *leading coefficient* of $f$;

$\mathrm{lm}(f) = X^{\mathrm{Deg}(f)}$ is the *leading monomial* of $f$;

$\mathrm{lt}(f) = \mathrm{lc}(f)\,\mathrm{lm}(f) = c_{\mathrm{Deg}(f)} X^{\mathrm{Deg}(f)}$ is the *leading term* of $f$.

If $f$, $g \in A$ are non-zero polynomials, then:

i) $\mathrm{Deg}(fg) = \mathrm{Deg}(f) + \mathrm{Deg}(g)$;
ii) if $f + g \neq 0$, then $\mathrm{Deg}(f+g) \leq \max_{>}\{\mathrm{Deg}(f), \mathrm{Deg}(g)\}$.

Moreover, we associate to every ideal $0 \neq I \subset A$ the set

$$\mathrm{Deg}(I) = \{\mathrm{Deg}(f) \colon f \in I,\ f \neq 0\}.$$

It is easy to verify that $\mathrm{Deg}(I)$ is an $\mathcal{E}$-subset. Therefore, by **T.2.5**, it has a staircase, called the *staircase* of $I$.

It is also possible to associate to $I$ the monomial ideal

$$\mathrm{Lt}(I) = (\mathrm{lt}(f) \colon f \in I,\ f \neq 0) = (\mathrm{lm}(f) \colon f \in I,\ f \neq 0),$$

which is called the *initial ideal* or *leading term ideal* of $I$ with respect to $>$.
By definition, the staircase of $I$ and the staircase of $\mathrm{Lt}(I)$ are the same.
In general, for any subset $F \subseteq A \setminus \{0\}$ we define $\mathrm{Lt}(F) = (\mathrm{lt}(f) \colon f \in F)$.
Clearly, if $G \neq \{0\}$ is a subset of an ideal $I$, then $\mathrm{Lt}(G) \subseteq \mathrm{Lt}(I)$.

**Gröbner Basis**

Let $>$ be a monomial ordering, and let $I$ be a non-zero ideal of $A$. A set $G = \{g_1, \ldots, g_t\} \subset I \setminus \{0\}$ is a *Gröbner basis of $I$* with respect to $>$ if

$$\mathrm{Lt}(G) = (\mathrm{lt}(g_1), \ldots, \mathrm{lt}(g_t)) = \mathrm{Lt}(I).$$

By convention, the empty set generates the zero ideal, and we say that $G = \emptyset$ is the Gröbner basis of $I = (0)$.

## 2.3 Division in $K[x_1, \ldots, x_n]$

In the remaining sections of this chapter we always assume to have fixed a monomial ordering $>$ on $K[x_1, \ldots, x_n]$.

Let $f$, $g \in A \setminus \{0\}$. We say that a polynomial $f$ *reduces to a polynomial* $h \in A$ *modulo* $g$ *(in one step)* if and only if there exists a term $t_{\mathbf{b}} = c_{\mathbf{b}}X^{\mathbf{b}}$ in $f$ such that $\mathrm{lt}(g) \mid t_{\mathbf{b}}$ and $h = f - \dfrac{t_{\mathbf{b}}}{\mathrm{lt}(g)}g$.

The reduction operation is denoted by $f \xrightarrow{g} h$.

Note that with the reduction operation, the term $t_{\mathbf{b}}$ is replaced by a sum of terms of multidegree strictly smaller than $\mathbf{b}$. Furthermore, this process can be iterated until none of the terms of $f$ is divisible by $\mathrm{lt}(g)$. In this case we write $f \xrightarrow{g}_* h$ and say that $h$ is *reduced with respect to* $g$.

For example, consider the ring $\mathbb{Q}[x, y, z]$ and the lex order with $x > y > z$. Take $f = 3x^2y^3z^2 + xy^2z + 2xy$, $g = 2xy - z$, and $t_{\mathbf{b}} = 3x^2y^3z^2$. Then, $\mathrm{lt}(g) = 2xy$ and

$$f \xrightarrow{g} h = f - \frac{3x^2y^3z^2}{2xy}g = xy^2z + 2xy + \tfrac{3}{2}xy^2z^3.$$

The term $t_{\mathbf{b}}$ has been replaced by the term $\frac{3}{2}xy^2z^3$ whose multidegree is strictly smaller, because $(1, 2, 3) < (2, 3, 2)$.
Repeating this process, we obtain

$$\begin{aligned} f &\xrightarrow{g} \tfrac{3}{2}xy^2z^3 + xy^2z + 2xy \xrightarrow{g} \tfrac{3}{2}xy^2z^3 + xy^2z + z \\ &\xrightarrow{g} xy^2z + \tfrac{3}{4}yz^4 + z \xrightarrow{g}_* \tfrac{3}{4}yz^4 + \tfrac{1}{2}yz^2 + z. \end{aligned}$$

We can work in an analogous way with respect to a set of polynomials.

Let $f, f_1, \dots, f_s \in A\setminus\{0\}$, and let $F = \{f_1, \dots, f_s\}$. We say that $f$ *reduces to* $r \in A$ *modulo* $F$ when there exist indices $i_1, \dots, i_k \in \{1, \dots, s\}$ and polynomials $h_1, \dots, h_{k-1} \in A$ such that

$$f \xrightarrow{f_{i_1}} h_1 \xrightarrow{f_{i_2}} h_2 \xrightarrow{f_{i_3}} \dots \xrightarrow{f_{i_{k-1}}} h_{k-1} \xrightarrow{f_{i_k}} r.$$

The reduction of a polynomial $f$ modulo $F$ is denoted by $f \xrightarrow{F} r$.

A polynomial $r \in A$ is *reduced with respect to* $F$ when $r$ cannot be further reduced modulo $F$, *i.e.*, if $r = 0$ or if no monomial of $r$ is divisible by $\mathrm{lt}(f_i)$, for $i = 1, \dots, s$.

If $f \xrightarrow{F} r$ and $r$ is reduced with respect to $F$, then the polynomial $r$ is called *a remainder* of $f$ with respect to $F$, and we write $f \xrightarrow{F}_* r$.

We say *a* remainder because the reduction process depends on the order in which we use the polynomials of the set $F$.
For example, let $f = x_1x_2 - x_2$, $f_1 = x_1 - 1$, and $f_2 = x_1x_2$ be elements of $K[x_1, x_2]$ ordered with the lex order with $x_1 > x_2$.
Then, $f \xrightarrow{f_1}_* 0$, while $f \xrightarrow{f_2}_* -x_2$.

However, the reduction operation allows to define a notion of division of a multivariate polynomial by a set of polynomials similar to that of polynomials in one variable.
We divide a polynomial $f$ by $\{f_1, \dots, f_s\}$ using the procedure described by the following algorithm, excluding the trivial cases in which $f$ or $f_i = 0$.

**Division Algorithm**

**Input** $f, f_1, \dots, f_s \in A \setminus \{0\}$; $>$ monomial ordering.
**Output** $u_1, \dots, u_s$, $r \in A$ such that:

$f = \sum_{i=1}^{s} u_i f_i + r$;
$r$ is reduced with respect to $\{f_1, \dots, f_s\}$;
if $u_i f_i \neq 0$, then $\mathrm{Deg}(u_i f_i) \leq \mathrm{Deg}(f)$.

**Initialization** $p := f$; $u_1 := 0$; $u_2 := 0; \dots$; $u_s := 0$; $r := 0$;
  **while** $p \neq 0$ **repeat**
    **if** there exists $j$ such that $\mathrm{lt}(f_j) \mid \mathrm{lt}(p)$ **then**

$i := \min\{j \colon \mathrm{lt}(f_j) \mid \mathrm{lt}(p)\}$

$$u_i := u_i + \frac{\mathrm{lt}(p)}{\mathrm{lt}(f_i)}$$

$$p := p - \frac{\mathrm{lt}(p)}{\mathrm{lt}(f_i)} f_i$$

    **else**

$r := r + \mathrm{lt}(p)$

$p := p - \mathrm{lt}(p)$

    **endif**
  **endwhile**
**Return** $\{u_1, \dots, u_s, r\}$

The polynomials $u_1, \dots, u_s$, and $r$ provided by the previous algorithm satisfy the following properties.

**Division Theorem**

**T. 2.8.** (→ p. 216) For each set of non-zero polynomials $F = \{f_1,\ldots,f_s\}$, and any $f \in A \setminus \{0\}$ there exist polynomials $u_1,\ldots,u_s$, and $r \in A$ such that:

i) $f = \sum_{i=1}^{s} u_i f_i + r$;
ii) $r$ is reduced with respect to $F$;
iii) if $u_i f_i \neq 0$, then $\mathrm{Deg}(u_i f_i) \leq \mathrm{Deg}(f)$.

By the previous theorem, the leading monomial of $f$ appears either in $r$ or in one of the leading monomials of $u_i f_i$.
More precisely,

$$\mathrm{lm}(f) = \begin{cases} \max\{\max_{u_i\neq 0}\{\mathrm{lm}(u_i)\mathrm{lm}(f_i)\},\, \mathrm{lm}(r)\} & \text{if } r \neq 0; \\ \max_{u_i \neq 0}\{\mathrm{lm}(u_i)\mathrm{lm}(f_i)\} & \text{otherwise.} \end{cases} \tag{2.1}$$

We also remark that the Division Algorithm just described depends on the order of the polynomials $f_1,\ldots,f_s$ when choosing the minimum index $i$ such that $\mathrm{lt}(f_i) \mid \mathrm{lt}(p)$. This choice may affect the remainder produced by the division, see **E.9.5**. We will show that this is not the case when $f_1,\ldots,f_s$ are a Gröbner basis for the ideal they generate. Indeed, this property characterizes Gröbner bases.

## 2.4 Gröbner Bases: First Properties

By definition, if $I \neq 0$ is an ideal of $A$, then a set $G = \{g_1,\ldots,g_t\} \subset I\setminus\{0\}$ is a Gröbner basis of $I$ with respect to $>$ when $\mathrm{Lt}(G) = (\mathrm{lt}(g_1),\ldots,\mathrm{lt}(g_t)) = \mathrm{Lt}(I)$. Equivalently, $G$ is a Gröbner basis when $\{\mathrm{Deg}(g_1),\ldots,\mathrm{Deg}(g_t)\}$ is a boundary of $\mathrm{Deg}(I)$, *i.e.*, when

$$\mathrm{Deg}(I) = \bigcup_{i=1}^{t}(\mathrm{Deg}(g_i) + \mathbb{N}^n).$$

We remark that, in general, a Gröbner basis with respect to one monomial ordering is not a Gröbner basis with respect to a different one, see **E.9.6**.

**T. 2.9.** (→ p. 216) Let $I \subset A$ be a non-zero ideal, and consider a set $G = \{g_1,\ldots,g_t\} \subset I \setminus \{0\}$. Then, the following properties are equivalent:

1. $G$ is a Gröbner basis of $I$ with respect to $>$;

2. $f \in I$ if and only if $f \xrightarrow{G}_* 0$;
3. $f \in I \setminus \{0\}$ if and only if

$$f = \sum_{i=1}^{t} u_i g_i \quad \text{with} \quad \mathrm{lm}(f) = \max_{u_i \neq 0}\{\mathrm{lm}(u_i)\mathrm{lm}(g_i)\}.$$

**T. 2.10.** ($\rightarrow$ p. 217) Let $G = \{g_1, \dots, g_t\}$ be a Gröbner basis of $I$ with respect to $>$. Then, $I = (g_1, \dots, g_t)$, *i.e.*, a Gröbner basis of $I$ is a set of generators of $I$.

We also say that a set $G \subset A$ is a *Gröbner basis* when $G$ is a Gröbner basis of the ideal $(G)$.
The following statement presents a fundamental property of Gröbner basis.

**Uniqueness of the Remainder**

**T. 2.11.** ($\rightarrow$ p. 217) Let $G = \{g_1, \dots, g_t\} \subset I \setminus \{0\}$ be a Gröbner basis of $I$ with respect to $>$. Then, for every $f \subset A$ the remainder of the division of $f$ by $G$ is unique.

The converse of the previous statement is also true, and this property characterizes Gröbner bases. For a complete proof see [1, Theorem 1.6.7].
Therefore, given a Gröbner basis $G$ and a polynomial $f \in A$, there exists *a unique $r$ reduced modulo $G$ such that $f \xrightarrow{G}_* r$.*
We call $r$ *the remainder* of the division of $f$ by $G$, and we denote it by $\overline{f}^G$.

It is important to note that, while the remainder of the division of a polynomial by a Gröbner basis $G$ is unique and does not depend on the order in which the polynomials are used, this is not true for the coefficients $u_i$ of the polynomials of $G$, that appear in the representation of $f$.

For example, let $G = \{g_1, g_2\} = \{x_1 + x_3, x_2 - x_3\}$ and $f = x_1x_2$. It is not difficult to verify that $G$ is a Gröbner basis with respect to the lex order with $x_1 > x_2 > x_3$. Nevertheless, if we divide $f$ first by $g_1$ and then by $g_2$, we obtain $f = x_2g_1 - x_3g_2 - x_3^2$, while, if we divide $f$ first by $g_2$ and then by $g_1$, we obtain $f = x_1g_2 + x_3g_1 - x_3^2$. Thus, the remainder is unique, but the coefficients of the polynomials $g_1$ and $g_2$ in the two expressions of $f$ are different.

**T. 2.12.** ($\rightarrow$ p. 217) Let $G = \{g_1, \dots, g_t\}$ and $G' = \{g'_1, \dots, g'_{t'}\}$ be two Gröbner bases of $I \subset A$ with respect to the same monomial ordering $>$, and let $r$ and $r'$ be the remainders of the division of a polynomial $f \in A$ by $G$ and $G'$, respectively. Then, $r = r'$.

As a consequence of the previous results, we obtain the following important theorem.

**Hilbert's Basis Theorem**

**T. 2.13.** Every ideal of $A$ is finitely generated.

**Proof T. 2.13.** Let $I \neq 0$ be an ideal of $A$. Then, the initial ideal $\mathrm{Lt}(I)$ with respect to some monomial ordering has a finite set of monomial generators, say $\mathrm{Lt}(I) = (m_1, \dots, m_t)$.
For all $i = 1, \dots, t$, there exists $f_i \in I$ such that $\mathrm{lt}(f_i) = m_i$. Hence, $\{f_1, \dots, f_t\}$ is a Gröbner basis of $I$, and $I = (f_1, \dots, f_t)$ by **T.2.10**. □

**T. 2.14.** Let $B$ be a ring. Then, every ideal of $B$ is finitely generated if and only if every ascending chain of ideals of $B$ is stationary.

We leave the proof of this fact to the reader as an exercise. For the general proof in the case of modules see **T.7.2**.1.

A ring is called *Noetherian* when every ascending chain of its ideals is stationary, see Chapter 7. Therefore, the ring of polynomials with finitely many variables and coefficients in $K$ is Noetherian.
For the general result see **T.7.4**.

## 2.5 Buchberger's Algorithm

If $G$ is a Gröbner basis of $I$ with respect to $>$, then $G$ is a set of generators of $I$. Moreover, for each polynomial $f \in I$, there exists $g_i \in G$ such that $\mathrm{lt}(g_i) \mid \mathrm{lt}(f)$. This property does not hold for a set of generators $\{f_1, \dots, f_k\}$ which is not a Gröbner basis, because of the cancellations that may occur between the terms of maximal multidegree in the expression $f = \sum_i u_i f_i$.

For example, let $I = (f_1, f_2)$ with $f_1 = x_1x_2^2 + x_1$ and $f_2 = x_1^2x_2 + x_2$. If $>$ is the lex order with $x_1 > x_2$, then $f = x_1f_1 - x_2f_2 = x_1^2 - x_2^2 \in I$, but $x_1^2 \notin (x_1x_2^2, x_1^2x_2) = (\mathrm{lt}(f_1), \mathrm{lt}(f_2))$.

Now, we address the problem of the construction of a Gröbner basis.
Let $f$, $g$ be non-zero polynomials with $\mathrm{Deg}(f) = \mathbf{a}$ and $\mathrm{Deg}(g) = \mathbf{b}$.
Let also $X^{\mathbf{c}}$ be the least common multiple of $\mathrm{lm}(f)$ and $\mathrm{lm}(g)$, that is, $\mathbf{c} = (\max\{a_1, b_1\}, \dots, \max\{a_n, b_n\})$.

***S*-polynomial**

The *$S$-polynomial of $f$ and $g$* is defined as

$$S(f, g) = \frac{X^{\mathbf{c}}}{\mathrm{lt}(f)} f - \frac{X^{\mathbf{c}}}{\mathrm{lt}(g)} g.$$

This definition leads to an important criterion to decide whether a set $G$ is a Gröbner basis. For its proof see [5, Chapter 2, §6].

**Buchberger's Criterion**

Let $I = (g_1, \ldots, g_t)$ be an ideal of $A$. Then, $\{g_1, \ldots, g_t\}$ is a Gröbner basis with respect to $>$ if and only if

$$\overline{S(g_i, g_j)}^G = 0 \text{ for all } i, j = 1, \ldots, t.$$

The previous criterion can be used to provide an algorithm for the construction of a Gröbner basis of an ideal $I = (f_1, \ldots, f_s)$.
For each pair of generators, we compute a remainder of their $S$-polynomial with respect to $\{f_1, \ldots, f_s\}$. If any of these remainders is non-zero, we add it to the current set of generators, obtaining a new set of generators $G$ of $I$. Repeating this procedure, new polynomials are added to $G$, until all of its $S$-polynomials reduce to zero with respect to the current set of generators. If this construction terminates in a finite number of steps, the criterion guarantees that the result is a Gröbner basis.

**Buchberger's Algorithm**

**Input** $I = (f_1, \ldots, f_s) \subseteq A$, $f_i \neq 0$ for $i = 1, \ldots, s$; $>$ a monomial ordering.
**Output** A Gröbner basis $G$ of $I$ with respect to $>$ containing $\{f_1, \ldots, f_s\}$.
**Initialization**

$F := \{f_1, \ldots, f_s\}$
$G := F$
$\Sigma := \{(f_i, f_j) \in G \times G \colon f_i \neq f_j\}$

**while** $\Sigma \neq \emptyset$ **repeat**
  **choose** $(f, g) \in \Sigma$
    $\Sigma := \Sigma \setminus \{(f, g)\}$
    $p := \overline{S(f, g)}^G$
    **if** $p \neq 0$ then
      $\Sigma := \Sigma \cup \{(h, p) \colon h \in G\}$
      $G := G \cup \{p\}$
    **endif**
**endwhile**
**Return** $G$

At each step of the algorithm, the set $G$ is contained in $I$ because when updating $G$, we add remainders of $S$-polynomials, and these remainders are elements of $I$, reduced with respect to the current $G$.

**T. 2.15.** Buchberger's Algorithm terminates and it is correct.

**Proof T. 2.15.** The termination condition, $\overline{S(f,g)}^G = 0$ for each pair $f, g \in G$, guarantees that the algorithm is correct and that the output is a Gröbner basis, by Buchberger's Criterion.

Since at each step the set $G_{i+1}$ is constructed adding to $G_i$ an element of $I$ that does not reduce to zero modulo $G_i$, the chain $\mathrm{Lt}(G_1) \subsetneq \mathrm{Lt}(G_2) \subsetneq \ldots$ is an ascending chain of ideals of $A$.
Since $A$ is Noetherian, see **T.2.14**, the chain is stationary, *i.e.*, the process stops and the algorithm terminates. $\square$

We end this section with a property that can be used to reduce the computation of a Gröbner basis from a set of generators.

**T. 2.16.** ($\rightarrow$ p. 217) Let $f$, $g \in A$ be such that $\gcd(\mathrm{lt}(f), \mathrm{lt}(g)) = 1$. Then, for any set $G$ containing $f$ and $g$, we have $\overline{S(f,g)}^G = 0$.
In particular, if $G = \{g_1, \ldots, g_t\} \subset A \setminus \{0\}$ and $\gcd(\mathrm{lt}(g_i), \mathrm{lt}(g_j)) = 1$ for all $i \neq j$, then $G$ is a Gröbner basis.

For an application on how to use it, see **E.9.8**.

## 2.6 Minimal and Reduced Gröbner Bases

We have seen that every ideal $I$ of $A$ has a Gröbner basis with respect to a fixed monomial ordering and how to construct it.

If we make all the elements of a Gröbner basis monic and choose a subset whose leading terms form a staircase of $\mathrm{Deg}(I)$, we obtain what is called a *minimal* Gröbner basis.
In other words, $G = \{g_1, \ldots, g_t\}$ is a minimal Gröbner basis of $I$ if and only if each $g_i$ is monic and $G(\mathrm{Lt}(I)) = \{\mathrm{lt}(g_1), \ldots, \mathrm{lt}(g_t)\}$.
Thus, by definition, a Gröbner basis $G$ of $I$, consisting of monic polynomials, is minimal exactly when $|G| = |G(\mathrm{Lt}(I))|$.

Given any Gröbner basis of $I$, it is always possible to extract a minimal Gröbner basis $G$. However minimality does not guarantee the uniqueness of the polynomials of $G$.
For example, let $I = (x_1, x_2) \subset K[x_1, x_2]$. Then, for every $a \in K$ and for every $k \in \mathbb{N}$, the sets $\{x_1 + a x_2^k, x_2\}$ are minimal Gröbner bases of $I$ with respect to the lex order with $x_1 > x_2$.

To obtain uniqueness, an additional condition on $G = \{g_1, \ldots, g_t\}$ is necessary: we require that any $g_i \in G$ is reduced with respect to $G \setminus \{g_i\}$, *i.e.*, none of the terms of $g_i$ is divisible by $\mathrm{lt}(g_j)$ for $j \neq i$.

**Reduced Gröbner Bases**

A Gröbner basis $G = \{g_1, \ldots, g_t\}$ of an ideal $I$ is *reduced* if

i) it is minimal;

ii) $\overline{g_i}^{G-\{g_i\}} = g_i$ for all $i = 1, \ldots, t$.

It is always possible to construct a reduced Gröbner basis as follows.

**Construction of a Reduced Gröbner Basis**

**T. 2.17.** (→ p. 218) Let $G = \{g_1, \ldots, g_t\}$ be a minimal Gröbner basis of an ideal $I \subset A$ with respect to $>$. Let the elements $g_1', \ldots, g_t'$ be defined via the following procedure:

$$g_1 \xrightarrow{G_1'}_* g_1', \qquad \text{with } G_1' = \{g_2, \ldots, g_t\};$$

$$\vdots$$

$$g_i \xrightarrow{G_i'}_* g_i', \qquad \text{with } G_i' = \{g_1', \ldots, g_{i-1}', g_{i+1}, \ldots, g_t\};$$

$$\vdots$$

$$g_t \xrightarrow{G_t'}_* g_t', \qquad \text{with } G_t' = \{g_1', \ldots, g_{t-1}'\}.$$

Then, $G' = \{g_1', \ldots, g_t'\}$ is a reduced Gröbner basis of $I$.

**Uniqueness of the Reduced Gröbner Basis**

**T. 2.18.** (→ p. 218) Let $G = \{g_1, \ldots, g_t\}$ and $G' = \{g_1', \ldots, g_t'\}$ be reduced Gröbner bases of an ideal $I$ with respect to the same monomial ordering $>$. Then, $G = G'$.

It should be noted that a minimal set of generators of $I$ is not necessarily a Gröbner basis of $I$. It should also be noted that a minimal or reduced Gröbner basis of $I$ is not necessarily a minimal set of generators of $I$.

For example, let $f_1 = x_1^3$, $f_2 = x_1^2x_2 - x_2^2$, and $I = (f_1, f_2) \subset K[x_1, x_2]$. Consider the lex order with $x_1 > x_2$. The reduced Gröbner basis of $I$ is $G = \{f_1, f_2, f_3, f_4\}$, where $f_3 = x_1x_2^2$ and $f_4 = x_2^3$. Hence, $G$ has 4 elements, while $I$ can be generated by 2 elements.

## 2.7 Some Applications

In this section, we use the previous results to find constructive solutions to the following problems:

1 . given a polynomial $f \in A$ and an ideal $I = (f_1, \ldots, f_s) \subseteq A$, decide whether $f \in I$;

1′. when $f \in I$, find $c_1, \ldots, c_s \in A$ such that $f = \sum_{i=1}^{s} c_i f_i$;

2 . decide whether two ideals $I$, $J \subset A$ are equal;

3 . find a canonical representation for the elements of the quotient $A/I$;

4 . find a basis for the $K$-vector space $A/I$;

5 . decide whether an element is invertible in $A/I$, and, when it is invertible, compute its inverse.

For the first of these problems, we compute a Gröbner basis $G$ of $I$, and then, verify whether the remainder of the division of $f$ by $G$ is 0. Applying **T.2.9** and **T.2.11**, we obtain the following result.

**Membership Test**

**T. 2.19.** Let $I \subset A$ be an ideal, and let $f \in A$ be a polynomial. Let also $G = \{g_1, \ldots, g_t\}$ be a Gröbner basis of $I$ with respect to $>$. Then,

$$f \in I \text{ if and only if } \overline{f}^G = 0.$$

During the execution of Buchberger's Algorithm, it is possible to keep track of the representation of the new generators $g_j$ as a linear combination of the original polynomials $f_i$. At each step, if $\{h_1, \ldots, h_l\}$ is the current basis and a new polynomial $g$ is added, then

$$g = S(h_\alpha, h_\beta) - \sum_{i=1}^{l} v_i h_i,$$

where the coefficients $v_i$ are explicitly computed by the Division Algorithm. In this way, Buchberger's Algorithm returns not only the Gröbner basis $G$ as output, but also a matrix $M$ of size $t \times s$ with entries in $A$ such that

$$M(f_1, \ldots, f_s)^{\mathrm{t}} = (g_1, \ldots, g_t)^{\mathrm{t}}. \tag{2.2}$$

Hence, when we divide a polynomial $f \in I$ by $G = \{g_1, \ldots, g_t\}$ and obtain polynomials $u_1, \ldots, u_t$ such that $f = \sum_{i=1}^{t} u_i g_i$ we can express $f \in I$ as a combination of the polynomials $f_1, \ldots, f_s$ using (2.2). This construction solves problem 1′, see **E.9.12** for an explicit example.

A solution to problem 2 is an immediate consequence of the uniqueness of the reduced Gröbner basis **T.2.18**.

**Equality Test for Ideals**

**T. 2.20.** Let $I,\ J \subset A$ be ideals, and let $G$ and $G'$ be the reduced Gröbner bases of $I$ and $J$, respectively. Then,

$$I = J \text{ if and only if } G = G'.$$

The following criterion is also often useful.

**T. 2.21.** ($\rightarrow$ p. 218) Let $I,\ J \subseteq A$ be ideals such that $I \subseteq J$, and $\mathrm{Lt}(I) = \mathrm{Lt}(J)$ for some monomial ordering $>$. Then, $I = J$.

**T. 2.22.** Let $G$ be a Gröbner basis of an ideal $I$ with respect to $>$.

1. The map $\overline{\phantom{f}}^G : A \longrightarrow A$ defined by $f \mapsto \overline{f}^G$ is $K$-linear.
2. For any $f,\ g \in A$, the following equivalence holds:

$$f \equiv g \mod I \text{ if and only if } \overline{f}^G = \overline{g}^G.$$

**Proof T. 2.22.** 1. To verify that the equality

$$\overline{af + bg}^G = a\overline{f}^G + b\overline{g}^G$$

holds for any $f,\ g \in A$ and any $a,\ b \in K$, we let $r = \overline{f}^G$ and $s = \overline{g}^G$, and then, prove that $ar + bs$ is the remainder of $af + bg$ with respect to $G$.
We can write

$$af + bg = a(f - r) + b(g - s) + ar + bs,$$

where $f - r$ and $g - s$ belong to $I$.
Since the monomials of $ar + bs$ are monomials of $r$ or of $s$, which are reduced modulo $G$, we have that $ar + bs$ is the remainder of $af + bg$ with respect to $G$.
2. We have $f \equiv g \mod I$ if and only if $f - g \in I$. By **T.2.9**, this is equivalent to $\overline{f - g}^G = 0$. Hence, $\overline{f}^G - \overline{g}^G = 0$ by part 1. □

Therefore,

$$\left\{\overline{f}^G : f \in A\right\}$$

is a set of representatives of $A/I$. Indeed, if $r = \overline{f}^G$, then $f - r \in I$, and therefore, $\overline{f} = \overline{r}$ in $A/I$, see problem 3.

The following result solves problem 4.

**$K$-basis of $A/I$**

**T. 2.23.** ($\to$ p. 218) Let $I \subset A$ be an ideal, and let $G = \{g_1, \ldots, g_t\}$ be a Gröbner basis of $I$ with respect to $>$.

1. Let $\mathcal{B} = \{X^{\mathbf{a}}\colon \operatorname{lm}(g) \nmid X^{\mathbf{a}} \text{ for all } g \in G\} = \{X^{\mathbf{a}}\colon X^{\mathbf{a}} \notin \operatorname{Lt}(I)\}$. Then,
$$\overline{\mathcal{B}} = \{\overline{m} \in A/I\colon m \in \mathcal{B}\}$$
is a $K$-basis of $A/I$.
2. Let $f \in A$, and let
$$\overline{f}^G = \sum_{X^{\mathbf{a}} \in \mathcal{B}} c_{\mathbf{a}} X^{\mathbf{a}}.$$
Then, the coordinates of $\overline{f}$ with respect to the basis $\overline{\mathcal{B}}$ are $(c_{\mathbf{a}})_{X^{\mathbf{a}} \in \mathcal{B}}$.

The following result solves problem 5, and it provides a procedure for computing the inverse of $f \mod I$.

**T. 2.24.** ($\to$ p. 219) An element $f \in A$ is invertible modulo $I = (f_1, \ldots, f_s)$ if and only if $(I, f) = 1$.

We compute a Gröbner basis $G$ of the ideal $(I, f)$ with respect to $>$.
If $(I, f)$ is a proper ideal, then $f$ is not invertible modulo $I$.
Otherwise, we write $1 = u_1 g_1 + \ldots + u_t g_t$ as a combination of the elements of the basis $G$. Applying the procedure used for the solution of problem 1′, *i.e.*, applying a matrix like the one in (2.2), we obtain a combination
$$1 = h_1 f_1 + \ldots + h_s f_s + fg.$$
Therefore, $g$ is the inverse of $f$ modulo $I$.

### 2.7.1 Lexicographic Orderings and Elimination

Gröbner bases with respect to lexicographic orderings satisfy an important property, known as *elimination property*, which we will study in this section. This property has several important and practical applications.

Throughout this section, we will assume that $>$ is the lexicographic order with $x_1 > x_2 > \ldots > x_n$. Let $I \subset A$ be an ideal. For $k = 1, \ldots, n-1$, we define the ideal
$$I_k = I \cap K[x_{k+1}, \ldots, x_n]$$
as the *$k$-th elimination ideal of $I$*. This ideal is the contraction of $I$ with respect to the inclusion homomorphism $K[x_{k+1}, \ldots, x_n] \longrightarrow A$.

**Variable Elimination Theorem**

**T. 2.25.** Let $G$ be a Gröbner basis of an ideal $I \subset A$ with respect to $>$. Then, for any $k = 1, \ldots, n-1$,

$$G_k = G \cap K[x_{k+1}, \ldots, x_n]$$

is a Gröbner basis of the $k$-th elimination ideal $I_k$ of $I$.

**Proof T. 2.25.** Let $1 \leq k \leq n-1$ be a fixed integer. Since $G_k \subset I_k$, we have to prove that $\mathrm{Lt}(I_k) \subseteq \mathrm{Lt}(G_k)$. We show that for every polynomial $f \in I_k \subseteq I$, the monomial $\mathrm{lt}(f)$ is divisible by $\mathrm{lt}(g)$ for some $g \in G_k$.
Since $f \in I$, then $\mathrm{lt}(f)$ is divisible by $\mathrm{lt}(g)$ for some $g \in G$. Moreover, since $f \in I_k \subseteq K[x_{k+1}, \ldots, x_n]$ only the variables $x_{k+1}, \ldots, x_n$ may appear in $\mathrm{lt}(g)$.

The crucial property of the lex order is that every monomial divisible by $x_i$ with $i = 1, \ldots, k$ is greater than any monomial in $K[x_{k+1}, \ldots, x_n]$. Therefore, all the terms of $g$ which are smaller than $\mathrm{lt}(g)$ belong to $K[x_{k+1}, \ldots, x_n]$ as well. Thus, $g \in G_k$. □

As an application of the previous result, we obtain a procedure to construct a generating set for the intersection and quotient of ideals.

**T. 2.26.** (→ p. 219) Let $I$, $J \subseteq A$ be ideals.

1. [**Intersection of Ideals**] Let $t$ be a new variable and consider the ideal $(tI, (1-t)J) \subset A[t]$. Then,

$$I \cap J = (tI, (1-t)J) \cap A.$$

2. [**Quotient of Ideals**] Let $J = (f_1, \ldots, f_s)$. Then,

$$I : J = \bigcap_{i=1}^{s} \left( \frac{1}{f_i}(I \cap (f_i)) \right).$$

By part 1, a Gröbner basis of $I \cap J$, is obtained by applying **T.2.25** to $(tI, (1-t)J)$ in $A[t]$ using a lex order where $t > x_i$ for all $i$.
By part 2, the computation of the quotient of ideals can be reduced to the computation of intersections, and we can repeatedly apply part 1 to compute them.

**T. 2.27.** (→ p. 219) [**Radical Membership Test**] Let $I \subsetneq A$ be an ideal, and let $t$ be a new variable. Then,

$$f \in \sqrt{I} \text{ if and only if } (I, 1-tf) = A[t].$$

# Chapter 3
# Affine Algebraic Varieties

In this chapter we introduce the concepts of affine algebraic varieties and their associated ideals, which consist of polynomials that vanish on these varieties. In addition, we explore the properties of these varieties and ideals by using the theory of ideals developed so far. Furthermore, we provide a proof of the Hilbert Nullstellensatz, which establishes a fundamental relation between geometry and algebra. We then apply this theorem, along with the Gröbner basis theory, to analyze and study solutions of systems of polynomial equations.

## 3.1 Definitions and First Properties

Consider a subset $F \subset A = K[x_1, \dots, x_n]$. We define the *affine algebraic variety associated with* $F$, denoted by $\mathbb{V}_K(F)$, or simply $\mathbb{V}(F)$, as follows:

$$\mathbb{V}(F) = \{(a_1, \dots, a_n) \in K^n : f(a_1, \dots, a_n) = 0 \text{ for all } f \in F\}.$$

By Hilbert's Basis Theorem **T.2.13**, every ideal of $A$ is finitely generated. Moreover, if $I = (f_1, \dots, f_s)$ is an ideal, then it can be easily seen that $(a_1, \dots, a_n) \in K^n$ verifies $f(a_1, \dots, a_n) = 0$ for every $f \in I$ if and only if $f_i(a_1, \dots, a_n) = 0$ for every $i = 1, \dots, s$. Thus, $\mathbb{V}(I) = \mathbb{V}(F)$ for each set of generators $F$ of $I$.
Let $V \subseteq K^n$ be an affine variety. The *ideal associated with* $V$ or, simply the *ideal of* $V$, is defined as:

$$\mathbb{I}(V) = \{f \in A : f(\alpha) = 0 \text{ for all } \alpha \in V\} \subseteq A.$$

Additionally, we define the *coordinate ring of* $V$ as the ring

$$A/\mathbb{I}(V).$$



A. Bandini et al., *Commutative Algebra through Exercises*, UNITEXT 159,
https://doi.org/10.1007/978-3-031-56910-4_3

Analogous definitions can be given for arbitrary subsets of $K^n$. However, our primary focus will be on varieties. Therefore, initially, we will always assume that $V$ is of the form $V = \mathbb{V}(I)$, where $I$ is an ideal of $A$. For further details, refer to the appendix at the end of this chapter.

We define the maps

$$\begin{aligned} \mathcal{Z} = \{V \subseteq K^n\colon\ V \text{ is an affine variety}\} &\longleftrightarrow \mathcal{I} = \{I \subseteq A\colon\ I \text{ is an ideal}\} \\ \mathbb{V}(I) &\longleftarrow I \\ V &\longmapsto \mathbb{I}(V). \end{aligned}$$

**Ideal-Variety Correspondence**

**T. 3.1.** ($\to$ p. 220) Let $I$, $J \subseteq A$ be ideals, and let $V$, $W \subseteq K^n$ be affine varieties. Then:

1. $I \subseteq J \Rightarrow \mathbb{V}(I) \supseteq \mathbb{V}(J)$;
2. $I \subseteq \mathbb{I}(\mathbb{V}(I))$;
3. $\mathbb{V}(\mathbb{I}(\mathbb{V}(I))) = \mathbb{V}(I)$;
4. $V \subseteq W \iff \mathbb{I}(V) \supseteq \mathbb{I}(W)$;
5. $\mathbb{V}(I + J) = \mathbb{V}(I) \cap \mathbb{V}(J)$;
6. $\mathbb{V}(IJ) = \mathbb{V}(I) \cup \mathbb{V}(J)$;
7. $\mathbb{V}(I \cap J) = \mathbb{V}(I) \cup \mathbb{V}(J)$;
8. $\mathbb{V}(I) = \mathbb{V}(\sqrt{I})$.

In particular, statements 5 and 6 show that finite intersections and finite unions of affine varieties are affine varieties.
If $V \subseteq W$, then we say that $V$ is a *subvariety* of $W$.

We will see that we can define a topology on $K^n$, namely the *Zariski topology*, whose closed sets are the affine varieties, see **T.3.20**. The next result shows that this topological space is Noetherian.

**T. 3.2.** ($\to$ p. 220) Every descending chain of affine varieties is stationary.

An affine variety $V$ is said to be *irreducible* if it cannot be expressed as the union of two proper subvarieties $V_1$ and $V_2$, *i.e.*, $V$ is irreducible if and only if

$$V = V_1 \cup V_2 \implies V = V_1 \ \text{ or } \ V = V_2.$$

**T. 3.3.** A variety $V$ is irreducible if and only if $\mathbb{I}(V)$ is a prime ideal.

**Proof T. 3.3.** Let $fg \in \mathbb{I}(V)$. We have to prove that if $V$ is irreducible, then $f \in \mathbb{I}(V)$ or $g \in \mathbb{I}(V)$. For this purpose, we define the varieties $V_1 = V \cap \mathbb{V}(f)$ and $V_2 = V \cap \mathbb{V}(g)$, and show that $V_1 \cup V_2 = V$.
From **T.3.1**.6 it follows that

$$\begin{aligned} V_1 \cup V_2 &= (V \cap \mathbb{V}(f)) \cup (V \cap \mathbb{V}(g)) \\ &= V \cap (\mathbb{V}(f) \cup \mathbb{V}(g)) = V \cap (\mathbb{V}(fg)) = V, \end{aligned}$$

where the last equality holds because $\mathbb{V}(fg) \supseteq \mathbb{V}(\mathbb{I}(V)) = V$, see **T.3.1**.1 and 3. Since $V$ is irreducible, we have either $V = V_1$ or $V = V_2$. This implies $\mathbb{V}(f) \supseteq V$ or $\mathbb{V}(g) \supseteq V$, and therefore, either $f \in \mathbb{I}(\mathbb{V}(f)) \subseteq \mathbb{I}(V)$ or $g \in \mathbb{I}(\mathbb{V}(g)) \subseteq \mathbb{I}(V)$ by **T.3.1**.4.

Conversely, assume that $\mathbb{I}(V)$ is prime and that $V = V_1 \cup V_2$.
In this case, the corresponding ideals satisfy $\mathbb{I}(V) = \mathbb{I}(V_1 \cup V_2) = \mathbb{I}(V_1) \cap \mathbb{I}(V_2)$, where the last equality can be easily verified. Since $\mathbb{I}(V)$ is prime, it is also irreducible, see **T.1.13**. Consequently, we have either $\mathbb{I}(V_1) = \mathbb{I}(V)$ or $\mathbb{I}(V_2) = \mathbb{I}(V)$. Since $\mathbb{V}(\mathbb{I}(V)) = V$, we obtain that either $V_1 = V$ or $V_2 = V$.
Therefore, $V$ is irreducible. □

Note that the variety associated to a prime ideal is not always irreducible. For example, let $K = \mathbb{Z}/(2)$, $A = K[x, y]$, and $I = (x + y)$. Then, $A/I \simeq K[x]$ is a domain, and therefore, $I$ is prime, while $\mathbb{V}(I) = \{(0,0),(1,1)\} \subsetneq K^2$ is reducible, being the union of

$$V_1 = \{(0,0)\} = \mathbb{V}((x,y)) \quad \text{and} \quad V_2 = \{(1,1)\} = \mathbb{V}((x+1, y+1)),$$

which are varieties strictly contained in $\mathbb{V}(I)$. We also note that

$$V = \mathbb{V}((x,y)(x+1,y+1)) = \mathbb{V}((x^2+x, xy+x, xy+y, y^2+y)),$$

therefore $\mathbb{I}(\mathbb{V}(I)) \supseteq (x^2+x, xy+x, xy+y, y^2+y) \supsetneq I$, and $\mathbb{I}(\mathbb{V}(I))$ is not prime, by the previous result.

**Decomposition into Irreducible Varieties**

**T. 3.4.** (→ p. 220) 1. Every affine variety $V$ can be decomposed as a finite union of irreducible varieties, *i.e.*, there exist a positive integer $t$ and irreducible varieties $V_1, \dots, V_t$ such that $V = \bigcup_{i=1}^{t} V_i$.

2. If the decomposition $V = \bigcup_{i=1}^{t} V_i$ is minimal, *i.e.*, if $V_i \not\subseteq V_j$ for all $i \neq j$, then it is unique up to permutation.

We call the varieties $V_i$ the *irreducible components* of $V$.

We conclude this section with a remark about varieties consisting of a single point.

**T. 3.5.** (→ p. 221) Let $\alpha = (a_1, \dots, a_n) \in K^n$. Then, $\{\alpha\}$ is a variety. Furthermore,

$$\mathbb{I}(\{\alpha\}) = (x_1 - a_1, \dots, x_n - a_n)$$

is a maximal ideal, denoted by $\mathfrak{m}_\alpha$.

Therefore, for every point $\alpha = (a_1, \dots, a_n) \in K^n$, there exists a corresponding maximal ideal of $K[x_1, \dots, x_n]$. However, it is important to note that the converse is not true in general. As an example consider the maximal ideal $(x^2 + 1) \subset \mathbb{R}[x]$. In this case there is no element $\alpha \in \mathbb{R}$ such that $\alpha^2 + 1 = 0$.

## 3.2 The Resultant

Let $A$ be a domain, and let $f = \sum_{i=0}^{m} a_i x^i$ and $g = \sum_{i=0}^{\ell} b_i x^i$ be polynomials in $A[x] \setminus \{0\}$ not both constant and of degree $m$ and $\ell$, respectively.
We define the *Sylvester matrix of $f$ and $g$* as the $(m+\ell) \times (m+\ell)$ square matrix

$$\mathrm{Syl}(f,g) = \left.\begin{pmatrix} a_m & a_{m-1} & \cdots & \cdots & a_0 & 0 & \cdots & 0 \\ 0 & a_m & a_{m-1} & \cdots & \cdots & a_0 & & \vdots \\ \vdots & & \ddots & \ddots & & & \ddots & 0 \\ 0 & \cdots & 0 & a_m & a_{m-1} & \cdots & \cdots & a_0 \\ b_\ell & \cdots & b_1 & b_0 & 0 & \cdots & \cdots & 0 \\ 0 & \ddots & & \ddots & \ddots & & & \vdots \\ \vdots & & \ddots & & & \ddots & & \vdots \\ 0 & \cdots & 0 & b_\ell & \cdots & b_1 & b_0 & 0 \\ 0 & \cdots & \cdots & 0 & b_\ell & \cdots & b_1 & b_0 \end{pmatrix}\begin{matrix} \left.\vphantom{\begin{matrix} a \\ a \\ a \\ a \end{matrix}}\right\} \ell \\ \left.\vphantom{\begin{matrix} a \\ a \\ a \\ a \\ a \end{matrix}}\right\} m \end{matrix}\right. .$$

For further details on the construction of this matrix the reader can refer to [5, Chapter 3, §6].

The *resultant of $f$ and $g$*, denoted by $\mathrm{Res}(f,g)$, is defined as the determinant of the Sylvester matrix $\mathrm{Syl}(f,g)$, *i.e.*,

$$\mathrm{Res}(f,g) = \det \mathrm{Syl}(f,g).$$

If $f, g \in A \setminus \{0\}$ we set $\mathrm{Res}(f,g) = 1$.

In particular, if $m > 0$ and $\ell = 0$, then the Sylvester matrix is the diagonal $m \times m$ matrix with $g = b_0$ on the diagonal, hence $\mathrm{Res}(f,g) = b_0^m$.
Similarly, if $m = 0$ and $\ell > 0$, then $\mathrm{Res}(f,g) = a_0^\ell$.

The following statements immediately follow from the properties of the determinant.

**Properties of the Resultant 1**

**T. 3.6.** Let $A$, $f$, and $g$ be as above. Then:

1. $\operatorname{Res}(f,g) \in A$;
2. $\operatorname{Res}(f,g) = (-1)^{m\ell} \operatorname{Res}(g,f)$;
3. for all $a$, $b \in A$,

$$\operatorname{Res}(af,g) = a^\ell \operatorname{Res}(f,g) \quad \text{and} \quad \operatorname{Res}(f,bg) = b^m \operatorname{Res}(f,g).$$

The resultant also has the following important property.

**T. 3.7.** Let $A$, $f$, and $g$ be as above. Then, there exist polynomials $h_1$, $h_2 \in A[x]$ with $\deg h_1 < \deg g$ and $\deg h_2 < \deg f$ such that

$$\operatorname{Res}(f,g) = h_1 f + h_2 g.$$

In particular, $\operatorname{Res}(f,g) \in (f,g) \cap A$.

**Proof T. 3.7.** By multiplying the $i$-th column of the Sylvester matrix by $x^{m+\ell-i}$ and adding it to the last column for $i = 1, \ldots, m+\ell-1$, we obtain

$$\operatorname{Res}(f,g) = \det \operatorname{Syl}(f,g) = \det \begin{pmatrix} a_m & \cdots & \cdots & a_0 & & x^{\ell-1} f(x) \\ & \ddots & & & & \vdots \\ & & a_m & \cdots & \cdots & f(x) \\ b_\ell & \cdots & b_1 & b_0 & & x^{m-1} g(x) \\ & \ddots & & & & \vdots \\ & & b_\ell & \cdots & \cdots & g(x) \end{pmatrix}.$$

By expanding the determinant with respect to the last column, it is evident that the resultant is the sum of $f$ multiplied by a polynomial of degree at most $\ell - 1$ and of $g$ multiplied by a polynomial of degree at most $m-1$. $\square$

As a simple example consider $f = ax^2 + bx + c \in \mathbb{Q}[x]$ with $a \neq 0$. In this case, the Sylvester matrix of $f$ and $f'$ is given by

$$\operatorname{Syl}(f,f') = \begin{pmatrix} a & b & c \\ 2a & b & 0 \\ 0 & 2a & b \end{pmatrix}$$

and $\operatorname{Res}(f,f') = -a(b^2 - 4ac) = 4a^2 f - (2a^2 x + ab) f'$.

Hence, $\text{Res}(f, f') = 0$ if and only if $f$ has a double root, *i.e.*, if and only if $f$ and $f'$ have a common root.
This property is crucial and holds in general, we will prove it using the following statement.

**T. 3.8.** Consider a positive integer $m$, indeterminates $Y = y_1, \ldots, y_m$, and the polynomial $f_m(x)$ defined by

$$f_m(x) = \prod_{i=1}^{m} (x - y_i) = \sum_{i=0}^{m} a_i^{(m)} x^i \in A[Y][x] = A[Y, x].$$

Then, the following properties hold:

1. the coefficients $a_i^{(m)}$ are linear in $y_j$ for each $j = 1, \ldots, m$;
2. $a_i^{(m)}(y_1, \ldots, y_{m-1}, 0) = a_{i-1}^{(m-1)}(y_1, \ldots, y_{m-1})$ for each $0 < i \leq m$, and $a_0^{(m)}(y_1, \ldots, y_{m-1}, 0) = 0$;
3. for every $g = \sum_{i=0}^{\ell} b_i x^i \in A[x]$,

$$\text{Res}(f_m(x), g(x)) = g(y_m) \, \text{Res}(f_{m-1}(x), g(x)) \in A[Y].$$

**Proof T. 3.8.** 1. The coefficients of $f_m$ are the elementary symmetric polynomials in the variables $y_1, \ldots, y_m$.
Therefore,

$$\begin{aligned}
a_m^{(m)} &= 1; \\
a_{m-1}^{(m)} &= -(y_1 + \ldots + y_m); \\
a_{m-2}^{(m)} &= y_1 y_2 + \ldots + y_{m-1} y_m; \\
&\vdots \\
a_0^{(m)} &= (-1)^m y_1 y_2 \cdots y_m.
\end{aligned}$$

It follows that the coefficients $a_i^{(m)}$ are linear in each variable $y_j$.
2. By writing $f_m(x) = f_{m-1}(x)(x - y_m)$, we can observe that for every $i > 0$, we have

$$a_i^{(m)}(y_1, \ldots, y_{m-1}, y_m) = a_{i-1}^{(m-1)}(y_1, \ldots, y_{m-1}) - y_m a_i^{(m-1)}(y_1, \ldots, y_{m-1}).$$

When we evaluate this expression at $y_m = 0$, we obtain

$$a_i^{(m)}(y_1, \ldots, y_{m-1}, 0) = a_{i-1}^{(m-1)}(y_1, \ldots, y_{m-1}).$$

The statement for $i = 0$ follows immediately from part 1.

3. Consider the Sylvester matrix

$$\operatorname{Syl}(f_m, g) = \begin{pmatrix} a_m^{(m)} & \cdots & a_0^{(m)} & & & \\ & \ddots & & & & \\ & & a_m^{(m)} & & \cdots & a_0^{(m)} \\ b_\ell & \cdots & b_1 & b_0 & & \\ & \ddots & & & \ddots & \\ & & b_\ell & & \cdots & b_0 \end{pmatrix}.$$

By multiplying the $i$-th column of the Sylvester matrix by $y_m^{m+\ell-i}$ and adding it to the last column for $i = 1, \ldots, m+\ell-1$, we obtain

$$\begin{pmatrix} a_m^{(m)} & \cdots & a_0^{(m)} & & & y_m^{\ell-1} f_m(y_m) \\ & \ddots & & & & \vdots \\ & & a_m^{(m)} & & \cdots & f_m(y_m) \\ b_\ell & \cdots & b_1 & b_0 & & y_m^{m-1} g(y_m) \\ & \ddots & & & & \vdots \\ & & b_\ell & & \cdots & g(y_m) \end{pmatrix}.$$

Since $f_m(y_m) = 0$, it follows from the properties of the determinant that

$$\operatorname{Res}(f_m(x), g(x)) = g(y_m) \det \begin{pmatrix} a_m^{(m)} & \cdots & a_0^{(m)} & & & 0 \\ & \ddots & & & & \vdots \\ & & a_m^{(m)} & & \cdots & 0 \\ b_\ell & \cdots & b_1 & b_0 & & y_m^{m-1} \\ & \ddots & & & & \vdots \\ & & b_\ell & & \cdots & 1 \end{pmatrix}.$$

Let $M$ denote the latter matrix. Then, we have proved that

$$\operatorname{Res}(f_m(x), g(x)) = g(y_m) \det M.$$

Since $A$ is a domain, we have

$$\deg_{y_m}(\operatorname{Res}(f_m(x), g(x))) = \deg_{y_m}(g(y_m)) + \deg_{y_m}(\det M).$$

By part 1, we have

$$\deg_{y_m}(\operatorname{Res}(f_m(x), g(x))) = \deg_{y_m}(\det(\operatorname{Syl}(f_m, g))) \leq \ell.$$

Moreover, $\deg_{y_m}(g(y_m)) = \ell$. Therefore, $\deg_{y_m}(\det M) = 0$, and $\det M$ does not change when evaluated at $y_m = 0$. Applying part 2, we obtain

$$
\begin{aligned}
\operatorname{Res}(f_m(x), g(x)) &= g(y_m) \det \begin{pmatrix}
a_{m-1}^{(m-1)} & \cdots & a_0^{(m-1)} & \cdots & \cdots & 0 \\
& \ddots & & \ddots & & \vdots \\
& & a_{m-1}^{(m-1)} & \cdots & a_0^{(m-1)} & 0 \\
b_\ell & \cdots & b_1 & b_0 & & \vdots \\
& \ddots & & \ddots & & 0 \\
& & b_\ell & \cdots & b_1 & 1
\end{pmatrix} \\
&= g(y_m) \det \begin{pmatrix}
a_{m-1}^{(m-1)} & \cdots & a_0^{(m-1)} & \cdots & 0 \\
& \ddots & & \ddots & \vdots \\
& & a_{m-1}^{(m-1)} & \cdots & a_0^{(m-1)} \\
b_\ell & \cdots & b_1 & b_0 & \\
& \ddots & & \ddots & \\
& & b_\ell & \cdots & b_0
\end{pmatrix} \\
&= g(y_m) \operatorname{Res}(f_{m-1}(x), g(x)).
\end{aligned}
$$
□

**Properties of the Resultant 2**

**T. 3.9.** (→ p. 221) Let $K$ be a field, and let

$$
f = \sum_{i=0}^{m} a_i x^i \quad \text{and} \quad g = \sum_{i=0}^{\ell} b_i x^i
$$

be polynomials in $K[x]$ such that $a_m$, $b_\ell \neq 0$.
Let also $\alpha_1, \ldots, \alpha_m$ and $\beta_1, \ldots, \beta_\ell$ be the roots of $f$ and $g$ in $\overline{K}$, respectively. Then:

1. $\operatorname{Res}(f, g) = a_m^\ell \prod_{i=1}^m g(\alpha_i) = (-1)^{m\ell} b_\ell^m \prod_{j=1}^\ell f(\beta_j)$;
2. $\operatorname{Res}(f, g) = a_m^\ell b_\ell^m \prod\limits_{j=1}^{\ell} \prod\limits_{i=1}^{m} (\alpha_i - \beta_j)$;
3. $\operatorname{Res}(f, g) = 0$ if and only if $f$ and $g$ have a common root in $\overline{K}$;
4. $\operatorname{Res}(f, g) = 0$ if and only if $f$ and $g$ have a common factor of positive degree in $K[x]$.

The following result will be a valuable tool for proving the Extension Theorem **T.3.11**.

**T. 3.10.** (→ p. 222) Let $f$ and $g$ be polynomials in $K[x_1, \ldots, x_n]$ written as

$$
f = \sum_{i=0}^{m} c_i(x_2, \ldots, x_n) x_1^i \quad \text{and} \quad g = \sum_{i=0}^{\ell} d_i(x_2, \ldots, x_n) x_1^i,
$$

of degree $m$ and $\ell$ in $x_1$, respectively. Moreover, let $\beta \in K^{n-1}$.
If $f(x_1, \beta)$ has degree $m$ in $x_1$, $g(x_1, \beta)$ is non-zero and has degree $\ell - r$ in $x_1$ for some integer $0 \le r \le \ell$, then

$$\operatorname{Res}_{x_1}(f, g)(\beta) = c_m(\beta)^r \operatorname{Res}_{x_1}(f(x_1, \beta), g(x_1, \beta)).$$

## 3.3 The Extension Theorem

Before proving the Extension Theorem, we recall the following facts.
It is easy to verify, by induction on the number of variables, that a polynomial $f$ in $n$ variables with coefficients in an infinite field $K$ is non-zero if and only if there exists $\alpha \in K^n$ such that $f(\alpha) \neq 0$.
Moreover, every algebraically closed field $K = \overline{K}$ is infinite, because if $K = \{a_1, \dots, a_s\}$ is finite, then the polynomial $\prod_{i=1}^{s}(x - a_i) + 1 \in K[x]$ would have no roots in $K$, leading to a contradiction.
Therefore, $f \in \overline{K}[x_1, \dots, x_n]$ is non-zero if and only if there exists $\alpha \in \overline{K}^n$ such that $f(\alpha) \neq 0$.

For the rest of this section, we let $A = K[x_1, \dots, x_n]$. Let $I = (f_1, \dots, f_s)$ be an ideal of $A$. The variety associated to $I$ corresponds to the set of solutions of the system $\Sigma = \{f_i = 0 \colon i = 1, \dots, s\}$. If $k$ is a fixed integer, with $1 \le k \le n - 1$, and $I_k$ is the $k$-th elimination ideal of $I$, we call $(a_{k+1}, \dots, a_n) \in \mathbb{V}(I_k)$ a *partial solution* of $\Sigma$, or of $\mathbb{V}(I)$.

**Extension Theorem**

**T. 3.11.** Consider an algebraically closed field $K = \overline{K}$.
Let $I = (f_1, \dots, f_s)$ be a proper ideal of $A$, and let $I_1 = I \cap K[x_2, \dots, x_n]$ be the first elimination ideal of $I$. For each $i = 1, \dots, s$, write

$$f_i = c_i(x_2, \dots, x_n) x_1^{N_i} + f_i',$$

with $\deg_{x_1} f_i' < N_i$ and $c_i \in K[x_2, \dots, x_n] \setminus \{0\}$.
Let $\beta = (a_2, \dots, a_n) \in \mathbb{V}(I_1)$ be a partial solution of $\mathbb{V}(I)$.
If $\beta \notin \mathbb{V}(c_1, \dots, c_s)$, then there exists $a_1 \in K$ such that

$$(a_1, \beta) = (a_1, a_2, \dots, a_n) \in \mathbb{V}(I).$$

In this case we say that $(a_1, \beta)$ *extends* $\beta$.

**Proof T. 3.11.** Since $\beta \notin \mathbb{V}(c_1, \dots, c_s)$, there exists $i \in \{1, \dots, s\}$ such that $c_i(\beta) \neq 0$. Therefore, $\deg_{x_1} f_i(x_1, \beta) = \deg_{x_1} f_i(x_1, \dots, x_n) = N_i \geq 0$.
If $N_i = 0$, then $f_i(x_1, \dots, x_n) = c_i(x_2, \dots, x_n) \in I_1$, and $c_i(\beta) \neq 0$ contradicts the hypothesis $\beta \in \mathbb{V}(I_1)$. Thus, $N_i \geq 1$.
Let $\varphi \colon A \longrightarrow K[x_1]$ be the evaluation homomorphism defined by

$$f(x_1, x_2, \dots, x_n) \mapsto f(x_1, \beta),$$

which is surjective.
The image $I^e = \varphi(I)$ is a principal ideal, hence there exists $g \in I$ such that

$$I^e = (\varphi(g)) = (g(x_1, \beta)).$$

In particular, for each $f \in I$ we have that $g(x_1, \beta)$ divides $f(x_1, \beta)$, and, since $f_i(x_1, \beta) \neq 0$, we have $g(x_1, \beta) \neq 0$. Thus, if we prove that there exists $a_1 \in K$ such that $g(a_1, \beta) = 0$, we obtain $f(a_1, \beta) = 0$ for each $f \in I$, that is, $(a_1, \beta) \in \mathbb{V}(I)$ extends $\beta$.

To find $a_1$, we define $h(x_2, \dots, x_n) = \operatorname{Res}_{x_1}(f_i, g)$.
Since, by **T.3.7**, $h \in I_1$, then $h(\beta) = 0$ by hypothesis, and by **T.3.10**, we have

$$h(\beta) = c_i(\beta)^r \operatorname{Res}_{x_1}(f_i(x_1, \beta), g(x_1, \beta)),$$

where $\deg_{x_1} g(x_1, \beta) = \deg_{x_1} g - r$.
If $r = \deg_{x_1} g$, *i.e.*, if $g(x_1, \beta)$ is a non-zero constant, then we obtain a contradiction since

$$h(\beta) = c_i(\beta)^{\deg_{x_1} g} g(x_1, \beta)^{N_i} \neq 0.$$

Thus, as $K$ is algebraically closed and $\deg_{x_1} g(x_1, \beta) \geq 1$, the polynomial $g(x_1, \beta)$ has at least one root $a_1 \in K$. □

## 3.4 Hilbert's Nullstellensatz

We recall that the *(total) degree* of a term $cX^{\mathbf{a}}$, where $\mathbf{a} = (a_1, \dots, a_n) \in \mathbb{N}^n$ and $c \in K^*$, is defined as $|\,\mathbf{a}\,| = \sum_{i=1}^n a_i$.

A polynomial is said to be *homogenous* if all its terms have the same degree. A non-zero polynomial $f = f(x_1, \dots, x_n)$ is *homogenous of degree d* if and only if $f(tx_1, tx_2, \dots, tx_n) = t^d f(x_1, \dots, x_n)$.
By collecting terms of the same degree, every polynomial $f \in A$ can be expressed as a finite sum of homogenous polynomials of different degrees. In this case, the degree of $f$ is the maximum of these degrees.

The following result holds for every infinite $K$, *i.e.*, not necessarily algebraically closed.

**T. 3.12.** (→ p. 222) Assume $K$ is an infinite field, and let $f \in A$ be a polynomial of degree $N \geq 1$. Then, there exist $\alpha_2, \dots, \alpha_n \in K$ and a linear change of coordinates $\varphi\colon A \longrightarrow K[y_1, \dots, y_n]$ defined by

$$\begin{aligned} x_1 &\mapsto y_1, \\ x_i &\mapsto y_i + \alpha_i y_1 \quad \text{for } i = 2, \dots, n, \end{aligned}$$

such that

$$\varphi(f) = cy_1^N + f', \quad \text{with} \quad c \in K^* \quad \text{and} \quad \deg_{y_1} f' < N.$$

The following important theorem establishes a fundamental relation between ideals and their varieties.

**Hilbert's Nullstellensatz**

**T. 3.13.** [**Nullstellensatz, weak form**] Let $K = \overline{K}$, and let $I \subseteq A$ be an ideal. Then,

$$\mathbb{V}(I) = \emptyset \text{ if and only if } I = (1).$$

**T. 3.14.** [**Nullstellensatz, strong form**] Let $K = \overline{K}$, and let $I \subseteq A$ be an ideal. Then,

$$\mathbb{I}(\mathbb{V}(I)) = \sqrt{I}.$$

**Proof T. 3.13.** Clearly, if $I = (1)$, then $\mathbb{V}(I) = \emptyset$.

Conversely, we proceed by induction on the number of variables $n$.
If $n = 1$, then $I = (f)$ because $K[x]$ is a PID. Moreover, since $K$ is algebraically closed, the only polynomials without roots are the non-zero constants, and hence, $I = (1)$.
Suppose the claim is true for all $i < n$, and let $I = (f_1, \ldots, f_s) \subset A$ be such that $\mathbb{V}(I) = \emptyset$. If $f_1 \in K^*$, then $I = (1)$.
Otherwise, if $f_1$ has total degree $N \geq 1$, by **T.3.12**, there exists a linear change of coordinates $\varphi$ such that $\varphi(f_1) = cy_1^N + f'$, where $c \in K^*$ and $\deg_{y_1} f' < N$. Since $\varphi$ is an isomorphism, $\varphi(I)$ is an ideal of $K[y_1, \ldots, y_n]$ and $\mathbb{V}(\varphi(I)) = \emptyset$. From the Extension Theorem, it follows that $\mathbb{V}(\varphi(I)_1) = \emptyset$. In fact, every element $\beta \in \mathbb{V}(\varphi(I)_1)$ could be extended to some $\alpha \in \mathbb{V}(\varphi(I))$, because $c \in K^*$ and $\mathbb{V}((c)) = \emptyset$, but this contradicts the hypothesis.
Therefore, by the inductive hypothesis, $1 \in \varphi(I)_1 \subseteq \varphi(I)$, and thus, $1 \in I$. □

**Proof T. 3.14.** The inclusion $\sqrt{I} \subseteq \mathbb{I}(\mathbb{V}(\sqrt{I})) = \mathbb{I}(\mathbb{V}(I))$ follows from **T.3.1**.2 and 8, and holds over every field.

We prove the opposite inclusion by way of the Rabinowitsch trick.
Let $f \in \mathbb{I}(\mathbb{V}(I))$, and let $J = (I, 1 - tf) \subseteq K[t, x_1, \ldots, x_n]$. If we prove that $\mathbb{V}(J) = \emptyset$, then, by the weak form of the Nullstellensatz, we obtain $1 \in J$. The conclusion will follow from the Radical Membership Test **T.2.27**.
Let $\alpha = (b, \beta) \in K^{n+1}$ with $\beta = (a_1, \ldots, a_n) \in K^n$.
If $\beta \in \mathbb{V}(I)$, then

$$f(\beta) = 0 \quad \text{and} \quad (1 - tf)(b, \beta) = 1 - bf(\beta) = 1.$$

This implies that $\alpha \notin \mathbb{V}(J)$.

On the other hand, if $\beta \notin \mathbb{V}(I)$, there exists $g \in I$ such that $g(\beta) \neq 0$. Viewing $g$ as a polynomial in $K[t, x_1, \ldots, x_n]$ which does not depend on $t$, we have $g(b, \beta) \neq 0$, that is, $\alpha \notin \mathbb{V}(J)$ in this case as well.
Therefore, $\mathbb{V}(J) = \emptyset$. □

The Nullstellensatz has several immediate consequences. One of them is the characterization of the maximal ideals of $K[x_1, \ldots, x_n]$ when $K = \overline{K}$, see also **T.3.5**.

**T. 3.15.** (→ p. 223) Let $K = \overline{K}$. Then, $\mathfrak{m} \subset A$ is a maximal ideal if and only if there exists $\alpha = (a_1, \ldots, a_n) \in K^n$ such that

$$\mathfrak{m} = \mathfrak{m}_\alpha = (x_1 - a_1, \ldots, x_n - a_n).$$

**T. 3.16.** (→ p. 223) Let $K = \overline{K}$. Then, for every ideal $I \subseteq A$, the set $\operatorname{Min} I$ consisting of the *minimal prime ideals containing* $I$ is finite.
In particular, $\sqrt{I}$ can be expressed as a finite intersection of primes of $A$.

We will see later that this property is true in general for every ideal in a Noetherian ring, see **T.7.6** and **T.7.10**. Therefore, the assumption $K = \overline{K}$ is unnecessary for polynomial rings.

## 3.5 Systems of Polynomial Equations

In this section we use the Nullstellensatz to address the problem of solving systems of polynomial equations in $A = K[x_1, \ldots, x_n]$.

Consider a system of polynomial equations $\Sigma = \{f_i = 0 \colon i = 1, \ldots, s\}$, let $I = (f_1, \ldots, f_s)$, and suppose $G$ is a Gröbner basis of $I$ with respect to a fixed monomial ordering.
The system $\Sigma$ has solutions if and only if $\mathbb{V}(I) \neq \emptyset$, which is equivalent to $I \neq (1)$ by the Nullstellensatz. We can determine whether $I \neq (1)$ by using Gröbner bases.

**Solvability Test for Polynomial Systems**

The polynomial system $\Sigma$ has a solution in $\overline{K}^n$ if and only if $I \neq (1)$, *i.e.*, if and only if $1 \notin G$.

Once we have established the existence of solutions for $\Sigma$, we are also interested in the number of solutions, *i.e.*, the cardinality of $\mathbb{V}(I)$.

**T. 3.17.** Assume $K = \overline{K}$, and let $I = (G) \subset A$, where $G$ is a Gröbner basis of $I$ with respect to a fixed monomial ordering. The following statements are equivalent:

1. $\mathbb{V}(I)$ is finite;
2. for every $i = 1, \ldots, n$, there exist $c_i \in \mathbb{N}$ and $g_{l_i} \in G$ such that $\text{lm}(g_{l_i}) = x_i^{c_i}$;
3. the $K$-vector space $A/I$ is finite dimensional.

**Proof T. 3.17.** We prove that $1 \Rightarrow 2 \Rightarrow 3 \Rightarrow 1$.
$1 \Rightarrow 2$. If $\mathbb{V}(I) = \emptyset$, then, by the weak form of the Nullstellensatz, we have $I = (1)$, that is, $x_i^0 = 1 \in G$ for each $i = 1, \ldots, n$.
Otherwise, let $\mathbb{V}(I) = \{\alpha_1, \ldots, \alpha_s\}$ with $\alpha_j = (a_{j1}, \ldots, a_{jn}) \in K^n$ for every $j = 1, \ldots, s$.
For each $i = 1, \ldots, n$, consider the polynomial

$$f_i(X) = \prod_{j=1}^{s} (x_i - a_{ji}).$$

Then, $f_i(\alpha_j) = 0$ for each $j = 1, \ldots, s$. Therefore, by the strong form of the Nullstellensatz, we have $f_i \in \mathbb{I}(\mathbb{V}(I)) = \sqrt{I}$. Thus, for every $i = 1, \ldots, n$, there exists $d_i \in \mathbb{N}$ such that $f_i^{d_i} \in I$, and its leading monomial $x_i^{sd_i}$ is in $\text{Lt}(I) = \text{Lt}(G)$.
As a consequence, for each $i = 1, \ldots, n$, there exists an element $g_{l_i}$ in $G$ whose leading monomial is $\text{lm}(g_{l_i}) = x_i^{c_i}$ for some $c_i \in \mathbb{N}$, $c_i \leq sd_i$.

$2 \Rightarrow 3$. By **T.2.23**.1, it suffices to prove that the set

$$\mathcal{B} = \{X^{\mathbf{a}} \colon \text{lm}(g) \nmid X^{\mathbf{a}} \text{ for every } g \in G\}$$

is finite. If $X^{\mathbf{a}} \in \mathcal{B}$, then the hypothesis implies that $a_1 < c_1, \ldots, a_n < c_n$. Therefore, $\mathcal{B}$ contains a finite number of elements.

$3 \Rightarrow 1$. Let $d = \dim_K A/I < \infty$.
For each $i = 1, \ldots, n$ the vectors $\{\overline{1}, \overline{x_i}, \ldots, \overline{x_i}^d\}$ are linearly dependent. Therefore, there exists a non-trivial $K$-linear combination $\sum_{j=0}^{d} b_{ij}\overline{x_i}^j = 0$. Thus,

$$\sum_{j=0}^{d} b_{ij} x_i^j \in I \cap K[x_i]$$

is a non-zero univariate polynomial of $I$, and if $\alpha = (a_1, \ldots, a_n) \in \mathbb{V}(I)$, then $a_i$ is one of its roots. Hence, for every $i$, there are only a finite number of possible values for $a_i$.
Thus, $\mathbb{V}(I)$ is finite. □

We remark that in the previous proof the assumption $K = \overline{K}$ is only necessary to prove that $1 \Rightarrow 2$. Furthermore, the implication $3 \Rightarrow 2$ remains valid even without this additional condition.

**T. 3.18.** ($\rightarrow$ p. 223) Suppose $K = \overline{K}$, and let $I$ be a radical ideal such that $\mathbb{V}(I)$ is a finite set. Then, $I$ is a 0-dimensional ideal, $|\,\mathbb{V}(I)| = \dim_K A/I$, and the ring $A/I$ is a direct sum of a finite number of fields.

In the above setting the ring $A/I$ has Krull dimension 0 and is Noetherian. This is because the ideals of $A/I$ are in one-to-one correspondence with ideals of $A$ containing $I$, and all ideals of $A$ are finitely generated. Therefore, $A/I$ is a finite direct sum of local Noetherian rings of dimension 0. This result holds in greater generality, see **T.7.16** and **T.7.17**.

**T. 3.19.** ($\rightarrow$ p. 224) If $K = \overline{K}$ and $\mathbb{V}(I)$ is finite, then

$$|\,\mathbb{V}(I)| = |\,\mathbb{V}(\sqrt{I})| = \dim_K(A/\sqrt{I}) \le \dim_K(A/I) < \infty.$$

In particular, $I$ is 0-dimensional.

If $K \subsetneq \overline{K}$ and $I = (f_1, \ldots, f_s)$, then we have

$$\begin{aligned}\mathbb{V}(I) &= \{\alpha \in K^n \colon f_i(\alpha) = 0 \text{ for each } i = 1, \ldots, s\}\\ &\subseteq \{\alpha \in \overline{K}^n \colon f_i(\alpha) = 0 \text{ for each } i = 1, \ldots, s\}\\ &= \mathbb{V}_{\overline{K}}(\{f_i \colon i = 1, \ldots, s\}) = \mathbb{V}_{\overline{K}}\left(I\,\overline{K}[X]\right).\end{aligned}$$

Therefore, if the variety of points in the algebraic closure of $K$ is finite, then the variety $\mathbb{V}(I)$ is finite as well. In this case we have

$$|\,\mathbb{V}(I)| \le |\,\mathbb{V}_{\overline{K}}\left(I\,\overline{K}[X]\right)| \le \dim_{\overline{K}}\left(\overline{K}[X]/I\,\overline{K}[X]\right) = \dim_K(K[X]/I) < \infty.$$

How is the last equality justified?
Can we still conclude that $A/I$ is 0-dimensional?

## 3.6 Appendix: the Zariski Topology

In this section, we will provide a brief introduction to the definition and initial properties of the Zariski topology. We will focus on two sets that we have already encountered in the previous chapters, namely $K^n$ and $\operatorname{Spec} A$.

**T. 3.20.** ($\rightarrow$ p. 224) Let $K^n$ be the affine space. We say that $\mathcal{Y} \subseteq K^n$ is closed when $\mathcal{Y}$ is an affine variety of $K^n$, and that $\mathcal{A} \subseteq K^n$ is open when its complement in $K^n$ is closed. Let $\tau$ be the family of the open sets of $K^n$, then $\tau$ is a topology on $K^n$ called the *Zariski topology.*

Let $S$ be a subset of $K^n$. If we define

$$\mathbb{I}(S) = \{f \in K[x_1, \dots, x_n] \colon f(\alpha) = 0 \text{ for each } \alpha \in S\},$$

then $\mathbb{I}(S)$ is an ideal. We refer to the variety $\overline{S} = \mathbb{V}(\mathbb{I}(S))$ as the *Zariski closure of* $S$.

**T. 3.21.** ($\rightarrow$ p. 224) The variety $\overline{S} = \mathbb{V}(\mathbb{I}(S))$ is the smallest affine variety containing $S$.

**Closure Theorem**

**T. 3.22.** Let $K = \overline{K}$, let $I \subsetneq K[x_1, \dots, x_n]$ be an ideal, and let $I_k$ be the $k$-th elimination ideal of $I$. Moreover, let $V = \mathbb{V}(I)$, and let $\pi_k \colon K^n \longrightarrow K^{n-k}$ be the projection $\pi_k(a_1, \dots, a_n) = (a_{k+1}, \dots, a_n)$. Then,

$$\overline{\pi_k(V)} = \mathbb{V}(I_k).$$

**Proof T. 3.22.** Let $f \in I_k$ and $\alpha = (a_1, \dots, a_n) \in V$.
Then, $f = f(x_{k+1}, \dots, x_n)$ and $f$ vanishes in $\alpha$, *i.e.*,

$$f(\alpha) = f(a_{k+1}, \dots, a_n) = f(\pi_k(\alpha)) = 0.$$

Therefore, $f$ vanishes on all points of $\pi_k(V)$. This shows that, $\pi_k(V) \subseteq \mathbb{V}(I_k)$. The first inclusion then follows from **T.3.21**.

To prove the opposite inclusion, let $f \in \mathbb{I}(\pi_k(V)) \subseteq K[x_{k+1}, \dots, x_n]$. Viewing $f$ as a polynomial in $K[x_1, \dots, x_n]$, we have $f(\alpha) = f(a_{k+1}, \dots, a_n) = 0$ for each $\alpha \in V$. Hence, $f \in \mathbb{I}(V) = \sqrt{I}$ by the strong form of the Nullstellensatz, and there exists $m \in \mathbb{N}_+$ such that $f^m \in I \cap K[x_{k+1}, \dots, x_n] = I_k$.
This yields $f \in \sqrt{I_k}$, and therefore, $\mathbb{I}(\pi_k(V)) \subseteq \sqrt{I_k}$.
By considering the corresponding varieties, we have

$$\mathbb{V}(I_k) = \mathbb{V}(\sqrt{I_k}) \subseteq \mathbb{V}(\mathbb{I}(\pi_k(V)) = \overline{\pi_k(V)}. \qquad \square$$

**T. 3.23.** Consider an affine variety $V$, a subvariety $W \subseteq V$, and two ideals $I$, $J$ of $K[x_1, \dots, x_n]$. Then:

1. $V = W \cup (\overline{V \setminus W})$;
2. $\mathbb{V}(I : J) \supseteq \overline{\mathbb{V}(I) \setminus \mathbb{V}(J)}$;
3. if $K = \overline{K}$ and $I = \sqrt{I}$, then

$$\mathbb{V}(I : J) = \overline{\mathbb{V}(I) \setminus \mathbb{V}(J)}.$$

**Proof T. 3.23.** 1. From $V \supseteq V \setminus W$, we obtain $V \supseteq \overline{V \setminus W}$. Thus, the inclusion $\supseteq$ is satisfied.

Since $V = W \cup (V \setminus W) \subseteq W \cup \overline{(V \setminus W)}$, the opposite inclusion also holds.

2. It is sufficient to prove that $I : J \subseteq \mathbb{I}(\mathbb{V}(I) \setminus \mathbb{V}(J))$. By passing to the corresponding varieties, the conclusion immediately follows.

Consider $f \in I : J$ and $\alpha \in \mathbb{V}(I) \setminus \mathbb{V}(J)$. Then, $fg \in I$ for every $g \in J$, and hence, $f(\alpha)g(\alpha) = fg(\alpha) = 0$ for each $g \in J$.

Since $\alpha \notin \mathbb{V}(J)$, there exists a $g \in J$ such that $g(\alpha) \neq 0$, and then, $f(\alpha) = 0$.

3. By part 2, it is sufficient to prove that $I : J \supseteq \mathbb{I}(\mathbb{V}(I) \setminus \mathbb{V}(J))$. The claim follows again by passing to the corresponding varieties.

Let $f \in \mathbb{I}(\mathbb{V}(I) \setminus \mathbb{V}(J))$ and $g \in J$. We have to show that $fg \in I$.

The polynomial $fg$ vanishes on every $\alpha \in \mathbb{V}(I)$. Indeed, $f(\alpha) = 0$ when $\alpha \in \mathbb{V}(I) \setminus \mathbb{V}(J)$, and $g(\alpha) = 0$ when $\alpha \in \mathbb{V}(J)$.

Thus, by the Nullstellensatz and the hypothesis, $fg \in \mathbb{I}(\mathbb{V}(I)) = \sqrt{I} = I$. □

Now, let $A$ be any commutative ring. We define a topology on $\operatorname{Spec} A$, the *spectrum* of $A$. Let $\mathcal{X} = \operatorname{Spec} A$, and let $E$ be a subset of $A$. We define

$$\mathcal{V}(E) = \{\mathfrak{p} \in \mathcal{X} \colon E \subseteq \mathfrak{p}\}.$$

**The Zariski Topology**

**T. 3.24.** (→ p. 224) Let $E$, $E' \subseteq A$ be subsets and let $I$, $J$, $I_\alpha \subseteq A$ be ideals, where $\alpha$ varies in a set of indices $\Lambda$. Then:

1. $\mathcal{V}(\{0\}) = \mathcal{X}$ and $\mathcal{V}(A) = \emptyset$;
2. $E \subseteq E' \implies \mathcal{V}(E) \supseteq \mathcal{V}(E')$;
3. $\mathcal{V}(E) = \mathcal{V}((E)) = \mathcal{V}\left(\sqrt{(E)}\right)$;
4. $\mathcal{V}(I) \cup \mathcal{V}(J) = \mathcal{V}(I \cap J) = \mathcal{V}(IJ)$ (finite union);
5. $\bigcap_{\alpha \in \Lambda} \mathcal{V}(I_\alpha) = \mathcal{V}(\bigcup_{\alpha \in \Lambda} I_\alpha)$ (arbitrary intersection).

The topology whose closed sets are precisely the subsets of the form $\mathcal{V}(E)$, where $E \subset A$, is called *Zariski topology on $\mathcal{X}$*.

**T. 3.25.** (→ p. 225) Let $\mathcal{X} = \operatorname{Spec} A$ be endowed with the Zariski topology. The sets

$$\mathcal{X}_f = \mathcal{X} \setminus \mathcal{V}(\{f\}) = \{\mathfrak{p} \in \mathcal{X} \colon f \notin \mathfrak{p}\},$$

where $f$ varies in $A$, are a basis of open sets for the topological space $\mathcal{X}$.

**T. 3.26.** (→ p. 225) Let $\mathcal{X} = \operatorname{Spec} A$ be endowed with the Zariski topology. Then, $\mathcal{X}$ is compact.

**T. 3.27.** (→ p. 225) Let $\mathcal{X} = \operatorname{Spec} A$ be endowed with the Zariski topology, and let $\mathcal{Y} \subset \mathcal{X}$. Then, $\overline{\mathcal{Y}} = \mathcal{V}(\bigcap_{\mathfrak{p} \in \mathcal{Y}} \mathfrak{p})$.

**T. 3.28.** (→ p. 226) Let $A$ be a ring of positive dimension, and let $\mathcal{X} = \operatorname{Spec} A$ be endowed with the Zariski topology.
Then, $\mathcal{X}$ is a $T_0$, but not a $T_1$, topological space. In particular, $\mathcal{X}$ is not a Hausdorff space.

**T. 3.29.** (→ p. 226) Let $\phi\colon A \longrightarrow B$ be a ring homomorphism, and let $\mathcal{X} = \operatorname{Spec} A$ and $\mathcal{Y} = \operatorname{Spec} B$ be endowed with the Zariski topology.
Then, $\phi^*\colon \mathcal{Y} \longrightarrow \mathcal{X}$, defined by $\phi^*(\mathfrak{q}) = \phi^{-1}(\mathfrak{q})$, is a continuous map.

**T. 3.30.** (→ p. 226) Let $\mathcal{X} = \operatorname{Spec} A$ be endowed with the Zariski topology, and let $\mathcal{N}(A)$ be the nilradical of $A$. Then, $\mathcal{X}$ and $\operatorname{Spec}(A/\mathcal{N}(A))$ are homeomorphic.

**T. 3.31.** Let $\mathcal{X} = \operatorname{Spec} A$ be endowed with the Zariski topology. Then, $\mathcal{X}$ is irreducible if and only if $\mathcal{N}(A)$ is prime, *i.e.*, if and only if $|\operatorname{Min} A| = 1$.

**Proof T. 3.31.** Recall that a topological space $\mathcal{X}$ is irreducible if and only if for every pair of non-empty open subsets $\mathcal{A}$, $\mathcal{B} \subset \mathcal{X}$ we have $\mathcal{A} \cap \mathcal{B} \neq \emptyset$ or, equivalently, if and only if for every open set $\mathcal{A} \neq \emptyset$ we have $\overline{\mathcal{A}} = \mathcal{X}$.
By the previous result, we can assume $A$ is reduced, and prove that $\mathcal{X}$ is irreducible if and only if $(0)$ is prime.
If $\mathcal{X}$ is irreducible and $f$, $g \in A$ are such that $fg = 0$, then

$$\begin{aligned} \mathcal{X}_f \cap \mathcal{X}_g &= (\mathcal{X} \setminus \mathcal{V}(f)) \cap (\mathcal{X} \setminus \mathcal{V}(g)) = \mathcal{X} \setminus (\mathcal{V}(f) \cup \mathcal{V}(g)) \\ &= \mathcal{X} \setminus \mathcal{V}(fg) = \mathcal{X} \setminus \mathcal{V}(0) = \emptyset. \end{aligned}$$

By the hypothesis, either $\mathcal{X}_f = \emptyset$ or $\mathcal{X}_g = \emptyset$, *i.e.*, either $\mathcal{V}(f) = \mathcal{X}$ or $\mathcal{V}(g) = \mathcal{X}$. This is equivalent to saying that either $f \in \mathfrak{p}$ for every prime ideal $\mathfrak{p}$, and therefore $f \in \mathcal{N}(A) = (0)$ or, analogously, $g = 0$.

Conversely, suppose $(0)$ is prime and let $\mathcal{X} \setminus \mathcal{V}(I)$ and $\mathcal{X} \setminus \mathcal{V}(J)$ be two non-empty open sets. Since $\mathcal{V}(I) \neq \mathcal{X}$ and $\mathcal{V}(J) \neq \mathcal{X}$, we have that $I \neq (0)$ and $J \neq (0)$. Therefore, $(0) \in (\mathcal{X} \setminus \mathcal{V}(I)) \cap (\mathcal{X} \setminus \mathcal{V}(J))$, *i.e.*, the intersection is non-empty. □

# Chapter 4
# Modules

In this chapter we study modules over a ring $A$ and their homomorphisms. We will highlight similarities and distinctions between modules and both Abelian groups and vector spaces, which serve as important examples when $A = \mathbb{Z}$ and when $A = K$ is a field, respectively.
Additionally, we will examine exact sequences and commutative diagrams, and we will conclude the chapter with a constructive proof of the Structure Theorem for finitely generated modules over a PID.

## 4.1 Modules and Submodules

Let $A$ be a ring. A set $M$ is an *A-module* if $(M, +)$ is an Abelian group together with an *outer product*, or *scalar multiplication*, $\cdot\colon A \times M \longrightarrow M$, $(a, m) \mapsto a \cdot m = am$ such that, for any $a$, $b \in A$ and any $m$, $n \in M$:

i) $(a + b)m = am + bm$;

ii) $a(m + n) = am + an$;

iii) $(ab)m = a(bm)$;

iv) $1_A m = m$.

Every ring $A$ has a natural structure of $A$-module with the outer product operation provided by the multiplication, or inner product, of $A$.
When $A = \mathbb{Z}$, an $A$-module is an Abelian group.
When $A = K$ is a field, an $A$-module is a $K$-vector space.
If $A$ is a subring of $B$, then $\cdot\colon A \times B \longrightarrow B$, defined by the multiplication of $a \in A$ by $b \in B$, both considered as elements of $B$, satisfies properties i)-iv). Hence, $B$ is an $A$-module.

For any $A$-module $M$, a subset $N \subseteq M$ is an *A-submodule* of $M$, or simply a *submodule* when there is no ambiguity about the ring $A$, if the following conditions are satisfied:



A. Bandini et al., *Commutative Algebra through Exercises*, UNITEXT 159,
https://doi.org/10.1007/978-3-031-56910-4_4

i) $(N, +)$ is a subgroup of $(M, +)$;

ii) $N$ is closed under the outer product defined on $M$, *i.e.*, for any $a \in A$ and any $n \in N$ we have $an \in N$.

For example, the module $\{0\}$ is a submodule of any module and will be denoted as 0.
The submodules of the $A$-module $A$ are precisely the ideals of $A$.
Let $M$ be an $A$-module. Then, for any ideal $I \subset A$, the set

$$IM = \left\{ \sum_{i=1}^{s} a_i m_i \colon a_i \in I,\ m_i \in M \text{ for some } s \in \mathbb{N} \right\},$$

consisting of all finite linear combinations of elements of $M$ with coefficients in $I$, is a submodule of $M$.
A non-zero module is *simple* if it has no non-zero proper submodules.

For any submodule $N$ of an $A$-module $M$, we can define a natural $A$-module structure on the quotients group $M/N$ by setting $a\,\overline{m} = \overline{am}$ for each $a \in A$ and $\overline{m} \in M/N$.
In particular, for any ideal $I \subset A$, there exists a natural $A$-module structure on the quotient $M/IM$.
Moreover, $M/IM$ also has an $A/I$-module structure, defined by $\bar{a}\,\overline{m} = \overline{am}$ for each $\bar{a} \in A/I$ and $\overline{m} \in M/IM$. The reader should verify that this product is well-defined and provides the required structure, see **E.10.1**.

### Restriction of Scalars

Let $f \colon A \longrightarrow B$ be a ring homomorphism, and let $M$ be a $B$-module. The outer product

$$A \times M \longrightarrow M, \quad (a, m) \mapsto f(a)m$$

defines an $A$-module structure on $M$ which is said to be obtained *by restriction of scalars via* $f$, see **E.10.2**.

For example, we have seen above that, when $A$ is a subring of $B$, then $B$ is an $A$-module under the inner product of $B$ restricted to the elements of $A$. This $A$-module structure on $B$ is obtained by restriction of scalars via the inclusion homomorphism of $A$ in $B$.

It is worth recalling that, in general, the union of submodules is not a submodule, unless the submodules form a chain of inclusions.

**Operations on Submodules**

Let $\{M_h\}_{h\in H}$ be any family of submodules of an $A$-module $M$.

**Intersection.** The set

$$\bigcap_{h\in H} M_h = \{m \in M\colon\ m \in M_h \ \text{ for all } h\}$$

is a submodule of $M$.

**Sum.** The set

$$\sum_{h\in H} M_h = \left\{\sum_{h\in H} a_h m_h\colon\ a_h \in A,\, m_h \in M_h,\, a_h = 0 \ \text{ for almost all } h\right\}$$

consisting of all finite linear combinations of elements of $\bigcup_{h\in H} M_h$ with coefficients in $A$, is a submodule of $M$.
It is the smallest submodule which contains $\bigcup_{h\in H} M_h$.

It is easy to verify that the subsets of $M$ defined above are submodules of $M$. When $N$ and $P$ are submodules of $M$ such that $N \cap P = 0$, then we denote their sum by $N \oplus P$ and call it the *direct sum of* $N$ *and* $P$.

Since the product of elements of a module is not defined, there is no analogue of the quotient of ideals. Instead, we consider the following set.

Let $M$ be an $A$-module, and let $N$, $P$ be submodules of $M$. We define the set

$$N : P = \{a \in A\colon\ aP \subseteq N\}.$$

In particular, if $N = 0$, then $0 : P$ is also denoted by $\operatorname{Ann} P$ and is called the *annihilator of* $P$.

It is easy to prove that $N : P$ and $\operatorname{Ann} P$ are ideals of $A$, see **E.10.5**.

**T. 4.1.** (→ p. 227) Let $M$ be an $A$-module, and let $I \subseteq A$ be an ideal. If $I \subseteq \operatorname{Ann} M$, then the outer product with $\overline{a}m = am$, for any $\overline{a} \in A/I$ and $m \in M$, defines a natural $A/I$-module structure on $M$.

This structure is particularly useful when $I$ is a maximal ideal because, in this case, $M$ has a natural structure of $A/I$-vector space.

## 4.2 Module Homomorphisms

Let $M$, $N$ be $A$-modules. A map $f\colon M \longrightarrow N$ is an *$A$-module homomorphism* if it is a group homomorphism and is *$A$-linear*, *i.e.*, if for all $m, n \in M$ and for all $a \in A$:

i) $f(m+n) = f(m) + f(n)$;
ii) $f(am) = af(m)$.

A homomorphism $f\colon M \longrightarrow M$ is called *endomorphism* of $M$.

**The Module $\mathrm{Hom}_A(M, N)$**

**T. 4.2.** (→ p. 227) The set $\mathrm{Hom}_A(M,N)$ of all $A$-module homomorphisms $M \longrightarrow N$, together with the operations defined as, for every $f$, $g \in \mathrm{Hom}_A(M,N)$,

i) $(f+g)(m) = f(m) + g(m)$;
ii) $(af)(m) = a(f(m))$,

for every $a \in A$, and for every $m \in M$, is an $A$-module.
Moreover, for any $A$-module $M$, we have $\mathrm{Hom}_A(A,M) \simeq M$.

The $A$-module $\mathrm{Hom}_A(M,M)$ of the endomorphisms of $M$ is denoted by $\mathrm{End}_A M$. We note that $\mathrm{End}_A M$ with the sum defined above and the inner product defined by map composition is also a non-commutative ring with identity $\mathrm{id}_M$.

Given an $A$-module homomorphism $f\colon M \longrightarrow N$, we can define:

i) $\mathrm{Ker}\, f = \{m \in M\colon\ f(m) = 0\}$, the *kernel of $f$*;
ii) $\mathrm{Im}\, f = f(M) \subseteq N$, the *image of $f$*;
iii) $\mathrm{Coker}\, f = N/\mathrm{Im}\, f$, the *cokernel of $f$*.

It is easy to verify that $f$ is injective if and only if $\mathrm{Ker}\, f = 0$, and that $f$ is surjective if and only if $\mathrm{Coker}\, f = 0$.

For example, if $M$ is an $A$-module and $a \in A$, then we can consider the multiplication map by $a$, denoted by

$$M \xrightarrow{a} M$$

and defined by $m \mapsto am$. It is an $A$-module homomorphism and therefore an endomorphism of $M$. Its kernel is $\{m \in M\colon am = 0\}$, its image is $aM = (a)M$, and its cokernel is $M/aM$.

**Module Homomorphism Theorems**

**T. 4.3.** (→ p. 227) Let $M$, $N$, and $P$ be $A$-modules.

1. A homomorphism $f\colon M \longrightarrow N$ induces an isomorphism

$$M/\operatorname{Ker} f \simeq \operatorname{Im} f.$$

2. Let $N$ and $P$ be submodules of $M$ with $P \subseteq N$. Then, $N/P$ is a submodule of $M/P$ and

$$(M/P)/(N/P) \simeq M/N.$$

3. Let $N$ and $P$ be submodules of $M$. Then,

$$(N+P)/P \simeq N/(N \cap P).$$

## 4.3 Free Modules

The following definitions generalize well-known concepts from the theory of vector spaces.

**Generators, Free Sets, and Bases**

Let $M$ be an $A$-module, and let $S \subseteq M$ be a subset. The set

$$\langle S \rangle = \langle S \rangle_A = \left\{ \sum_{i=1}^{k} a_i s_i \colon \; a_i \in A, \; s_i \in S, \quad \text{for some } k \in \mathbb{N} \right\} \subseteq M,$$

consisting of all finite linear combinations of elements of $S$, with coefficients in $A$, is a submodule of $M$ called the *submodule generated* by $S$.

We say that $S$ is a *set of generators*, or a *generating set*, of $M$, or that $S$ *generates* $M$, if $\langle S \rangle = M$, *i.e.*, if for every $m \in M$ there exist $a_1, \ldots, a_k \in A$ and $s_1, \ldots, s_k \in S$ such that $m = \sum_{i=1}^{k} a_i s_i$.

We say that $S$ is *free* if for every $a_1, \ldots, a_k \in A$ and distinct $s_1, \ldots, s_k \in S$ such that $\sum_{i=1}^{k} a_i s_i = 0$, then $a_i = 0$ for each $i = 1, \ldots, k$.

In this case the elements of $S$ are said to be *linearly independent*.

A free set of generators of a module $M$ is called a *basis* of $M$.

By convention, the empty set is a basis of the module 0.

From the definition of submodule it immediately follows that $\langle S\rangle$ is the smallest submodule of $M$ containing $S$.
Let $M = \langle S\rangle$. If $S$ is finite, then $M$ is said to be *finitely generated.*
If $S = \{s\}$, then $M$ is called *cyclic.*
If $S$ is free, then $M$ is said to be *free.* Therefore, by definition, a module is free when it has a basis.

For example, each ring $A$ is a free $A$-module with basis $\{1\}$.
For any integer $n \neq 0, \pm 1$, the ring $\mathbb{Z}/(n)$ is free as a module over itself, but it is not a free $\mathbb{Z}$-module.
The ring $A^n$ is a free $A$-module, and the set

$$\{e_1 = (1,0,\dots,0),\dots,e_n = (0,\dots,0,1)\}$$

is a basis of $A^n$, called *the canonical basis.*

**T. 4.4.** ($\rightarrow$ p. 227) Let $M$ be an $A$-module generated by a set $S$. Then, $M$ is free with basis $S$ if and only if every element of $M$ can be written in a unique way as a linear combination of elements of $S$.

Not all properties of generating sets and of free sets that hold for vector spaces generalize to modules. For further details, see **E.10.4**.
However, when $M$ is a free module, all of its bases are *equipotent*, *i.e.*, they have the same cardinality.

**Rank of a Free Module**

**T. 4.5.** Let $M$ be a free $A$-module. Then, all bases of $M$ are equipotent.

The cardinality of any basis is called the *rank* of $M$, denoted by $\operatorname{rank}_A M$, or simply by $\operatorname{rank} M$ if there is no ambiguity about the ring $A$.

**Proof T. 4.5.** We claim that if $\mathcal{B} = \{m_h\}_{h\in H}$ is a basis of $M$ as an $A$-module, then, for any maximal ideal $\mathfrak{m}$ of $A$, the set $\overline{\mathcal{B}} = \{\overline{m_h}\}_{h\in H}$ is a basis for the $A/\mathfrak{m}$-vector space $M/\mathfrak{m}M$. The statement will then follow from the corresponding result for vector spaces, since all bases of a vector space are equipotent.
Every $m \in M$ can be written as a finite sum $\sum_h a_h m_h$ for some $a_h \in A$. This implies that every element $\overline{m} \in M/\mathfrak{m}M$ can be written as

$$\overline{m} = \overline{\sum_h a_h m_h} = \sum_h \overline{a_h m_h} = \sum_h \overline{a_h}\,\overline{m_h}.$$

Therefore, $\overline{\mathcal{B}}$ is a set of generators.
If $0 = \sum_h \overline{a_h}\,\overline{m_h} = \overline{\sum_h a_h m_h}$, then $\sum_h a_h m_h \in \mathfrak{m}M$. Hence, there exist elements $c_j \in \mathfrak{m}$ such that $\sum_h a_h m_h = \sum_j c_j m_j$, where, by definition, all sums are finite. Reordering the indices if necessary, we can write $\sum_h (a_h - c_h) m_h = 0$.

Since $\mathcal{B}$ is a basis of $M$, this implies that $a_h = c_h \in \mathfrak{m}$ for any $h$, that is, $\overline{a_h} = 0$ for all $h$.
Therefore, $\overline{\mathcal{B}}$ is a free set. □

## 4.4 Direct Sum and Direct Product of Modules

For any family $\{M_h\}_{h\in H}$ of $A$-modules, we define the *direct sum*

$$\bigoplus_{h\in H} M_h = \{(m_h)_{h\in H}\colon\ m_h \in M_h \text{ for all } h \text{ and } m_h = 0 \text{ for almost all } h\}.$$

Without requiring that the number of non-zero elements is finite, we obtain the definition of *direct product* of modules

$$\prod_{h\in H} M_h = \{(m_h)_{h\in H}\colon\ m_h \in M_h\}.$$

An $A$-module structure can be defined on these sets by defining sum and outer product componentwise. We note that the first module is a submodule of the second one. Moreover, when there are only finitely many components, *i.e.*, when $|H| < +\infty$, the two modules coincide.
Also note that the direct sum of infinitely many commutative rings with identity is still a commutative ring but without identity. Indeed the identity belongs only to the corresponding infinite direct product.

For a finite number of rings $A_1, \ldots, A_k$, we do not distinguish between the notation $A_1 \times \cdots \times A_k$ and $A_1 \oplus \cdots \oplus A_k$, or the corresponding compact notation $\prod_{i=1}^{k} A_i$ and $\bigoplus_{i=1}^{k} A_i$.

Consider a set $S$, which can be finite or infinite. The free $A$-module

$$A^S = \bigoplus_{s\in S} A$$

is the direct sum of as many copies of $A$ as there are elements in $S$, and has canonical basis $\{e_s\colon s \in S\}$.
In particular, if $S$ is a subset of an $A$-module $M$, then the map defined by $e_s \mapsto s$ is a surjective homomorphism

$$A^S \longrightarrow \bigoplus_{s\in S} \langle s\rangle,$$

where $\langle s\rangle$ is the cyclic $A$-module $\langle s\rangle_A = As$ generated by an element $s \in S$.

**Universal Property of Direct Sum and Direct Product**

**T. 4.6.** Let $\{M_h\}_{h\in H}$ be a family of $A$-modules, and let $N$ be an $A$-module.

1. For every $h \in H$, let $i_h\colon M_h \longrightarrow \bigoplus_{h\in H} M_h$ be the inclusion homomorphism. Assume that, for any $h \in H$, there exists a homomorphism $\varphi_h\colon M_h \longrightarrow N$.
   Then, there exists a unique homomorphism $\varphi\colon \bigoplus_{h\in H} M_h \longrightarrow N$ such that the following diagram commutes for all $h \in H$:

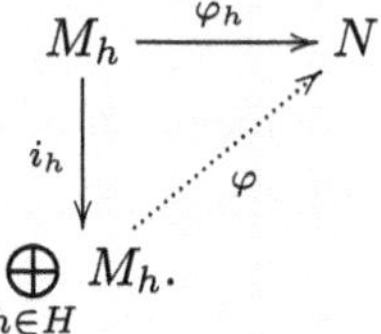


2. For every $h \in H$, let $\pi_h\colon \prod_{h\in H} M_h \longrightarrow M_h$ be the projection homomorphism. Assume that, for any $h \in H$, there exists a homomorphism $\psi_h\colon N \longrightarrow M_h$.
   Then, there exists a unique homomorphism $\psi\colon N \longrightarrow \prod_{h\in H} M_h$ such that the following diagram commutes for all $h \in H$:

$$\begin{array}{ccc} & & \prod_{h\in H} M_h \\ & {\scriptstyle\psi}\nearrow & \big\downarrow{\scriptstyle\pi_h} \\ N & \xrightarrow[\psi_h]{} & M_h. \end{array}$$

**Proof T. 4.6.** 1. The homomorphism $\varphi\colon \bigoplus_{h\in H} M_h \longrightarrow N$ defined by

$$m = (m_h)_{h\in H} \mapsto \sum_{h\in H} \varphi_h(m_h)$$

is well-defined, because for any such $m$ there are only finitely many components $m_h \neq 0$. Therefore, the sum of the $\varphi_h(m_h)$ is a finite sum of elements of $N$. The equality $\varphi \circ i_h = \varphi_h$ for all $h \in H$ can be easily verified.
Moreover, if $\varphi'$ is another homomorphism such that $\varphi' \circ i_h = \varphi_h$ for all $h \in H$, then, for every $m \in \bigoplus_{h\in H} M_h$,

$$\varphi'(m) = \varphi'\Big(\sum_{h\in H} i_h(m_h)\Big) = \sum_{h\in H} \varphi'(i_h(m_h)) = \sum_{h\in H} \varphi_h(m_h) = \varphi(m).$$

2. Define the homomorphism $\psi\colon N \longrightarrow \prod_{h\in H} M_h$ by $\psi(n) = (\psi_h(n))_{h\in H}$.
It can be easily verified that $\pi_h \circ \psi = \psi_h$ for all $h \in H$.
Moreover, if $\psi'$ is another homomorphism such that $\pi_h \circ \psi' = \psi_h$, with $\psi'(n) = (n_h)_{h\in H}$ and $n \in N$, then $n_h = \pi_h \circ \psi'(n) = \psi_h(n)$ for all $n \in N$.
Therefore, $\psi = \psi'$. □

For example, let $A$ be a ring and let $x$ be a variable.
For any $i \in \mathbb{N}$, consider the cyclic module $M_i = \langle x^i \rangle = Ax^i$. Then, by the universal properties **T.4.6**.1 and 2, we obtain canonical $A$-module isomorphisms

$$\bigoplus_{i\in\mathbb{N}} M_i \simeq A[x] \quad \text{and} \quad \prod_{i\in\mathbb{N}} M_i \simeq A[[x]].$$

To conclude this section, we present two characterizations of free modules.

**T. 4.7.** An $A$-module $M = \langle S \rangle$ is free with basis $S$ if and only if for any $A$-module $N$ and any map $f\colon S \longrightarrow N$, there exists a unique $A$-module homomorphism $\tilde{f}\colon M \longrightarrow N$ such that $\tilde{f}_{|S} = f$.

**Proof T. 4.7.** In **T.4.4** it has been proved that if $M$ is free with basis $S$, then every element $m \in M$ can be uniquely written as a linear combination $m = \sum_i a_i s_i$ of elements of $S$.
In order for $\tilde{f}$ to be a homomorphism such that $\tilde{f}_{|S} = f$, it must be defined as

$$\tilde{f}(m) = \tilde{f}\Big(\sum_i a_i s_i\Big) = \sum_i a_i \tilde{f}(s_i) = \sum_i a_i f(s_i)$$

for all $m \in M$, and therefore, $\tilde{f}$ is uniquely determined.

Conversely, suppose $N = A^S$ and consider the map $f\colon S \longrightarrow A^S$ that takes an element $s \in S$ and maps it to the corresponding element $e_s$ of the canonical basis of the free $A$-module $A^S$. According to our assumption, this map lifts to a homomorphism $\tilde{f}$ such that $\tilde{f}(s) = f(s) = e_s$ for all $s \in S$.
Since any element of $A^S$ is of the form $\sum_i a_i e_{s_i}$, which is the image of $\sum_i a_i s_i$, the map $\tilde{f}$ is clearly surjective.
It is injective as well, since for any $m = \sum_i a_i s_i$, we have

$$0 = \tilde{f}(m) = \sum_i a_i \tilde{f}(s_i) = \sum_i a_i e_{s_i}$$

if and only if $a_i = 0$ for any $i$.
Thus, $M$ is isomorphic to the free $A$-module $A^S$ and has basis $S$. □

**T. 4.8.** (→ p. 228) Every $A$-module $M$ is a quotient of a free $A$-module.
In particular, if $M$ is generated by $n$ elements, then $M$ is a quotient of $A^n$.

**T. 4.9.** (→ p. 228) An $A$-module $M$ is free if and only if there exist $A$-modules $\{M_h\}_{h\in H}$ such that $M \simeq \bigoplus_{h\in H} M_h$, where $M_h \simeq A$ for all $h$.

## 4.5 Nakayama's Lemma

In this section we will introduce a very useful tool in the development of the theory of modules, known as Nakayama's Lemma. It is a direct consequence of the Cayley-Hamilton Theorem.

**Cayley-Hamilton Theorem**

**T. 4.10.** Consider a finitely generated $A$-module $M$ with $n$ generators. Let $I$ be an ideal of $A$, and let $\varphi \in \mathrm{End}_A M$ be an endomorphism of $M$, such that $\varphi(M) \subseteq IM$.
Then, there exist $a_0, \ldots, a_{n-1} \in I$ such that

$$\varphi^n + \sum_{i=0}^{n-1} a_i \varphi^i = 0_{\mathrm{End}_A M}.$$

**Proof T. 4.10.** Let $M = \langle m_1, \ldots, m_n \rangle$. Since $\varphi(m_i) \in IM$ for every $i$, it follows that $\varphi(m_i) = \sum_{j=1}^{n} c_{ij} m_j$ for some $c_{ij} \in I$.
Hence, we obtain $n$ linear equations of the form

$$\varphi(m_i) - \sum_{j=1}^{n} c_{ij} m_j = \sum_{j=1}^{n} (\delta_{ij}\varphi - c_{ij}) m_j = 0,$$

where $\delta_{ij}$ denotes Kronecker's delta.
This system can be represented in matrix form

$$T_\varphi \begin{pmatrix} m_1 \\ m_2 \\ \vdots \\ m_n \end{pmatrix} = \begin{pmatrix} \varphi - c_{11} & -c_{12} & \cdots & -c_{1n} \\ -c_{21} & \varphi - c_{22} & & \vdots \\ \vdots & & \ddots & \\ -c_{n1} & \cdots & & \varphi - c_{nn} \end{pmatrix} \begin{pmatrix} m_1 \\ m_2 \\ \vdots \\ m_n \end{pmatrix} = \begin{pmatrix} 0 \\ 0 \\ \vdots \\ 0 \end{pmatrix},$$

where $T_\varphi$ is a $n \times n$ matrix with coefficients in $A[\varphi] \subset \mathrm{End}_A M$.
Even if the ring $\mathrm{End}_A M$ is not commutative, we note that the subring $A[\varphi]$ is commutative, hence we can consider the adjoint matrix $T_\varphi^*$ and the determinant of $T_\varphi$. Moreover, we have

$$\det T_\varphi = \varphi^n + \sum_{i=0}^{n-1} a_i \varphi^i \in A[\varphi], \text{ with } a_i \in I \text{ for all } i,$$

because $c_{ij} \in I$ for all $i$, $j$.
Multiplying on the left by $T_\varphi^*$, we obtain

$$\begin{pmatrix}0\\0\\\vdots\\0\end{pmatrix} = T_\varphi^* \begin{pmatrix}0\\0\\\vdots\\0\end{pmatrix} = T_\varphi^* T_\varphi \begin{pmatrix}m_1\\m_2\\\vdots\\m_n\end{pmatrix} = \begin{pmatrix}\det T_\varphi & 0 & \cdots & 0\\ 0 & \det T_\varphi & & \vdots\\ \vdots & & \ddots & \\ 0 & \cdots & & \det T_\varphi\end{pmatrix}\begin{pmatrix}m_1\\m_2\\\vdots\\m_n\end{pmatrix},$$

which yields $\det T_\varphi(m_i) = 0$ for all $i = 1, \ldots, n$.
Thus, $\det T_\varphi$ is the zero endomorphism, since it annihilates each of the generators of $M$. □

**Nakayama's Lemma**

**T. 4.11.** Let $M$ be a finitely generated $A$-module.

1. Let $I$ be an ideal of $A$ such that $M = IM$. Then, there exists an element $a \in A$ such that $a \equiv 1 \mod I$ and $aM = 0$.
2. Let $\mathcal{J}(A)$ be Jacobson radical of $A$, and let $I \subseteq \mathcal{J}(A)$ be an ideal of $A$ such that $IM = M$. Then, $M = 0$.
3. Consider $N \subseteq M$ a submodule, and let $I \subseteq \mathcal{J}(A)$ be an ideal of $A$ such that $M = IM + N$. Then, $M = N$.

**Proof T. 4.11.** 1. By applying the Cayley-Hamilton Theorem to the identity endomorphism $\mathrm{id}_M$, we find elements $a_i \in I$ such that

$$\mathrm{id}_M + \sum_{i=0}^{n-1} a_i \mathrm{id}_M = 0.$$

Hence, $(1 + \sum_{i=0}^{n-1} a_i)m = 0$ for any $m \in M$. Letting $a = 1 + \sum_{i=0}^{n-1} a_i$, we have $aM = 0$ and $a \equiv 1 \mod I$.

2. By part 1, there exists an element $a$ such that $a \equiv 1 \mod I$ and $aM = 0$. Therefore, $1 - a \in I \subseteq \mathcal{J}(A)$ and $a \in A^*$, see **T.1.15**.
Thus, $aM = 0$ yields $M = 0$.

3. By part 2, it is sufficient to prove that $M/N = I(M/N)$.
Since

$$I\,(M/N) = (IM + N)/N = M/N,$$

the proof is complete. □

It should be noted that Nakayama's Lemma fails to hold for modules which are not finitely generated.
For example, consider $\mathbb{Q}$ as a $\mathbb{Z}$-module. Clearly, $\mathbb{Q}$ is not finitely generated (prove it!). Moreover, for any prime $p$, we have $p\,\mathbb{Q} = \mathbb{Q}$. However, there exists no $n \equiv 1 \mod (p)$ such that $n\mathbb{Q} = 0$.

By using Nakayama's Lemma, we can establish a connection between the cardinality of a minimal set of generators of a module $M$, which is finitely generated over a local ring $(A, \mathfrak{m}, K)$, and the dimension of $M/\mathfrak{m}M$ as a $K$-vector space.

**T. 4.12.** ($\rightarrow$ p. 228) Consider a local ring $(A, \mathfrak{m}, K)$ and a finitely generated $A$-module $M$. Let $\overline{m_1}, \ldots, \overline{m_k}$ be a basis of the $K$-vector space $M/\mathfrak{m}M$, and let $\pi\colon M \longrightarrow M/\mathfrak{m}M$ denote the canonical projection.
If $n_1, \ldots, n_k \in M$ are such that $\pi(n_i) = \overline{m_i}$ for all $i$, then $M = \langle n_1, \ldots, n_k \rangle_A$.

**T. 4.13.** Consider a local ring $(A, \mathfrak{m}, K)$, and let $M$ be a finitely generated $A$-module. Then, all minimal sets of generators of $M$ have the same cardinality, namely $\mu(M) = \dim_K M/\mathfrak{m}M$.

**Proof T. 4.13.** Let $d = \dim_K M/\mathfrak{m}M$, and let $\{m_1, \ldots, m_k\}$ be a minimal generating set of $M$. Since $\overline{m_1}, \ldots, \overline{m_k}$ generate $M/\mathfrak{m}M$, it follows that $k \geq d$. If $k > d$, the elements $\overline{m_1}, \ldots, \overline{m_k}$ are linearly dependent, and there exists a subset of $d$ elements which is a basis of $M/\mathfrak{m}M$. Then, by the previous result, a proper subset of $\{m_1, \ldots, m_k\}$ generates $M$. This contradicts the hypothesis of minimality. □

In the case of finitely generated vector spaces, it is well-known that any injective or surjective endomorphism is an isomorphism. This property does not hold for finitely generated modules.
For example, the injective endomorphism $f\colon \mathbb{Z} \longrightarrow \mathbb{Z}$, defined by $f(1) = 2$, is not surjective.
However the following statement holds.

**T. 4.14.** Consider a finitely generated $A$-module $M$, and let $f \in \mathrm{End}_A M$ be a surjective endomorphism. Then, $f$ is also injective.

**Proof T. 4.14.** Let $x$ be a variable. We define an $A[x]$-module structure on $M$ as follows:
for any polynomial $p(x) = \sum_i a_i x^i$ and any $m \in M$, we set

$$p(x)m = \sum_i a_i f^i(m),$$

where $f^i$ denotes the composition of $f$ with itself $i$ times.
Since $f$ is surjective, we have $M = f(M) = xM$. Applying Nakayama's Lemma 1, we obtain a polynomial $p(x) \in A[x]$, such that $p(x) \equiv 1 \mod (x)$ and $p(x)M = 0$. Let $p(x) = 1 + xq(x)$ for some $q(x) \in A[x]$.
Then, for every $m \in \mathrm{Ker}\, f$ we have

$$m = (p(x) - xq(x))m = p(x)m - xq(x)m = 0,$$

hence $\mathrm{Ker}\, f = 0$. □

We conclude this section by highlighting another property of free modules.

**T. 4.15.** (→ p. 228) Let $M$ be a free $A$-module of rank $r$. Then, every generating set with $r$ elements is a basis of $M$.

## 4.6 Categories and Functors

In this brief section, we present some essential terminology and definitions related to categories and functors. This concepts play a central role in category theory and homological algebra, and have gained widespread significance in many branches of mathematics since the publication of the book by H. Cartan and S. Eilenberg [4]. Essentially, category theory provides a language that is more suited to higher levels of abstraction.
To assist the reader in understanding this new perspective, a few basic concepts will be outlined here.

A *category* is defined as a pair, denoted by $\mathcal{C} = (\mathrm{Obj}(\mathcal{C}), \mathrm{Mor}(\mathcal{C}))$. It consists of $\mathrm{Obj}(\mathcal{C})$, which represents the *objects* in the category, and $\mathrm{Mor}(\mathcal{C})$, which represents the *morphisms* in the category.
As examples, $\mathcal{S}et$, $\mathcal{R}ing$, and $\mathcal{T}op$, correspond to categories where the objects are sets, rings, and topological spaces, and the morphisms are functions between sets, ring homomorphisms, and continuous functions, respectively.

A category is equipped with a *composition law*, denoted by $\circ$, for the elements of $\mathrm{Mor}(\mathcal{C})$.
This composition law must satisfy the following properties:

i) *associativity*: for any objects $A$, $B$, $C$, $D$ in $\mathrm{Obj}(\mathcal{C})$ and any morphisms $f$, $g$, $h$ in $\mathrm{Mor}(\mathcal{C})$, with $f\colon A \longrightarrow B$, $g\colon B \longrightarrow C$, $h\colon C \longrightarrow D$, it holds

$$h \circ (g \circ f) = (h \circ g) \circ f;$$

ii) *existence of identity elements*: for any object $A \in \mathrm{Obj}(\mathcal{C})$, there exists a morphism $\mathrm{id}_A\colon A \longrightarrow A$ in $\mathrm{Mor}(\mathcal{C})$, such that for any object $B$ in $\mathrm{Obj}(\mathcal{C})$ and any morphisms $g$, $h \in \mathrm{Mor}(\mathcal{C})$, with $g\colon A \longrightarrow B$ and $h\colon B \longrightarrow A$, it holds

$$g \circ \mathrm{id}_A = g \quad \text{and} \quad \mathrm{id}_A \circ h = h.$$

A map between two categories is called a *functor* if it preserves the structure of the category by mapping objects and morphisms in a consistent manner. More precisely, $F : \mathcal{C} \longrightarrow \mathcal{D}$ is a functor if:

i) for any $A \in \mathrm{Obj}(\mathcal{C})$, $F(A) \in \mathrm{Obj}(\mathcal{D})$;
ii) for any $f \in \mathrm{Mor}(\mathcal{C})$, $F(f) \in \mathrm{Mor}(\mathcal{D})$;
iii) for any $A \in \mathrm{Obj}(\mathcal{C})$, $F(\mathrm{id}_A) = \mathrm{id}_{F(A)}$;
iv) $F$ must be consistent with the composition laws defined on the categories.

Regarding point iv), a functor $F\colon \mathcal{C} \longrightarrow \mathcal{D}$ can act on the morphisms in two ways. Let $f$, $g \in \mathrm{Mor}(\mathcal{C})$. Then:

a) $F(f)\colon F(A) \longrightarrow F(B)$. In this case, we require that

$$F(g \circ f) = F(g) \circ F(f)$$

holds, and we call $F$ *covariant.*

b) $F(f)\colon F(B) \longrightarrow F(A)$. In this case, we require that

$$F(g \circ f) = F(f) \circ F(g)$$

holds, and we call $F$ *contravariant.*

As easy and basic examples of functors we recall the *constant functor* and the *forgetful functor.*
For any two categories $\mathcal{C}$ and $\mathcal{D}$, the constant functor associates each object of $\mathcal{C}$ to a fixed object $X$ of $\mathcal{D}$, and each element of $\mathrm{Mor}(\mathcal{C})$ to $\mathrm{id}_X \in \mathrm{Mor}(\mathcal{D})$. A forgetful functor instead forgets or drops some or all of the input's structure or properties while mapping to the output. For example, if $\mathcal{C} = \mathcal{R}ing$ and $\mathcal{D} = \mathcal{G}rp$ are the categories of rings with ring homomorphisms and of groups with group homomorphisms, respectively, the functor $F\colon \mathcal{C} \longrightarrow \mathcal{D}$ which fixes objects and morphisms, simply forgetting the inner product of rings and the extra structure they have because of it, is a forgetful functor.

In the following we will examine some notable examples of functors in commutative algebra, such has $\mathrm{Hom}(M, \bullet)$, $\mathrm{Hom}(\bullet, N)$, $\bullet \otimes N$, and $S^{-1}\bullet$. These functors are all defined within the category of modules.

## 4.7 Exact Sequences

Let $\{M_i\}_{i\in\mathbb{N}}$ be a family of $A$-modules, and let $\{\varphi_i\colon M_{i-1} \longrightarrow M_i\}_{i\in\mathbb{N}}$ be a family of homomorphisms. A *sequence* of $A$-modules

$$\ldots \longrightarrow M_{i-1} \xrightarrow{\varphi_i} M_i \xrightarrow{\varphi_{i+1}} M_{i+1} \longrightarrow \ldots$$

is called *complex of $A$-modules* if the composition of any two consecutive homomorphisms is zero, *i.e.*, if $\varphi_{i+1} \circ \varphi_i = 0$ for all $i$, or, equivalently, if and only if $\mathrm{Im}\,\varphi_i \subseteq \mathrm{Ker}\,\varphi_{i+1}$ for all $i$.
If the complex is bounded on the left and/or on the right, *i.e.*, if the modules $M_i$ are definitely equal to 0 on the left and/or on the right, only the non-trivial parts of the complex are considered. In such cases, a 0 is inserted on the left and/or on the right. In the following, we assume that all complexes have only a finite number of non-trivial modules.

A sequence is *exact at $M_i$* if $\mathrm{Im}\,\varphi_i = \mathrm{Ker}\,\varphi_{i+1}$. A sequence is *exact* when it is exact at any $M_i$. An exact sequence of the form

$$0 \longrightarrow M \xrightarrow{f} N \xrightarrow{g} P \longrightarrow 0 \tag{4.1}$$

is called *short exact sequence.*

It immediately follows from the definition that the sequence $0 \longrightarrow M \xrightarrow{f} N$ is exact if and only if $f$ is injective, the sequence $N \xrightarrow{g} P \longrightarrow 0$ is exact if and only if $g$ is surjective, and the sequence (4.1) is exact if and only if $f$ is injective, $g$ is surjective, and $\operatorname{Coker} f = N/\operatorname{Im} f = N/\operatorname{Ker} g \simeq P$.

Short exact sequences are frequently employed as follows. Given a short exact sequence like (4.1) involving three modules $M$, $N$, and $P$, we aim to investigate an invariant $\rho$ or a property $\mathcal{P}$ associated with them. If we know $\rho$ for two of the modules, in many cases, it is possible to determine, or estimate $\rho$ for the third module. Similarly, if we know the property $\mathcal{P}$ holds for two of the modules, we can often establish that $\mathcal{P}$ also holds for the third module. For example, if the modules are $K$-vector spaces, we can consider $\rho = \dim_K$. See **E.10.29** for an example where $\mathcal{P}$ is defined as "being finitely generated".

### 4.7.1 The Functors $\operatorname{Hom}_A(\bullet, N)$ and $\operatorname{Hom}_A(M, \bullet)$

Consider $A$-modules $M$, $M_1$, $N$, and $N_1$, and homomorphisms $f\colon M_1 \longrightarrow M$ and $g\colon N \longrightarrow N_1$. We define the following homomorphisms:

$$f^*\colon \operatorname{Hom}_A(M, N) \longrightarrow \operatorname{Hom}_A(M_1, N), \qquad f^*(\varphi) = \varphi \circ f$$
$$g_*\colon \operatorname{Hom}_A(M, N) \longrightarrow \operatorname{Hom}_A(M, N_1), \qquad g_*(\varphi) = g \circ \varphi.$$

With these definitions, the following diagrams are commutative:

Moreover, the following properties hold:

i) if $f = \mathrm{id}_M$, then $f^* = \mathrm{id}_{\operatorname{Hom}(M,N)}$ for all $N$;
ii) for any $A$-module $M_2$ and any homomorphism $f'\colon M_2 \longrightarrow M_1$,

$$(f \circ f')^* = f'^* \circ f^*;$$

i′) if $g = \mathrm{id}_N$, then $g_* = \mathrm{id}_{\operatorname{Hom}(M,N)}$ for all $M$;
ii′) for any $A$-module $N_2$ and any homomorphism $g'\colon N_1 \longrightarrow N_2$,

$$(g' \circ g)_* = g'_* \circ g_*.$$

For any $A$-module $N$, the functor $\operatorname{Hom}_A(\bullet, N)$ maps an $A$-module $M$ to the $A$-module $\operatorname{Hom}_A(M, N)$, and an $A$-module homomorphism $f\colon M_1 \longrightarrow M$

to the $A$-module homomorphism $\mathrm{Hom}_A(f, N) = f^*$. It also sends identity homomorphisms to identity homomorphisms and behaves well with respect to composition by i) and ii). Therefore, $\mathrm{Hom}_A(\bullet, N)$ is a functor from the category of $A$-modules to itself. Since

$$\mathrm{Hom}_A(f \circ f', N) = \mathrm{Hom}_A(f', N) \circ \mathrm{Hom}_A(f, N),$$

the order in the composition is reversed, and $\mathrm{Hom}_A(\bullet, N)$ is a contravariant functor.

In a similar manner, it can be shown that, for any $A$-module $M$, the functor $\mathrm{Hom}_A(M, \bullet)$ is a covariant functor.

**Left-exactness of Hom($\bullet$, $N$) and of Hom($M$, $\bullet$)**

**T. 4.16.** 1. Let $M_1 \xrightarrow{f} M \xrightarrow{g} M_2 \longrightarrow 0$ be an exact sequence of $A$-modules. Then, for any $A$-module $N$, the sequence

$$0 \longrightarrow \mathrm{Hom}_A(M_2, N) \xrightarrow{g^*} \mathrm{Hom}_A(M, N) \xrightarrow{f^*} \mathrm{Hom}_A(M_1, N)$$

is exact.

2. Let $0 \longrightarrow N_1 \xrightarrow{f} N \xrightarrow{g} N_2$ be an exact sequence of $A$-modules. Then, for any $A$-module $M$, the sequence

$$0 \longrightarrow \mathrm{Hom}_A(M, N_1) \xrightarrow{f_*} \mathrm{Hom}_A(M, N) \xrightarrow{g_*} \mathrm{Hom}_A(M, N_2)$$

is exact.

When referring to these properties, we will say that the functors $\mathrm{Hom}_A(\bullet, N)$ and $\mathrm{Hom}_A(M, \bullet)$ are *left-exact.*

**Proof T. 4.16.** 1. We prove that the sequence is exact for any $A$-module $N$.

At $\mathrm{Hom}_A(M_2, N)$ In other words, $g^*$ is injective.

Consider $\varphi \in \mathrm{Hom}_A(M_2, N)$, and assume $g^*(\varphi) = \varphi \circ g = 0$. Since $g$ is surjective by hypothesis, for any $m_2 \in M_2$, there exists $m \in M$ such that $m_2 = g(m)$. Consequently, $\varphi(m_2) = \varphi(g(m)) = 0$.
Therefore, $\varphi = 0$ and $\mathrm{Ker}\, g^* = 0$.

At $\mathrm{Hom}_A(M, N)$ In other words, $\mathrm{Im}\, g^* = \mathrm{Ker}\, f^*$.

Since $f^* \circ g^* = (g \circ f)^* = 0$, it immediately follows that $\mathrm{Im}\, g^* \subseteq \mathrm{Ker}\, f^*$.

To prove the opposite inclusion, take $\psi \in \mathrm{Ker}\, f^*$.
We have to find $\varphi \in \mathrm{Hom}(M_2, N)$ such that $g^*(\varphi) = \varphi \circ g = \psi$. Since $g$ is surjective, for any $m_2 \in M_2$, there exists $m \in M$ such that $m_2 = g(m)$, and we can define $\varphi(m_2) = \psi(m)$.

Now, we only need to check that $\varphi$ is well-defined. Clearly, $\varphi$ has values in $N$. Let $n$ be another element of $g^{-1}(m_2)$, then $m - n \in \operatorname{Ker} g = \operatorname{Im} f$ and there exists $m_1 \in M_1$ such that $m - n = f(m_1)$.
Hence, $\psi(m-n) = \psi(f(m_1)) = f^*(\psi)(m_1) = 0$, because $\psi \in \operatorname{Ker} f^*$.
Therefore, $\varphi$ is well-defined, because the definition does not depend on the choice of the element in the preimage of $m_2$.

2. The proof is similar to the previous one. We have to prove that the second sequence is exact for any $A$-module $M$.

At $\operatorname{Hom}_A(M, N_1)$ In other words, $f_*$ is injective.

Let $\varphi \in \operatorname{Hom}_A(M, N_1)$ be such that $f_*(\varphi) = f \circ \varphi = 0$. Then, $\operatorname{Im} \varphi \subseteq \operatorname{Ker} f$, and $\varphi = 0$ follows by the injectivity of $f$.

At $\operatorname{Hom}_A(M, N)$ In other words, $\operatorname{Im} f_* = \operatorname{Ker} g_*$.

Since $g_* \circ f_* = (g \circ f)_* = 0$, we have that $\operatorname{Im} f_* \subseteq \operatorname{Ker} g_*$.

To prove the opposite inclusion, take $\psi \in \operatorname{Ker} g_*$, *i.e.*, such that $g \circ \psi = 0$. We have to find $\varphi \in \operatorname{Hom}_A(M, N_1)$ such that $f_*(\varphi) = \psi$. We define $\varphi$ by $\varphi(m) = f^{-1}(\psi(m))$ for any $m \in M$.
Note that such map is well-defined because $f$ is injective, and, by hypothesis, $\operatorname{Im} \psi \subseteq \operatorname{Ker} g = \operatorname{Im} f$. □

We note that the converse of the statements of **T.4.16** also holds.

**T. 4.17.** (→ p. 228) 1. Let $M_1 \xrightarrow{f} M \xrightarrow{g} M_2 \longrightarrow 0$ be a sequence of $A$-modules such that the sequence

$$0 \longrightarrow \operatorname{Hom}_A(M_2, N) \xrightarrow{g^*} \operatorname{Hom}_A(M, N) \xrightarrow{f^*} \operatorname{Hom}_A(M_1, N)$$

is exact for any $A$-module $N$. Then,

$$M_1 \xrightarrow{f} M \xrightarrow{g} M_2 \longrightarrow 0$$

is exact.

2. Let $0 \longrightarrow N_1 \xrightarrow{f} N \xrightarrow{g} N_2$ be a sequence of $A$-modules such that the sequence

$$0 \longrightarrow \operatorname{Hom}_A(M, N_1) \xrightarrow{f_*} \operatorname{Hom}_A(M, N) \xrightarrow{g_*} \operatorname{Hom}_A(M, N_2)$$

is exact for any $A$-module $M$. Then,

$$0 \longrightarrow N_1 \xrightarrow{f} N \xrightarrow{g} N_2$$

is exact.

### 4.7.2 Split Sequences

For any pair of $A$-modules $M$ and $N$, the definition of direct sum of modules yields an exact sequence

$$0 \longrightarrow M \xrightarrow{i_M} M \oplus N \xrightarrow{\pi_N} N \longrightarrow 0,$$

where $i_M(m) = (m, 0)$ and $\pi_N(m, n) = n$.
In general, a short exact sequence

$$0 \longrightarrow M \xrightarrow{f} N \xrightarrow{g} P \longrightarrow 0$$

does not provide a decomposition of $N$ in terms of $M$ and $P$. The sequences for which such decomposition holds are characterized by the following proposition.

**T. 4.18.** Let $0 \longrightarrow M \xrightarrow{f} N \xrightarrow{g} P \longrightarrow 0$ be an exact sequence of $A$-modules. The following statements are equivalent:

1. there exists an isomorphism $N \xrightarrow{\varphi} M \oplus P$ that makes the following diagram commutative

$$\begin{array}{ccccccccc} 0 & \longrightarrow & M & \xrightarrow{f} & N & \xrightarrow{g} & P & \longrightarrow & 0 \\ & & \downarrow{\scriptstyle \mathrm{id}_M} & & \downarrow{\scriptstyle \varphi} & & \downarrow{\scriptstyle \mathrm{id}_P} & & \\ 0 & \longrightarrow & M & \xrightarrow{i_M} & M \oplus P & \xrightarrow{\pi_P} & P & \longrightarrow & 0; \end{array}$$

2. there exists a homomorphism $r \colon N \longrightarrow M$ such that $r \circ f = \mathrm{id}_M$, in other words, $r$ is a left-inverse of $f$;
3. there exists a homomorphism $s \colon P \longrightarrow N$ such that $g \circ s = \mathrm{id}_P$, in other words, $s$ is a right-inverse of $g$.

If any of the previous equivalent conditions holds, we say that the sequence *splits*, that $r$ is a *retraction* of $f$, and $s$ a *section* of $g$.

It is worth noting that in a short exact sequence such as

$$0 \longrightarrow M \xrightarrow{i_M} M \oplus P \xrightarrow{\pi_P} P \longrightarrow 0$$

the canonical inclusion and projection play the role of a section and a retraction, respectively, meaning that $\pi_P \circ i_P = \mathrm{id}_P$ and $\pi_M \circ i_M = \mathrm{id}_M$.

**Proof T. 4.18.** Consider the following diagram:

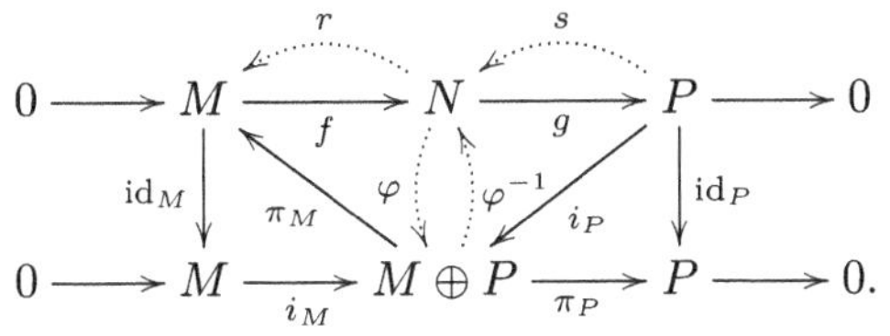

$$
\begin{array}{ccccccccc}
 & & & \overset{r}{\swarrow\cdots} & & \overset{s}{\swarrow\cdots} & & & \\
0 & \longrightarrow & M & \xrightarrow{f} & N & \xrightarrow{g} & P & \longrightarrow & 0 \\
 & & \Big\downarrow \mathrm{id}_M & \nwarrow \pi_M & \varphi \,\vdots\, \varphi^{-1} & \swarrow i_P & \Big\downarrow \mathrm{id}_P & & \\
0 & \longrightarrow & M & \xrightarrow[i_M]{} & M \oplus P & \xrightarrow[\pi_P]{} & P & \longrightarrow & 0.
\end{array}
$$

$1 \Rightarrow 2$. To define $r$, it is sufficient to set $r(n) = (\pi_M \circ \varphi)(n)$, that is, if $n = \varphi^{-1}(m,p) \in N$, then $r(n) = \pi_M(m,p) = m$.
Indeed,

$$r(f(m)) = \pi_M(\varphi(f(m))) = \pi_M(i_M(\mathrm{id}_M(m))) = (\pi_M \circ i_M)(m) = m.$$

$1 \Rightarrow 3$. Since $g$ is surjective, for any $p \in P$ there exists $n \in N$ such that $g(n) = p$. We define $s(p) = (\varphi^{-1} \circ i_P)(p)$. In this way, we have that $\varphi(s(p)) = i_P(p)$. Therefore,

$$p = \pi_P(i_P(p)) = \pi_P(\varphi(s(p))) = g(s(p)).$$

$2 \Rightarrow 1$. Every $n \in N$ can be expressed as $n = f(r(n)) + (n - f(r(n)))$.
Since $r(n - f(r(n))) = r(n) - (r \circ f)(r(n)) = 0$, we have $N = \operatorname{Im} f + \operatorname{Ker} r$.
We show that this sum is a direct sum. If $u \in \operatorname{Ker} r \cap \operatorname{Im} f$, then $u = f(m)$ for some $m \in M$. Moreover, since $r \circ f = \mathrm{id}_M$ and $u \in \operatorname{Ker} r$, it follows that $0 = r(u) = r(f(m)) = m$, and thus, $u = 0$.
Now, recall that $f$ is injective, $\operatorname{Im} f = \operatorname{Ker} g$, and $g$ is surjective.
Therefore, $\operatorname{Im} f \simeq M$, $g_{|\operatorname{Ker} r}$ is an isomorphism, and we have

$$N = \operatorname{Im} f \oplus \operatorname{Ker} r \simeq \operatorname{Im} f \oplus \operatorname{Im} g \simeq M \oplus P.$$

The isomorphism $\varphi$ is defined by

$$\varphi(n) = (r(n), g(n)).$$

It is straightforward to verify that the diagram is commutative.
Indeed,

$$\varphi(f(m)) = (r(f(m)), g(f(m))) = (m, 0) = i_M(m),$$

and

$$\pi_P(\varphi(n)) = \pi_P(r(n), g(n)) = g(n).$$

$3 \Rightarrow 1$. This is similar to the previous one.
For any $n \in N$ write $n = (n - s(g(n))) + s(g(n))$.
Since $n - s(g(n)) \in \operatorname{Ker} g = \operatorname{Im} f$ and $\operatorname{Im} f \cap \operatorname{Im} s = 0$, it follows that $N = \operatorname{Im} f \oplus \operatorname{Im} s \simeq M \oplus P$.
To conclude the proof, define

$$\varphi(n) = (f^{-1}(n - s(g(n))), g(n)).$$

This function is well-defined, because $n - s(g(n)) \in \operatorname{Im} f$ and $f$ is injective. □

As shown in the proof above, if the initial short exact sequence splits, then also the sequence $0 \longrightarrow P \xrightarrow{s} N \xrightarrow{r} M \longrightarrow 0$ is exact and splits.

### 4.7.3 Snake Lemma

We conclude this section with a highly useful result.

**Snake Lemma**

**T. 4.19.** Given a commutative diagram of $A$-modules with exact rows

$$\begin{array}{ccccccc} & & M & \xrightarrow{f} & N & \xrightarrow{g} & P & \longrightarrow 0 \\ & & \downarrow{\alpha} & & \downarrow{\beta} & & \downarrow{\gamma} \\ 0 & \longrightarrow & M' & \xrightarrow{f'} & N' & \xrightarrow{g'} & P', \end{array}$$

there exists an exact sequence

$$\operatorname{Ker}\alpha \xrightarrow{\tilde{f}} \operatorname{Ker}\beta \xrightarrow{\tilde{g}} \operatorname{Ker}\gamma \xrightarrow{\delta} \operatorname{Coker}\alpha \xrightarrow{\overline{f'}} \operatorname{Coker}\beta \xrightarrow{\overline{g'}} \operatorname{Coker}\gamma.$$

Moreover, if $f$ is injective, then $\tilde{f}$ is also injective, and if $g'$ is surjective, then $\overline{g'}$ is also surjective.

The homomorphism $\delta$ is usually called the *connecting homomorphism.*

**Proof T. 4.19.** Consider the induced diagram

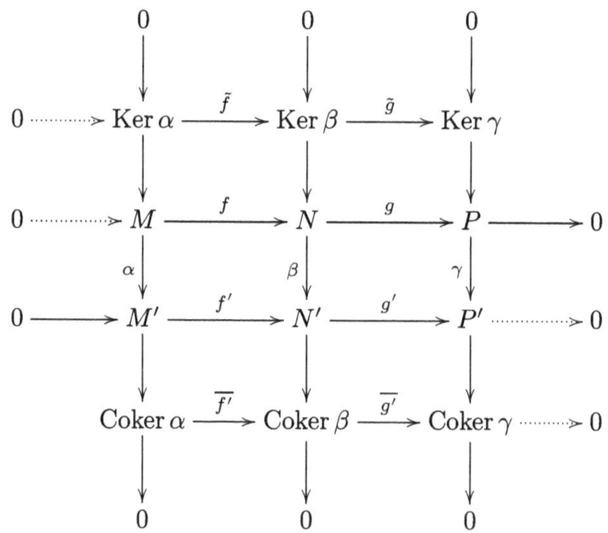

where the middle squares are commutative by assumption, and the columns are exact sequences, since the maps are the natural inclusions and projections.

We begin by verifying that all morphisms are well-defined.
The homomorphisms $\tilde{f}$ and $\tilde{g}$ are the restrictions $f_{|\operatorname{Ker}\alpha}$ and $g_{|\operatorname{Ker}\beta}$.
Take $m \in \operatorname{Ker}\alpha$. Then,

$$\beta(f(m)) = f'(\alpha(m)) = 0,$$

meaning $f(m) \in \operatorname{Ker}\beta$.
As a result, $\tilde{f}$ is well-defined. The proof that $\tilde{g}$ is well-defined is similar.
The homomorphisms $\overline{f'}$ and $\overline{g'}$ are defined by $\overline{f'}(\overline{m}) = \overline{f'(m)}$ and $\overline{g'}(\overline{n}) = \overline{g'(n)}$.
Take $\overline{m} = \overline{m'} \in \operatorname{Coker}\alpha$.
Then, $m - m' = \alpha(u)$ for some $u \in M$, and therefore,

$$f'(m - m') = f'(\alpha(u)) = \beta(f(u)) \in \operatorname{Im}\beta.$$

It follows that $\overline{f'(m)} = \overline{f'(m')} \in \operatorname{Coker}\beta$, meaning that $\overline{f'}$ is well-defined. The proof that $\overline{g'}$ is well-defined is similar.

Now, we construct $\delta$, which plays a crucial role in the proof.
We aim to find a homomorphism $\delta\colon \operatorname{Ker}\gamma \longrightarrow \operatorname{Coker}\alpha$. Its definition, like the proofs of the other essential parts of the statement, is obtained by using a strategy called *diagram chasing*, which we now explain.
Take $p \in \operatorname{Ker}\gamma \subseteq P$, and let $n \in N$ be such that $g(n) = p$. This $n$ exists because $g$ is surjective.
Since $0 = \gamma(g(n)) = g'(\beta(n))$, we have $\beta(n) \in \operatorname{Ker} g' = \operatorname{Im} f'$. Therefore, there exists a unique $m \in M'$ such that $f'(m) = \beta(n)$, because $f'$ is injective.
We define

$$\delta(p) = \overline{m} \in \operatorname{Coker}\alpha.$$

To prove that $\delta$ is well-defined, we must show that if $n'$ is an element of $N$ such that $g(n') = p$ and $m'$ is the element of $M'$ such that $f'(m') = \beta(n')$, then $\overline{m} = \overline{m'}$ in $\operatorname{Coker}\alpha$, *i.e.*, $m - m' \in \operatorname{Im}\alpha$.
We have $n - n' \in \operatorname{Ker} g = \operatorname{Im} f$, thus there exists an element $u \in M$ such that $f(u) = n - n'$.
As a result,

$$f'(m - m') = \beta(n - n') = \beta(f(u)) = f'(\alpha(u)).$$

Since $f'$ is injective, we have $m - m' = \alpha(u)$. This shows that $\delta$ is well-defined. It is straightforward to verify that $\delta$ is a homomorphism.

We can now proceed to demonstrate that the final sequence in the statement is exact.
At $\operatorname{Ker}\beta$ In other words, $\operatorname{Im}\tilde{f} = \operatorname{Ker}\tilde{g}$.

It is clear that, since $g \circ f = 0$, we also have $\tilde{g} \circ \tilde{f} = 0$. Therefore, $\operatorname{Im}\tilde{f} \subseteq \operatorname{Ker}\tilde{g}$.

To prove the opposite inclusion, we consider an element $n \in \operatorname{Ker}\beta$ such that $\tilde{g}(n) = 0$. Then, $g(n) = 0$, which implies that $n \in \operatorname{Ker} g = \operatorname{Im} f$. Therefore,

there exists $m \in M$ such that $f(m) = n$. To complete the proof, we need to prove that $m \in \operatorname{Ker} \alpha$.
Since

$$f'(\alpha(m)) = \beta(f(m)) = \beta(n) = 0$$

and $f'$ is injective, we have $m \in \operatorname{Ker} \alpha$.

At $\operatorname{Ker} \gamma$ In other words, $\operatorname{Im} \tilde{g} = \operatorname{Ker} \delta$.

To show that $\delta \circ \tilde{g} = 0$, let $n \in \operatorname{Ker} \beta$. We first observe that since $\beta(n) = 0$, we have $\beta(n) = f'(0)$. Therefore, by the definition of $\delta$, we see that $\delta(\tilde{g}(n)) = 0$.

To prove the opposite inclusion, consider an element $p \in \operatorname{Ker} \delta$.
We can express $p$ as $g(n)$ for some $n \in N$. Since $\delta(p) = 0$, we know that $f'(m) = \beta(n)$ for some $m \in \operatorname{Im} \alpha$.
By choosing $u \in M$ such that $\alpha(u) = m$, we have

$$\beta(f(u)) = f'(\alpha(u)) = \beta(n).$$

Hence, $n - f(u) \in \operatorname{Ker} \beta$.
Therefore, we can conclude that

$$\tilde{g}(n - f(u)) = g(n) - g(f(u)) = g(n) = p,$$

which means $p \in \operatorname{Im} \tilde{g}$.

At $\operatorname{Coker} \alpha$ In other words, $\operatorname{Im} \delta = \operatorname{Ker} \overline{f'}$.

Take $p \in \operatorname{Ker} \gamma$. There exist $n \in N$ and $m \in M'$ such that $p = g(n)$ and $\beta(n) = f'(m)$. Therefore, $\delta(p) = \overline{m} \in \operatorname{Coker} \alpha$.
Moreover, since $f'(m) \in \operatorname{Im} \beta$, we have $\overline{f'}(\delta(p)) = \overline{f'(m)} = 0$ in $\operatorname{Coker} \beta$.
Hence, $\operatorname{Im} \delta \subseteq \operatorname{Ker} \overline{f'}$.

To prove the opposite inclusion, take $\overline{m} \in \operatorname{Ker} \overline{f'}$. Thus, $f'(m) \in \operatorname{Im} \beta$.
Let $n \in N$ be such that $f'(m) = \beta(n)$, and let $p = g(n)$.
Then,

$$\gamma(g(n)) = g'(\beta(n)) = g'(f'(m)) = 0,$$

hence $p \in \operatorname{Ker} \gamma$ and $\overline{m} = \delta(p)$ by definition of $\delta$, as desired.

At $\operatorname{Coker} \beta$ In other words, $\operatorname{Im} \overline{f'} = \operatorname{Ker} \overline{g'}$.

It is clear that $\overline{g'} \circ \overline{f'} = 0$. Therefore, $\operatorname{Im} \overline{f'} \subseteq \operatorname{Ker} \overline{g'}$.

To prove the opposite inclusion, take $\overline{n'} \in \operatorname{Ker} \overline{g'} \subseteq \operatorname{Coker} \beta$.
Then, we have $g'(n') \in \operatorname{Im} \gamma$, and there exists $p \in P$ such that $g'(n') = \gamma(p)$.
Let $n \in N$ be such that $g(n) = p$ and consider the element $n' - \beta(n)$.
We have

$$g'(n' - \beta(n)) = g'(n') - g'(\beta(n)) = g'(n') - \gamma(g(n)) = \gamma(p) - \gamma(p) = 0.$$

Therefore, $n' - \beta(n) \in \operatorname{Ker} g' = \operatorname{Im} f'$, and there exists $m' \in M'$ such that $f'(m') = n' - \beta(n)$.
Thus, we obtain

$$\overline{n'} = \overline{n' - \beta(n)} = \overline{f'(m')} = \overline{f'}(\overline{m'}),$$

and both inclusions are verified.

Moreover, if $f$ is injective, the sequence is exact

**at Ker $\alpha$** It is clear since $\tilde{f}$ is a restriction of $f$.

Finally, if $g'$ is surjective, the sequence is exact

**at Coker $\gamma$** For any $\overline{p} \in \operatorname{Coker}\gamma$, there exists $n' \in N'$ such that $g'(n') = p$. Therefore, $\overline{p} = \overline{g'}(\overline{n'})$ and $\overline{g'}$ is also surjective. □

**T. 4.20.** (→ p. 230) In a commutative diagram of $A$-modules with exact rows

$$\begin{array}{ccccccccc} 0 & \longrightarrow & M & \xrightarrow{f} & N & \xrightarrow{g} & P & \longrightarrow & 0 \\ & & \downarrow{\scriptstyle\alpha} & & \downarrow{\scriptstyle\beta} & & \downarrow{\scriptstyle\gamma} & & \\ 0 & \longrightarrow & M' & \xrightarrow{f'} & N' & \xrightarrow{g'} & P' & \longrightarrow & 0, \end{array}$$

if any two of the homomorphisms $\alpha$, $\beta$, and $\gamma$ are isomorphisms, then the third one is an isomorphism as well.

## 4.8 Projective Modules

In **T.4.16** we have seen that for any exact sequence

$$0 \longrightarrow M_1 \xrightarrow{f} M \xrightarrow{g} M_2 \longrightarrow 0$$

and for any $A$-module $N$, the following sequences are also exact:

$$0 \longrightarrow \operatorname{Hom}_A(M_2, N) \xrightarrow{g^*} \operatorname{Hom}_A(M, N) \xrightarrow{f^*} \operatorname{Hom}_A(M_1, N)$$

$$0 \longrightarrow \operatorname{Hom}_A(N, M_1) \xrightarrow{f_*} \operatorname{Hom}_A(N, M) \xrightarrow{g_*} \operatorname{Hom}_A(N, M_2).$$

However, in general, these sequences are not right-exact, *i.e.*, the maps $f^*$ and $g_*$ are not surjective.

For example, consider the exact sequence of $\mathbb{Z}$-modules

$$0 \longrightarrow \mathbb{Z} \xrightarrow{n} \mathbb{Z} \xrightarrow{\pi} \mathbb{Z}/(n) \longrightarrow 0,$$

where the first map is the multiplication by $n \neq 0, \pm 1$ and $\pi$ is the canonical projection. Take the $\mathbb{Z}$-module $\mathbb{Z}/(n)$ and consider the functors $\operatorname{Hom}_{\mathbb{Z}}(\bullet, \mathbb{Z}/(n))$ and $\operatorname{Hom}_{\mathbb{Z}}(\mathbb{Z}/(n), \bullet)$.

These functors induce, respectively, two sequences

$$0 \longrightarrow \operatorname{Hom}_{\mathbb{Z}}(\mathbb{Z}/(n), \mathbb{Z}/(n)) \xrightarrow{\pi^*} \operatorname{Hom}_{\mathbb{Z}}(\mathbb{Z}, \mathbb{Z}/(n)) \xrightarrow{n^*} \operatorname{Hom}_{\mathbb{Z}}(\mathbb{Z}, \mathbb{Z}/(n))$$

$$0 \longrightarrow \mathrm{Hom}_{\mathbb{Z}}(\mathbb{Z}/(n), \mathbb{Z}) \xrightarrow{n_*} \mathrm{Hom}_{\mathbb{Z}}(\mathbb{Z}/(n), \mathbb{Z}) \xrightarrow{\pi_*} \mathrm{Hom}_{\mathbb{Z}}(\mathbb{Z}/(n), \mathbb{Z}/(n)).$$

For any $g \in \mathrm{Hom}_{\mathbb{Z}}(\mathbb{Z}, \mathbb{Z}/(n))$, we have

$$n^*(g)(m) = g(nm) = ng(m) = 0$$

for any $m \in \mathbb{Z}$.
By **T.4.2**, we have $\mathrm{Hom}_{\mathbb{Z}}(\mathbb{Z}, \mathbb{Z}/(n)) \simeq \mathbb{Z}/(n) \neq 0$. Hence, $n^* = 0$ is not surjective.
Moreover, since $\mathrm{Hom}_{\mathbb{Z}}(\mathbb{Z}/(n), \mathbb{Z}) = 0$, the map $\pi_*$ cannot be surjective either.

A functor is called *exact* if it preserves short exact sequences, *i.e.*, for any short exact sequence, the induced sequence is also exact.
To characterize the modules $M$ for which the functor $\mathrm{Hom}_A(M, \bullet)$ is also right-exact, and as a result exact, we introduce a new definition.

**Projective Module**

An $A$-module $P$ is *projective* if, for any pair of $A$-modules $M$ and $N$, and any homomorphisms $g\colon M \longrightarrow N$ and $f\colon P \longrightarrow N$ with $g$ surjective, there exists a homomorphism $\tilde{f}\colon P \longrightarrow M$ which makes the following diagram commute:

$$\begin{array}{ccccc} & & P & & \\ & \tilde{f} \swarrow & \downarrow f & & \\ M & \xrightarrow[g]{} & N & \longrightarrow & 0. \end{array} \qquad \boxed{g \circ \tilde{f} = f}$$

We remark that the zero module is projective.

**T. 4.21.** (→ p. 230) Every free module is projective.

**Characterization of Projective Modules**

**T. 4.22.** Let $P$ be an $A$-module. The following conditions are equivalent:

1. $P$ is projective;
2. for any short exact sequence of $A$-modules

$$0 \longrightarrow N_1 \xrightarrow{f} N \xrightarrow{g} N_2 \longrightarrow 0,$$

the sequence

$$0 \longrightarrow \mathrm{Hom}_A(P, N_1) \xrightarrow{f_*} \mathrm{Hom}_A(P, N) \xrightarrow{g_*} \mathrm{Hom}_A(P, N_2) \longrightarrow 0$$

is exact;

3. every exact sequence $0 \longrightarrow M \longrightarrow N \longrightarrow P \longrightarrow 0$ splits;
4. $P$ is a direct summand of every module $M$ that has a surjective map $\pi\colon M \longrightarrow P$;
5. $P$ is a direct summand of a free module.

**Proof T. 4.22.** $1 \Leftrightarrow 2$. The equivalence immediately follows from the definition of projective module.
For any surjective map $g\colon N \longrightarrow N_2$, the induced homomorphism $g_*$ is surjective if and only if for any $\varphi \in \mathrm{Hom}(P, N_2)$ there exists $\psi \in \mathrm{Hom}_A(P, N)$ such that the following diagram commutes:

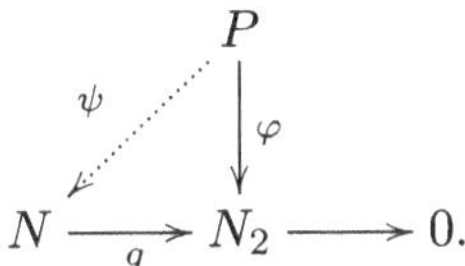

$1 \Rightarrow 3$. Consider the exact sequence $0 \longrightarrow M \xrightarrow{\alpha} N \xrightarrow{\beta} P \longrightarrow 0$.
As $P$ is projective, there exists a homomorphism $s$ that makes the following diagram commutative:

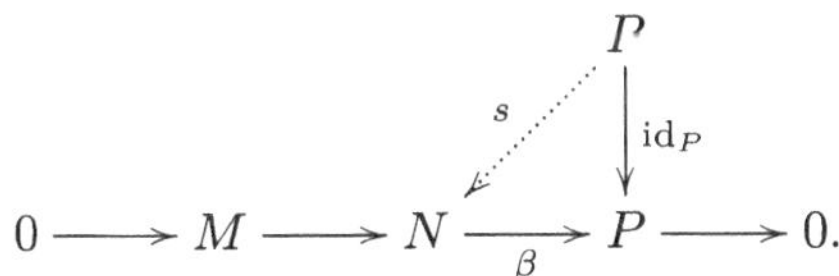

Therefore, $s\colon P \longrightarrow N$ is a section of $\beta$ and the sequence splits, see **T.4.18**.

$3 \Rightarrow 4$. If $M$ is a module with a surjective map $\pi\colon M \longrightarrow P$, then, the sequence $0 \longrightarrow \operatorname{Ker}\pi \longrightarrow M \xrightarrow{\pi} P \longrightarrow 0$ is exact. By hypothesis, this sequence splits. As a result, $M \simeq \operatorname{Ker}\pi \oplus P$.

$4 \Rightarrow 5$. By **T.4.8**, every $A$-module is a quotient of a free module.

$5 \Rightarrow 1$. Let $f\colon P \longrightarrow N$ and $g\colon M \longrightarrow N$ be homomorphisms, where $g$ is surjective. We need to find a homomorphism $\tilde{f}\colon P \longrightarrow M$ such that $g \circ \tilde{f} = f$. By assumption, there exist $A$-modules $Q$ and $F$, with $F$ free, such that $F = Q \oplus P$.
Let $i_P\colon P \longrightarrow F$ be the inclusion homomorphism. By using the universal property of direct sum **T.4.6**.1, and considering the zero homomorphism $f'\colon Q \longrightarrow N$, we can extend $f$ to a unique homomorphism $\varphi\colon F \longrightarrow N$ such that

$$\varphi \circ i_P = f \quad \text{and} \quad \varphi \circ i_Q = f' = 0.$$

Since $F$ is free, it is also projective by **T.4.21**. Thus, there exists $\widetilde{\varphi}\colon F \longrightarrow M$ such that $\varphi = g \circ \widetilde{\varphi}$.

We have constructed a diagram with commutative triangles

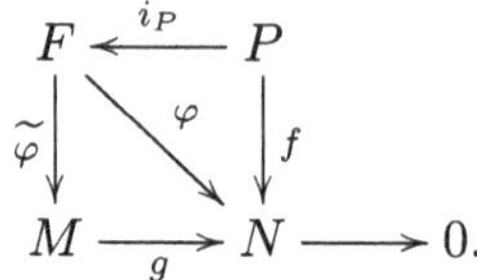

Now it is sufficient to define $\tilde{f}$ as $\widetilde{\varphi} \circ i_P$. □

For example, consider the ring $A = \mathbb{Z}[\sqrt{-6}]$ and the ideals $I = (2, \sqrt{-6})$ and $J = (3, \sqrt{-6})$. It is easy to verify that $I + J = A$ and $IJ = (\sqrt{-6}) \simeq A$. Additionally, $I$ and $J$ are not principal ideals and they are not free $A$-modules, see also **E.10.40**.
The sequence of $A$-modules

$$\begin{array}{ccccccccc} 0 \longrightarrow & IJ & \longrightarrow & I \oplus J & \longrightarrow & A & \longrightarrow 0 \\ & a & \mapsto & (a, -a) & & & \\ & & & (a, b) & \mapsto & a + b & \end{array}$$

is exact, see also **E.10.35**.
Since $A$ is free, and thus projective, the sequence splits by **T.4.22**, and we obtain

$$I \oplus J \simeq A \oplus IJ \simeq A \oplus A = A^2.$$

As a consequence, $I$ and $J$ are projective, again by **T.4.22**.

We conclude this section by briefly noting that the notion of injective module can be introduced in a similar manner to characterize the modules $M$ for which the functor $\mathrm{Hom}_A(\bullet, M)$ is exact. Further information can be found in **E.10.42**, **E.10.43**, and **E.10.44**.

**Injective Modules**

An $A$-module $E$ is *injective* if, for any pair of $A$-modules $M$ and $N$, and any homomorphisms $g\colon M \longrightarrow E$ and $f\colon M \longrightarrow N$, with $f$ injective, there exists a homomorphism $\tilde{g}\colon N \longrightarrow E$ that makes the following diagram commutative:

$$\begin{array}{ccccc} 0 \longrightarrow & M & \xrightarrow{f} & N \\ & \downarrow g & \swarrow \tilde{g} & \\ & E. & & \end{array}$$

$$\boxed{\tilde{g} \circ f = g}$$

## 4.9 Modules over a PID

In general, submodules of a free or finitely generated module are not necessarily free or finitely generated, see **E.10.4**. Nevertheless, these properties hold for modules over a principal ideal domain $A$. In this section, we will focus on this specific scenario.

We recall that all bases of a free module $M$ are equipotent, see **T.4.5**, and their common cardinality is by definition $\operatorname{rank} M$.

**Properties of Submodules over a PID**

**T. 4.23.** Let $A$ be a principal ideal domain, let $M$ be an $A$-module, and let $N \subseteq M$ be a non-zero submodule of $M$.

1. If $M$ is free, then $N$ is free and $\operatorname{rank} N \leq \operatorname{rank} M$.
2. If $M$ is finitely generated, then $N$ is finitely generated.

**Proof T. 4.23.** 1. We prove the statement under the assumption that $M$ is finitely generated. For a general proof see [6, Chapter III, Theorem 7.1 and Appendix 2, §2].

The proof proceeds by induction on the rank of $M$.
If $M$ is cyclic, its submodules are isomorphic to ideals of $A$. Since $A$ is a PID and $N \neq 0$, there exists $0 \neq a \in A$ such that $N \simeq (a)$.
Let $c \in A$. Then, $ca = 0$ implies $c = 0$ because $A$ is a domain.
Thus, $N$ is free. Moreover, $\operatorname{rank} N = \operatorname{rank} M = 1$.

Now, assume that $\operatorname{rank} M = r + 1$, and that the claim is true for all free modules of rank less than or equal to $r$.
Let $\{m_1, \ldots, m_{r+1}\}$ be a basis of $M$, and define

$$N_r = N \cap \langle m_1, \ldots, m_r \rangle.$$

If $N = N_r$, then $N$ is free and $\operatorname{rank} N \leq \operatorname{rank}\langle m_1, \ldots, m_r \rangle = r < \operatorname{rank} M$.
Otherwise, $N_r \subsetneq N$ and $N_r$ is free by the inductive assumption. Recalling that for any $n \in N$, there exist unique $b_1, \ldots, b_r$ and $a_n \in A$ such that $n = b_1 m_1 + \ldots + b_r m_r + a_n m_{r+1}$, we define

$$I = \{a_n \in A \colon n \in N\}.$$

Since $a_{n_1} + a_{n_2} = a_{n_1+n_2}$ and $ca_n = a_{cn}$, we have that $I$ is an ideal of $A$. Moreover, since $N_r \neq N$, it follows that $I \neq 0$. As $A$ is a PID, there exist $b_1, \ldots, b_r$, $a \in A$, with $a \neq 0$, and $n_0 \in N \setminus N_r$ such that $I = (a)$ and $n_0 = b_1 m_1 + \ldots + b_r m_r + a m_{r+1}$.
To conclude the proof, we show that

$$N \simeq N_r \oplus \langle n_0 \rangle.$$

Let $n \in N$, then, $a_n = ka$ for some $k \in A$. Therefore, $n - kn_0 \in N_r$, and thus,

$$n = (n - kn_0) + kn_0 \in N_r + \langle n_0 \rangle.$$

Since $m_1, \ldots, m_r, n_0$ are linearly independent, $N_r \cap \langle n_0 \rangle = 0$. The conclusion follows from the inductive assumption, once we observe that $\langle n_0 \rangle$ is free.

Finally, we have $\operatorname{rank} N = \operatorname{rank} N_r + 1 \leq r + 1 = \operatorname{rank} M$.

2. Given a generating set of $M$ consisting of $r$ elements, it is possible to define a surjective homomorphism $f : A^r \longrightarrow M$. Hence, $f^{-1}(N)$ is a free and finitely generated submodule of $A^r$. Consequently, $N$ is finitely generated. □

**T. 4.24.** (→ p. 230) Let $A$ be a principal ideal domain, and let $M$ be an $A$-module. Then, $M$ is projective if and only if it is free.

Next, we would like to provide a constructive proof of the structure theorem of finitely generated modules over a PID.
Before introducing some facts on matrices with entries in a PID and Smith normal form, we make the following remarks.
Let $\mathcal{B}_n = \left\{ e_1^{(n)}, \ldots, e_n^{(n)} \right\}$ be the canonical basis of the module $A^n$, and let $M = \langle m_1, \ldots, m_r \rangle$ be a finitely generated $A$-module. Then, $M$ is isomorphic to a quotient of $A^r$, via the homomorphism $f\colon A^r \longrightarrow M$ defined by $f(e_i^{(r)}) = m_i$, and the following short exact sequence holds:

$$0 \longrightarrow \operatorname{Ker} f \longrightarrow A^r \stackrel{f}{\longrightarrow} M \longrightarrow 0.$$

Since $\operatorname{Ker} f \subseteq A^r$, it is free of rank $s \leq r$ by **T.4.23**.1. Let $w_1, \ldots, w_s$ be a basis of $\operatorname{Ker} f$, and define a homomorphism $\varphi\colon A^s \longrightarrow A^r$ by letting $\varphi(e_i^{(s)}) = w_i$. In this way, we have $\operatorname{Ker} f \simeq \operatorname{Im} \varphi$, and consequently,

$$M \simeq A^r / \operatorname{Ker} f \simeq \operatorname{Coker} \varphi.$$

Let $\mathcal{B}_s$ and $\mathcal{B}_r$ be the canonical bases of $A^s$ and $A^r$, respectively. We associate with $\varphi$ a matrix $X = (x_{ij}) \in M_{rs}(A)$ whose columns generate the relations among the generators of $M$.
In other words, $(a_1, \ldots, a_r) \in A^r$ is such that

$$a_1 m_1 + \ldots + a_r m_r = 0, \quad \textit{i.e.}, \quad (a_1, \ldots, a_r) \in \operatorname{Ker} f,$$

if and only if there exists $u \in A^s$ such that $Xu^{\mathrm{t}} = (a_1, \ldots, a_r)^{\mathrm{t}}$, that is, $(a_1, \ldots, a_r) \in \operatorname{Im} \varphi$.
Therefore, in the light of the above observations, we will study finitely generated $A$-modules using matrices with entries in $A$.

### 4.9.1 Smith Normal Form

Given any $r \times s$ matrix with entries in a principal ideal domain $A$, as in the case when $A$ a field, we can perform the following *elementary operations*.

**Elementary Operations**

i) Swap two rows.
ii) Add a multiple of one row to another row.
iii) Multiply a row by an *invertible element* of $A$.

Similar elementary operations can be defined with columns.

We define an *elementary matrix* as a matrix obtained performing one of the above operations on an identity matrix.
In this way, any elementary operation on a matrix can be achieved by multiplying it, either on the left side if working with rows or on the right side if working with columns, by the corresponding elementary matrix.

We recall that a matrix is invertible if and only if its determinant is in $A^*$. Note that all elementary matrices are invertible.
Two matrices $X$, $Y \in M_{rs}(A)$ are said to be *equivalent* if there exist invertible matrices $R \in M_r(A)$ and $S \in M_s(A)$, such that $Y = RXS$.
It is easy to verify that this defines an equivalence relation on $M_{rs}(A)$.
Additionally, a matrix $D = (d_{ij}) \in M_{rs}(A)$, not necessarily square, is said to be *diagonal* if $d_{ij} = 0$ for all $i \neq j$.

Given a matrix $X$ with entries in $A$, we consider the ideals

$$\Delta_i(X) = (\det X_i \colon X_i \text{ is an } i \times i \text{ submatrix of } X) \subseteq A.$$

**T. 4.25.** (→ p. 230) Let $X$ and $Y$ be equivalent matrices. Then, we have $\Delta_i(X) = \Delta_i(Y)$ for all $i$.

**T. 4.26.** Every matrix with entries in a principal ideal domain $A$ is equivalent to a diagonal matrix.

**Proof T. 4.26.** The proof is divided in several steps.

Step 1 We start by showing the following basic fact. Given a non-zero $2 \times 2$ matrix $X$, we can find a triangular matrix that is equivalent to $X$ and has a specific entry in position $(1, 1)$.
Consider a matrix $X = \begin{pmatrix} a & c \\ b & d \end{pmatrix}$ with coefficients in $A$, where $a$ and $b$ are not both zero.
Let $0 \neq x = \gcd(a, b)$. Then, there exist $s$, $t \in A$ such that $sa + tb = x$.

Next, define $R = \begin{pmatrix} s & t \\ -bx^{-1} & ax^{-1} \end{pmatrix}$. This matrix is invertible, since it has determinant 1.
By multiplying $X$ on the left-hand side by $R$, we obtain

$$RX = \begin{pmatrix} s & t \\ -bx^{-1} & ax^{-1} \end{pmatrix} \begin{pmatrix} a & c \\ b & d \end{pmatrix} = \begin{pmatrix} x & * \\ 0 & * \end{pmatrix},$$

which is an upper-triangular matrix with the greatest common divisor of the elements of the first column of $X$ in position $(1, 1)$.

Similarly, if $a$ and $c$ are not both zero, and $0 \neq y = \gcd(a, c) = ua + vc$, for some $u$, $v \in A$, then, multiplying $X$ on the right by $S = \begin{pmatrix} u & -cy^{-1} \\ v & ay^{-1} \end{pmatrix}$, we obtain a lower-triangular matrix $\begin{pmatrix} y & 0 \\ * & * \end{pmatrix}$.

**Step 2** Let $X$ be an $r \times s$ matrix. We now show that $X$ is equivalent to a diagonal matrix.
If $X$ is zero we are done.
Otherwise, performing elementary operations, if necessary, we can assume that the element in position $(1, 1)$ of $X$ is non-zero.
Applying Step 1 to the $2 \times 2$ submatrix in the upper-left corner of $X$, we can construct an invertible matrix

$$R = \left(\begin{array}{c|c} R_2 & 0 \\ \hline 0 & I_{r-2} \end{array}\right),$$

where $I_{r-2}$ is the identity matrix of size $r - 2$, such that

$$RX = \left(\begin{array}{cc|c} x & * & \\ 0 & * & * \\ \hline \multicolumn{2}{c|}{*} & * \end{array}\right).$$

Now, we fix the first row of $RX$ and swap the remaining rows with the second one, if necessary, in order to apply repeatedly Step 1, and eventually obtain a new matrix $X_1$ with the following properties:

i) the element $x_1$ in position (1,1) is the greatest common divisor of all elements in the first column of $X$;
ii) all the elements of the first column except $x_1$ are zero;
iii) it is equivalent to $X$.

**Step 3** Now, fix the first column of $X_1$. Applying repeatedly Step 1 to columns, we obtain a matrix $X_2$ with the following properties:

i) the element $x_2$ in position (1,1) is the greatest common divisor of all elements in the first row of $X_1$;

ii) all the elements of the first row except $x_2$ are zero;

iii) it is equivalent to $X_1$, and hence, to $X$.

After this last operation, new non-zero elements can appear in the first column, hence we may need to apply again Step 2 to the first column.

Step 4 Iterating this process, we construct a sequence of elements

$$x_1, x_2, \ldots, x_i, \ldots,$$

all in position $(1,1)$, which are the greatest common divisor of the elements in the first column or of the elements in the first row of some matrix $X_i$ equivalent to $X$. Therefore, $x_{i+1} \mid x_i$ for all $i$. Since $A$ is a PID, there exists an index $n$ such that $x_n = x_{n+1}$, see **T.1.24**. This means that $x_n$ is the greatest common divisor of all the elements in both the first column and the first row of $X_n$.

Step 5 Using the entry $x_n$, we can annihilate all the remaining elements in both the first column and the first row of $X_n$, and obtain the matrix

$$\left(\begin{array}{c|ccc} x_n & 0 & \cdots & 0 \\ \hline 0 & & & \\ \vdots & & Y & \\ 0 & & & \end{array}\right).$$

Step 6 We apply again Steps 2 to 5 to the matrix $Y$. By iterating this whole procedure, we eventually diagonalize $X$. □

**Smith Normal Form**

Let $A$ be a principal ideal domain. A matrix $D = (d_{ij}) \in M_{rs}(A)$ is *in Smith (normal) form* if:

i) $D$ is diagonal;

ii) $d_{11} \mid d_{22} \mid \ldots \mid d_{tt}$, where $t = \min\{r, s\}$.

Note that, in general, there exist integers $0 \leq k_1 \leq k_2 \leq t$ such that $d_{ii} \in A^*$ for all $0 < i \leq k_1$ and $d_{ii} = 0$ for all $k_2 < i \leq t$.

**T. 4.27.** Every matrix $X$ with entries in $A$ is equivalent to a matrix in Smith normal form.

**Proof T. 4.27.** By **T.4.26**, we can assume that $X = (x_{ij})$ is diagonal. If the matrix is not already in Smith normal form, let $i$ be the smallest index for which there exists $j > i$ such that $x_{ii} \nmid x_{jj}$. We then choose $j$ as the smallest index with this property. By swapping rows and columns if necessary, we can assume, without loss of generality, that $i = 1$ and $j = 2$.
Let $x = \gcd(x_{11}, x_{22})$ and $s$, $t \in A$ such that $sx_{11} + tx_{22} = x$.

Consider the matrices

$$R = \left(\begin{array}{cc|c} s & t & \\ -x_{22}x^{-1} & x_{11}x^{-1} & 0 \\ \hline \multicolumn{2}{c|}{0} & I_{r-2} \end{array}\right) \quad \text{and} \quad S = \left(\begin{array}{cc|c} 1 & -tx_{22}x^{-1} & \\ 1 & sx_{11}x^{-1} & 0 \\ \hline \multicolumn{2}{c|}{0} & I_{s-2} \end{array}\right).$$

The product $RXS = (y_{ij})$ is still a diagonal matrix, with

$$y_{jj} = x_{jj} \ \text{ if } \ j > 2, \ y_{11} = x, \ \text{ and } \ y_{22} = x_{11}x_{22}x^{-1},$$

and hence, such that $y_{11} \mid y_{22}$.
By swapping rows and columns again, we obtain a diagonal matrix $(z_{ij})$, where $z_{hh} = x_{hh}$ for $h \neq i, j$, $z_{ii} = y_{11}$ and $z_{jj} = y_{22}$.
Furthermore, for all $1 \leq h \leq i-1$, it holds that $z_{hh} \mid z_{h+\ell,h+\ell}$ for all $\ell > 0$, while for all $i < h \leq j$ we have that $z_{ii} \mid z_{hh}$.

Repeating this procedure, we achieve the conclusion. □

**T. 4.28.** (→ p. 231) Let $X$ be a matrix equivalent to a matrix $D = (d_{ij})$ in Smith form. Then:

1. $\Delta_1(X) = (d_{11})$;
2. $\Delta_i(X) = (d_{ii})\Delta_{i-1}(X)$ for all $i > 1$.

Therefore, if $\Delta_i(X) = (\delta_i)$, we can define $d_{11} = \delta_1$ and $d_{ii} = \delta_i/\delta_{i-1}$ for all $i > 1$ such that $\delta_{i-1} \neq 0$. Note that the elements $d_{ii}$ are unique only up to invertible elements of $A$. This leads to an essentially unique form $D$, which, with a slight abuse of notation, is referred to as *the Smith (normal) form* of $X$. The elements $d_{ii}$ are called the *invariant factors* of $X$.
Thus, two matrices are equivalent if and only if the invariant factors of their Smith forms are pairwise associate.

Recall that elementary operations do not change $\Delta_i(X)$.
This is not the case when we multiply on the left or right by matrices whose determinant is not invertible in $A$.
For instance, $\begin{pmatrix} 2 & 4 \\ 3 & 8 \end{pmatrix}$ reduces to $\begin{pmatrix} 2 & 0 \\ 3 & 2 \end{pmatrix}$, by subtracting twice the first column from the second column. However, it cannot be reduced neither to $\begin{pmatrix} 2 & 4 \\ 0 & 2 \end{pmatrix}$, by subtracting $\frac{3}{2}$ times the first row from the second row, since $\frac{3}{2} \notin \mathbb{Z}$, nor to $\begin{pmatrix} 2 & 4 \\ 0 & 4 \end{pmatrix}$, by subtracting 3 times the first row from twice the second row.
Therefore, when calculating the Smith form, make sure to use only elementary operations.

### 4.9.2 Structure Theorems for Finitely Generated Modules

From the Smith normal form and its properties, we deduce a classification of finitely generated modules over principal ideal domains.

We first introduce some notation.
Let $N = \langle w_1, \ldots, w_s \rangle$ be a submodule of a module $L = \langle l_1, \ldots, l_r \rangle$, and let $x_{ij} \in A$ be such that $w_h = \sum_{k=1}^{r} x_{kh} l_k$ for all $h = 1, \ldots, s$.
Using the $r \times s$ matrix $X = (x_{ij})$, we can write

$$(w_1, \ldots, w_s) = (l_1, \ldots, l_r)X.$$

**T. 4.29.** Let $L$ be a free $A$-module of rank $r$, and let $N \subseteq L$ be a non-zero submodule. Then, there exist a basis $\{v_1, \ldots, v_r\}$ of $L$ and coefficients $d_1, \ldots, d_s \in A$, where $s \leq r$, such that $\{d_1 v_1, \ldots, d_s v_s\}$ is a basis of $N$.

**Proof T. 4.29.** Since $A$ is a PID, we know that $N$ is also a free module of rank $s \leq r$, see **T.4.23**.1.
Let $\{l_1, \ldots, l_r\}$ be a basis of $L$ and $\{w_1, \ldots, w_s\}$ a basis of $N$. There exists an $r \times s$ matrix $X$ with entries in $A$ such that $(w_1, \ldots, w_s) = (l_1, \ldots, l_r)X$.
By **T.4.27**, the matrix $X$ is equivalent to a matrix $D$ in Smith normal form. This means that there exist invertible matrices $R$ and $S$, of sizes $r \times r$ and $s \times s$, respectively, such that $RXS = D$.
Let $d_1, \ldots, d_k$ be the non-zero elements of the diagonal of $D$.
Then,

$$(w_1, \ldots, w_s)S = (l_1, \ldots, l_r)XS = (l_1, \ldots, l_r)R^{-1}D.$$

By defining $(v_1, \ldots, v_r) = (l_1, \ldots, l_r)R^{-1}$, we obtain that $\{v_1, \ldots, v_r\}$ is a basis of $L$. Moreover,

$$(w_1, \ldots, w_s)S = (v_1, \ldots, v_r)D = (d_1 v_1, \ldots, d_k v_k, 0, \ldots, 0),$$

with $k \leq s$. Since $S$ is invertible, it follows that $k = s$ and $\{d_1 v_1, \ldots, d_s v_s\}$ is a basis of $N$. □

Observe that, in the previous proof, the scalars $d_i$ are the diagonal entries of the Smith normal form of $X$.

Also note that, for all $i = 1, \ldots, s$, we have $av_i \in N$ if and only if $d_i \mid a$. Therefore, $(d_i) = N : \langle v_i \rangle$. On the other hand, if $s < i \leq r$, then $av_i \in N$ if and only if $a = 0$, *i.e.*, $(0) = N : \langle v_i \rangle$.

**T. 4.30.** (→ p. 231) Using the same notation as in **T.4.29**, we have:

1. $L/N \simeq \langle \overline{v_1} \rangle \oplus \ldots \oplus \langle \overline{v_r} \rangle$;
2. $L/N \simeq A/(d_1) \oplus \ldots \oplus A/(d_s) \oplus A^{r-s}$, where $(d_i) = 0 : \langle \overline{v_i} \rangle$ for all $i = 1, \ldots, s$.

**Structure Theorem 1**

**T. 4.31.** (→ p. 231) Let $M = \langle m_1, \ldots, m_r \rangle$ be a finitely generated $A$-module. Then, there exists a chain of ideals $I_1 \supseteq I_2 \supseteq \ldots \supseteq I_r$ such that

$$M \simeq \bigoplus_{i=1}^{r} A/I_i.$$

It is important to note that, based on what we have seen so far, every matrix with entries in $A$ is equivalent to a matrix in Smith normal form, which is essentially unique. This implies that, if we choose a different basis $w_1', \ldots, w_s'$ of the module of relations $\operatorname{Ker} f$, the expression of $M$ as a direct sum of cyclic modules will remain unchanged. However, it is not immediately clear what would happen if we choose a generating set $\{n_1, \ldots, n_{r'}\}$ of $M$, other than $\{m_1, \ldots, m_r\}$. The following general result examines this situation.

**T. 4.32.** Let $A$ be a ring, and let

$$I_1 \supseteq I_2 \supseteq \ldots \supseteq I_r \quad \text{and} \quad J_1 \supseteq J_2 \supseteq \ldots \supseteq J_{r'}$$

be chains of ideals of $A$, with $r \leq r'$. Assume that

$$M \simeq \bigoplus_{i=1}^{r} A/I_i \simeq \bigoplus_{h=1}^{r'} A/J_h.$$

Then:

1. $J_1 = J_2 = \ldots = J_{r'-r} = A$;
2. $J_{r'-r+i} = I_i$ for all $i = 1, \ldots, r$.

**Proof T. 4.32.** 1. We assume $r < r'$ and consider $B = A/J_1$. We prove that $B = 0$. First, note that $J_1 + J_h = J_1$ for all $h > 1$, due to the hypothesis. By **E.10.7** we have that

$$B^{r'} \simeq \bigoplus_{h=1}^{r'} A/J_1 \simeq \bigoplus_{h=1}^{r'} A/(J_1 + J_h) \simeq M/J_1M \simeq \bigoplus_{i=1}^{r} A/(J_1 + I_i).$$

Projecting $B^r$ onto $\bigoplus_{i=1}^{r} A/(J_1 + I_i)$, we obtain a surjective homomorphism of $B^r$ onto $B^{r'}$ by composition. Since $r < r'$, this implies $B = 0$, see **E.10.9**. By repeating this reasoning, we obtain $J_2 = \cdots = J_{r'-r} = A$.

2. By part 1, we can assume that $r = r'$. By symmetry, it suffices to show that $I_h \subseteq J_h$ for all $h = 1, \ldots, r$.

Let $a \in I_h$. By **E.10.8**, we have that

$$aM \simeq a\left(\bigoplus_{i=1}^{r} A/I_i\right) \simeq \bigoplus_{i=1}^{r} A/(I_i : (a)).$$

Since $I_i \subseteq I_{i-1}$, we also have $a \in I_i$ for all $i \leq h$, and hence, $I_i : (a) = A$ for all $i \leq h$.
Accordingly,

$$\bigoplus_{i=h+1}^{r} A/(I_i : (a)) \simeq aM \simeq a\left(\bigoplus_{i=1}^{r} A/J_i\right) \simeq \bigoplus_{i=1}^{r} A/(J_i : (a)).$$

Thus, by part 1, $J_i : (a) = A$ for all $i \leq h$, and therefore, $a \in J_h$. □

We now develop a more refined version of the above structure theorem using properties of torsion submodules.

**Torsion Submodule**

Let $A$ be a domain, and let $M$ be an $A$-module. The *torsion submodule of $M$* is the set

$$T(M) = \{m \in M : am = 0 \text{ for some } a \in A \setminus \{0\}\},$$

see **E.10.46**.
Its elements are called *torsion elements.*
The module $M$ is said to be *torsion* if $M = T(M)$ and *torsion-free* if $T(M) = 0$.

**T. 4.33.** (→ p. 231) Let $A$ be a PID, and let $M$ be a finitely generated $A$-module. Then:

1. $T(M)$ is finitely generated;
2. $M \simeq T(M) \oplus A^k$, for some $k \geq 0$;
3. $\operatorname{Ann} T(M) \neq 0$.

The previous result shows that every finitely generated module $M$ over a PID can be decomposed into the direct sum of a torsion module and a free module. This modules are referred to as the *torsion part* and the *free part* of $M$, respectively.

To further decompose $M$, we introduce the following definition.
Let $M$ be an $A$-module, and let $a \in A$. The *$a$-component* of $M$ is defined as

$$M_{[a]} = \{m \in M : a^k m = 0 \text{ for some } k \in \mathbb{N}\} = \bigcup_{k \in \mathbb{N}} 0 :_M (a^k).$$

This set is a submodule of $M$, see **E.10.47**.

**Structure Theorem 2**

Let $A$ be a principal ideal domain.

**T. 4.34.** (→ p. 232) Let $p \in A$ be a prime element, and let $M = M_{[p]}$ be a finitely generated $A$-module. Then, $M$ can be expressed as

$$M \simeq A/(p^{k_1}) \oplus A/(p^{k_2}) \oplus \ldots \oplus A/(p^{k_s}),$$

where $k_1 \leq k_2 \leq \ldots \leq k_s$.

**T. 4.35.** (→ p. 232) Let $M$ be a finitely generated $A$-module. Then, $M$ can be expressed as a direct sum of cyclic modules and of a free module

$$M \simeq \bigoplus_{i=1}^{h} A/(q_i) \oplus A^k,$$

where $(q_i) \subset A$ are primary ideals for all $i$, and $h,\, k \geq 0$.

This decomposition is unique up to the order of the components. In particular, the set of primary ideals that appears in this decomposition is unique, and an ideal $(q_i)$ can appear multiple times.
The elements $q_i$ are unique up to invertible elements and are called the *elementary divisors* of $M$.

We conclude this section with the following observation: given an $A$-module $M$ generated by $r$ elements, with $M \simeq \operatorname{Coker}\varphi$ and $\varphi$ associated with a matrix $X \in M_{rr'}(A)$, the rank of the free part of $M$ is equal to the number of zero-rows in the Smith form of $X$, see also **T.4.30**.

## 4.10 Appendix: the Rational Canonical Form and the Jordan Form

In the previous section we have seen how to generalize the structure theorems of finitely generated Abelian groups to finitely generated modules over a PID. In this appendix we will focus on vector spaces. When we are given an endomorphism of a vector space $V$, we will show how it is possible to use it to obtain a decomposition of $V$.
Let $V$ be a non-zero $K$-vector space of finite dimension, and let $\varphi \in \operatorname{End}_K(V)$ be a fixed endomorphism of $V$.

We define a $K[x]$-module structure on $V$ using $\varphi$ as follows:
given a $p(x) = \sum_{i=0}^{t} a_i x^i$ and a vector $v \in V$, we set

$$p(x)v = \sum_{i=0}^{t} a_i \varphi^i(v),$$

see also the proof of **T.4.14**.
We refer to this structure on $V$ as the $K[x]$-module structure *under* $\varphi$. Additionally, for any subset $W \subseteq V$, we use $\langle W \rangle$ to denote the $K[x]$-submodule of $V$ generated by $W$, and we denote the annihilator of any $K[x]$-module $T$ as $\operatorname{Ann} T$.

**T. 4.36.** Let $0 \neq V$ be a $K$-vector space with $\dim_K V = n$, together with its $K[x]$-module structure under $\varphi$. Then, the following hold:

1. $\operatorname{Ann} V$ is a non-zero proper ideal of $K[x]$.
   Its monic generator is called the *minimal polynomial* of $\varphi$.
2. For all $0 \neq v \in V$, $\operatorname{Ann}\langle v \rangle$ is non-zero proper ideal of $K[x]$.
3. Any basis $\mathcal{B}$ of a vector subspace $W$ of $V$ is a generating set of $\langle W \rangle$, but it is never a basis of $\langle W \rangle$.
4. Any $K[x]$-submodule $T \subseteq V$ is a *$\varphi$-invariant* subspace of $V$, *i.e.*, such that $\varphi(T) \subseteq T$.
5. Let $0 \neq v \in V$. Let $f_v(x)$ be the monic generator of $\operatorname{Ann}\langle v \rangle$, and let $d = \deg f_v(x)$. Then, $\langle v \rangle$ is a $K$-vector space of dimension $d$ and has a basis
$$\{v, \varphi(v), \ldots, \varphi^{d-1}(v)\}.$$
   Moreover, $\langle v \rangle$ is the smallest $\varphi$-invariant subspace of $V$ containing $v$.

**Proof T. 4.36.** 1. Since $V \neq 0$, we have $\operatorname{Ann} V \subsetneq K[x]$.
For any $f(x) \in K[x]$, it follows from the definition that $f(x)v = 0$ if and only if $f(\varphi)(v) = 0$. Therefore, $f(x) \in \operatorname{Ann} V$ if and only if $f(\varphi)$ is the zero endomorphism.
Since the $n^2 + 1$ vectors $\mathrm{id}_V, \varphi, \ldots, \varphi^{n^2} \in \operatorname{End}_K(V)$ are linearly dependent, it is possible to find a non-trivial $K$-linear combination $\sum_{i=0}^{n^2} a_i \varphi^i = 0$.
Thus, $f(x) = \sum_{i=0}^{n^2} a_i x^i$ is a non-zero element of $\operatorname{Ann} V$.
2. Since $\langle v \rangle$ is a submodule of $V$, from part 1 it immediately follows that

$$0 \neq \operatorname{Ann} V \subseteq \operatorname{Ann}\langle v \rangle \subsetneq K[x].$$

3. The first statement is a consequence of the outer product definition, since for any $a \in K$ and $v \in V$, the multiplication $av$ in $V$ is the same, whether $V$ is considered as a $K$-vector space or as a $K[x]$-module.

On the other hand, by part 1, $V$ is torsion as a $K[x]$-module.
Therefore, $V$ and all its submodules $\langle W \rangle$ are not free, and hence, do not have a basis.

4. For all $w \in T$, it holds that $p(x)w \in T$.
In particular, $\varphi(w) = xw \in T$ for all $w \in T$, *i.e.*, $\varphi(T) \subseteq T$.

5. Observe that, by part 2, $d > 0$.
Moreover,

$$\langle v \rangle = \{p(x)v \colon p(x) \in K[x]\}$$

is generated by $v$ over $K[x]$, hence by $\{\varphi^i(v) \colon i \in \mathbb{N}\}$ over $K$, since $\varphi^i(v) = x^i v$ for all $i$.

Furthermore, since $f_v(x)v = 0$ and $f_v(x)$ is monic, we can express $\varphi^d(v)$ and, consequently, $\varphi^{d+h}(v)$ for all $h \in \mathbb{N}$, as a $K$-linear combination of $v$, $\varphi(v), \ldots, \varphi^{d-1}(v)$. This proves that every element of $\langle v \rangle$ can be expressed as a $K$-linear combination of $v, \varphi(v), \ldots, \varphi^{d-1}(v)$, and thus, these vectors generate $\langle v \rangle$ as a $K$-vector space.
To prove that these elements are linearly independent, assume, by contradiction, that there exists a non-trivial $K$-linear combination $\sum_{i=0}^{d-1} a_i \varphi^i(v)$ that equals to 0.
Thus, there exists a non-zero polynomial

$$f(x) = \sum_{i=0}^{d-1} a_i x^i \in \operatorname{Ann}\langle v \rangle = (f_v(x)),$$

and this contradicts the fact that $\deg f_v = d$.

By what we proved above, it follows easily that $\langle v \rangle$ is $\varphi$-invariant.
If $T$ is another $\varphi$-invariant subspace containing $v$, then it follows that

$$\varphi^i(v) \in \varphi^i(T) \subseteq \varphi(T) \subseteq T \text{ for all } i,$$

*i.e.*, $\langle v \rangle \subseteq T$. □

Since $K[x]$ is a PID, the finitely generated $K[x]$-module $V$, which has been shown to be torsion in **T.4.36**.1, can be decomposed into a direct sum of cyclic modules

$$V = \langle v_1 \rangle \oplus \cdots \oplus \langle v_s \rangle \simeq K[x]/\operatorname{Ann}\langle v_1 \rangle \oplus \cdots \oplus K[x]/\operatorname{Ann}\langle v_s \rangle,$$

by the structure theorem **T.4.31**.
Consider, for all $i = 1, \ldots, s$, the monic polynomial $f_i = f_{v_i}$ generating $\operatorname{Ann}\langle v_i \rangle$, see **T.4.36**.5. Let $d_i = \deg f_i > 0$, and write

$$f_i = \sum_{j=0}^{d_i} a_j^{(i)} x^j \text{ for some } a_j^{(i)} \in K.$$

**The Rational Canonical Form**

**T. 4.37.** (→ p. 232) With the notation introduced above, let

$$V = \langle v_1 \rangle \oplus \cdots \oplus \langle v_s \rangle$$

be a decomposition of $V$ into a direct sum of cyclic modules.
Then:

1. the set $\{v_1, \varphi(v_1), \ldots, \varphi^{d_1-1}(v_1), \ldots, v_s, \varphi(v_s), \ldots, \varphi^{d_s-1}(v_s)\}$ is a basis of $V$;
2. the matrix $M$ associated with $\varphi$ with respect to this basis is a block matrix
$$M = \begin{pmatrix} M_{f_1} & 0 & \ldots & 0 \\ 0 & M_{f_2} & \ldots & 0 \\ \vdots & & \ddots & \\ 0 & 0 & \ldots & M_{f_s} \end{pmatrix},$$
where each block is given by the *companion matrix* of $f_i$
$$M_{f_i} = \begin{pmatrix} 0 & 0 & \ldots & 0 & -a_0^{(i)} \\ 1 & 0 & \ldots & 0 & -a_1^{(i)} \\ 0 & 1 & \ldots & 0 & -a_2^{(i)} \\ \vdots & & \ddots & & \vdots \\ 0 & & \ldots & 1 & -a_{d_i-1}^{(i)} \end{pmatrix}.$$
$M$ is called the *rational canonical form* of $\varphi$.

By the structure theorem **T.4.31**, the polynomials $f_i$ defining the rational canonical form of $\varphi$ satisfy the divisibility conditions $f_1 \mid f_2 \mid \ldots \mid f_s$.

As a working example, consider $V = \mathbb{Q}^4$, and let $\varphi \in \mathrm{End}_{\mathbb{Q}}(V)$ be the endomorphism given by $\varphi(v) = Av$, where

$$A = (a_{ij}) = \begin{pmatrix} 1 & 0 & 1 & 0 \\ 0 & 1 & -2 & 0 \\ -2 & -1 & 2 & 0 \\ -2 & -1 & -2 & 1 \end{pmatrix}. \tag{4.2}$$

First we decompose $V$ into a direct sum of cyclic $\mathbb{Q}[x]$-modules, and then, we determine the rational canonical form of $\varphi$.

Let $\mathcal{B} = \{e_1, e_2, e_3, e_4\}$ be the canonical basis of $V$. By **T.4.36**.3, $\mathcal{B}$ is a generating set of $V$ as a $\mathbb{Q}[x]$-module. Let $f\colon \mathbb{Q}[x]^4 \longrightarrow V$ be the homomorphism defined by

$$f(\mathbf{p}) = p_1e_1 + p_2e_2 + p_3e_3 + p_4e_4,$$

for all $\mathbf{p} = (p_1, p_2, p_3, p_4) \in \mathbb{Q}[x]^4$.
This homomorphism $f$ is surjective and the sequence

$$0 \longrightarrow \operatorname{Ker} f \longrightarrow \mathbb{Q}[x]^4 \xrightarrow{f} V \longrightarrow 0$$

is exact.
In order to compute $\operatorname{Ker} f$, we consider the relation

$$xe_1 = \varphi(e_1) = Ae_1 = \sum_{h=1}^{4} a_{h1}e_h,$$

which implies that

$$f(x - a_{11}, -a_{21}, -a_{31}, -a_{41}) = 0.$$

Hence, $\mathbf{r}_1 = (x - a_{11}, -a_{21}, -a_{31}, -a_{41}) \in \operatorname{Ker} f$.
Similarly, we define $\mathbf{r}_2$, $\mathbf{r}_3$, and $\mathbf{r}_4$ and consider the submodule

$$\langle \mathbf{r}_i \colon i = 1, \ldots, 4 \rangle \subseteq \operatorname{Ker} f.$$

To prove the opposite inclusion, let $\mathbf{p} = (p_1, p_2, p_3, p_4) \in \mathbb{Q}[x]^4$. Dividing by the polynomials $x - a_{ii}$, we can express $p_i$ as $p_i = h_i(x - a_{ii}) + c_i$, where $c_i \in \mathbb{Q}$ and $\deg h_i < \deg p_i$ for all $i$.
Therefore,

$$\begin{aligned}
\mathbf{p}^{\mathrm{t}} &= \begin{pmatrix} h_1(x - a_{11}) + c_1 \\ h_2(x - a_{22}) + c_2 \\ h_3(x - a_{33}) + c_3 \\ h_4(x - a_{44}) + c_4 \end{pmatrix} \\
&= h_1\mathbf{r}_1^{\mathrm{t}} + h_2\mathbf{r}_2^{\mathrm{t}} + h_3\mathbf{r}_3^{\mathrm{t}} + h_4\mathbf{r}_4^{\mathrm{t}} + \begin{pmatrix} h_2a_{12} + h_3a_{13} + h_4a_{14} + c_1 \\ h_1a_{21} + h_3a_{23} + h_4a_{24} + c_2 \\ h_1a_{31} + h_2a_{32} + h_4a_{34} + c_3 \\ h_1a_{41} + h_2a_{42} + h_3a_{43} + c_4 \end{pmatrix} \\
&= h_1\mathbf{r}_1^{\mathrm{t}} + h_2\mathbf{r}_2^{\mathrm{t}} + h_3\mathbf{r}_3^{\mathrm{t}} + h_4\mathbf{r}_4^{\mathrm{t}} + \begin{pmatrix} q_1 \\ q_2 \\ q_3 \\ q_4 \end{pmatrix},
\end{aligned}$$

where the polynomials $q_1, \ldots, q_4$ are such that $\max_i\{\deg q_i\} < \max_i\{\deg p_i\}$.
Repeating this procedure, since the $x - a_{ii}$ are linear polynomials, we can eventually write

$$\mathbf{p}^{\mathrm{t}} = g_1\mathbf{r}_1^{\mathrm{t}} + g_2\mathbf{r}_2^{\mathrm{t}} + g_3\mathbf{r}_3^{\mathrm{t}} + g_4\mathbf{r}_4^{\mathrm{t}} + \begin{pmatrix} d_1 \\ d_2 \\ d_3 \\ d_4 \end{pmatrix}$$

for some $g_1, \ldots, g_4 \in \mathbb{Q}[x]$ and $d_1, \ldots, d_4 \in \mathbb{Q}$.
Therefore, when $\mathbf{p} \in \operatorname{Ker} f$ it follows that

$$0 = f(\mathbf{p}) = f(d_1, d_2, d_3, d_4) = d_1 e_1 + d_2 e_2 + d_3 e_3 + d_4 e_4.$$

This implies $d_i = 0$ for all $i$, since $\mathcal{B}$ is a basis of $V$, *i.e.*, $\mathbf{p} \in \langle \mathbf{r}_1, \mathbf{r}_2, \mathbf{r}_3, \mathbf{r}_4 \rangle$, and thus $\langle \mathbf{r}_1, \mathbf{r}_2, \mathbf{r}_3, \mathbf{r}_4 \rangle = \operatorname{Ker} f$.

To obtain the desired decomposition of $V$, we find the Smith form of the matrix of relations whose columns are the vectors $\mathbf{r}_i$, *i.e.*, the matrix $xI - A$. We thus consider

$$A - xI = \begin{pmatrix} a_{11} - x & a_{12} & a_{13} & a_{14} \\ a_{21} & a_{22} - x & a_{23} & a_{24} \\ a_{31} & a_{32} & a_{33} - x & a_{34} \\ a_{41} & a_{42} & a_{43} & a_{44} - x \end{pmatrix}$$

$$= \begin{pmatrix} 1-x & 0 & 1 & 0 \\ 0 & 1-x & -2 & 0 \\ -2 & -1 & 2-x & 0 \\ -2 & -1 & -2 & 1-x \end{pmatrix},$$

*i.e.*, the *characteristic matrix* of $A$.
The Smith form of $A - xI$ is

$$\begin{pmatrix} 1 & 0 & 0 & 0 \\ 0 & 1 & 0 & 0 \\ 0 & 0 & x-1 & 0 \\ 0 & 0 & 0 & x^3 - 4x^2 + 5x - 2 \end{pmatrix},$$

therefore,

$$V \simeq \mathbb{Q}[x]/(x-1) \oplus \mathbb{Q}[x]/(x^3 - 4x^2 + 5x - 2).$$

The rational canonical form $M$ of $A$ consists of two blocks, given by the companion matrices of

$$f_1 = a_1^{(1)} x + a_0^{(1)} = x - 1$$

and

$$f_2 = a_3^{(2)} x^3 + a_2^{(2)} x^2 + a_1^{(2)} x + a_0^{(2)} = x^3 - 4x^2 + 5x - 2.$$

Hence,

$$M = \left(\begin{array}{c|c} M_{f_1} & 0 \\ \hline 0 & M_{f_2} \end{array}\right) = \left(\begin{array}{c|ccc} 1 & 0 & 0 & 0 \\ \hline 0 & 0 & 0 & 2 \\ 0 & 1 & 0 & -5 \\ 0 & 0 & 1 & 4 \end{array}\right), \tag{4.3}$$

and this concludes the first part of our example.

Consider again the decomposition $V = \bigoplus_{i=1}^{s} \langle v_i \rangle \simeq \bigoplus_{i=1}^{s} K[x]/(f_i)$.

Using unique factorization in $K[x]$ and the Chinese Remainder Theorem, we can further decompose the summands as

$$\langle v_i \rangle \simeq K[x]/(f_i) \simeq \bigoplus_{j=1}^{t_i} K[x]/(f_{ij}^{e_{ij}}),$$

where the $f_{ij}$ are the distinct irreducible factors of $f_i$.
The last isomorphism in the above expression represents the decomposition of $\langle v_i \rangle$ as a sum of $f_{ij}$-components, see **E.10.49**. Given the divisibility relations among the polynomials $f_i$, it is clear that non-trivial $f_{ij}$-components also appear in the decomposition of $\langle v_k \rangle$ for $k \geq i$.

In the special case when the field $K$ is algebraically closed, each

$$f_i = \prod_{j=1}^{t_i} \left(x - \lambda_j^{(i)}\right)^{e_{ij}}$$

factors into a product of powers of linear polynomials, where $\lambda_j^{(i)} \in K$ and $\lambda_j^{(i)} \neq \lambda_h^{(i)}$ if $j \neq h$. Note that $e_{ij}$ is the multiplicity of $\lambda_j^{(i)}$ as a root of $f_i$. As previously remarked, $\lambda_j^{(i)}$ also appears as a root of $f_k$ for $k \geq i$.
Therefore, for all $i$, we can write

$$\langle v_i \rangle = \langle w_1^{(i)} \rangle \oplus \ldots \oplus \langle w_{t_i}^{(i)} \rangle,$$

where $\operatorname{Ann}\langle w_j^{(i)} \rangle = \left(x - \lambda_j^{(i)}\right)^{e_{ij}}$, and obtain a decomposition of $V$ as a direct sum of cyclic $K[x]$-modules

$$V = \langle w_1^{(1)} \rangle \oplus \ldots \oplus \langle w_{t_1}^{(1)} \rangle \oplus \ldots \oplus \langle w_1^{(s)} \rangle \oplus \ldots \oplus \langle w_{t_s}^{(s)} \rangle.$$

**The Jordan Form**

**T. 4.38.** Let $K$ be an algebraically closed field, and let

$$V = \langle w_1^{(1)} \rangle \oplus \ldots \oplus \langle w_{t_1}^{(1)} \rangle \oplus \ldots \oplus \langle w_1^{(s)} \rangle \oplus \ldots \oplus \langle w_{t_s}^{(s)} \rangle.$$

be the decomposition of $V$ obtained above.
Let also $\psi = \bigoplus_{i,j} \psi_{ij}$, with $\psi_{ij} = (\varphi - \lambda_j^{(i)} \mathrm{id}_V)_{|\langle w_j^{(i)} \rangle}$. Then:

1. a basis of $V$ is given by

$$\begin{aligned}&\Big\{w_1^{(1)}, \psi_{11}(w_1^{(1)}), \dots, \psi_{11}^{e_{11}-1}(w_1^{(1)}), \dots,\\ &w_{t_1}^{(1)}, \psi_{1t_1}(w_{t_1}^{(1)}), \dots, \psi_{1t_1}^{e_{1t_1}-1}(w_{t_1}^{(1)}), \dots\\ &\vdots\\ &w_1^{(s)}, \psi_{s1}(w_1^{(s)}), \dots, \psi_{s1}^{e_{s1}-1}(w_1^{(s)}), \dots,\\ &w_{t_s}^{(s)}, \psi_{st_s}(w_{t_s}^{(s)}), \dots, \psi_{st_s}^{e_{st_s}-1}(w_{t_s}^{(s)})\Big\};\end{aligned}$$

2. the matrix associated with $\varphi$ with respect to such basis is a diagonal block matrix

$$\begin{pmatrix} J_{\lambda_1^{(1)}} & & & & & & & & & \\ & J_{\lambda_2^{(1)}} & & & & & & & & \\ & & \ddots & & & & & & 0 & \\ & & & J_{\lambda_{t_1}^{(1)}} & & & & & & \\ & & & & J_{\lambda_1^{(2)}} & & & & & \\ & & & & & \ddots & & & & \\ & & & & & & J_{\lambda_{t_{s-1}}^{(s-1)}} & & & \\ & 0 & & & & & & J_{\lambda_1^{(s)}} & & \\ & & & & & & & & \ddots & \\ & & & & & & & & & J_{\lambda_{t_s}^{(s)}} \end{pmatrix},$$

where each block $J_{\lambda_j^{(i)}}$ is a Jordan matrix of size $e_{ij}$,

$$J_{\lambda_j^{(i)}} = \begin{pmatrix} \lambda_j^{(i)} & 0 & \dots & & 0 \\ 1 & \lambda_j^{(i)} & 0 & \dots & \vdots \\ 0 & 1 & \lambda_j^{(i)} & \ddots & \vdots \\ \vdots & & \ddots & \ddots & 0 \\ 0 & 0 & & 1 & \lambda_j^{(i)} \end{pmatrix}.$$

**Proof T. 4.38.** Since $V$ is expressed as a direct sum, we can focus on the action of $\psi$ on each of the subspaces $\langle w_j^{(i)} \rangle$.
By definition, this action is given by $\psi_{ij}$.

To simplify the notation, let $w_j^{(i)} = w$, $\lambda_j^{(i)} = \lambda$, and $\mathrm{Ann}\langle w\rangle = (x-\lambda)^e$. Then, we have $\psi = \varphi - \lambda \mathrm{id}_V$, and we can work with the cyclic $K[x]$-module $\langle w\rangle \simeq K[x]/(x-\lambda)^e$, which is a $K$-vector space of dimension $e$.

1. According to **T.4.36**.5, the set $\{w, \varphi(w), \ldots, \varphi^{e-1}(w)\}$ is a basis of $\langle w\rangle$. From the definition of $\psi$, it follows that

$$\varphi^i(w) = (\psi + \lambda\,\mathrm{id}_V)^i(w) \in \langle w, \psi(w), \ldots, \psi^{e-1}(w)\rangle,$$

and therefore, $\{w, \psi(w), \ldots, \psi^{e-1}(w)\}$ is also a basis of $\langle w\rangle$.

2. Regarding the action of $\varphi$ on $\langle w\rangle$, we have

$$\varphi(\psi^i(w)) = (\psi + \lambda\,\mathrm{id}_V)(\psi^i(w)) = \begin{cases} \psi^{i+1}(w) + \lambda\psi^i(w) & \text{if } 0 \le i < e-1 \\ \lambda\psi^{e-1}(w) & \text{if } i = e-1, \end{cases}$$

where the last equality is given by $\psi^e(w) = (x-\lambda)^e w = 0$. Therefore, the matrix representing $\varphi$ with respect to the basis $\{w, \psi(w), \ldots, \psi^{e-1}(w)\}$ is

$$J_\lambda = \begin{pmatrix} \lambda & 0 & \ldots & \ldots & 0 \\ 1 & \lambda & 0 & \ldots & 0 \\ 0 & 1 & \lambda & \ldots & 0 \\ \vdots & & \ddots & \ddots & \\ 0 & 0 & \ldots & 1 & \lambda \end{pmatrix},$$

and the proof is complete. □

Let us revisit our previous example in order to find the Jordan form of the matrix $A$ in (4.2) over $\mathbb{C}$. We computed the rational canonical form $M$ of $A$ in (4.3).

Since $f_2 = x^3 - 4x^2 + 5x - 2 = (x-1)^2(x-2)$, we immediately have that the Jordan form of $A$ is

$$\left(\begin{array}{c|cc|c} 1 & 0 & 0 & 0 \\ \hline 0 & 1 & 0 & 0 \\ 0 & 1 & 1 & 0 \\ \hline 0 & 0 & 0 & 2 \end{array}\right).$$

Finally, we note that the $(x-1)$-component is represented by the $3 \times 3$ upper-left block of the matrix. This block consists of two smaller blocks: the first block, of size 1, is derived from $M_{f_1}$ and the second block, of size 2, is derived from $M_{f_2}$.

# Chapter 5
# Tensor Product

We introduce a new product operation between modules called tensor product. We provide a detailed description of its construction and fundamental properties. We then interpret the tensor product with a given module as a functor. Finally, we highlight an important application of this operation, namely, extension of scalars.

## 5.1 Universal Property of Tensor Product

Let $A$ be a ring, and let $M$, $N$, $P$ be $A$-modules. A function $b\colon M \times N \longrightarrow P$ is *$A$-bilinear* if for any $m \in M$ the map

$$b_{(m,\cdot)}\colon N \longrightarrow P, \quad b_{(m,\cdot)}(n) \mapsto b(m,n),$$

and for any $n \in N$ the map

$$b_{(\cdot,n)}\colon M \longrightarrow P, \quad b_{(\cdot,n)}(m) \mapsto b(m,n)$$

are $A$-linear.
Let $\mathrm{Bil}(M,N;P)$ denote the set of all $A$-bilinear maps from $M \times N$ to $P$. We define an $A$-module structure on it by setting, for any $b, b' \in \mathrm{Bil}(M,N;P)$ and $\alpha \in A$,

i) $(b+b')(m,n) = b(m,n) + b'(m,n)$;
ii) $(\alpha b)(m,n) = \alpha b(m,n)$,

for all $m \in M$ and $n \in N$, see **E.11.1**.

**T. 5.1.** Let $A$ be a ring, and let $M$, $N$, and $P$ be $A$-modules. Then,

$$\mathrm{Bil}(M,N;P) \simeq \mathrm{Hom}_A(M, \mathrm{Hom}_A(N,P)).$$



A. Bandini et al., *Commutative Algebra through Exercises*, UNITEXT 159,
https://doi.org/10.1007/978-3-031-56910-4_5

**Proof T. 5.1.** Define

$$\Phi\colon \operatorname{Bil}(M,N;P) \longrightarrow \operatorname{Hom}_A(M,\operatorname{Hom}_A(N,P))$$

as the function that sends an $A$-bilinear map $b$ to $\varphi_b\colon M \longrightarrow \operatorname{Hom}_A(N,P)$ defined by $\varphi_b(m)(n) = b_{(m,\cdot)}(n) = b(m,n)$.
Additionally, let

$$\Psi\colon \operatorname{Hom}_A(M,\operatorname{Hom}_A(N,P)) \longrightarrow \operatorname{Bil}(M,N;P)$$

be the function that sends a homomorphism $\varphi$ to the $A$-bilinear map $b_\varphi$ defined by $b_\varphi(m,n) = \varphi(m)(n)$ for any $m \in M$ and $n \in N$.
To complete the proof, it is sufficient to verify that:

i) $\Phi$ and $\Psi$ are well-defined, *i.e.*, $\varphi_b$ is a homomorphism and $b_\varphi$ is $A$-bilinear;
ii) $\Phi$ and $\Psi$ are $A$-modules homomorphisms;
iii) $\Phi$ and $\Psi$ are mutually inverse. □

We define the tensor product using the following universal property, and then we prove its existence and uniqueness.

**Tensor Product and its Universal Property**

Let $A$ be a ring, and let $M$, $N$ be $A$-modules.
We define the *tensor product of* $M$ *and* $N$, denoted by $(T,\tau)$, as an $A$-module $T$ along with an $A$-bilinear map $\tau\colon M \times N \longrightarrow T$.
This pair satisfies the following universal property:
for any $f \in \operatorname{Bil}(M,N;P)$, there exists a unique $A$-module homomorphism $\tilde{f}\colon T \longrightarrow P$ which makes the following diagram commutative:

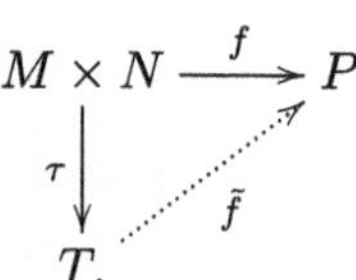


**T. 5.2.** Let $A$ be a ring, and let $M$, $N$ be $A$-modules. Then, the tensor product of $M$ and $N$ exists and is unique up to isomorphism.

**Proof T. 5.2.** Uniqueness Assume that $(T_1,\tau_1)$ and $(T_2,\tau_2)$ are two tensor products of $M$ and $N$. Using the universal property twice, first with $P = T_1$ and then with $P = T_2$, we obtain a diagram

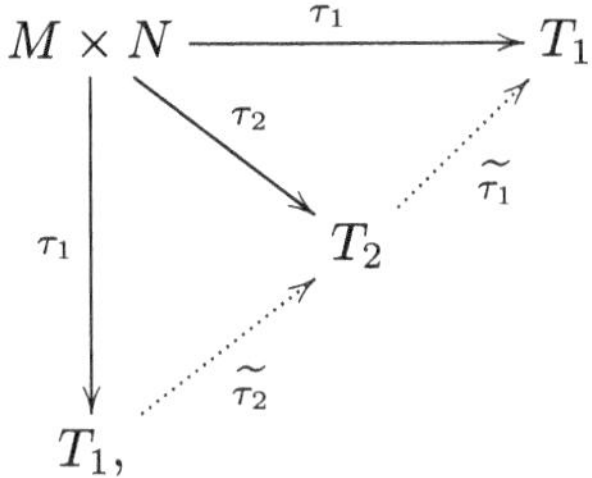


where the homomorphisms $\widetilde{\tau_1}$ and $\widetilde{\tau_2}$ are provided by the universal property. Hence, they are unique, and such that $\tau_2 = \widetilde{\tau_2} \circ \tau_1$ and $\tau_1 = \widetilde{\tau_1} \circ \tau_2$.
The composition $\widetilde{\tau_1} \circ \widetilde{\tau_2}$ is an endomorphism of $T_1$ such that $(\widetilde{\tau_1} \circ \widetilde{\tau_2}) \circ \tau_1 = \tau_1$.
Since $\mathrm{id}_{T_1} \circ \tau_1 = \tau_1$, the uniqueness in the universal property yields

$$\widetilde{\tau_1} \circ \widetilde{\tau_2} = \mathrm{id}_{T_1}.$$

Similarly we have $\widetilde{\tau_2} \circ \widetilde{\tau_1} = \mathrm{id}_{T_2}$, and we can conclude that $T_1 \simeq T_2$.

Existence Consider the free $A$-module $F = A^{M \times N}$, which has canonical basis $\mathcal{B} = \{e_{(m,n)} : (m,n) \in M \times N\}$, and the map $\mathrm{i}: M \times N \longrightarrow F$ which sends $(m,n)$ to $e_{(m,n)}$.
Moreover, consider the submodule of $F$

$$\begin{aligned} D = \langle &\mathrm{i}(m_1 + m_2, n) - \mathrm{i}(m_1, n) - \mathrm{i}(m_2, n),\ \mathrm{i}(am, n) - a\mathrm{i}(m, n), \\ &\mathrm{i}(m, n_1 + n_2) - \mathrm{i}(m, n_1) - \mathrm{i}(m, n_2),\ \mathrm{i}(m, an) - a\mathrm{i}(m, n) : \\ &m,\ m_1,\ m_2 \in M,\ n,\ n_1,\ n_2 \in N,\ a \in A\rangle, \end{aligned}$$

and define the map

$$\tau = \pi \circ \mathrm{i}: M \times N \longrightarrow F \longrightarrow F/D,$$

where $\pi: F \longrightarrow F/D$ is the canonical projection.
Note that $\tau$ is $A$-bilinear by construction, since $D$ contains all the relations that $\tau$ must satisfy to be $A$-bilinear.

Next, we prove that the pair $(T = F/D, \tau)$ is the tensor product of $M$ and $N$ by showing that it satisfies the universal property.
Suppose $P$ is an $A$-module, and let $f \in \mathrm{Bil}(M, N; P)$.
Consider the diagram

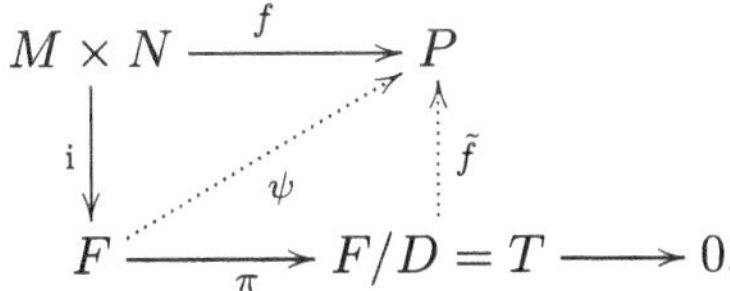


We have to prove that there exists a unique homomorphism $\tilde{f}$ such that

$$\tilde{f} \circ \tau = \tilde{f} \circ \pi \circ \mathrm{i} = f.$$

Since $F$ is a free module, there exists a unique homomorphism $\psi$ which makes the upper left triangle commute, *i.e.*, $f = \psi \circ \mathrm{i}$.
Next, we define $\tilde{f}$ such that $\tilde{f} \circ \pi = \psi$, and thus,

$$\tilde{f} \circ \tau = \tilde{f} \circ \pi \circ \mathrm{i} = \psi \circ \mathrm{i} = f.$$

We set

$$\tilde{f} \colon F/D \longrightarrow P, \quad \overline{x} \mapsto \psi(x) \text{ for any } x \in F.$$

If $\tilde{f}$ is well-defined, then it is a homomorphism, since $\psi$ is a homomorphism, and the diagram commutes by construction.
For $\tilde{f}$ to be well-defined, we need $\psi_{|D} = 0$. It is sufficient to prove that this is true for the generators of $D$.
For example, since $f$ is bilinear, we have

$$\begin{aligned}
&\psi(\mathrm{i}(m+m',n) - \mathrm{i}(m,n) - \mathrm{i}(m',n)) \\
&\qquad = (\psi \circ \mathrm{i})(m+m',n) - (\psi \circ \mathrm{i})(m,n) - (\psi \circ \mathrm{i})(m',n) \\
&\qquad = f(m+m',n) - f(m,n) - f(m',n) = 0.
\end{aligned}$$

To show that $\tilde{f}$ is unique, assume that there exist two homomorphisms $\tilde{f}_1, \tilde{f}_2$ such that $\tilde{f}_1 \circ \pi \circ \mathrm{i} = \tilde{f}_2 \circ \pi \circ \mathrm{i}$. This implies that $\tilde{f}_1 \circ \pi$ and $\tilde{f}_2 \circ \pi$ coincide on the elements of the canonical basis of $F$, and thus, on $F$. Therefore, for any $\overline{x} \in F/D$, there exists $x \in F$ such that

$$\tilde{f}_1(\overline{x}) = \tilde{f}_1(\pi(x)) = \tilde{f}_2(\pi(x)) = \tilde{f}_2(\overline{x}). \qquad \square$$

The tensor product of $M$ and $N$ is denoted by $M \otimes_A N$, or simply by $M \otimes N$ when there is no ambiguity on the ring $A$.
Additionally, we denote the element $\tau(m,n)$ by $m \otimes_A n$, or simply $m \otimes n$. This element is referred to as a *simple tensor*, or *elementary tensor*. Any element of $M \otimes N$ is called a *tensor*.
By construction, we have that for all $m$, $m_1$, $m_2 \in M$, $n$, $n_1$, $n_2 \in N$ and $a \in A$, the following relations hold:

i) $(m_1 + m_2) \otimes n = m_1 \otimes n + m_2 \otimes n$;
ii) $m \otimes (n_1 + n_2) = m \otimes n_1 + m \otimes n_2$;
iii) $a(m \otimes n) = am \otimes n = m \otimes an$.

**T. 5.3.** (→ p. 233) Let $A$ be a ring, and let $M$, $N$ be $A$-modules.

1. For all $m \in M$ and $n \in N$, we have $m \otimes 0 = 0 \otimes n = 0$.
2. The set $\{m \otimes n \colon m \in M, n \in N\}$ of simple tensors is a generating set of $M \otimes N$.

3. Let $\mathcal{G}_1$ and $\mathcal{G}_2$ be sets of generators of $M$ and $N$, respectively. Then, the set of simple tensors

$$\mathcal{G}_1 \otimes \mathcal{G}_2 = \{m \otimes n \colon m \in \mathcal{G}_1,\ n \in \mathcal{G}_2\}$$

is a set of generators of $M \otimes N$.

4. If $M$ and $N$ are finitely generated, then $M \otimes N$ is finitely generated.

In particular, **T.5.3**.2 implies that a tensor is a finite linear combination of simple tensors.

It should be noted that $M \otimes N$ can be zero even if $M \neq 0$ and $N \neq 0$. As an example, we show that $\mathbb{Z}/(5) \otimes_{\mathbb{Z}} \mathbb{Z}/(7) = 0$.
Using **T.5.3**.1, we see that

$$5(\overline{a} \otimes \overline{b}) = 5\overline{a} \otimes \overline{b} = \overline{5a} \otimes \overline{b} = 0 \otimes \overline{b} = 0.$$

Similarly, we obtain $7(\overline{a} \otimes \overline{b}) = 0$. Thus, for any simple tensor $m \otimes n$ we have $m \otimes n = 1(m \otimes n) = (21 - 20)(m \otimes n) = 0$.
The conclusion follows from **T.5.3**.2.

As a non-trivial example, we compute the tensor product $A[x] \otimes_A A[y]$ of the $A$-modules $M = A[x]$ and $N = A[y]$, which are generated by $\{x^i \colon i \in \mathbb{N}\}$ and $\{y^j : j \in \mathbb{N}\}$, respectively.
Clearly, $M \otimes_A N$ is generated by $\{x^i \otimes y^j : (i, j) \in \mathbb{N}^2\}$.
Consider the diagram

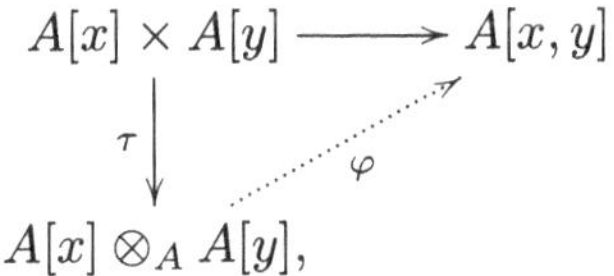


where the horizontal map $(x^i, y^j) \mapsto x^i y^j$ is the usual multiplication in $A[x, y]$, which is $A$-bilinear, and the homomorphism $\varphi$ is given by

$$\varphi\Big(\sum_{i,j} a_{ij}(x^i \otimes y^j)\Big) = \sum_{i,j} a_{ij} x^i y^j.$$

Then, $\varphi$ is surjective, since, for any polynomial $p = \sum_{i,j} a_{ij} x^i y^j \in A[x, y]$, we have $\varphi(\sum_{i,j} a_{ij}(x^i \otimes y^j)) = p$.
Furthermore, if

$$\varphi(\sum_{i,j} a_{ij}(x^i \otimes y^j)) = \sum_{i,j} a_{ij} x^i y^j = 0,$$

then $a_{ij} = 0$ for all $i$, $j$, hence $\varphi$ is injective as well.

**Properties of Tensor Product**

**T. 5.4.** (→ p. 233) Let $A$ be a ring, and let $M$, $N$, and $P$ be $A$-modules. Then:

1. $A \otimes M \simeq M$;
2. $M \otimes N \simeq N \otimes M$;
3. $(M \otimes N) \otimes P \simeq M \otimes (N \otimes P)$;
4. $(M \oplus N) \otimes P \simeq (M \otimes P) \oplus (N \otimes P)$;
5. if $I$ is an ideal of $A$, then $M \otimes (A/I) \simeq M/IM$;
6. if $M$ and $N$ are free of rank $m$ and $n$, respectively, then $M \otimes N$ is free of rank $mn$.

In general, a tensor product of free modules is free, see **E.11.4**.
Note that part 4 establishes that tensor product and direct sum commute. For a similar statement regarding tensor product and direct product, see **ToF.14.91**.

**T. 5.5.** (→ p. 235) Let $A$ be a ring, and let $M$, $N$, and $P$ be $A$-modules. Then,

$$\mathrm{Hom}_A(M \otimes_A N, P) \simeq \mathrm{Bil}(M, N; P) \simeq \mathrm{Hom}_A(M, \mathrm{Hom}_A(N, P)).$$

The isomorphism between $\mathrm{Hom}_A(M \otimes_A N, P)$ and $\mathrm{Hom}_A(M, \mathrm{Hom}_A(N, P))$ is called the *adjunction formula* Hom-$\otimes$.

## 5.2 Tensor Product as a Functor

Let $f\colon M \longrightarrow M'$ and $g\colon N \longrightarrow N'$ be $A$-module homomorphisms.
They induce a well-defined homomorphism

$$f \otimes g\colon M \otimes N \longrightarrow M' \otimes N', \quad (f \otimes g)(m \otimes n) = f(m) \otimes g(n),$$

see **E.11.6**.
Using **E.11.7**, it is easy to see that if $f$ and $g$ are isomorphisms, then $f \otimes g$ is also an isomorphism. In particular, if $M \simeq M'$, then $M \otimes N \simeq M' \otimes N$ for every $A$-module $N$.
Moreover, according to **E.11.7**, for any $A$-module $N$, the map $\bullet \otimes_A N$, sending an $A$-module $M$ to $M \otimes_A N$ and an $A$-module homomorphism $f\colon M \longrightarrow M'$ to $f \otimes_A \mathrm{id}_N\colon M \otimes_A N \longrightarrow M' \otimes_A N$, defines a covariant functor from the category of $A$-modules to itself.

Similarly, we can define the functor $N \otimes_A \bullet$, and, by **T.5.4**.2, we can identify the two functors.

**Right Exactness of $\bullet \otimes N$**

**T. 5.6.** Let $M_1 \xrightarrow{f} M \xrightarrow{g} M_2 \longrightarrow 0$ be an exact sequence. Then, for any $A$-module $N$, the sequence

$$M_1 \otimes N \xrightarrow{f \otimes \mathrm{id}_N} M \otimes N \xrightarrow{g \otimes \mathrm{id}_N} M_2 \otimes N \longrightarrow 0$$

is exact.

**Proof T. 5.6.** By **T.4.16**.1, the exactness of $M_1 \xrightarrow{f} M \xrightarrow{g} M_2 \to 0$ implies the exactness of

$$0 \longrightarrow \mathrm{Hom}(M_2, \mathrm{Hom}(N, Q)) \xrightarrow{g^*} \mathrm{Hom}(M, \mathrm{Hom}(N, Q)) \xrightarrow{f^*} \mathrm{Hom}(M_1, \mathrm{Hom}(N, Q))$$

for any $A$-module $Q$.
Using the isomorphisms defined in **T.5.1** and **T.5.5**, we obtain the commutative diagram

$$\begin{array}{ccc}
\mathrm{Hom}(M, \mathrm{Hom}(N, Q)) & \xrightarrow[\varphi \mapsto \varphi \circ f]{f^*} & \mathrm{Hom}(M_1, \mathrm{Hom}(N, Q)) \\
\simeq \downarrow & & \downarrow \simeq \\
\mathrm{Bil}(M, N; Q) & \xrightarrow[b_\varphi \mapsto b_{\varphi \circ f}]{} & \mathrm{Bil}(M_1, N; Q) \\
\simeq \downarrow & & \downarrow \simeq \\
\mathrm{Hom}(M \otimes N, Q) & \xrightarrow[\widetilde{b_\varphi} \mapsto \widetilde{b_{\varphi \circ f}}]{} & \mathrm{Hom}(M_1 \otimes N, Q).
\end{array}$$

Since for any $\widetilde{b_\varphi} \in \mathrm{Hom}(M \otimes N, Q)$ and for any simple tensor $m_1 \otimes n$ we have

$$\begin{aligned}
\Big(f \otimes \mathrm{id}_N\Big)^* \big(\widetilde{b_\varphi}\big)(m_1 \otimes n) &= \big(\widetilde{b_\varphi} \circ (f \otimes \mathrm{id}_N)\big)(m_1 \otimes n) \\
&= \widetilde{b_\varphi}(f(m_1) \otimes n) = b_\varphi(f(m_1), n) = \varphi(f(m_1))(n) \\
&= b_{\varphi \circ f}(m_1, n) = \widetilde{b_{\varphi \circ f}}(m_1 \otimes n),
\end{aligned}$$

the last horizontal homomorphism coincides with $(f \otimes \mathrm{id}_N)^*$.
In a similar way, it can be verified that the induced homomorphism on $\mathrm{Hom}(M_2 \otimes N, Q)$ is $(g \otimes \mathrm{id}_N)^*$.

As a result, the sequence

$$0 \longrightarrow \mathrm{Hom}(M_2 \otimes N, Q) \xrightarrow{(g\otimes \mathrm{id}_N)^*} \mathrm{Hom}(M \otimes N, Q) \xrightarrow{(f\otimes \mathrm{id}_N)^*} \mathrm{Hom}(M_1 \otimes N, Q)$$

is exact for any $Q$.
Using **T.4.17**.1, we can conclude that the sequence

$$M_1 \otimes N \xrightarrow{f\otimes \mathrm{id}_N} M \otimes N \xrightarrow{g\otimes \mathrm{id}_N} M_2 \otimes N \longrightarrow 0$$

is exact. □

As we have seen for the functors $\mathrm{Hom}_A(M, \bullet)$ and $\mathrm{Hom}_A(\bullet, N)$, see **T.4.16** and **T.4.17**, the converse of the previous statement also holds.

**T. 5.7.** (→ p. 235) Let $M_1 \xrightarrow{f} M \xrightarrow{g} M_2 \longrightarrow 0$ be a sequence of $A$-modules. If, for every $A$-module $N$, the sequence

$$M_1 \otimes N \xrightarrow{f\otimes \mathrm{id}_N} M \otimes N \xrightarrow{g\otimes \mathrm{id}_N} M_2 \otimes N \longrightarrow 0$$

is exact, then $M_1 \xrightarrow{f} M \xrightarrow{g} M_2 \longrightarrow 0$ is exact.

An $A$-module $Q$ is said to be *flat* if for every injective $A$-module homomorphism $f : M_1 \longrightarrow M$, the induced map $f \otimes \mathrm{id}_Q : M_1 \otimes Q \longrightarrow M \otimes Q$ is also injective. By **T.5.6**, the functor $\bullet \otimes Q$ preserves short exact sequences if and only if $Q$ is flat.
Not all modules are flat. For example, let $A = \mathbb{Z} = M = N$, $Q = \mathbb{Z}/(2)$, and consider the short sequence $0 \longrightarrow \mathbb{Z} \xrightarrow{2} \mathbb{Z} \longrightarrow \mathbb{Z}/(2) \longrightarrow 0$.
Tensoring with $Q$, we observe that the sequence

$$0 \longrightarrow \mathbb{Z} \otimes_{\mathbb{Z}} \mathbb{Z}/(2) \longrightarrow \mathbb{Z} \otimes_{\mathbb{Z}} \mathbb{Z}/(2)$$

is not exact since the homomorphism $2 \otimes \mathrm{id}_{\mathbb{Z}/(2)}$ is zero.
Therefore, since $\mathbb{Z} \otimes_{\mathbb{Z}} \mathbb{Z}/(2) \simeq \mathbb{Z}/(2) \neq 0$, this homomorphism cannot be injective.

## 5.3 Extension of Scalars

In Section 4.1, it was shown that given rings $A$ and $B$, and a homomorphism $f\colon A \longrightarrow B$, we can define an $A$-module structure on any $B$-module $M$ by restricting scalars via $f$.

This is achieved by defining the scalar multiplication as follows

$$A \times M \longrightarrow M, \quad (a, m) \mapsto f(a)m.$$

We can view restriction of scalars as a covariant functor from the category of $B$-modules to the category of $A$-modules.
Indeed, any $B$-module homomorphism $u \colon M \longrightarrow N$ induces an $A$-module homomorphism $\tilde{u}$ between $M$ and $N$, since for any $m \in M$ and $a \in A$, we have

$$\tilde{u}(am) = u(f(a)m) = f(a)u(m) = a\tilde{u}(m).$$

In this way, the ring $B$ has two module structures. It is not only a module over itself, but also an $A$-module with the structure defined via $f$.
The two scalar multiplications commute, since

$$(ab)b' = (f(a)b)b' = f(a)bb' = a(bb')$$

hold for any $a \in A$ and any $b, b' \in B$.
We call $B$ an $(A, B)$-*bimodule.*

On the other hand, suppose that $M$ is an $A$-module. The tensor product $M_B = B \otimes_A M$ has a natural $A$-module structure. Since $B$ is a $B$-module, we can also define a $B$-module structure on $M_B$ as follows

$$B \times M_B \longrightarrow M_B, \quad (b, b' \otimes m) \mapsto bb' \otimes m.$$

We say that the structure of $B$-module on $M_B$ is defined by *extension of scalars*. Note that the module $M_B$ obtained via extension of scalars is an $(A, B)$-bimodule.
One classical example of extension of scalars is given by a domain $A$ and its field of fractions $B = Q(A)$. Another example is provided by a field extension $K \subset L$, where extension of scalars transforms a $K$-vector space into an $L$-vector space.

As previously noted for restriction of scalars, we can view extension of scalars as a covariant functor from the category of $A$-modules to the category of $B$-modules. This functor is essentially defined by $B \otimes_A \bullet$.

The following fact is very useful. For a complete proof, see [2, Theorem 8.8].

**T. 5.8.** Let $A$ and $B$ be rings. Let also $M$ be an $A$-module, $N$ an $(A, B)$-bimodule, and $P$ a $B$-module. Then, there exists an isomorphism of $(A, B)$-bimodules

$$(M \otimes_A N) \otimes_B P \simeq M \otimes_A (N \otimes_B P).$$

# Chapter 6
# Localization

In this chapter we discuss the construction of rings of fractions, an operation which generalizes the construction of the rational numbers $\mathbb{Q}$ from the ring $\mathbb{Z}$, or, in greater generality, the construction of the field of fractions of a domain. We will define fractions using as denominators elements of a chosen subset of the ring. We will see that, when we choose the denominators in the complement of a prime ideal, this construction produces a local ring whose maximal ideal is the extension of that prime with respect to a canonical homomorphism.

## 6.1 Rings of Fractions

Let $A$ be a ring. We say that a subset $S \subset A$ is *multiplicative*, or *multiplicatively closed* if $1 \in S$ and $st \in S$ for any $s, t \in S$.

**T. 6.1.** (→ p. 236) Let $S \subset A$ be a multiplicative subset. For $a$, $b \in A$ and $s$, $t \in S$, the relation

$$(a, s) \sim (b, t) \iff \text{there exists } u \in S \text{ such that } u(at - bs) = 0$$

defines an equivalence relation on $A \times S$.

We denote $(A \times S)/\sim$ by $S^{-1}A$, and we denote by $\frac{a}{s}$ the equivalence class of the element $(a, s)$.

**Ring of Fractions**

**T. 6.2.** (→ p. 236) Let $S$ be a multiplicative subset of a ring $A$. Then, the set $S^{-1}A$ with the sum and product operations defined by



A. Bandini et al., *Commutative Algebra through Exercises*, UNITEXT 159,
https://doi.org/10.1007/978-3-031-56910-4_6

$$\frac{a}{s}+\frac{b}{t}=\frac{at+bs}{st} \quad \text{and} \quad \frac{a}{s}\cdot\frac{b}{t}=\frac{ab}{st},$$

for all $a$, $b \in A$ and $s$, $t \in S$, is a commutative ring with identity, where $0 = \frac{0}{1}$ and $1 = \frac{1}{1}$.

The ring $S^{-1}A$ is called the *localization of $A$ at $S$* or the *ring of fractions of $A$ with respect to $S$*.
It is easy to verify that the map

$$\sigma = \sigma_S\colon A \longrightarrow S^{-1}A \ \text{ given by } \ \sigma(a) = \tfrac{a}{1}$$

is a canonical ring homomorphism, usually called *localization homomorphism*.
**T. 6.3.** (→ p. 236) Let $S$ be a multiplicative subset of $A$, and let $\sigma$ be the localization homomorphism $A \longrightarrow S^{-1}A$. Then:

1. $\sigma$ is injective if and only if $S \cap \mathcal{D}(A) = \emptyset$;
2. $S^{-1}A = 0$ if and only if $S \cap \mathcal{N}(A) \neq \emptyset$.

The following is the fundamental and defining property of localization.

**Universal Property of Ring of Fractions**

**T. 6.4.** Let $g\colon A \longrightarrow B$ be a ring homomorphism such that $g(S) \subseteq B^*$. Then, there exists a unique ring homomorphism $\tilde{g}\colon S^{-1}A \longrightarrow B$ such that the following diagram commutes

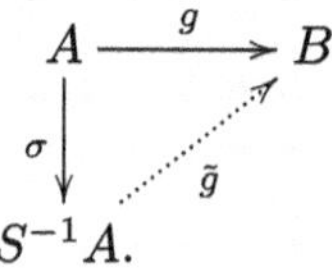


**Proof T. 6.4.** If $\tilde{g}$ exists, then, for every $a \in A$ and $s \in S$

$$\tilde{g}(\tfrac{a}{1}) = \tilde{g}\sigma(a) = g(a) \ \text{ and } \ \tilde{g}(\tfrac{1}{s}) = \tilde{g}((\tfrac{s}{1})^{-1}) = \tilde{g}(\tfrac{s}{1})^{-1} = g(s)^{-1}.$$

Hence, $\tilde{g}$ is defined as $\tilde{g}\left(\frac{a}{s}\right) = \tilde{g}\left(\frac{a}{1}\right)\tilde{g}\left(\frac{1}{s}\right) = g(a)g(s)^{-1}$, and therefore, $\tilde{g}$ is uniquely determined by $g$.
To complete the proof, we show that $\tilde{g}\colon S^{-1}A \longrightarrow B$ is well-defined.
Let $\frac{a}{s} = \frac{b}{t}$, and let $u \in S$ be such that $u(at - bs) = 0$. Then, we have $g(u)(g(a)g(t) - g(b)g(s)) = 0$.
Since $g(u) \in B^*$, we obtain $\tilde{g}\left(\frac{a}{s}\right) = g(a)g(s)^{-1} = g(b)g(t)^{-1} = \tilde{g}\left(\frac{b}{t}\right)$.
Since $g$ is a homomorphism, $\tilde{g}$ is also a homomorphism. □

**T. 6.5.** (→ p. 236) Using the same notation as in **T.6.4**, assume that:

i) if $g(a) = 0$, then there exists $s \in S$ such that $as = 0$;
ii) for any $b \in B$ there exist $a \in A$ and $s \in S$ such that $b = g(a)g(s)^{-1}$.

Then, $\tilde{g} \colon S^{-1}A \longrightarrow B$ is an isomorphism.

It is worth noting that if $g$ is injective, then $\tilde{g}$ is also injective.

Two examples of rings of fractions which are particularly important are the following:

i) *Localization at the powers of an element.* Let $S = S_f = \{f^n\}_{n \in \mathbb{N}}$ be the set of powers of an element $f \in A$.
In this case $S^{-1}A$ is denoted by $A_f$.
It is clear that $A_f \neq 0$ if and only if $f \notin \mathcal{N}(A)$;
ii) *Localization at the complement of a prime ideal.* Let $S = A \setminus \mathfrak{p}$ be the complement of a prime ideal $\mathfrak{p} \in \operatorname{Spec} A$.
In this case $S^{-1}A$ is denoted by $A_{\mathfrak{p}}$ and is called the *localization of $A$ at $\mathfrak{p}$*.

It is important to remark that in the last example the result of localization is a local ring.

**T. 6.6.** (→ p. 236) The ring $A_{\mathfrak{p}}$ is a local ring with maximal ideal

$$\mathfrak{p}A_{\mathfrak{p}} = \left\{ \frac{a}{s} \colon a \in \mathfrak{p},\ s \in S \right\}.$$

**Extension and Contraction of Ideals with respect to $\sigma_S$**

**T. 6.7.** (→ p. 237) Let $A$ be a ring, let $S \subset A$ be a multiplicative subset, and let $\sigma = \sigma_S$ be the localization homomorphism.

1. Let $I \subset A$ be an ideal, and let $I^e = (\sigma(I))$ be its extension in $S^{-1}A$. Then:

   a. $I^e = \left\{ \frac{a}{s} \colon a \in I,\ s \in S \right\} = S^{-1}I$;
   b. $S^{-1}I = S^{-1}A$ if and only if $I \cap S \neq \emptyset$;
   c. $I^{ec} = \{a \in A \colon as \in I \text{ for some } s \in S\} = \bigcup_{s \in S} I : (s)$.

2. Let $J \subset S^{-1}A$ be an ideal, and let $J^c = \sigma^{-1}(J)$ be its contraction in $A$. Then, $J = J^{ce}$, *i.e.*, every ideal of $S^{-1}A$ is the extension of an ideal in $A$.

**T. 6.8.** (→ p. 237) Let $A$ be a ring, and let $S \subset A$ be a multiplicative subset.
If $\mathfrak{p} \subset A$ is a prime ideal such that $\mathfrak{p} \cap S = \emptyset$, then $\mathfrak{p} = \mathfrak{p}^{ec}$. Furthermore, $\mathfrak{p}^e$ is a prime ideal of $S^{-1}A$.
In particular, there exists a one-to-one correspondence between prime ideals of $S^{-1}A$ and prime ideals of $A$ that do not intersect $S$:

$$\operatorname{Spec} S^{-1}A \longleftrightarrow \{\mathfrak{p} \in \operatorname{Spec} A \colon \mathfrak{p} \cap S = \emptyset\}.$$

It is straightforward to verify that

$$S = A \setminus \bigcup_{h \in H} \mathfrak{p}_h$$

is a multiplicative subset for any set of primes $\{\mathfrak{p}_h \in \operatorname{Spec} A \colon h \in H\}$. Furthermore, if $H$ is finite, the Prime Avoidance Lemma implies that $S^{-1}A$ is semilocal.

**Localization and Operations on Ideals**

**T. 6.9.** (→ p. 237) Let $I$, $J \subset A$ be ideals. Then:

1. $S^{-1}(I + J) = S^{-1}I + S^{-1}J$;
2. $S^{-1}(IJ) = S^{-1}I \cdot S^{-1}J$;
3. $S^{-1}(I \cap J) = S^{-1}I \cap S^{-1}J$;
4. $S^{-1}\sqrt{I} = \sqrt{S^{-1}I}$.
   In particular, $S^{-1}\mathcal{N}(A) = \mathcal{N}(S^{-1}A)$.

It is worth noting that in a localization, the inverses of all elements of $S$ are added, but there may be additional invertible elements. For example, consider the set $S = \{6^n\}_{n \in \mathbb{N}} \subset \mathbb{Z}$. Since $\frac{2}{1}\frac{3}{6} = \frac{1}{1}$, also the element $\frac{2}{1}$ is invertible in $S^{-1}\mathbb{Z}$, as well as all elements of the set $T = \{2^n 3^m\}_{n,m \in \mathbb{N}}$.
A more detailed analysis of this phenomenon can be found in Section 6.5.

## 6.2 Modules of Fractions

The construction we just described for rings can be generalized to modules. Given an $A$-module $M$, we can localize it with respect to a multiplicative subset $S \subset A$ to obtain an $S^{-1}A$-module $S^{-1}M$.

We define an equivalence relation on $M \times S$ by letting

$$(m, s) \sim (n, t) \iff \text{there exists } u \in S \text{ such that } u(tm - sn) = 0,$$

where $m, n \in M$ and $s, t \in S$.
We denote by $S^{-1}M$ the set $(M \times S)/\sim$ and by $\frac{m}{s}$ the equivalence class of an element $(m, s)$.

**Module of Fractions**

Let $M$ be an $A$-module, and let $S \subset A$ be a multiplicative subset. The set $S^{-1}M$ with the operations defined by

$$\frac{m}{s} + \frac{n}{t} = \frac{tm + sn}{st} \quad \text{and} \quad \frac{a}{s} \cdot \frac{m}{t} = \frac{am}{st},$$

where $m, n \in M$, $s, t \in S$, and $a \in A$, is an $S^{-1}A$-module.

The module $S^{-1}M$ is the *localization of $M$ at $S$* or the *module of fractions of $M$ with respect to $S$*.

Note that if $\operatorname{Ann} M \cap S \neq \emptyset$, then $S^{-1}M = 0$.
Indeed, if there exists $s \in \operatorname{Ann} M \cap S$, then for every element $\frac{m}{t} \in S^{-1}M$, we have $sm = 0$, hence $\frac{m}{t} = \frac{0}{1}$.

As in the case of rings, we denote $S^{-1}M$ by $M_f$ when $S = \{f^n\}_{n \in \mathbb{N}}$, and by $M_{\mathfrak{p}}$ when $S = A \setminus \mathfrak{p}$. In particular, when $M = I$ is an ideal of $A$ we write $I_{\mathfrak{p}} = IA_{\mathfrak{p}}$.

Let $\sigma\colon A \longrightarrow S^{-1}A$ be the localization homomorphism. By using the restriction of scalars via $\sigma$, any $S^{-1}A$-module is endowed with an $A$-module structure. This structure is defined by $am = \frac{a}{1}m$ for all $a \in A$ and $m \in M$.

By abuse of notation, we continue to refer to the canonical localization map for modules as $\sigma = \sigma_S$. We can easily verify that $\sigma\colon M \longrightarrow S^{-1}M$, defined by $\sigma(m) = \frac{m}{1}$, is an $A$-module homomorphism.

To establish the universal property of module of fractions, as was previously done for rings, it is essential to take into account that $M$ and $S^{-1}M$ are modules over distinct rings. Therefore, if we wish to factor a homomorphism $f : M \longrightarrow N$ through $S^{-1}M$, we need $N$ to be endowed with both an $A$-module and an $S^{-1}A$-module structure. Additionally, these two structures must be compatible, meaning that $an = \frac{a}{1}n$ holds for all $a \in A$ and $n \in N$.

**T. 6.10.** (→ p. 238) Let $A$ be a ring, let $S \subset A$ be a multiplicative subset, and let $N$ be an $A$-module.
It is possible to define an $S^{-1}A$-module structure on $N$ compatible with the $A$-module structure if and only if for every $s \in S$ the multiplication

$$\mu_s\colon N \xrightarrow{s} N$$

is an isomorphism.
In this case, the $S^{-1}A$-module structure on $N$ is unique.

**Universal Property of Module of Fractions**

**T. 6.11.** (→ p. 238) Let $S \subset A$ be a multiplicative subset, and let $M$ be an $A$-module. Let also $N$ be an $A$-module such that the multiplication $\mu_s$ is an isomorphism on $N$ for all $s \in S$.
Then, for any $A$-module homomorphism $f\colon M \longrightarrow N$ there exists a unique $S^{-1}A$-module homomorphism

$$\tilde{f}\colon S^{-1}M \longrightarrow N \ \text{ such that } \ \tilde{f}\left(\frac{m}{s}\right) = \frac{1}{s}f(m)$$

and the following diagram commutes:

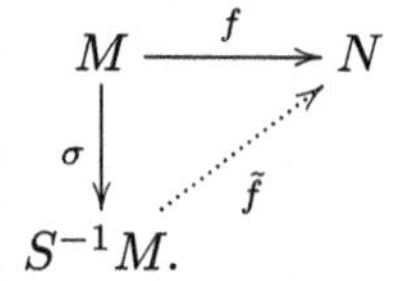

$$\begin{array}{ccc} M & \xrightarrow{f} & N \\ {\scriptstyle\sigma}\downarrow & \nearrow{\scriptstyle\tilde{f}} & \\ S^{-1}M. & & \end{array}$$

Note that, by **T.6.10**, $N$ is an $(A, S^{-1}A)$-bimodule.
Moreover, the homomorphism $\tilde{f}$ is also an $A$-module homomorphism by restriction of scalars.

## 6.3 The Functor $S^{-1}$

In order to view localization as a functor, we need to understand its effect on homomorphisms. Let $f\colon M \longrightarrow N$ be an $A$-module homomorphism. We can construct the following diagram:

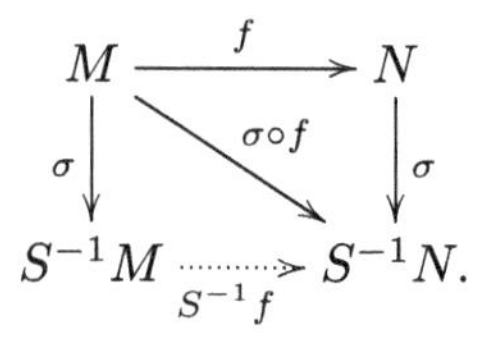

$$\begin{array}{ccc} M & \xrightarrow{f} & N \\ {\scriptstyle\sigma}\downarrow & \searrow^{\sigma\circ f} & \downarrow{\scriptstyle\sigma} \\ S^{-1}M & \xrightarrow[S^{-1}f]{} & S^{-1}N. \end{array}$$

It can be easily verified that the map given by

$$S^{-1}f\left(\frac{m}{s}\right) = \frac{f(m)}{s}$$

is an $S^{-1}A$-module homomorphism that makes the diagram commute.

Alternatively, we recall that $S^{-1}N$ is an $(A, S^{-1}A)$-bimodule, and therefore, by the universal property, we have $S^{-1}f = \widetilde{\sigma \circ f}$.

Moreover, for any $A$-module homomorphism $g\colon N \longrightarrow P$, we have

$$S^{-1}(g \circ f) = S^{-1}g \circ S^{-1}f.$$

Consequently, localization can be viewed as a covariant functor from the category of $A$-modules to the category of $S^{-1}A$-modules.

**Exactness of $S^{-1}$**

**T. 6.12.** ($\rightarrow$ p. 239) The functor $S^{-1}$ is exact, *i.e.*, for any exact sequence $M \xrightarrow{f} N \xrightarrow{g} P$ the sequence $S^{-1}M \xrightarrow{S^{-1}f} S^{-1}N \xrightarrow{S^{-1}g} S^{-1}P$ is also exact.
In particular, if $f$ is an injective homomorphism, then $S^{-1}f$ is also injective, and if $g$ is a surjective homomorphism, then $S^{-1}g$ is also surjective.

It can be easily verified that for any $A$-module $M$ and any localization $S^{-1}M$, all submodules of $S^{-1}M$ are of the form $S^{-1}N$ where $N$ is a submodule of $M$, see also **T.6.7**.2.

**Localization and Operations on Modules**

**T. 6.13.** Let $S \subset A$ be a multiplicative subset. Let also $M$, $N$, and $P$ be $A$-modules.

1. If $M$ and $N$ are submodules of $P$, then $S^{-1}(M{+}N) = S^{-1}M{+}S^{-1}N$.
2. If $M$ and $N$ are submodules of $P$, then $S^{-1}(M{\cap}N) = S^{-1}M{\cap}S^{-1}N$.
3. If $N$ is a submodule of $P$, then $S^{-1}N \subseteq S^{-1}P$ and

$$S^{-1}(P/N) \simeq S^{-1}P/S^{-1}N.$$

4. If $M$ is finitely generated, then $S^{-1}\operatorname{Ann}_A M = \operatorname{Ann}_{S^{-1}A} S^{-1}M$.
5. If $M$ and $N$ are submodules of $P$ and $N$ is finitely generated, then

$$S^{-1}(M : N) = S^{-1}M : S^{-1}N.$$

**Proof T. 6.13.** 1. For any $m \in M$, $n \in N$, and $s \in S$, we have $\frac{m+n}{s} = \frac{m}{s} + \frac{n}{s}$. As a consequence, $S^{-1}(M + N) \subseteq S^{-1}M + S^{-1}N$.

To prove the opposite inclusion, recall that $\frac{m}{s} + \frac{n}{t} = \frac{tm+sn}{st}$.

2. Since $M \cap N$ is contained in both $M$ and $N$, we immediately have the inclusion $S^{-1}(M \cap N) \subseteq S^{-1}M \cap S^{-1}N$.

To prove the opposite inclusion, let $\alpha \in S^{-1}M \cap S^{-1}N$.
Then, there exist $m \in M$, $n \in N$, and $s$, $t$, $u \in S$ such that $\alpha = \frac{m}{s} = \frac{n}{t}$ with $u(tm - sn) = 0$. This implies that $utm = usn \in M \cap N$.

Therefore,

$$\frac{m}{s}=\frac{utm}{uts}=\frac{usn}{uts}\in S^{-1}(M\cap N).$$

3. By applying $S^{-1}$ to the exact sequence $0\longrightarrow N\xrightarrow{i}P\xrightarrow{\pi}P/N\longrightarrow 0$, where $i$ is the inclusion homomorphism and $\pi$ the canonical projection, we obtain the exact sequence $0\longrightarrow S^{-1}N\xrightarrow{S^{-1}i}S^{-1}P\xrightarrow{S^{-1}\pi}S^{-1}(P/N)\longrightarrow 0$, since $S^{-1}$ is an exact functor by **T.6.12**.
Consider the diagram

$$\begin{array}{ccccccccc}
0 & \longrightarrow & S^{-1}N & \xrightarrow{S^{-1}i} & S^{-1}P & \xrightarrow{S^{-1}\pi} & S^{-1}(P/N) & \longrightarrow & 0\\
 & & \Big\downarrow \mathrm{id}_{S^{-1}N} & & \Big\downarrow \mathrm{id}_{S^{-1}P} & & \Big\downarrow \gamma & & \\
0 & \longrightarrow & S^{-1}N & \xrightarrow[j]{} & S^{-1}P & \xrightarrow[\eta]{} & S^{-1}P/S^{-1}N & \longrightarrow & 0,
\end{array}$$

where $j$ is the inclusion homomorphism, $\eta$ is the canonical projection defined by $\eta\left(\frac{p}{s}\right)=\frac{p}{s}+S^{-1}N$, and $\gamma(\frac{\overline{p}}{s})=\frac{p}{s}+S^{-1}N$, where $p\in P$ is any element congruent to $\overline{p}$ modulo $N$. The map $\gamma$ is well-defined, since, when $\frac{\overline{p}}{s}=\frac{\overline{q}}{t}$, there exists $u\in S$ such that $ut\overline{p}=us\overline{q}$, that is, $utp-usq=n$ for some $n\in N$. Therefore, we have

$$\begin{aligned}\gamma\left(\frac{\overline{p}}{s}\right)&=\frac{p}{s}+S^{-1}N=\frac{utp}{uts}+S^{-1}N=\frac{usq}{uts}+\frac{n}{uts}+S^{-1}N\\&=\frac{q}{t}+S^{-1}N=\gamma\left(\frac{\overline{q}}{t}\right).\end{aligned}$$

It is clear that the square on the left commutes.
For the square on the right, we have

$$\gamma\circ(S^{-1}\pi)\left(\frac{p}{s}\right)=\gamma\left(\frac{\overline{p}}{s}\right)=\eta\left(\frac{p}{s}\right).$$

The conclusion follows from **T.4.20**.

4. We proceed by induction on the number of generators of $M$.
If $M=0$, then there is nothing to prove.
Now, assume that $M=\langle m\rangle_A\neq 0$. Then, $S^{-1}M=\langle\frac{m}{1}\rangle_{S^{-1}A}$.
If $\frac{a}{s}\in S^{-1}\operatorname{Ann}_A M$, then $\frac{a}{s}\frac{m}{1}=\frac{am}{s}=0$, *i.e.*, $\frac{a}{s}\in\operatorname{Ann}_{S^{-1}A}S^{-1}M$.

To prove the opposite inclusion, let $\frac{a}{s}\in\operatorname{Ann}_{S^{-1}A}S^{-1}M$.
Then, $\frac{am}{s}=\frac{a}{s}\frac{m}{1}=0$, and there exists $u\in S$ such that $uam=0$. Hence, $ua\in\operatorname{Ann}_A M$ and $\frac{a}{s}=\frac{ua}{us}\in S^{-1}\operatorname{Ann}_A M$.

Proceeding to the inductive step, we write $M=N_1+N_2$ with $N_1$ and $N_2$ $A$-modules each having fewer generators than $M$.
By applying parts 1 and 2, along with the properties of the annihilator, see **E.8.21**, we obtain

$$\begin{aligned} S^{-1} \operatorname{Ann}_A M &= S^{-1} \operatorname{Ann}_A(N_1 + N_2) = S^{-1}(\operatorname{Ann}_A N_1 \cap \operatorname{Ann}_A N_2) \\ &= S^{-1} \operatorname{Ann}_A N_1 \cap S^{-1} \operatorname{Ann}_A N_2 \\ &= \operatorname{Ann}_{S^{-1}A} S^{-1}N_1 \cap \operatorname{Ann}_{S^{-1}A} S^{-1}N_2 \\ &= \operatorname{Ann}_{S^{-1}A}(S^{-1}N_1 + S^{-1}N_2) \\ &= \operatorname{Ann}_{S^{-1}A}(S^{-1}(N_1 + N_2)) = \operatorname{Ann}_{S^{-1}A} S^{-1}M, \end{aligned}$$

where the fourth equality follows from the inductive hypothesis.

5. By definition, $a \in M : N$ if and only if $aN \subseteq M$, and this is equivalent to saying that $a\,((N+M)/M) = 0$, *i.e.*,

$$M : N = \operatorname{Ann}_A\left((N+M)/M\right).$$

If $N$ is finitely generated, then $(N+M)/M$ is also finitely generated, and it is sufficient to use parts 1, 3 and 4 to obtain

$$\begin{aligned} S^{-1}(M : N) &= S^{-1} \operatorname{Ann}_A\left((N+M)/M\right) \\ &= \operatorname{Ann}_{S^{-1}A}\left(S^{-1}\left((N+M)/M\right)\right) \\ &= \operatorname{Ann}_{S^{-1}A}\left(S^{-1}(N+M)/S^{-1}M\right) \\ &= \operatorname{Ann}_{S^{-1}A}\left((S^{-1}N + S^{-1}M)/S^{-1}M\right) = S^{-1}M : S^{-1}N. \quad \square \end{aligned}$$

Note that part 4 is a special case of part 5, and neither of these statements holds in general.

For example, take $A = K\left[t, \frac{x}{t^n} : n \in \mathbb{N}\right]$ with $x$ and $t$ indeterminates, $M = \langle x \rangle_A$, $N = \langle \frac{x}{t^n} : n \in \mathbb{N} \rangle_A$, and $S = \{t^n : n \in \mathbb{N}\}$. Here, $M$ and $N$ are ideals of $A$, and we have

$$M : N = \left\{a \in A : \frac{ax}{t^n} \in M \text{ for all } n \in \mathbb{N}\right\} = \left(\frac{x}{t^n} : n \in \mathbb{N}\right) = N.$$

Hence, $S^{-1}(M : N) = S^{-1}N = S^{-1}(x) = S^{-1}M$, while

$$S^{-1}M : S^{-1}N = S^{-1}A.$$

We conclude this section by presenting some connections between localization and tensor product.

**Localization and Tensor Product**

**T. 6.14.** (→ p. 239) Let $S$ be a multiplicative subset of a ring $A$, and let $M$ be an $A$-module. Then, there exist canonical isomorphisms:

1. $S^{-1}M \simeq S^{-1}A \otimes_A M$;
2. $S^{-1}(M \otimes_A N) \simeq S^{-1}M \otimes_{S^{-1}A} S^{-1}N$.

## 6.4 Local Properties

Let $\mathcal{P}$ be a property of a ring $A$. We say that $\mathcal{P}$ is a *local property* if

$$\mathcal{P} \text{ holds for } A \iff \mathcal{P} \text{ holds for } A_{\mathfrak{p}} \text{ for all } \mathfrak{p} \in \operatorname{Spec} A.$$

Similarly, a property $\mathcal{P}$ of an $A$-module $M$ is a *local property* if

$$\mathcal{P} \text{ holds for } M \iff \mathcal{P} \text{ holds for the } A_{\mathfrak{p}}\text{-module } M_{\mathfrak{p}} \text{ for all } \mathfrak{p} \in \operatorname{Spec} A.$$

The following is a fundamental local property.

**T. 6.15.** (→ p. 239) Let $M$ be an $A$-module. Then, the following statements are equivalent:

1. $M = 0$;
2. $M_{\mathfrak{p}} = 0$ for all ideals $\mathfrak{p} \in \operatorname{Spec} A$;
3. $M_{\mathfrak{m}} = 0$ for all ideals $\mathfrak{m} \in \operatorname{Max} A$.

If $S = A \setminus \mathfrak{p}$ with $\mathfrak{p}$ prime, and $f\colon M \longrightarrow N$ is an $A$-module homomorphism, we denote the homomorphism $S^{-1}f$ by $f_{\mathfrak{p}}\colon M_{\mathfrak{p}} \longrightarrow N_{\mathfrak{p}}$.

**T. 6.16.** Let $f\colon M \longrightarrow N$ be a homomorphism. Then:

1. $f$ is injective if and only if $f_{\mathfrak{p}}$ is injective for all $\mathfrak{p} \in \operatorname{Spec} A$, equivalently, if and only if $f_{\mathfrak{m}}$ is injective for all $\mathfrak{m} \in \operatorname{Max} A$;
2. $f$ is surjective if and only if $f_{\mathfrak{p}}$ is surjective for all $\mathfrak{p} \in \operatorname{Spec} A$, equivalently, if and only if $f_{\mathfrak{m}}$ is surjective for all $\mathfrak{m} \in \operatorname{Max} A$.

**Proof T. 6.16.** 1. We claim that $(\operatorname{Ker} f)_{\mathfrak{p}} = \operatorname{Ker} f_{\mathfrak{p}}$ for any prime $\mathfrak{p}$.
We have that $\frac{m}{s} \in \operatorname{Ker} f_{\mathfrak{p}}$ if and only if $\frac{f(m)}{s} = f_{\mathfrak{p}}\left(\frac{m}{s}\right) = 0$, *i.e.*, if and only if there exists $u \in S = A \setminus \mathfrak{p}$ such that $f(um) = uf(m) = 0$.
Hence, $\frac{m}{s} = \frac{um}{us} \in (\operatorname{Ker} f)_{\mathfrak{p}}$.

To prove the opposite inclusion, let $\frac{m}{s} \in (\operatorname{Ker} f)_{\mathfrak{p}}$.
Then, $\frac{m}{s} = \frac{n}{t}$ for some $n \in \operatorname{Ker} f$. Hence, there exists $u \in S$ such that $utm = usn$, and

$$f_{\mathfrak{p}}\left(\frac{m}{s}\right) = f_{\mathfrak{p}}\left(\frac{utm}{uts}\right) = f_{\mathfrak{p}}\left(\frac{usn}{uts}\right) = \frac{f(usn)}{uts} = 0,$$

*i.e.*, $\frac{m}{s} \in \operatorname{Ker} f_{\mathfrak{p}}$.
The conclusion follows from **T.6.15**.

2. Proceeding as in part 1, we can prove that $\operatorname{Im} f_{\mathfrak{p}} = (\operatorname{Im} f)_{\mathfrak{p}}$. Therefore, $\operatorname{Coker} f_{\mathfrak{p}} = (\operatorname{Coker} f)_{\mathfrak{p}}$. The conclusion follows again from **T.6.15**. □

**T. 6.17.** (→ p. 240) Let $A$ be a ring, and let $M$ be an $A$-module. The following are local properties:

1. $A$ is reduced;
2. $M$ is flat.

An example of a property that is not local is $\mathcal{P}$ = "being Noetherian". Consider

$$A = (\mathbb{Z}/(2))\,[x_i\colon\ i \in \mathbb{N}]/I \quad \text{with} \quad I = (x_i^2 - x_i\colon\ i \in \mathbb{N}).$$

It is clear that $A$ is not Noetherian.
However, for any prime ideal $\mathfrak{p}$, the ring $A_\mathfrak{p}$ is local by **T.6.6**. The definition of $I$ and the fact that we are in characteristic 2 imply that every element of $A_\mathfrak{p}$ is idempotent. Since in a local ring the only idempotents are 0 and 1, see **E.8.17**, it follows that $A_\mathfrak{p} \simeq \mathbb{Z}/(2)$ is Noetherian for all $\mathfrak{p} \in \operatorname{Spec} A$.

## 6.5 Appendix: the Saturation of a Multiplicative Subset

As previously observed, in a localization $S^{-1}A$, it is possible for the set $\{\frac{t}{s} \in S^{-1}A\colon\ t \in S\}$ to be a proper subset of $(S^{-1}A)^*$. In order to explain this phenomenon, we will describe the saturation of a multiplicative subset $S \subset A$ and the set of invertible elements of $S^{-1}A$.

Consider a multiplicative subset $S \subset A$. Then, $\frac{a}{s} \in (S^{-1}A)^*$ if and only if there exists $\frac{b}{t} \in S^{-1}A$ such that $\frac{ab}{st} = \frac{1}{1}$, hence, if and only if there exist $u \in S$ and $b \in A$ such that $uab = ust \in S$.

A multiplicative subset $S \subset A$ is called *saturated* if for any $s$, $t \in A$ such that $st \in S$, both $s$ and $t$ are elements of $S$.

For example, the set of invertible elements $A^*$ of $A$ and the set $S = A \setminus \mathcal{D}(A)$ are saturated multiplicative subsets of $A$.
It should be noted that if $S$ is saturated, then $S = \{a \in A\colon\ \frac{a}{1} \in (S^{-1}A)^*\}$. Indeed, as previously shown, $\frac{a}{1} \in (S^{-1}A)^*$ if and only if $uab \in S$ for some $b \in A$ and $u \in S$, therefore, by definition of saturated set, $a \in S$.

**T. 6.18.** ($\rightarrow$ p. 240) A multiplicative subset $S$ is saturated if and only if

$$S = A \setminus \bigcup_{\substack{\mathfrak{p} \in \operatorname{Spec} A \\ \mathfrak{p} \cap S = \emptyset}} \mathfrak{p}.$$

The *saturation* of a subset $S \subseteq A$ is defined as

$$\overline{S} = \{t \in A\colon\ \text{there exists } a \in A \text{ such that } at \in S\}.$$

It is straightforward to verify that the saturation of a multiplicative subset is a saturated multiplicative subset, see **E.12.34**.

The following result provides a characterization of saturated sets and a description of the invertible elements of $S^{-1}A$.

**Properties of Saturation**

**T. 6.19.** Let $S \subset A$ be a multiplicative subset, and let $\overline{S}$ be its saturation. Then:

1. $S \subseteq \overline{S}$;
2. if $T \subset A$ is a saturated multiplicative subset such that $S \subseteq T$, then $\overline{S} \subseteq T$;
3. $\overline{S} = A \setminus \bigcup_{\substack{\mathfrak{p} \in \operatorname{Spec} A \\ \mathfrak{p} \cap S = \emptyset}} \mathfrak{p}$;
4. $(S^{-1}A)^* = \left\{\frac{a}{s} \colon a \in \overline{S},\ s \in S\right\}$;
5. $\overline{S} = \sigma_S^{-1}((S^{-1}A)^*)$, where $\sigma_S \colon A \longrightarrow S^{-1}A$ is the localization homomorphism;
6. $\overline{S}^{-1}A = S^{-1}A$.

**Proof T. 6.19.** 1. If $s \in S$, then $s \cdot 1 \in S$, hence $s \in \overline{S}$.

2. If $s \in \overline{S}$, then there exists $a \in A$ such that $as \in S \subseteq T$. Therefore, $s \in T$ since $T$ is saturated.

3. Let $T = A \setminus \bigcup_{\substack{\mathfrak{p} \in \operatorname{Spec} A \\ \mathfrak{p} \cap S = \emptyset}} \mathfrak{p}$.

Clearly, $T$ is a multiplicative subset containing $S$, and by **T.6.18**, it is saturated. Therefore, part 2 implies that $\overline{S} \subseteq T$.

To prove the opposite inclusion, observe that $\overline{S}$ is saturated and contains $S$. Hence, again by **T.6.18**,

$$\overline{S} = A \setminus \bigcup_{\substack{\mathfrak{p} \in \operatorname{Spec} A \\ \mathfrak{p} \cap \overline{S} = \emptyset}} \mathfrak{p} \supseteq A \setminus \bigcup_{\substack{\mathfrak{p} \in \operatorname{Spec} A \\ \mathfrak{p} \cap S = \emptyset}} \mathfrak{p} = T.$$

4. Consider the element $\frac{a}{s}$, with $a \in \overline{S}$. Then, there exists $t \in A$ such that $ta \in S$. Hence, $\frac{a}{s}\frac{st}{at} = \frac{1}{1}$, that is, $\frac{a}{s} \in (S^{-1}A)^*$.

To prove the opposite inclusion, let $\frac{a}{s}, \frac{b}{t} \in (S^{-1}A)^*$ be such that $\frac{a}{s}\frac{b}{t} = 1$. Then, there exists $u \in S$ such that $uab = ust \in S$, hence $a \in \overline{S}$.

5. By part 4, we immediately have $\sigma_S(\overline{S}) \subseteq (S^{-1}A)^*$.

To prove the opposite inclusion, take $u \in \sigma_S^{-1}((S^{-1}A)^*)$. Then, $\frac{u}{1}$ belongs to $(S^{-1}A)^*$, and there exists $\frac{b}{t}$ such that $\frac{u}{1}\frac{b}{t} = \frac{1}{1}$.
Thus, there exists $s \in S$ such that $sub = st \in S$, hence, $u \in \overline{S}$.

6. It is clear that $A \times S \subseteq A \times \overline{S}$.

Let $\sim_S$ and $\sim_{\overline{S}}$ be the equivalence relations associated to $S$ and $\overline{S}$, respectively, see **T.6.1**.
Then, for any $(a, s)$ and $(b, t)$ in $A \times S$, we have

$$(a, s) \sim_S (b, t) \text{ if and only if } (a, s) \sim_{\overline{S}} (b, t).$$

This implies that $S^{-1}A \subseteq \overline{S}^{-1}A$.

To prove the opposite inclusion, take $\frac{b}{t} \in \overline{S}^{-1}A$. Then, there exists $a \in A$ such that $at \in S$. Therefore, $\frac{b}{t} = \frac{ab}{at} \in S^{-1}A$. □

In general, for any two multiplicative subsets $S \subset T$ of $A$, the existence of a set inclusion between $S^{-1}A$ and $T^{-1}A$ depends on the compatibility between the equivalence relations $\sim_S$ and $\sim_T$, see **T.6.19**.6.
For example, take $A = \mathbb{Z}/(6)$ with

$$S = \{\overline{1}, \overline{5}\} = A^* \subset T = \{\overline{1}, \overline{5}, \overline{2}, \overline{4}\} = \mathbb{Z}/(6) \setminus (\overline{3}).$$

We have $S^{-1}A \simeq \mathbb{Z}/(6)$ and $T^{-1}A \simeq \mathbb{Z}/(3)$.

**T. 6.20.** (→ p. 240) Let $S$ and $T$ be multiplicative subsets of $A$. Then,

$$S^{-1}A = T^{-1}A \text{ if and only if } \overline{S} = \overline{T}.$$

# Chapter 7
# Noetherian and Artinian Rings. Primary Decomposition

In this final chapter we introduce Noetherian and Artinian rings through chain conditions on ideals. We prove that every ideal in a Noetherian ring is finitely generated and can be decomposed as a finite intersection of primary ideals. Additionally, we prove that $A$ is Noetherian if and only if $A[x_1, \ldots, x_n]$ is Noetherian, and that $A$ is Artinian if and only if, up to isomorphism, it is a finite direct sum of local Artinian rings. This result generalizes what we have seen in Chapters 2 and 3, where $A$ was a quotient of the polynomial ring $K[x_1, \ldots, x_n]$.

## 7.1 Noetherian and Artinian Modules

Let $(\Sigma, \leq)$ be a *poset*, *i.e.*, a non-empty set $\Sigma$ equipped with a partial order relation $\leq$. We recall that a chain of $\Sigma$ is a totally ordered subset of $\Sigma$.

**T. 7.1.** ($\rightarrow$ p. 241) The following conditions are equivalent:

1. every chain $\{s_\alpha\}_{\alpha \in \Lambda}$ of elements of $\Sigma$ is *stationary*, or *stabilizes*, *i.e.*, there exists $\alpha_0 \in \Lambda$ such that $s_\alpha = s_{\alpha_0}$ for all $\alpha \geq \alpha_0$;
2. every non-empty subset of $\Sigma$ has a maximal element with respect to $\leq$.

An important example to consider is the family $\Sigma$ of all ideals of a ring $A$. The natural order relations to be considered are the set containments $\subseteq$ and $\supseteq$. When the relation $\subseteq$ is used, condition 1 is referred to as *ascending chain condition* (a.c.c.), and when it is verified, we say that *$A$ satisfies a.c.c.*. When we use the relation $\supseteq$, condition 1 is called *descending chain condition* (d.c.c.), and when it is verified, we say that *$A$ satisfies d.c.c.*.

We can provide analogous definitions for the case of modules. Given an $A$-module $M$, we consider the posets $(\Sigma, \subseteq)$ and $(\Sigma, \supseteq)$ of all submodules of $M$. As before, we say that $M$ satisfies a.c.c. or d.c.c. depending on whether the order relation is $\subseteq$ or $\supseteq$.



A. Bandini et al., *Commutative Algebra through Exercises*, UNITEXT 159,
https://doi.org/10.1007/978-3-031-56910-4_7

Now, we extend to both rings and modules the notion of Noetherianity, already introduced for polynomial rings, see **T.2.13** and **T.2.14**.
Let $M$ be an $A$-module, and let $\Sigma$ be the family of submodules of $M$. The module $M$ is called *Noetherian* or *Artinian*, depending on whether $M$ satisfies a.c.c. or d.c.c., respectively.
A ring $A$ is said to be Noetherian or Artinian, if it is Noetherian or Artinian as an $A$-module.

**Noetherian Modules: First Properties**

**T. 7.2.** Let $I$ be an ideal of $A$, and let $M$, $N$, and $P$ be $A$-modules.

1. $M$ is Noetherian if and only if every submodule of $M$ is finitely generated.
2. Let $0 \longrightarrow N \xrightarrow{f} M \xrightarrow{g} P \longrightarrow 0$ be a short exact sequence. Then, $M$ is Noetherian if and only if $N$ and $P$ are Noetherian.
3. Let $M = \bigoplus_{i=1}^{n} M_i$. Then, $M$ is Noetherian if and only if $M_i$ is Noetherian for all $i$.
4. Let $A$ be Noetherian. Then, $A/I$ is Noetherian both as an $A$-module and as an $A/I$-module.
5. Let $A$ be Noetherian. Then, $M$ is Noetherian if and only if $M$ is finitely generated.
6. Let $I \subseteq \operatorname{Ann} M$. Then, $M$ is Noetherian as an $A$-module if and only if it is Noetherian as an $A/I$-module.
7. Let $M$ be Noetherian, and let $S \subset A$ be a multiplicative subset. Then, $S^{-1}M$ is Noetherian as an $S^{-1}A$-module.

**Proof T. 7.2.** 1. Assume that $M$ is Noetherian, and consider a submodule $N \subset M$. Let $\Sigma$ be the set of finitely generated submodules of $N$, ordered by $\subseteq$. Since $0 \in \Sigma$, the set $\Sigma$ is non-empty, and there exists a maximal element $N_0 \in \Sigma$ by **T.7.1**.

To prove that $N$ is finitely generated, it is sufficient to show that $N = N_0$. Assume, by contradiction, $N \neq N_0$, and let $n \in N \setminus N_0$. Then, we have $N_0 \subsetneq N_0 + \langle n \rangle \subseteq N$, where $N_0 + \langle n \rangle$ is finitely generated.
Thus, $N_0 + \langle n \rangle \in \Sigma$, which contradicts the maximality of $N_0$.

Conversely, assume $\{M_\alpha\}_{\alpha \in \Lambda}$ is an ascending chain of submodules of $M$. Then, $\widetilde{M} = \bigcup_\alpha M_\alpha$ is a submodule of $M$, and therefore, is finitely generated. Let $\{m_1, \ldots, m_n\}$ be a generating set of $\widetilde{M}$.
Then, there exists $\alpha_0 \in \Lambda$ such that

$$m_1, \ldots, m_n \in M_{\alpha_0} \subseteq \widetilde{M} = \langle m_1, \ldots, m_n \rangle.$$

Hence, the chain is stationary at $M_{\alpha_0}$.

2. Let $M$ be Noetherian, and let $\{N_\alpha\}_\alpha$ be an ascending chain of submodules of $N$. Then, $\{f(N_\alpha)\}_\alpha$ is an ascending chain in $M$, and hence, it is stationary. Since $f$ is injective, the original chain is also stationary in $N$, which implies that $N$ is Noetherian.
Now, consider an ascending chain $\{P_\alpha\}_\alpha$ of submodules of $P$.
The ascending chain $\{g^{-1}(P_\alpha)\}_\alpha$ of submodules of $M$ is stationary. Since $g$ is surjective, the chain $\{P_\alpha = g(g^{-1}(P_\alpha))\}_\alpha$ of submodules of $P$ is also stationary.

Conversely, assume $N$ and $P$ are Noetherian. Suppose $\{M_\alpha\}_\alpha$ is an ascending chain in $M$. Consider the chains $\{N_\alpha = f^{-1}(M_\alpha)\}_\alpha$ and $\{P_\alpha = g(M_\alpha)\}_\alpha$ in $N$ and $P$, respectively. Since both chains are stationary, there exists $\beta$ such that $N_\alpha = N_\beta$ and $P_\alpha = P_\beta$ for all $\alpha \geq \beta$.
Now, we show that

$$0 \longrightarrow N_\alpha \xrightarrow{f_\alpha} M_\alpha \xrightarrow{g_\alpha} P_\alpha \longrightarrow 0,$$

where $f_\alpha = f_{|N_\alpha}$ and $g_\alpha = g_{|M_\alpha}$, is exact for all $\alpha$.
Clearly, $f_\alpha$ is injective and $g_\alpha$ surjective. We only need to verify that the sequence is exact at $M_\alpha$.
It is clear that $g_\alpha \circ f_\alpha = 0$, hence $\operatorname{Im} f_\alpha \subseteq \operatorname{Ker} g_\alpha$.

To prove the opposite inclusion, let $m_\alpha \in \operatorname{Ker} g_\alpha \subseteq \operatorname{Ker} g = \operatorname{Im} f$. Then, there exists $n \in N$ such that $f(n) = m_\alpha$. Hence, $n \in N_\alpha$, as desired.
Finally, for all $\alpha \geq \beta$ consider the commutative diagram

$$\begin{array}{ccccccccc}
0 & \longrightarrow & N_\beta & \xrightarrow{f_\beta} & M_\beta & \xrightarrow{g_\beta} & P_\beta & \longrightarrow & 0 \\
 & & \downarrow{\scriptstyle \mathrm{id}_{N_\beta}} & & \downarrow{\scriptstyle i_\beta} & & \downarrow{\scriptstyle \mathrm{id}_{P_\beta}} & & \\
0 & \longrightarrow & N_\alpha & \xrightarrow[f_\alpha]{} & M_\alpha & \xrightarrow[g_\alpha]{} & P_\alpha & \longrightarrow & 0,
\end{array}$$

where $i_\beta$ is the inclusion homomorphism. To prove the claim, it suffices to apply **T.4.20** to this diagram.

3. Both implications follow from the previous part.
If $M = \bigoplus_{i=1}^n M_i$ is Noetherian, then every summand $M_i$ of $M$ is Noetherian.

Conversely, we observe that the sequence

$$0 \longrightarrow M_1 \longrightarrow M \longrightarrow \bigoplus_{i=2}^{n} M_i \longrightarrow 0$$

is exact, and then, proceed by induction on $n$.

4. Consider the exact sequence

$$0 \longrightarrow I \longrightarrow A \longrightarrow A/I \longrightarrow 0.$$

By part 2, $A$ is Noetherian implies that $A/I$ is a Noetherian $A$-module. Since $A/I$-submodules of $A/I$ are in one-to-one correspondence with ideals of $A$ containing $I$, it follows that ascending chains are stationary.
Therefore, $A/I$ is also Noetherian as an $A/I$-module.

5. If $M$ is Noetherian, then the conclusion follows directly from part 1.
Conversely, let $M = \langle m_1, \dots, m_n \rangle$ be finitely generated.
Consider the exact sequence

$$0 \longrightarrow \operatorname{Ker}\varphi \longrightarrow A^n \xrightarrow{\varphi} M \longrightarrow 0,$$

where $\varphi(e_i) = m_i$ for all $i$.
Since $A$ is Noetherian, $A^n$ and $M$ are also Noetherian, by parts 3 and 2.

6. We observe that $N \subseteq M$ implies $I \subseteq \operatorname{Ann} M \subseteq \operatorname{Ann} N$.
Hence, $N$ is an $A$-submodule of $M$ if and only if $N$ is an $A/I$-submodule of $M$, by **T.4.1**.

7. Recall that all submodules of $S^{-1}M$ are of the form $S^{-1}N$, for some submodule $N$ of $M$. Let $N = \langle n_1, \dots, n_r \rangle_A$ be a submodule of $M$.
Then,

$$S^{-1}N = \langle \tfrac{n_1}{1}, \dots, \tfrac{n_r}{1} \rangle_{S^{-1}A}.$$ □

Some properties that hold for Noetherian rings and modules have corresponding formulations for Artinian rings and modules, although the situation is not completely symmetric, see also **T.7.16**.

**T. 7.3.** (→ p. 241) Let $0 \longrightarrow N \longrightarrow M \longrightarrow P \longrightarrow 0$ be a short exact sequence of $A$-modules. Then, $M$ is Artinian if and only if $N$ and $P$ are Artinian.
In particular, if $A$ is Artinian and $I$ is an ideal of $A$, then $A/I$ is also Artinian.

## 7.2 Noetherian Rings and Primary Decomposition

We will prove now that a ring $A$ is Noetherian if and only if the polynomial ring $A[x_1, \dots, x_n]$ is Noetherian. This is a fundamental result, with one direction following from **T.7.2**.2 and the opposite direction proven below.

**Hilbert's Basis Theorem**

**T. 7.4.** If $A$ is a Noetherian ring, then $A[x_1, \dots, x_n]$ is also Noetherian.

**Proof T. 7.4.** The proof presented here is not constructive but it is more general than the one of **T.2.13**.
Since $A[x_1, \dots, x_n] = A[x_1, \dots, x_{n-1}][x_n]$, it is sufficient to prove the theorem for $n = 1$.
Assume, by contradiction, that there exists an ideal $I$ of $A[x]$ that is not finitely generated. Recursively define a sequence of elements of $I$ by setting

$f_0 = 0$ and selecting an element $f_i$ from $I \setminus (f_0, \ldots, f_{i-1})$ with minimal degree. This choice is always possible, otherwise $I$ would be finitely generated.
Let $\deg f_i = d_i$ and $a_i = \mathrm{lc}(f_i)$ for all $i \geq 1$. By construction, $d_i \leq d_{i+1}$ for every $i$.
Moreover, the ascending chain

$$(a_1) \subseteq \ldots \subseteq (a_1, \ldots, a_i) \subseteq \ldots$$

of ideals of $A$ is stationary by hypothesis.
Therefore, there exists an integer $k$ such that $a_{k+1} \in (a_1, \ldots, a_k)$, that is, $a_{k+1} = \sum_{i=1}^{k} b_i a_i$ for some $b_i \in A$.
Let now $g \in A[x]$ be the polynomial

$$g = f_{k+1} - \sum_{i=1}^{k} b_i x^{d_{k+1} - d_i} f_i \, .$$

Then, $g \in I \setminus (f_0, f_1, \ldots, f_k)$ and, by construction, $\deg g < d_{k+1} = \deg f_{k+1}$. This contradicts the choice of $f_{k+1}$. □

Fields, principal ideal domains, polynomial rings with coefficients in Noetherian rings, their finite direct sums, quotients, and localizations are all examples of Noetherian rings.

**Finiteness Theorem 1**

**T. 7.5.** (→ p. 241) Let $A$ be a Noetherian ring, and let $I$ be an ideal of $A$. Then, there exists a positive integer $k$ such that $\sqrt{I}^k \subseteq I$.

In $\mathbb{Z}$, the Fundamental Theorem of Arithmetic holds, and it can be easily rephrased in terms of ideals: every proper ideal can be expressed as a finite product of powers of distinct prime ideals, which in this case are primary ideals. The same holds in any principal ideal domain A: given a proper ideal $(a)$ of $A$, since $A$ is a UFD, we can factorize $a$, and obtain a decomposition of $(a)$ as a finite intersections of primary ideals by **E.8.25**.2 and **E.8.64**.
The results we are about to examine can be viewed as a generalization of these statements.
We remark that similar results also hold for submodules, see [2, Chapter 18].

An ideal $I$ of a ring $A$ is said to be *decomposable* if it can be expressed as a finite intersection of primary ideals of $A$. This expression is known as *primary decomposition of* $I$. A primary ideal appearing in a primary decomposition of $I$ is called *primary component of* $I$. The next result guarantees that in a Noetherian ring every ideal is decomposable.

**T. 7.6.** Let $A$ be a Noetherian ring, and let $I \subsetneq A$ be a proper ideal.

1. If $I$ is irreducible, then $I$ is primary.
2. $I$ is a finite intersection of irreducible ideals.
3. $I$ is decomposable.

**Proof T. 7.6.** 1. Assume $I$ is an irreducible ideal that is not primary. Then, there exist $a, b \in A$ such that $ab \in I$, with $a \notin \sqrt{I}$ and $b \notin I$.
We consider the ascending chain of ideals

$$I \subsetneq I : (a) \subseteq I : (a)^2 \subseteq \dots.$$

Since by hypothesis the chain is stationary, we have $I : (a)^n = I : (a)^{n+1}$ for some $n \in \mathbb{N}_+$. We want to prove that $I = (I, b) \cap (I, a^n)$. Since $I$ is irreducible by assumption, and properly contained in $(I, b)$ and $(I, a^n)$, we will obtain a contradiction.
It is clear that $I$ is contained in $(I, b) \cap (I, a^n)$.

To prove the opposite inclusion, let $c \in (I, b) \cap (I, a^n)$.
Then, we can write $c = i + a^n d \in (I, b)$, for some $i \in I$ and $d \in A$.
Therefore,

$$a^{n+1}d + ai = ac \in (aI, ab) \subseteq I,$$

hence $d \in I : (a)^{n+1} = I : (a)^n$, which yields $c \in I$.

2. Let $\Sigma$ be the set of all proper ideals of $A$ that cannot be expressed as a finite intersection of irreducible ideals.
Assume, by contradiction, that $\Sigma$ is non-empty. Since $A$ is Noetherian, $\Sigma$ has maximal elements, which are not irreducible. Hence, if $J$ is a maximal element, then it can be written as $J_1 \cap J_2$ with $J \subsetneq J_1$, and $J \subsetneq J_2$.
Moreover, both $J_1$ and $J_2$ are not in $\Sigma$. Therefore, both $J_1$ and $J_2$ can be expressed as a finite intersection of irreducible ideals. Thus, $J$ also has such a decomposition.
This contradicts the assumption that $J \in \Sigma$.

3. This follows directly from parts 1 and 2. □

A primary decomposition is not necessarily unique. To achieve some level of uniqueness, we first reduce the decomposition to a minimal one.

**Minimal Primary Decomposition**

A primary decomposition $I = \bigcap_{i=1}^{t} \mathfrak{q}_i$ is called *minimal* if it verifies:

i) $\mathfrak{p}_i = \sqrt{\mathfrak{q}_i} \neq \sqrt{\mathfrak{q}_j} = \mathfrak{p}_j$ for all $i \neq j$;
ii) $\mathfrak{q}_i \not\supseteq \bigcap_{j \neq i} \mathfrak{q}_j$ for all $i$.

As an initial step, we can remove from the collection of primary ideals any ideal that contains the intersection of the remaining ones. This eliminates redundant primary ideals and gives us a decomposition that satisfies ii).
We say that an ideal $I$ of a ring $A$ is $\mathfrak{p}$-*primary* if it is primary and $\mathfrak{p}$ is its radical, see **T.1.7**.1.
To also satisfy condition i), we use **E.8.63**, which allows us to replace all $\mathfrak{p}$-primary ideals with their intersection.

The next simple result is very useful.

**T. 7.7.** ($\rightarrow$ p. 241) Let $a \in A$, and let $\mathfrak{q}$ be a $\mathfrak{p}$-primary ideal of $A$.

1. If $a \in \mathfrak{q}$, then $\mathfrak{q} : (a) = A$.
2. If $a \notin \mathfrak{q}$, then $\mathfrak{q} : (a)$ is $\mathfrak{p}$-primary.
3. If $a \notin \mathfrak{p}$, then $\mathfrak{q} : (a) = \mathfrak{q}$.

Given a minimal primary decomposition of an ideal $I = \bigcap_{i=1}^{n} \mathfrak{q}_i$, where $\mathfrak{q}_i$ are $\mathfrak{p}_i$-primary ideals, we can prove that the prime ideals $\mathfrak{p}_i$ are independent of the chosen decomposition and can be characterized as follows.

**Uniqueness Theorem 1**

**T. 7.8.** Let $I = \bigcap_{i=1}^{t} \mathfrak{q}_i$ be a minimal primary decomposition of an ideal $I \subseteq A$. Then,

$$\left\{\mathfrak{p}_i \in \operatorname{Spec} A\colon\ \mathfrak{p}_i = \sqrt{\mathfrak{q}_i}\right\} = \left\{\sqrt{I : (a)}:\ a \in A\right\} \cap \operatorname{Spec} A.$$

**Proof T. 7.8.** Note that

$$I : (a) = \left(\bigcap_{i=1}^{t} \mathfrak{q}_i\right) : (a) = \bigcap_{i=1}^{t} (\mathfrak{q}_i : (a))$$

and, if $a \in \bigcap_{i=1}^{t} \mathfrak{q}_i$, then $\sqrt{I : (a)} = A$ is not prime.
We show now the two inclusions. Suppose that $a \in A$ and $\sqrt{I : (a)}$ is prime. We can write

$$\sqrt{I : (a)} = \sqrt{\bigcap_{i=1}^{t} (\mathfrak{q}_i : (a))} = \bigcap_{i=1}^{t} \sqrt{\mathfrak{q}_i : (a)} = \bigcap_{i\colon a \notin \mathfrak{q}_i} \mathfrak{p}_i,$$

where the last equality follows from **T.7.7**.
Since $\sqrt{I : (a)}$ is prime, there exists an index $j$ such that $\sqrt{I : (a)} = \mathfrak{p}_j$.

To prove the opposite inclusion, we will find for any $\mathfrak{p}_i = \sqrt{\mathfrak{q}_i}$ an element $a_i \in A$ such that $\mathfrak{p}_i = \sqrt{I : (a_i)}$. Since the given primary decomposition of $I$ is minimal, we can choose $a_i \in \bigcap_{j \neq i} \mathfrak{q}_j \setminus \mathfrak{q}_i$.

Then, we have

$$\sqrt{I:(a_i)} = \bigcap_{j=1}^{t} \sqrt{\mathfrak{q}_j:(a_i)} = \bigcap_{j\neq i} \sqrt{\mathfrak{q}_j:(a_i)} \cap \sqrt{\mathfrak{q}_i:(a_i)} = \mathfrak{p}_i,$$

where the last equality is obtained again using **T.7.7**. □

Note that the previous result holds for all decomposable ideals $I$, even if $A$ is not Noetherian.

**T. 7.9.** (→ p. 241) Using the same notation as in **T.7.8**, if $A$ is Noetherian, then

$$\{\mathfrak{p}_i \in \operatorname{Spec} A\colon\ \mathfrak{p}_i = \sqrt{\mathfrak{q}_i}\} = \{I:(a):\ a \in A\} \cap \operatorname{Spec} A.$$

We call the prime ideals $\mathfrak{p}_i$ the *associated primes of* $I$, and we denote the set of these primes by $\operatorname{Ass} I$. The minimal elements of $\operatorname{Ass} I$ are called the *minimal primes of* $I$. The associated primes which are not minimal are called *embedded primes of* $I$.

This terminology has its origin in geometry. For an ideal $I \subset K[x_1, \dots, x_n]$, where $K = \overline{K}$, a primary decomposition of $I$ induces a decomposition of $\mathbb{V}(I)$ as a union of irreducible varieties, as discussed in **T.3.4**.

The minimal primes of $I$ correspond to the subvarieties in a minimal decomposition of $\mathbb{V}(I)$, while the embedded primes correspond to the *embedded* subvarieties, *i.e.*, contained in one of those appearing in a minimal decomposition of $\mathbb{V}(I)$.

In fact, if $\mathfrak{q}_i$ is such that $\sqrt{\mathfrak{q}_i} = \mathfrak{p}_i \in \operatorname{Ass} I$, and $\mathfrak{p}_i$ is not a minimal prime of $I$, then there exists a minimal prime $\mathfrak{p}_j$ of $I$ such that $\mathfrak{p}_j \subsetneq \mathfrak{p}_i$, and consequently,

$$\mathbb{V}(\mathfrak{q}_i) = \mathbb{V}(\mathfrak{p}_i) \subsetneq \mathbb{V}(\mathfrak{p}_j) = \mathbb{V}(\mathfrak{q}_j).$$

**T. 7.10.** (→ p. 242) 1. Let $I$ be a decomposable ideal. Then, the set of minimal primes of $I$ is $\operatorname{Min} I$, see **E.8.75**.

2. [**Finiteness Theorem 2**] The set of minimal primes of a Noetherian ring is finite.
3. Let $(0) = \bigcap_{i=1}^{t} \mathfrak{q}_i$ be a minimal primary decomposition of the zero ideal of $A$, with $\sqrt{\mathfrak{q}_i} = \mathfrak{p}_i$. Then,

$$\mathcal{N}(A) = \bigcap_{\mathfrak{p}_i \in \operatorname{Min}(0)} \mathfrak{p}_i \quad \text{and} \quad \mathcal{D}(A) = \bigcup_{\mathfrak{p}_i \in \operatorname{Ass}(0)} \mathfrak{p}_i.$$

Part 1 of the previous result also provides justification for the notation $\operatorname{Min} I$.

**T. 7.11.** (→ p. 242) Let $S$ be a multiplicative subset of $A$, and let $\mathfrak{q}$ be a $\mathfrak{p}$-primary ideal of $A$.

1. The ideal $S^{-1}\mathfrak{q}$ is $S^{-1}\mathfrak{p}$-primary and $(S^{-1}\mathfrak{q})^c = \mathfrak{q}$, if $S \cap \mathfrak{p} = \emptyset$, and $S^{-1}\mathfrak{q} = S^{-1}A$, otherwise.
2. Let $I = \bigcap_{i=1}^t \mathfrak{q}_i$ be a minimal primary decomposition of $I$, with $\sqrt{\mathfrak{q}_i} = \mathfrak{p}_i$. Let also $k \leq t$ be such that $\mathfrak{p}_i \cap S = \emptyset$ for all $i = 1, \ldots, k$ and $\mathfrak{p}_i \cap S \neq \emptyset$, otherwise. Then,
$$(S^{-1}I)^c = \bigcap_{i=1}^k \mathfrak{q}_i.$$
3. Let $S = A \setminus \bigcup_{\mathfrak{p} \in \operatorname{Min} I} \mathfrak{p}$. Then,
$$(S^{-1}I)^c = \bigcap_{\sqrt{\mathfrak{q}} \in \operatorname{Min} I} \mathfrak{q}.$$

Let $I$ be a decomposable ideal of a ring $A$, and let $I = \bigcap_{i=1}^t \mathfrak{q}_i$ be a minimal primary decomposition. After reordering indices if necessary, we can write

$$I = \mathfrak{q}_1 \cap \ldots \cap \mathfrak{q}_k \cap \mathfrak{q}_{k+1} \cap \ldots \cap \mathfrak{q}_t,$$

with $\operatorname{Ass} I = \{\mathfrak{p}_i = \sqrt{\mathfrak{q}_i} : i = 1, \ldots, t\}$, and $\operatorname{Min} I = \{\mathfrak{p}_1, \ldots, \mathfrak{p}_k\}$ with $k \leq t$.

**Uniqueness Theorem 2**

**T. 7.12.** (→ p. 242) Let $I = \bigcap_{i=1}^t \mathfrak{q}_i$ be a minimal primary decomposition of the ideal $I \subseteq A$, as written above. Then, the primary ideals $\mathfrak{q}_1, \ldots, \mathfrak{q}_k$, whose radical is a minimal prime, are uniquely determined.
In particular, for all $i = 1, \ldots, k$,

$$\mathfrak{q}_i = (IA_{\mathfrak{p}_i})^c.$$

## 7.3 Artinian Rings

We will now present some of the key properties of Artinian rings, which are rings satisfying d.c.c..

**T. 7.13.** Let $A$ be an Artinian ring. Then:

1. $\operatorname{Spec} A = \operatorname{Max} A$, *i.e.*, all prime ideals of $A$ are maximal;
2. $A$ has a finite number of maximal ideals;
3. $\mathcal{N}(A)$ is nilpotent, *i.e.*, there exists $k \in \mathbb{N}$ such that $\mathcal{N}(A)^k = 0$.

**Proof T. 7.13.** 1. Let $\mathfrak{p} \subset A$ be a prime ideal, and let $B = A/\mathfrak{p}$.
We show that the domain $B$ is actually a field.
Consider a non-zero element $b \in B$ and the descending chain of ideals

$$(b) \supseteq (b^2) \supseteq \dots.$$

By hypothesis, there exists $k \in \mathbb{N}$ such that $(b^k) = (b^{k+1})$. Therefore, $b^k \in (b^{k+1})$, and there exists $c \in B$ such that $b^k = cb^{k+1}$.
Hence, $b^k(1 - cb) = 0$ and since $B$ is a domain, $cb = 1$.

2. Consider the family $\Sigma$ of all ideals of $A$ that can be expressed as an intersection of a finite number of maximal ideals. Such a family is non-empty (why?), thus, there exists a minimal element $I_0 = \bigcap_{i=1}^k \mathfrak{m}_i \in \Sigma$, see **T.7.1**.
For every maximal ideal $\mathfrak{m}$ of $A$, we have $I_0 \cap \mathfrak{m} = I_0$ by the minimality of $I_0$.
Therefore, $\mathfrak{m} \supseteq \bigcap_{i=1}^k \mathfrak{m}_i$, and hence, $\mathfrak{m} \supseteq \mathfrak{m}_i$ for some $i$, see **T.1.12**.2. Since $\mathfrak{m}_i$ is maximal, we have $\mathfrak{m} = \mathfrak{m}_i$.
Thus, $\operatorname{Max} A = \{\mathfrak{m}_1, \dots, \mathfrak{m}_k\}$.

3. By d.c.c., the descending chain

$$\mathcal{N}(A) \supseteq \mathcal{N}(A)^2 \supseteq \dots$$

is stationary, *i.e.*, there exists an integer $k$ such that $\mathcal{N}(A)^k = \mathcal{N}(A)^{k+1}$.
Let $I = \mathcal{N}(A)^k$. We assume, by contradiction, that $I \neq 0$.
Then, the family $\Sigma$ of all ideals $J$ such that $JI \neq 0$, is non-empty (why?), and thus, has a minimal element $J_0$. Hence, there exists $b \in J_0 \setminus \{0\}$ such that $bI \neq 0$ and $(b) \subseteq J_0$. Therefore, $J_0 = (b)$ by the minimality of $J_0$.
Since $(b)I \in \Sigma$, and

$$((b)I)I = (b)I^2 = (b)I \neq 0,$$

the minimality of $(b)$ implies $(b)I = (b)$.
Hence, $b = bc$, with $c \in I$, from which it follows

$$b = bc = bc^2 = \dots = bc^h = \dots.$$

Since $c \in I$ is nilpotent, there exists $s \in \mathbb{N}_+$ such that $c^s = 0$, and we have $b = bc^s = 0$. This contradicts the hypothesis on $b$, and hence, $I = 0$. □

**T. 7.14.** (→ p. 243) Let $(A, \mathfrak{m})$ be an Artinian local ring. Then, every element of $A$ is either invertible or nilpotent.

**T. 7.15.** Let $\mathfrak{m}_1, \dots, \mathfrak{m}_t$ be not necessarily distinct maximal ideals of a ring $A$. Then,

$$A/\prod_{i=1}^t \mathfrak{m}_i \text{ is Artinian if and only if is Noetherian.}$$

In particular, if $\prod_{i=1}^t \mathfrak{m}_i = 0$, then $A$ is Artinian if and only if is Noetherian.

**Proof T. 7.15.** We prove the assertion by induction on $t$.
If $t = 1$, then $A/\mathfrak{m}_1$ is a field, which is both Artinian and Noetherian.
Let $t > 1$. Consider $N = \left(\prod_{i=1}^{t-1} \mathfrak{m}_i\right) / \left(\prod_{i=1}^{t} \mathfrak{m}_i\right)$ and the short exact sequence

$$0 \longrightarrow N \longrightarrow A/\prod_{i=1}^{t} \mathfrak{m}_i \longrightarrow A/\prod_{i=1}^{t-1} \mathfrak{m}_i \longrightarrow 0.$$

By **T.4.1**, $N$ is an $A/\mathfrak{m}_t$-vector space, hence Artinian if and only if Noetherian by **E.13.2**. Moreover, $A/\prod_{i=1}^{t-1} \mathfrak{m}_i$ is Artinian if and only if is Noetherian, by the inductive hypothesis.
Hence, the conclusion follows from **T.7.2**.2 and **T.7.3**. □

**Characterization of Artinian Rings**

**T. 7.16.** (→ p. 243) A ring is Artinian if and only if is Noetherian and has dimension 0.

The last statement generalizes the discussion after the proof of **T.3.18**.

**Structure Theorem of Artinian Rings**

**T. 7.17.** A ring $A$ is Artinian if and only if it is isomorphic to a finite direct sum of Artinian local rings.

**Proof T. 7.17.** In the proof of **T.7.16** we have shown that $(0) = \prod_{i=1}^{s} \mathfrak{m}_i^k$ is a product of powers of distinct maximal ideals. Since these powers are pairwise comaximal, by the Chinese Remainder Theorem, we obtain

$$A \simeq A/(0) \simeq A/\prod_{i=1}^{s} \mathfrak{m}_i^k \simeq \prod_{i=1}^{s} A/\mathfrak{m}_i^k,$$

where the rings $A/\mathfrak{m}_i^k$ are local and Artinian.

The converse is a straightforward application of **T.7.3**. □

# Part II
# Exercises

# Chapter 8
# Rings and Ideals

**E. 8.1.** (→ p. 245) Let $S$ be a subset of a ring $A$.
Prove that $(S)$ is the intersection of all ideals of $A$ containing $S$.

**E. 8.2.** (→ p. 245) Let $A$ be a ring.
Prove that the following are equivalent:

1. $A$ is a domain;
2. [**Cancellation Law**] for any $a$, $b$, $c \in A$, with $a \neq 0$,
$$ab = ac \iff b = c;$$
3. $A \setminus \{0\}$ is closed under multiplication.

**E. 8.3.** (→ p. 245) Let $A = \mathbb{Z}/(n)$, with $n \neq 0, \pm 1$.

1. Let $n = 24$. Describe $\mathcal{D}(A)$, $A^*$, $\operatorname{Spec} A$, and $\operatorname{Max} A$.
2. Same task for $n = 17$.
3. For which values of $n$ is the ring $A$ a domain?
For which values of $n$ is the ring $A$ a field?

**E. 8.4.** (→ p. 246) Let $a \in A$ be nilpotent.
Prove:

1. $1 - a$ is invertible in $A$;
2. the sum of an invertible element and of a nilpotent is invertible.

**E. 8.5.** (→ p. 246) Let $f(x) = \sum_{i=0}^{n} a_i x^i$ be a polynomial of $A[x]$.
Prove:

1. $f$ is invertible if and only if $a_0$ is invertible and $a_1, \ldots, a_n$ are nilpotent;
2. $f$ is nilpotent if and only if $a_0, \ldots, a_n$ are nilpotent;
3. $f$ is a zero-divisor if and only if there exists $0 \neq a \in A$ such that $af = 0$.



A. Bandini et al., *Commutative Algebra through Exercises*, UNITEXT 159,
https://doi.org/10.1007/978-3-031-56910-4_8

**E. 8.6.** (→ p. 247) Prove that, for any ring $A$,

$$\mathcal{J}(A[x]) = \mathcal{N}(A[x]).$$

**E. 8.7.** (→ p. 247) Let $I \subset A$ be an ideal.
Prove that if $\bigcap_{n\in\mathbb{N}} I^n = 0$, then $1 + a \notin \mathcal{D}(A)$ for all $a \in I$.

**E. 8.8.** (→ p. 247) Let $f\colon A \longrightarrow B$ be a ring homomorphism.
Prove:

1. $\operatorname{Ker} f$ is an ideal of $A$;
2. $f$ is injective if and only if $\operatorname{Ker} f = (0)$;
3. $\operatorname{Im} f$ is a subring of $B$.

**E. 8.9.** (→ p. 247) Prove that there exists a unique ring homomorphism $f\colon \mathbb{Z} \longrightarrow \mathbb{Z}$.

**E. 8.10.** (→ p. 248) Let $A$ be a ring, let $I \subseteq A$ be an ideal, and fix $a \in A$.
Consider the set

$$J = \{f \in A[x]\colon\ f(a) \in I\}.$$

1. Prove that $J$ is an ideal.
2. Prove that $J$ is prime if and only if $I$ is prime.
3. Set $A = \mathbb{Q}[y],\ a = y - 1$, and $I = (y - 2)$.
   Describe $J$.

**E. 8.11.** (→ p. 248) Let $A \neq 0$ be a ring.
Prove that the following are equivalent:

1. $A$ is a field;
2. the only ideals of $A$ are (0) and (1);
3. every homomorphism $A \longrightarrow B$, where $B$ is a non-trivial ring, is injective.

**E. 8.12.** (→ p. 248) Let $A$ be a ring, and let $I$ be an ideal of $A$.
Prove that the set

$$I[x] = \Big\{f(x) = \sum_i a_i x^i \in A[x]\colon\ a_i \in I \text{ for all } i\Big\},$$

of all polynomials of $A[x]$ with coefficients in $I$ is an ideal of $A[x]$.
Moreover, prove that

$$A[x]/I[x] \simeq (A/I)\,[x].$$

**E. 8.13.** (→ p. 248) [**Gauss' Lemma**] Let $f = \sum_{i=0}^{n} f_i x^i \in A[x]$.
We say that $f$ is *primitive* if and only if $(f_0, \ldots, f_n) = (1)$.
Prove that $f,\ g \in A[x]$ are primitive if and only if $fg$ is primitive.

**E. 8.14.** (→ p. 249) Let $A$ be a ring that satisfies the following conditions:

i) the Jacobson radical $\mathcal{J}(A)$ is a non-zero prime ideal;
ii) every ideal $I \supseteq \mathcal{J}(A)$ is principal;
iii) $\mathcal{D}(A) \subseteq \mathcal{J}(A)$.

Prove that $A$ is a local ring with maximal ideal $\mathcal{J}(A)$.

**E. 8.15.** (→ p. 249) Let $A$ be a local ring with principal maximal ideal $\mathfrak{m} = (m)$.
Prove:

1. for any $0 \neq a, b \in \mathfrak{m}$, we have $(a) = (b) \iff a = bu$ for some $u \in A^*$;
2. if $\mathfrak{m} \neq (0)$, then $m$ is an irreducible element of $A$.

**E. 8.16.** (→ p. 249) Let $I$ and $J$ be ideals of a ring $A$.
Prove that if $I \subseteq \mathcal{J}(A)$ and $(I, J) = (1)$, then $J = (1)$.

**E. 8.17.** (→ p. 249) Prove that in a local ring the only idempotent elements are 0 and 1.

**E. 8.18.** (→ p. 250) Let $A$ be a Boolean ring.
Prove:

1. $2a = 0$ for all $a \in A$;
2. every prime ideal $\mathfrak{p}$ is maximal, and $A/\mathfrak{p}$ is a field with two elements;
3. every finitely generated ideal is principal.

**E. 8.19.** (→ p. 250) Prove that if every proper ideal of $A$ is prime, then $A$ is a field.

**E. 8.20.** (→ p. 250) [**Operations in $\mathbb{Z}$**] Let $I = (m)$, $J = (n)$, and $H = (h)$ be ideals of $\mathbb{Z}$.
Prove:

1. $I + J = (\gcd(m, n))$;
2. $I \cap J = (\operatorname{lcm}(m, n))$;
3. $IJ = (mn)$;
4. $I : J = (m/\gcd(m, n))$;
5. $I \cap (J + H) = (I \cap J) + (I \cap H)$;
6. $(I + J)(I \cap J) = IJ$.

**E. 8.21.** (→ p. 251) [**Properties of Ideal Quotients**] Let $I$, $J$, $H$, and $I_\alpha$ be ideals of a ring $A$, where $\alpha$ varies in a set of indices $\Lambda$.
Prove:

1. $I \subseteq I : J$;

2. $(I : J)J \subseteq I$;
3. $(I : J) : H = I : JH = (I : H) : J$;
4. $\left(\bigcap_{\alpha\in\Lambda} I_\alpha\right) : J = \bigcap_{\alpha\in\Lambda}(I_\alpha : J)$;
5. $I : \sum_{\alpha\in\Lambda} I_\alpha = \bigcap_{\alpha\in\Lambda} I : I_\alpha$.

**E. 8.22.** (→ p. 251) Let $I \subset A$ be an ideal, and let $f \in A$.
Prove that

$$I : (f) = \frac{1}{f}(I \cap (f)).$$

**E. 8.23.** (→ p. 251) Let $I \subset A$ be an ideal, and let $f, g \in A$.
Prove that

$$(I, f, g) = (1) \implies (I, f) \cap (I, g) = (I, fg).$$

**E. 8.24.** (→ p. 251) Let $I$, $J$, and $H$ be ideals of a ring $A$.
Prove:

1. if $I + H = A$ and $J + H = A$, then $I \cap J + H^n = A$ for every $n \in \mathbb{N}$;
2. if $I \subseteq H$, $I \cap J = H \cap J$, and $I/(I \cap J) = H/(H \cap J)$, then $I = H$.

**E. 8.25.** (→ p. 252) Let $I$, $J$, $H_1, \ldots, H_k \subseteq A$ be ideals.
Prove:

1. if $I + H_i = A$ for every $i$, then $I + H_1H_2 \cdots H_k = A$;
2. if $I + J = A$, then $I^m + J^n = A$ for all $n$, $m \in \mathbb{N}$.

**E. 8.26.** (→ p. 252) Let $I$ and $J$ be ideals of $A$.
Prove:

1. if $\sqrt{IJ} = A$, then $I = A$ and $J = A$;
2. if $\mathfrak{p} \subset A$ is a prime ideal such that $IJ = \mathfrak{p}$, then either $I = \mathfrak{p}$ or $J = \mathfrak{p}$.

**E. 8.27.** (→ p. 252) Let $I$, $J \subseteq A$ be ideals.
Prove:

1. $I + J = (1)$ if and only if $\sqrt{I} + \sqrt{J} = (1)$;
2. $\sqrt{I + \sqrt{J}} = \sqrt{I + J}$.

**E. 8.28.** (→ p. 252) Provide an example where

$$\sqrt{I} + \sqrt{J} \neq \sqrt{I + J}.$$

**E. 8.29.** (→ p. 252) Let $A = K[x, y]$ and $I = (x^2, xy)$.
Prove that $\sqrt{I}$ is prime and $I$ is not primary.

**E. 8.30.** (→ p. 253) Let $A$ be a ring satisfying the following property: every ideal $I \not\subseteq \mathcal{N}(A)$ contains a non-zero idempotent.
Prove that $\mathcal{J}(A) = \mathcal{N}(A)$.

**E. 8.31.** (→ p. 253) Let $A$ be a ring.
Prove that the following are equivalent:

1. $A$ has a unique prime ideal;
2. every element of $A$ is either invertible or nilpotent;
3. $A/\mathcal{N}(A)$ is a field.

**E. 8.32.** (→ p. 253) The *radical* of a subset $E$ of a ring $A$ is defined as

$$\sqrt{E} = \{a \in A\colon\ a^n \in E \text{ for some } n \in \mathbb{N}\}.$$

Let $\{E_\alpha\}_{\alpha \in \Lambda}$ be a family of subsets of $A$.
Prove that

$$\sqrt{\bigcup_\alpha E_\alpha} = \bigcup_\alpha \sqrt{E_\alpha}.$$

**E. 8.33.** (→ p. 253) Prove that

$$\mathcal{D}(A) = \bigcup_{0 \neq a \in A} \sqrt{\operatorname{Ann} a}$$

holds in any ring $A$.

**E. 8.34.** (→ p. 253) Let $A$ be a ring.
Prove:

1. $a \in A$ is invertible if and only if $\overline{a}$ is invertible in $A/\mathcal{J}(A)$;
2. if $a \in \mathcal{J}(A)$ is idempotent, then $a = 0$.

**E. 8.35.** (→ p. 254) Let $A$ be a ring, and let $a \in \mathcal{J}(A)$.
Prove that if $a$ is idempotent modulo an ideal $I \subset A$, then $a \in I$.

**E. 8.36.** (→ p. 254) Let $A$ be an infinite domain with only finitely many invertible elements.
Prove that $A$ has infinitely many maximal ideals.

**E. 8.37.** (→ p. 254) Let $A$ and $B$ be rings.
Describe all ideals, prime ideals, and maximal ideals of $A \times B$.

**E. 8.38.** (→ p. 255) Prove that every ideal of $A = \prod_{i=1}^n A_i$ is of the form

$$I = \prod_{i=1}^n I_i,$$

where $I_i$ is an ideal of the ring $A_i$ for each $i$.
Describe all prime and all maximal ideals of $A$.

**E. 8.39.** (→ p. 255) 1. A ring $A$ is finite direct product of fields if and only if it contains only finitely many ideals and $\mathcal{J}(A) = (0)$.

2. A finite ring $A$ is a direct product of fields if and only if $\mathcal{N}(A) = (0)$.

**E. 8.40.** (→ p. 255) Let $A$ be the ring $\mathbb{Z} \times \mathbb{Z}/(36) \times \mathbb{Q}$.
Find:

1. the nilradical of $A$;
2. the idempotent elements of $A$;
3. the ideals of $A$, and verify whether they are principal or not;
4. all prime and all maximal ideals of $A$.

**E. 8.41.** (→ p. 256) Let $p$, $p_1, \dots, p_n$ be distinct primes of $\mathbb{Z}$.

1. Prove that the ring

$$A_{(p)} = \left\{ \frac{a}{b} \in \mathbb{Q} \colon b \not\equiv 0 \mod p \right\}$$

is local, and describe its maximal ideal and residue field.

2. Prove that the ring

$$A_{(p_1, \dots, p_n)} = \left\{ \frac{a}{b} \in \mathbb{Q} \colon b \not\equiv 0 \mod p_j \,,\ 1 \leq j \leq n \right\}$$

is semilocal, and describe its maximal ideals.

**E. 8.42.** (→ p. 257) 1. Prove that the direct product of finitely many local rings is a semilocal ring.

2. Prove that the direct product of finitely many semilocal rings is semilocal.
3. Does the converse of part 1 also holds?

**E. 8.43.** (→ p. 257) Let $A$ be a ring satisfying the following property: for any $a \in A$, there exists $n > 1$ such that $a^n = a$.
Prove that every prime ideal of $A$ is maximal.

**E. 8.44.** (→ p. 257) [**$\mathcal{D}(A)$ is a Union of Primes**] Let $A$ be a ring, and let

$$\Sigma = \{I \subset A \colon I \text{ ideal}, I \subseteq \mathcal{D}(A)\}$$

be partially ordered by set inclusion $\subseteq$.
Prove:

1. $\Sigma$ has maximal elements and each of them is a prime ideal of $A$;

2. $\mathcal{D}(A)$ is a union of prime ideals.

**E. 8.45.** ($\rightarrow$ p. 258) Let $A$ be a ring where all prime ideals are principal. Prove that $A$ is a PIR.

**E. 8.46.** ($\rightarrow$ p. 258) [**Properties of Extension and Contraction of Ideals**] Let $f\colon A \longrightarrow B$ be a ring homomorphism, and let $I_1, I_2 \subset A$ and $J_1, J_2 \subset B$ be ideals.
Prove:

1. $(I_1 + I_2)^e = I_1^e + I_2^e$;
2. $(I_1 I_2)^e = I_1^e I_2^e$;
3. $(I_1 \cap I_2)^e \subseteq I_1^e \cap I_2^e$;
4. $(J_1 + J_2)^c \supseteq J_1^c + J_2^c$;
5. $(J_1 J_2)^c \supseteq J_1^c J_2^c$;
6. $(J_1 \cap J_2)^c = J_1^c \cap J_2^c$.

Provide examples to show that the inclusions in parts 3, 4 and 5 may be strict.

**E. 8.47.** ($\rightarrow$ p. 259) Let $i\colon A \longrightarrow A[x]$ be the inclusion homomorphism, and let $I[x] = \{f(x) = \sum_j a_j x^j \in A[x]\colon\ a_j \in I \text{ for all } j\}$, see **E.8.12**.
Prove:

1. $I[x]$ is the extension of $I$ with respect to $i$;
2. $I$ is prime if and only if $I[x]$ is prime.
3. Is it true that when $I$ is maximal, $I[x]$ is also maximal?

**E. 8.48.** ($\rightarrow$ p. 259) Let $f\colon A \longrightarrow B$ be a ring homomorphism, and let $I \subset A$ be an ideal.
Prove:

1. $(f(\sqrt{I})) \subseteq \sqrt{(f(I))}$, that is, $(\sqrt{I}\,)^e \subseteq \sqrt{I^e}$;
2. if $f$ is surjective and $\operatorname{Ker} f \subseteq I$, then $(\sqrt{I}\,)^e = \sqrt{I^e}$;
3. if $J$ is an ideal of $B$, then $\sqrt{J^c} = (\sqrt{J}\,)^c$.

**E. 8.49.** ($\rightarrow$ p. 260) Consider the inclusion homomorphism $\mathbb{Z} \longrightarrow \mathbb{Z}[i]$. Find all non-zero primes $p \in \mathbb{Z}$ such that $(p)^e$ is a prime ideal.

**E. 8.50.** ($\rightarrow$ p. 260) For any odd prime $p$, let $\zeta_p$ be a primitive $p$-th root of unity, and consider the inclusion homomorphism $\mathbb{Z} \longrightarrow \mathbb{Z}[\zeta_p]$.
Prove that

$$\frac{1-\zeta_p^a}{1-\zeta_p} \in \mathbb{Z}[\zeta_p]^* \quad \text{for all } a = 1, \ldots, p-1.$$

As a consequence, prove that in $\mathbb{Z}[\zeta_p]$ we have

$$(p)^e = (1-\zeta_p)^{p-1}.$$

**E. 8.51.** (→ p. 261) Let $f\colon A \longrightarrow B$ be a ring homomorphism.
Prove:

1. $f(\mathcal{N}(A)) \subseteq \mathcal{N}(B)$;
2. if $f$ is surjective, then $f(\mathcal{J}(A)) \subseteq \mathcal{J}(B)$;
3. if $f$ is not surjective, then the previous statement does not hold in general;
4. the inclusion in part 2 may be strict;
5. if $f$ is surjective and $A$ is semilocal, then $f(\mathcal{J}(A)) = \mathcal{J}(B)$.

**E. 8.52.** (→ p. 261) Let $A$ be a ring, and let $I \subset \mathcal{N}(A)$ be an ideal.
Prove that $A$ is local if and only if $A/I$ is local.

**E. 8.53.** (→ p. 262) Let $A$ be a ring such that $\mathcal{D}(A) \subseteq \mathcal{J}(A)$.
Prove that two elements $a,\, b \in A$ such that $(a) = (b)$ are associate.

**E. 8.54.** (→ p. 262) 1. Let $A$ be a PID.
Prove that if $d = \gcd(a,b)$, then there exist $u,\, v \in A$ such that $ua+vb = d$.
In particular, the equality $(a,b) = (\gcd(a,b))$ holds.

2. Find an example of a ring $A$ in which the previous property does not hold.

**E. 8.55.** (→ p. 262) Let $A$ be a PID, and let $I,\, J \subset A$ be ideals.
Prove that $(I+J)^2 = I^2 + J^2$.

**E. 8.56.** (→ p. 262) Let $A$ be a PIR.
Prove that if $\mathcal{J}(A) = \mathcal{D}(A) \neq (0)$, then $A$ is a local ring.

**E. 8.57.** (→ p. 262) Let $A$ be a PID.

1. Prove that $a \in A$ is irreducible if and only if $A/(a)$ is a field.
2. Prove that the maximal ideals of $A$ are those generated by irreducible elements.
3. Set $A = K[x]$, and let $f \in A$ be a polynomial of positive degree.
   Prove that $(f)$ is prime if and only if $f$ is irreducible.

**E. 8.58.** (→ p. 262) Let $K$ be a field, and let $f \in K[x] \setminus \{0\}$ be a monic polynomial. We say that $f$ is *squarefree* if all irreducible elements appearing in its factorization are distinct.
Prove that the ring $A = K[x]/(f)$ is reduced if and only if $f$ is squarefree.

**E. 8.59.** (→ p. 263) We say that a field $K$ is *perfect* if $\operatorname{char} K = 0$ or

$$\operatorname{char} K = p > 0 \text{ and } K = K^{(p)}, \quad \text{where } K^{(p)} = \{\alpha^p \colon \alpha \in K\}.$$

Let $K$ be a perfect field, let $f \in K[x] \setminus \{0\}$ be a monic polynomial, and denote by $f'$ its derivative.

1. Prove that $f$ is squarefree if and only if $\gcd(f, f') = 1$.
2. Let $K \subseteq L$ be a field extension.
   Prove that $f$ is squarefree in $K[x]$ if and only if $f$ is squarefree in $L[x]$.

In particular, the above statements hold when $K$ is finite or algebraically closed.

**E. 8.60.** (→ p. 263) Let $r \leq n$ be positive integers, and let $K$ be a perfect field. Moreover, let $h_{i_j}(x_{i_j}) \in K[x_{i_j}] \setminus K$ with $j = 1, \ldots, r$ be squarefree univariate polynomials.
Prove that $(h_{i_1}, \ldots, h_{i_r}) \subset K[x_1, \ldots, x_n]$ is radical.

**E. 8.61.** (→ p. 264) [**Lagrange Interpolation**] Let $\alpha_1, \ldots, \alpha_n$ be distinct elements of a field $K$, and let $\beta_1, \ldots, \beta_n \in K$.
Prove that there exists a polynomial $f(x) \in K[x]$ of degree $n-1$ such that $f(\alpha_i) = \beta_i$ for all $i$.

**E. 8.62.** (→ p. 264) [**Berlekamp's Algorithm**] Let $p$ be a prime in $\mathbb{Z}$, let $A = \mathbb{Z}/(p)[x]$, and let $f \in A$ be a squarefree monic polynomial.
Moreover, let $B = A/(f)$, and let $\varphi_p \colon B \longrightarrow B$ be the Frobenius homomorphism given by

$$\varphi_p(\overline{g}) = \overline{g}^p.$$

Prove:

1. $\dim_{\mathbb{Z}/(p)} \operatorname{Ker}(\varphi_p - \mathrm{id}_B) = n$, where $n$ is the number of irreducible factors of $f$;
2. for any $\overline{g} \in \operatorname{Ker}(\varphi_p - \mathrm{id}_B)$, we have

$$f(x) = \prod_{a \in \mathbb{Z}/(p)} \gcd(f(x), g(x) - a),$$

   where $g(x) \in A$ is any representative of $\overline{g}$.

**E. 8.63.** (→ p. 265) Let $\mathfrak{q}_1, \mathfrak{q}_2 \subset A$ be primary ideals with $\sqrt{\mathfrak{q}_1} = \sqrt{\mathfrak{q}_2}$.
Prove that $\mathfrak{q} = \mathfrak{q}_1 \cap \mathfrak{q}_2$ is also primary with radical $\sqrt{\mathfrak{q}_1}$.

**E. 8.64.** (→ p. 265) Let $A$ be a PID.
Prove that the primary ideals of $A$ are generated by powers of prime elements.

**E. 8.65.** (→ p. 265) 1. Prove that if $A$ is a UFD and $p \in A \setminus \{0\}$ is prime, then $(p^i)$ is a primary ideal for all $i \in \mathbb{N}_+$.
2. Find a primary ideal which is not a power of a prime ideal.
3. Let $A = K[x, y, z]$, $I = (xy - z^2) \subset A$, $B = A/I$, and $\mathfrak{p} = (\bar{x}, \bar{z}) \subset B$.
   Prove that $\mathfrak{p}$ is prime and $\mathfrak{p}^2$ is not primary.

**E. 8.66.** (→ p. 266) Let $A$ be a ring.
Prove that $A[x]$ is a PID if and only if $A$ is a field.

**E. 8.67.** (→ p. 266) Prove that in the ring $\mathbb{Z}[\sqrt{-5}]$ the element 3 is irreducible, but the ideal (3) is not irreducible.

**E. 8.68.** (→ p. 266) Let $A$ be a domain.
Prove that if every ascending chain of principal ideals of $A$ is stationary, then (UFD1) holds in $A$.

**E. 8.69.** (→ p. 266) Show that a domain which is a quotient of a UFD is not necessarily a UFD.

**E. 8.70.** (→ p. 267) Let $A$ be a ring and let $a \in A$. We define

$$I_a = \{ab - b \colon b \in A\},$$

and say that $a$ is a *quasi-regular* element when $I_a = A$.
Prove:

1. $I_a$ is an ideal for any $a \in A$;
2. $a$ is quasi-regular if and only if there exists $c \in A$ such that $a + c - ac = 0$;
3. every nilpotent element of $A$ is quasi-regular;
4. if every element in $A$ except 1 is quasi-regular, then $A$ is a field.

**E. 8.71.** (→ p. 267) Let $A$ be a domain which is not a field and such that every proper ideal of $A$ is a finite product of maximal ideals.
Prove:

1. if $\mathfrak{m} \in \operatorname{Max} A$, then, for every $a \in \mathfrak{m} \setminus \{0\}$, there exists an ideal $I$ such that $I\mathfrak{m} = (a)$;
2. if $J$ and $H$ are ideals of $A$ and $\mathfrak{m} \in \operatorname{Max} A$, then $J\mathfrak{m} = H\mathfrak{m}$ implies that $J = H$.

**E. 8.72.** (→ p. 267) Let $A$ be a ring, and let $I$, $J$ be ideals of $A$.
Prove:

1. if $I$ is primary and $J \nsubseteq \sqrt{I}$, then $\sqrt{I : J^m} = \sqrt{I}$ for all $m \geqslant 1$;
2. if $I = \sqrt{I}$ and $h \notin I$, then $I : (h)$ is radical.

**E. 8.73.** (→ p. 268) Let $p = \sum_{i \in \mathbb{N}} a_i x^i \in A[[x]]$.
Prove:

1. $p$ is invertible if and only if $a_0$ is invertible;
2. if $p$ is nilpotent, then $a_i$ is nilpotent for all $i$, but the converse does not hold in general;
3. $p \in \mathcal{J}(A[[x]])$ if and only if $a_0 \in \mathcal{J}(A)$;
4. the contraction of a maximal ideal $\mathfrak{m}$ of $A[[x]]$, with respect to the inclusion homomorphism $A \longrightarrow A[[x]]$, is a maximal ideal of $A$.
   Moreover, show that $\mathfrak{m}$ is generated by $\mathfrak{m}^c$ and $x$.

**E. 8.74.** (→ p. 269) Let $A$ be a ring, and let $I \subset A$ be an ideal.
Prove that if an element $g \in A$ satisfies $I : (g^m) = I : (g^{m+1})$ for some $m \in \mathbb{N}$, then:

1. $I : (g^{m+s}) = I : (g^m)$ for all $s \in \mathbb{N}$;
2. $I = (I : (g^m)) \cap (I, g^m)$.

**E. 8.75.** (→ p. 269) [**Existence of Minimal Primes**] Prove that the set of prime ideals of a ring $A \neq 0$ contains minimal elements with respect to set inclusion.
Moreover, prove that, for any ideal $I$, the set

$$\mathcal{V}(I) = \{\mathfrak{p} \in \operatorname{Spec} A \colon \ \mathfrak{p} \supseteq I\}$$

contains minimal elements with respect to set inclusion.

The set of minimal elements of $\mathcal{V}(I)$ is denoted by $\operatorname{Min} I$. If $I = 0$, we also use the notation $\operatorname{Min} A$.

**E. 8.76.** (→ p. 270) Let $A$ be a reduced ring, and let $a \in A$ be a zero-divisor. Prove that $a$ belongs to one of the minimal primes of $A$.

**E. 8.77.** (→ p. 270) Consider the ring $A = \mathbb{Z}[x, y]$ and the ideal

$$I = (9x^2 - y,\ 7y^2 + 2x + y,\ 63).$$

1. Prove that every prime ideal of $A/I$ is maximal.
2. Prove that $A/I \simeq \mathbb{Z}/(9) \times (\mathbb{Z}/(7))^2$.
3. Decompose $I$ as an intersection of primary ideals.
4. Prove that if $B$ is a domain and there exists an injective ring homomorphism $f \colon B \longrightarrow A/I$, then $B$ is a field.

**E. 8.78.** (→ p. 271) Prove that

$$\operatorname{Spec} \mathbb{Z}[x] = \{(0),\ (p),\ (f(x)),\ (p, g(x))\} \ \text{ and } \ \operatorname{Max} \mathbb{Z}[x] = \{(p,\ g(x))\},$$

where $p$ varies over the primes of $\mathbb{Z}$, $f(x)$ over the irreducible polynomials of $\mathbb{Z}[x]$, and $g(x)$ over the polynomials of $\mathbb{Z}[x]$ which are irreducible modulo $p$.

# Chapter 9
# Polynomials, Gröbner Bases, Resultant, and Varieties

**E. 9.1.** (→ p. 272) Prove that a partial order relation $>$ is a well-order on $\mathbb{N}^n$ if and only if every descending chain in $\mathbb{N}^n$ is stationary.

**E. 9.2.** (→ p. 272) Prove that the partial order relations lex, deglex, degrevlex are monomial orderings.

**E. 9.3.** (→ p. 272) Let $>$ be a total order on $\mathbb{N}^n$ such that if $\mathbf{a} > \mathbf{b}$, then $\mathbf{a}+\mathbf{c} > \mathbf{b}+\mathbf{c}$ for all $\mathbf{a}, \mathbf{b}, \mathbf{c} \in \mathbb{N}^n$.
Prove that $>$ is a monomial ordering if and only if $\mathbf{a} \geq \mathbf{0}$ for all $\mathbf{a} \in \mathbb{N}^n$.

**E. 9.4.** (→ p. 273) Let $K \subset K'$ be a field extension, and let $I \subset K[x_1, \ldots, x_n]$ be an ideal. Let also $I^e$ be the ideal generated by $I$ in $K'[x_1, \ldots, x_n]$.
Prove that every Gröbner basis of $I$ with respect to a fixed monomial ordering is also a Gröbner basis of $I^e$ with respect to the same ordering.

In particular, given any monomial ordering, the staircase of $I$ and the staircase of $I^e$ are the same.

**E. 9.5.** (→ p. 274) Let $A = K[x_1, x_2]$ be equipped with the lex order with $x_1 > x_2$. Let also

$$f = x_1^4 x_2 \quad \text{and} \quad F = \{f_1 = x_1^3,\ f_2 = x_1^2 x_2 - x_2^2\}.$$

Show that the remainder of division of $f$ by $F$ is not unique.

**E. 9.6.** (→ p. 274) Consider $g_1 = z + x,\ g_2 = y - x \in \mathbb{Q}[x, y, z],\ G = \{g_1, g_2\}$, and $I = (g_1, g_2)$. Let $>_1$ denote the lex order with $z > y > x$, and let $>_2$ denote the lex order with $x > y > z$.
Show that $G$ is a Gröbner basis of $I$ with respect to $>_1$ but $G$ is not a Gröbner basis with respect to $>_2$.

**E. 9.7.** (→ p. 274) [**Monomiality Test**] Prove that an ideal $I$ is monomial if and only if its reduced Gröbner basis with respect to any fixed monomial ordering consists of monomials.



A. Bandini et al., *Commutative Algebra through Exercises*, UNITEXT 159,
https://doi.org/10.1007/978-3-031-56910-4_9

**E. 9.8.** ($\rightarrow$ p. 275) Let $I = (x^2 - xy,\, xz - y^2,\, yz^2 - z^4) \subseteq \mathbb{R}[x, y, z]$.
Compute a minimal Gröbner basis of $I$ with respect to the lex order with $x > y > z$, and reduce it.

**E. 9.9.** ($\rightarrow$ p. 276) Let $I = (yz - y,\, xy + 2z^2,\, y - z) \subset \mathbb{Q}[x, y, z]$, and let $f = x^3z - y^2$.
Determine whether $f \in I$.

**E. 9.10.** ($\rightarrow$ p. 276) Let $I = (x^2 + xy + y^2,\, xy^2 + 1) \subset A = (\mathbb{Z}/(2))[x, y]$.
Let also

$$f_1 = x^3 + y^5 + xy^2 \quad \text{and} \quad f_2 = y(x^2 + x + y).$$

Determine whether $\overline{f_1} = \overline{f_2}$ in $A/I$.

**E. 9.11.** ($\rightarrow$ p. 276) Let $I = (x^2y - y + x,\, y^2 - yx - x^2,\, x^3 + y - 2x) \subset \mathbb{Q}[x, y]$ and $f = y + x + 1$.
Compute the inverse of $\overline{f}$ in $A = \mathbb{Q}[x, y]/I$.

**E. 9.12.** ($\rightarrow$ p. 277) Let $I = (x^2y + z,\, xz + y) \subset \mathbb{Q}[x, y, z]$.

1. Compute the reduced Gröbner basis $G$ of $I$ with respect to the deglex order with $x > y > z$.
2. Compute the transition matrix from $G$ to the given generators.
3. Verify that $f = xy^2z + y^3 \in I$, and write $f$ as a linear combination of the elements of $G$ as well as of the original generators of $I$.

**E. 9.13.** ($\rightarrow$ p. 278) Let $I = (x^2 + 2y^2 - 3,\, x^2 + xy + y^2 - 3) \subset \mathbb{C}[x, y]$.
Compute $I \cap \mathbb{C}[y]$.

**E. 9.14.** ($\rightarrow$ p. 278) Let $I,\, J \subset K[x, y]$, with

$$I = (x(x + y)^2,\, y) \quad \text{and} \quad J = (x^2,\, x + y).$$

Compute $I : J$.

**E. 9.15.** ($\rightarrow$ p. 278) Let $K$ be a field of characteristic different from 2, and let

$$I = (x^2 + y^2,\, x^3y^3 + y^4) \subset K[x, y] \quad \text{and} \quad f = x^2 + 5x.$$

Determine whether $f \in \sqrt{I}$.

**E. 9.16.** ($\rightarrow$ p. 278) Let $I = (x^2y^2z^4,\, x^2 + y^2 + z^2 - 1,\, 2 - xy) \subset \mathbb{C}[x, y, z]$.

1. Prove that $\dim_{\mathbb{C}} \mathbb{C}[x, y, z]/I$ is finite and compute it.
2. Let $J = (3x^3 + xz - 2,\, xy + z^2 - 2) \subset \mathbb{C}[x, y, z]$.
   Show that $I + J = (1)$.

**E. 9.17.** (→ p. 279) Find the values of the parameter $a \in \mathbb{C}$ for which the system of polynomial equations

$$\begin{cases} x + y = a \\ x^2 + y^2 = a^2 \\ x^3 + y^3 = a^5 \end{cases}$$

has solutions in $\mathbb{C}^2$.
If there are solutions, compute them.

**E. 9.18.** (→ p. 280) Let $I = (x^2y + xz + yz,\ y^2z)$ be an ideal of $\mathbb{R}[x, y, z]$, and let $A = \mathbb{R}[x, y, z]/I$.

1. Compute the reduced Gröbner basis of $I$ with respect to the lex order with $x > y > z$.
2. Find all the nilpotents of $A$.
3. Show that $(x^2y^3,\ y^3z) \subsetneq I \subsetneq (x^2,\ z)$.

**E. 9.19.** (→ p. 280) Let $f,\ g_1,\ g_2 \in K[x] \setminus \{0\}$, and let $a_m = \mathrm{lc}(f)$.
Prove:

1. $\mathrm{Res}(f, g_1g_2) = \mathrm{Res}(f, g_1)\,\mathrm{Res}(f, g_2)$;
2. if $g_1f + g_2 \neq 0$ and $N = \deg(g_1f + g_2)$, then

$$\mathrm{Res}(f, g_1f + g_2) = a_m^{N-\deg(g_2)}\,\mathrm{Res}(f, g_2).$$

**E. 9.20.** (→ p. 281) [**Construction of Polynomials with Given Roots**]
Let $f,\ g \in K[x]$ be polynomials of degrees $m,\ n > 0$.
Let also $\alpha_1, \ldots, \alpha_m$ and $\beta_1, \ldots, \beta_n$ be the roots in $\overline{K}$ of $f$ and $g$, respectively.
Prove:

1. the polynomial $r(x) = \mathrm{Res}_y(f(x - y),\ g(y))$ has roots $\alpha_i + \beta_j$;
2. the polynomial $r(x) = \mathrm{Res}_y(f(x + y),\ g(y))$ has roots $\alpha_i - \beta_j$;
3. the polynomial $r(x) = \mathrm{Res}_y\left(y^m f(\frac{x}{y}),\ g(y)\right)$ has roots $\alpha_i\beta_j$;
4. if $g(0) \neq 0$, then the polynomial $r(x) = \mathrm{Res}_y(f(xy),\ g(y))$ has roots $\frac{\alpha_i}{\beta_j}$.

**E. 9.21.** (→ p. 281) Let $f,\ g \in \mathbb{Q}[x]$ be polynomials of positive degree.
Prove that if $f(0) = 1$, then $\mathrm{Res}(f, x^k g) = \mathrm{Res}(f, g)$ for all $k \in 2\mathbb{N}$.

**E. 9.22.** (→ p. 282) Let $A = \mathbb{Z}[x]$.

1. Let $f, g \in A$ be monic polynomials such that $\mathrm{Res}(f, g) = p$, where $p \in \mathbb{Z}$ is prime.
   Prove that $(f, g) \cap \mathbb{Z} = (p)$.
2. Let $I = (x^2 - 4x + 1,\ x^2 - x) \subset A$.
   Compute $I \cap \mathbb{Z}$, and describe the quotient ring $A/I$.

**E. 9.23.** (→ p. 282) Let $f \in \mathbb{Z}[x]$ be a polynomial such that $\gcd(f, f') = 1$. Prove that the set of primes $p \in \mathbb{Z}$ such that the ring

$$(\mathbb{Z}/(p))[x]/(\overline{f})$$

is not reduced is finite.

**E. 9.24.** (→ p. 282) Let $A = K[x, y, z]$ and $I = (xz - y,\ yz - x) \subset A$. Decompose $\mathbb{V}(I)$ as a union of irreducible varieties.

**E. 9.25.** (→ p. 283) Let $I = (x^2 - yzt,\ t - yt,\ zt - y)$ be an ideal of $\mathbb{C}[x, y, z, t]$.

1. Find the irreducible components of $\mathbb{V}(I)$.
2. Decide whether $f = xt + y \in I$.

**E. 9.26.** (→ p. 283) Let $I = (x^2 + y^2 + z^2 - 1,\ x + y + z - 1) \subset \mathbb{C}[x, y, z]$. Compute:

1. $\dim_{\mathbb{C}} \mathbb{C}[x, y, z]/I$;
2. $\mathbb{V}(I) \cap \mathbb{V}(z - 1)$.

**E. 9.27.** (→ p. 283) Let $I = (xy^3,\ xy + y^2,\ y^2 - z^2) \subset \mathbb{C}[x, y, z]$.

1. Compute $I_1 = I \cap \mathbb{C}[y, z]$ and $I_2 = I \cap \mathbb{C}[z]$.
2. Let $\pi_1 : \mathbb{C}^3 \longrightarrow \mathbb{C}^2$ be the projection on the last two components given by

$$\pi_1(a_1, a_2, a_3) = (a_2, a_3).$$

   Verify whether $\pi_1(\mathbb{V}(I)) = \mathbb{V}(I_1)$.

**E. 9.28.** (→ p. 284) Let $I = (t^2 - x,\ t^3 - y,\ t^4 - z) \subset \mathbb{C}[x, y, z, t]$.

1. Compute $J = I \cap \mathbb{C}[x, y]$.
2. Determine whether every element of $\mathbb{V}(J) \subset \mathbb{C}^2$ can be extended to an element of $\mathbb{V}(I) \subset \mathbb{C}^4$.

**E. 9.29.** (→ p. 284) In $\mathbb{C}[x, y, z]$ consider the ideals $I = (x^2 - y^2 - yz,\ xy - y^2 z)$ and $J = (x^2 - y^2 - yz,\ xy - y^2 z,\ y^3 z^2 - y^3 - y^2 z)$.
Determine whether $I = J$ and whether $\mathbb{I}(\mathbb{V}(I)) = I$.

**E. 9.30.** (→ p. 284) Let $I = (x + y + z,\ xy + yz + zx,\ xyz - 1) \subseteq \mathbb{C}[x, y, z]$. Prove:

1. the set $\mathbb{V}(I)$ consists of the points of $\mathbb{C}^3$ obtained from all the permutations of the coordinates of $(1, \alpha, \alpha^2)$, where $\alpha = \frac{1}{2}(-1 + \sqrt{-3})$;
2. $I$ is radical.

**E. 9.31.** (→ p. 285) Let $I = (x^3 + y^3 + z^3 + 1,\ x^2 + y^2 + z^2 + 1,\ x + y + z + 1)$ be an ideal of $(\mathbb{Z}/(2))\,[x, y, z]$.

1. Is $\mathbb{V}(I) \subset \overline{\mathbb{Z}/(2)}^{\,3}$ finite?
2. Decompose $\mathbb{V}(I)$ as a union of irreducible varieties.

**E. 9.32.** (→ p. 285) Let $I = (x^2 - yz + y^2,\ xyz - x) \subset \mathbb{Q}[x, y, z]$.

1. Find a reduced Gröbner basis of $I$.
2. Compute $I^c$ with respect to the inclusion $\mathbb{Q}[y, z] \longrightarrow \mathbb{Q}[x, y, z]$.
3. Determine whether $\mathbb{V}(I) \subset \mathbb{Q}^3$ is finite.
4. Compute $\operatorname{Min} I$.

**E. 9.33.** (→ p. 286) Let $I = (yt^2 + x^3z^2t^3,\ z^2 + yt^2,\ x^2t^2) \subset K[x, y, z, t]$, and let $A$ be its quotient ring.

1. Verify that $I$ is a monomial ideal.
2. Decompose $I$ as an irredundant intersection of primary ideals.
3. Find $\mathcal{N}(A)$ and write it as an irredundant intersection of prime ideals.
4. Determine whether $\mathbb{V}(I)$ is finite.

**E. 9.34.** (→ p. 286) Let $I = (y^2 - xz,\ x^2 - y^2,\ x^2 - yz) \subset \mathbb{Q}[x, y, z]$, and let $A = \mathbb{Q}[x, y, z]/I$.

1. Determine the irreducible components of $\mathbb{V}(I)$ and whether $\mathbb{V}(I)$ is finite.
2. Determine whether $f = y(x^2 + x + y) \in \sqrt{I}$.

**E. 9.35.** (→ p. 286) Let $I = (x^2z,\ y^2z^2 - yz,\ y^2 - z^2) \subset \mathbb{C}[x, y, z]$. Determine whether $\mathbb{V}(I)$ is finite and $I \subseteq (x^2,\ y + 1,\ z - 1)$.

**E. 9.36.** (→ p. 287) Let $I = (xyz - 2,\ y^2z - x,\ 3x^2z^2 - y) \subset K[x, y, z]$.

1. Prove that if $K = \mathbb{C}$, then $\mathbb{V}(I)$ is finite.
2. Find, if possible, primes $p \in \mathbb{Z}$ such that when $K = \overline{\mathbb{Z}/(p)}$ then $\mathbb{V}(I)$ is empty or infinite.

**E. 9.37.** (→ p. 287) Let $I = (x^2 + y^2 + z^2 - 2,\ y^2 - z^2 + 1,\ xz - 1) \subset \mathbb{Q}[x, y, z]$, and let $A$ be its quotient ring.

1. Prove that $A$ is a finite dimensional $\mathbb{Q}$-vector space and find a basis of $A$.
2. Compute the coordinates of $f = x^2 + y^2z + 2y + 1$ with respect to the basis found in part 1.
3. Verify whether $\dim_{\mathbb{Q}} A = |\,\mathbb{V}_{\mathbb{C}}(I)|$.
4. Decompose $\sqrt{I}$ as an intersection of maximal ideals in $\mathbb{Q}[x, y, z]$, if possible.

**E. 9.38.** (→ p. 288) Let $\Sigma$ be the system of polynomial equations

$$\begin{cases} f_1 = x^2 - 3xy + y^2 = 0 \\ f_2 = x^3 - 8x + 3y = 0 \\ f_3 = x^2y - 3x + y = 0. \end{cases}$$

1. Verify that $\Sigma$ has a finite number of solutions in $\mathbb{C}^2$.
2. Find all solutions $\beta$ of $\Sigma$ such that $\beta \in \mathbb{Q}^2$.
3. Decompose the variety $\mathbb{V}_{\mathbb{R}}(f_1, f_2, f_3)$ as a union of irreducible varieties.

**E. 9.39.** (→ p. 288) Let $I = (xz - yz, y^2 - z, xyz - 1) \subset \mathbb{Q}[x, y, z]$.

1. Find, if possible, a non-zero univariate polynomial $p(y) \in I$.
2. Describe the set

$$S = \{q(y) \in \mathbb{Q}[y] \colon \mathbb{V}_{\mathbb{Q}}((q, I)) \neq \emptyset\},$$

and verify whether $S$ is an ideal.

**E. 9.40.** (→ p. 289) Let $V = \{\alpha_1, \ldots, \alpha_m\} \subset \mathbb{C}^n$, where $\alpha_i \neq \alpha_j$ for all $i \neq j$, and let

$$A = \mathbb{C}[x_1, \ldots, x_n]/\mathbb{I}(V)$$

be the coordinate ring of $V$.

1. Prove that there exist $m$ non-zero elements $a_1, \ldots, a_m \in A$, such that

$$a_i^2 = a_i \text{ for all } i, \; a_i a_j = 0 \text{ for } i \neq j, \text{ and } \sum_{i=1}^{m} a_i = 1.$$

2. Determine the number of idempotents of $A$, and find all such elements.

# Chapter 10
# Modules

## 10.1 Modules, Submodules and Homomorphisms

**E. 10.1.** (→ p. 290) Prove that, for any ideal $I \subset A$ and any $A$-module $M$, defining $\overline{a}\,\overline{m} = \overline{am}$ for all $\overline{a} \in A/I$ and $\overline{m} \in M/IM$, we obtain an $A/I$-module structure on $M/IM$.

**E. 10.2.** (→ p. 290) Prove that the restriction of scalars via a ring homomorphism $f\colon A \longrightarrow B$ induces an $A$-module structure on any $B$-module.

**E. 10.3.** (→ p. 290) Let $f\colon A \longrightarrow B$ be a surjective ring homomorphism and consider $B$ as an $A$-module by restriction of scalars via $f$.
Prove that the ideals of $B$ coincide with the $A$-submodules of $B$.

**E. 10.4.** (→ p. 290) We say that a generating set of an $A$-module $M$ is *minimal* if none of its proper subsets is still a generating set of $M$.
Moreover, we say that a free generating set is *maximal* if any set properly containing it is no longer free.
Prove:

1. finite minimal generating sets of a module may not have the same number of elements;
2. a minimal generating set of a module may not be a basis;
3. a maximal free subset of a module may not be a basis;
4. a submodule of a finitely generated module may not be finitely generated;
5. not all modules have a basis;
6. a submodule of a free module may not be free.

**E. 10.5.** (→ p. 290) Let $M$ be an $A$-module. Let $N$, $P$ be submodules of $M$. Prove that $N : P = \{a \in A\colon aP \subseteq N\}$, and in particular, $\operatorname{Ann} P = 0 : P$ are ideals of $A$.



A. Bandini et al., *Commutative Algebra through Exercises*, UNITEXT 159,
https://doi.org/10.1007/978-3-031-56910-4_10

**E. 10.6.** (→ p. 291) Let $M$ be an $A$-module. For any $m \in M$, let

$$\operatorname{Ann} m = \{a \in A \colon am = 0\}$$

be the *annihilator of* $m$.
Prove that $\operatorname{Ann} m = \operatorname{Ann}\langle m \rangle$.

**E. 10.7.** (→ p. 291) Let $I, J_1, J_2 \subset A$ be ideals and let $M = A/J_1 \oplus A/J_2$.
Prove that

$$M/IM \simeq A/(J_1 + I) \oplus A/(J_2 + I).$$

**E. 10.8.** (→ p. 291) Consider an ideal $J \subset A$, and an element $a \in A$. Let also $M = A/J$.
Prove that

$$aM \simeq A/(J : (a)).$$

**E. 10.9.** (→ p. 291) Let $A$ be a ring, and let $f \colon A^m \longrightarrow A^n$ be a surjective $A$-module homomorphism.
Prove that if $n > m$, then $A = 0$.

**E. 10.10.** (→ p. 291) Let $M$ be a free $A$-module of rank $r$.
Prove that every set of generators of $M$ has cardinality at least $r$.

**E. 10.11.** (→ p. 291) Consider a ring $A$, a nilpotent ideal $I \subset A$, and an $A$-module homomorphism $\varphi \colon M \longrightarrow N$.
Prove that if the induced homomorphism

$$\overline{\varphi} \colon M/IM \longrightarrow N/IN$$

is surjective, then $\varphi$ is also surjective.

**E. 10.12.** (→ p. 292) Let $A \neq 0$ be a ring. Consider positive integers $m$ and $n$, and let $f \colon A^m \longrightarrow A^n$ be an $A$-module homomorphism.
Prove:

1. if $f$ is surjective, then $m \geq n$;
2. if $f$ is injective, then $m \leq n$;
3. if $f$ is an isomorphism, then $m = n$.

**E. 10.13.** (→ p. 292) Let $M$ be a finitely generated $A$-module, and let $N \subseteq M$ be a non-trivial submodule.

1. Prove that $M \not\simeq M/N$.
2. Provide a counterexample to the previous statement when $M$ is not finitely generated.

**E. 10.14.** (→ p. 292) Let $A$ be a ring, and let $M$, $N$ be two $A$-modules.
Prove:

1. $M \neq 0$ is simple if and only if $M \simeq A/\mathfrak{m}$, where $\mathfrak{m} \in \operatorname{Max} A$;
2. if $M$, $N \neq 0$ are simple, and $\varphi\colon M \longrightarrow N$ is a homomorphism, then $\varphi$ is either the zero homomorphism or an isomorphism;
3. if $M$ is simple, then $\mathcal{J}(A)M = 0$.

**E. 10.15.** (→ p. 293) 1. Prove that an $A$-module $M \neq 0$ is simple if and only if for any $0 \neq m \in M$ we have $\langle m \rangle = M$.

2. Describe all simple $\mathbb{Z}$-modules $M$.

**E. 10.16.** (→ p. 293) Let $M$ be a cyclic $\mathbb{Z}$-module, and let $N$ and $P$ be submodules of $M$.
Prove that if there exist coprime $p$, $q \in \mathbb{Z}$ such that $\operatorname{Ann} N = (p)$, $\operatorname{Ann} P = (q)$, and $\operatorname{Ann} M = (pq)$, then $M = N \oplus P$.

**E. 10.17.** (→ p. 293) Let $M$, $N$, and $P$ be $A$-modules.
Prove that

$$\operatorname{Hom}_A(M, P) \oplus \operatorname{Hom}_A(N, P) \simeq \operatorname{Hom}_A(M \oplus N, P),$$

and

$$\operatorname{Hom}_A(P, M) \oplus \operatorname{Hom}_A(P, N) \simeq \operatorname{Hom}_A(P, M \oplus N).$$

**E. 10.18.** (→ p. 294) Let $n \neq 1$ be a positive integer.
Prove:

1. $\operatorname{Hom}_{\mathbb{Z}}(\mathbb{Q}, \mathbb{Z}) = 0$;
2. $\operatorname{Hom}_{\mathbb{Z}}(\mathbb{Z}/(n), \mathbb{Z}) = 0$;
3. $\operatorname{Hom}_{\mathbb{Z}}(\mathbb{Z}/(n), \mathbb{Q}/\mathbb{Z}) \neq 0$.

**E. 10.19.** (→ p. 294) Consider a ring $A$ and ideals $I$, $J \subsetneq A$, and let $M \neq 0$ be an $A$-module.
Prove:

1. $\operatorname{Hom}_A(A/I, M) \simeq 0 :_M I = \{m \in M : Im = 0\}$;
2. $\operatorname{Hom}_A(A/I, M)$ has an $A/I$-module structure;
3. if $M = A/J$, then $\operatorname{Hom}_A(A/I, M) \simeq (J : I)/J$.

**E. 10.20.** (→ p. 295) Let $A = K[x, y, z]$, $I = (x^3, x^2y, yz)$, and $J = (x^2, yz)$.
Compute the dimension of $\operatorname{Hom}_A(A/I, A/J)$ as a $K$-vector space.

**E. 10.21.** (→ p. 295) Let $(A, \mathfrak{m})$ be a local ring, and let $M \neq 0$ be a finitely generated $A$-module.
Prove that $\operatorname{Hom}_A(M, A/\mathfrak{m}) \neq 0$.

**E. 10.22.** (→ p. 296) Let $K$ be a field, and let $B \subset K$ be a local ring that is not a field.
Prove that $K$ is not finitely generated as a $B$-module.

**E. 10.23.** (→ p. 296) Let $M$ be a finitely generated $A$-module, and let $I \subset A$ be an ideal.
Prove that
$$\sqrt{\operatorname{Ann}(M/IM)} = \sqrt{\operatorname{Ann} M + I}.$$

**E. 10.24.** (→ p. 296) Let $A$ be a ring, and let $M$ be an $A$-module.
Prove that if $\mathcal{N}(A)$ is finitely generated and $\mathcal{N}(A)M = M$, then $M = 0$.

**E. 10.25.** (→ p. 296) Find a counterexample to the statement of Nakayama's Lemma 2 when the module $M$ is not finitely generated.

## 10.2 Exact Sequences and Projective Modules

**E. 10.26.** (→ p. 297) [**Five Lemma**] Consider the following commutative diagram with exact rows:

$$\begin{array}{ccccccccc} M_1 & \longrightarrow & M_2 & \longrightarrow & M_3 & \longrightarrow & M_4 & \longrightarrow & M_5 \\ \downarrow{\scriptstyle \alpha_1} & & \downarrow{\scriptstyle \alpha_2} & & \downarrow{\scriptstyle \alpha_3} & & \downarrow{\scriptstyle \alpha_4} & & \downarrow{\scriptstyle \alpha_5} \\ N_1 & \longrightarrow & N_2 & \longrightarrow & N_3 & \longrightarrow & N_4 & \longrightarrow & N_5. \end{array}$$

Prove:

1. if $\alpha_1$ is surjective and $\alpha_2,\ \alpha_4$ are injective, then $\alpha_3$ is injective;
2. if $\alpha_5$ is injective and $\alpha_2,\ \alpha_4$ are surjective, then $\alpha_3$ is surjective.

In particular, if $\alpha_1$ is surjective, $\alpha_5$ is injective, and $\alpha_2,\ \alpha_4$ are isomorphisms, then $\alpha_3$ is an isomorphism.

**E. 10.27.** (→ p. 297) Let $M$, $N$ and $P$ be $\mathbb{Z}$-modules, and let $p$, $q$ be distinct primes in $\mathbb{Z}$.
Prove that if $pN = qP = 0$, then every exact sequence

$$0 \longrightarrow N \xrightarrow{f} M \xrightarrow{g} P \longrightarrow 0$$

splits.

**E. 10.28.** (→ p. 298) Let $M = \mathbb{Z}/(2)$, $N = \mathbb{Z}/(4)$, and $P = \mathbb{Z}/(8)$.
Find, if possible, short exact sequences of $\mathbb{Z}$-modules of the form:

1. $0 \longrightarrow M \longrightarrow N \longrightarrow M \longrightarrow 0$;

2. $0 \longrightarrow M \longrightarrow N \oplus M \longrightarrow M \oplus M \longrightarrow 0$;
3. $0 \longrightarrow M \longrightarrow P \longrightarrow M \oplus M \longrightarrow 0$.

**E. 10.29.** (→ p. 298) Let $0 \longrightarrow N \xrightarrow{f} M \xrightarrow{g} P \longrightarrow 0$ be a short exact sequence of $A$-modules .
Prove that if $N$ and $P$ are finitely generated, then $M$ is also finitely generated.

**E. 10.30.** (→ p. 298) Let $f\colon M \longrightarrow N$ and $g\colon N \longrightarrow M$ be $A$-module homomorphisms such that $g \circ f = \mathrm{id}_M$.
Prove that $N \simeq \operatorname{Ker} g \oplus \operatorname{Im} f$.

**E. 10.31.** (→ p. 299) Let

$$0 \longrightarrow M \xrightarrow{\varphi} N \xrightarrow{f} P \longrightarrow 0$$

and

$$0 \longrightarrow P \xrightarrow{g} T \xrightarrow{\psi} W \longrightarrow 0$$

be two exact sequences of $A$-modules.
Prove that

$$0 \longrightarrow M \xrightarrow{\varphi} N \xrightarrow{g\circ f} T \xrightarrow{\psi} W \longrightarrow 0$$

is exact.

**E. 10.32.** (→ p. 299) Let $n \neq 0, \pm 1$.
Prove that $\mathbb{Z}/(n)$ is a projective $\mathbb{Z}/(n)$-module, but it is not projective as a $\mathbb{Z}$-module.

**E. 10.33.** (→ p. 299) Let $A = \mathbb{Z}/(12)$.
Prove that $\mathbb{Z}/(4)$ is a projective $A$-module, but not a free $A$-module.

**E. 10.34.** (→ p. 299) Let $A = \mathbb{Z}/(4)$ and $B = \mathbb{Z}/(6)$.
Find all non-trivial submodules of $A$ and $B$, and determine whether they are projective or not as $A$-modules and as $B$-modules, respectively.

**E. 10.35.** (→ p. 300) Let $I$ and $J$ be comaximal ideals of a ring $A$.
Prove:

1. $I \oplus J \simeq IJ \oplus A$;
2. if $A$ is a domain and $IJ$ is principal, then $I$ and $J$ are projective.

**E. 10.36.** (→ p. 300) Let $0 \longrightarrow M \xrightarrow{f} N \xrightarrow{g} P \longrightarrow 0$ be an exact sequence of $A$-modules.

1. Assume $A = \mathbb{Z}$ and two out of the three modules in the sequence are isomorphic to $\mathbb{Z}$.
   What can we deduce about the third module in all possible cases?

2. Assume $A = \mathbb{Z}$ and one out of the three modules in the sequence is isomorphic to $\mathbb{Z}$.
   What can we deduce about the remaining two modules in all possible cases?
3. Assume $A$ is a PID.
   Do the conclusions of parts 1 and 2 still hold?

**E. 10.37.** (→ p. 301) Let $N \subset M$, $N' \subset M'$ be $A$-modules such that

$$M/N \simeq M'/N' \simeq A.$$

Prove:

1. the sequences
$$0 \longrightarrow N \longrightarrow M \longrightarrow M/N \longrightarrow 0$$
   and
$$0 \longrightarrow N' \longrightarrow M' \longrightarrow M'/N' \longrightarrow 0$$
   split;
2. if $N \simeq N'$, then $M \simeq M'$.

**E. 10.38.** (→ p. 301) Let $A$ be a ring, and let $M \neq 0$ be an $A$-module. Prove:

1. if $\varphi \in \operatorname{End}_A M$ and $\varphi^2 = \varphi$, then
$$M \simeq \varphi(M) \oplus (\mathrm{id}_M - \varphi)(M);$$
2. if $M$ is finitely generated, then $M$ is projective if and only if there exist $n \in \mathbb{N}_+$ and $f \in \operatorname{End}_A(A^n)$ such that $f^2 = f$ and $M \simeq f(A^n)$.

**E. 10.39.** (→ p. 301) Prove that a domain $A$ is a field if and only if every $A$-module is projective.

**E. 10.40.** (→ p. 301) Consider the ring $A = \mathbb{Z}[\sqrt{-5}]$, and let

$$I = (3, 1 - \sqrt{-5}) \text{ and } J = (3, 1 + \sqrt{-5})$$

be ideals of $A$.

1. Prove that $I$ and $J$ are distinct maximal ideals and are not principal. Hence, $A$ is not a PID.
2. Prove that $I \cap J = IJ = (3)$.
   Moreover, $I$ and $J$ are projective $A$-modules that are not free.

**E. 10.41.** (→ p. 302) An $A$-module $M$ is called *finitely presented* if there exists an exact sequence $A^m \longrightarrow A^n \longrightarrow M \longrightarrow 0$ for some $m, n \in \mathbb{N}$.
Prove that a projective $A$-module is finitely presented if and only if is finitely generated.

**E. 10.42.** (→ p. 302) [**Characterization of Injective Modules**] Let $E$ be an $A$-module.
Prove that the following are equivalent:

1. $E$ is injective;
2. $\mathrm{Hom}_A(\bullet, E)$ is an exact functor;
3. every exact sequence
$$0 \longrightarrow E \longrightarrow M \longrightarrow N \longrightarrow 0$$
splits;
4. if $E$ is isomorphic to a submodule of $M$, then $E$ is a direct summand of $M$, *i.e.*, there exists a submodule $L$ of $M$ such that $M \simeq E \oplus L$.

**E. 10.43.** (→ p. 304) Let $A$ be a ring, and let $F$ be a free $A$-module.

1. Prove that if $A$ is a field, then $F$ is injective.
2. Does the previous statement hold for all rings?

**E. 10.44.** (→ p. 304) [**Baer's Criterion**] Prove that an $A$-module $E$ is injective if and only if for every ideal $I \subset A$, every homomorphism $f\colon I \longrightarrow E$ can be extended to a homomorphism $\tilde{f}\colon A \longrightarrow E$.

## 10.3 Modules over a PID and Smith Normal Form

**E. 10.45.** (→ p. 305) Let $A$ be a PID.
Prove that every submodule of a projective module is projective.

**E. 10.46.** (→ p. 305) Let $A$ be a domain, and let $M$ be an $A$-module.
Verify that the subset $T(M)$ of the torsion elements of $M$ is a submodule.

**E. 10.47.** (→ p. 306) Let $A$ be a ring, and let $M$ be an $A$-module.
For any $a \in A$, let $M_{[a]} = \{m \in M\colon a^k m = 0 \text{ for some } k \in \mathbb{N}\}$ be the $a$-component of $M$.
Prove that $M_{[a]}$ is a submodule of $M$.

**E. 10.48.** (→ p. 306) Let $A$ be a PID, and consider a finitely generated $A$-module $M$. Let $0 \neq a, b \in A$ be such that $\gcd(a, b) = 1$.

1. Prove that $M_{[ab]} = M_{[a]} \oplus M_{[b]}$.

2. Let $\pi_a$ and $\pi_b$ be the projections of $M_{[ab]}$ onto $M_{[a]}$ and $M_{[b]}$, respectively. Prove that there exist $c, d \in A$ such that, for all $m \in M_{[ab]}$, we have

$$\pi_a(m) = cm \quad\text{and}\quad \pi_b(m) = dm.$$

3. Prove that $M_{[ab]}$ is cyclic if and only if $M_{[a]}$ and $M_{[b]}$ are cyclic.

**E. 10.49.** (→ p. 306) Let $A$ be a PID, and let $M \neq 0$ be a cyclic $A$-module such that $M \not\cong A$.
Prove that there exist primes $p_1, \ldots, p_h \in A$ such that

$$M = \bigoplus_{i=1}^{h} M_{[p_i]},$$

where every $M_{[p_i]}$ is cyclic.

**E. 10.50.** (→ p. 307) Let $A$ be a domain, and let $M$ be an $A$-module. Prove:

1. every free $A$-module is torsion-free;
2. if $A$ is a PID, and $M$ is finitely generated and torsion-free, then $M$ is free.
3. Does the above conclusion still hold when $A$ is not a PID? Or when $A$ is a PID but $M$ is not finitely generated?

**E. 10.51.** (→ p. 307) Let $M$ be a $\mathbb{Z}$-module such that the sequence

$$0 \longrightarrow \mathbb{Z}^3 \xrightarrow{f} \mathbb{Z}^4 \longrightarrow M \longrightarrow 0,$$

is exact, where

$$f(x, y, z) = (x + y + z,\ -3x + y + z,\ x - 3y - 3z,\ x + 3y + z).$$

Express $M$ as a direct sum of cyclic $\mathbb{Z}$-modules.

**E. 10.52.** (→ p. 307) Let $\varphi\colon \mathbb{Q}[x]^4 \longrightarrow \mathbb{Q}[x]^4$ be the homomorphism given by

$$\varphi(a, b, c, d) = (a + 3c,\ b + 2xc + 3d,\ (x^2 - x)(a + 3c) + 2xd,\ (x^2 - x)(b + 3d)).$$

Find the dimension of $\operatorname{Coker}\varphi$ over $\mathbb{Q}$.

**E. 10.53.** (→ p. 308) Let $a \in \mathbb{Z}$ and $\varphi\colon \mathbb{Z}^3 \longrightarrow \mathbb{Z}^3$ be the homomorphism given by

$$\varphi(x, y, z) = (6x + 2y + 4z,\ ay + 4z,\ 2x + 2y + 2z).$$

Determine the isomorphism classes of $\operatorname{Coker}\varphi$ as $a$ varies.
Are there any values of $a$ for which $\operatorname{Coker}\varphi$ is infinite?

**E. 10.54.** (→ p. 308) Let $\varphi\colon \mathbb{Z}^3 \longrightarrow \mathbb{Z}^3$ be the homomorphism defined by the matrix

$$\begin{pmatrix} a & b & c \\ 0 & a & b \\ 0 & 0 & a \end{pmatrix} \quad \text{with } a, b, c \in \mathbb{Z}.$$

Prove:

1. $\operatorname{Coker}\varphi$ has at most two generators if and only if $\gcd(a, b, c) = 1$;
2. $\operatorname{Coker}\varphi$ is cyclic if and only if $\gcd(a, b) = 1$.

**E. 10.55.** (→ p. 308) Let $\varphi\colon \mathbb{Z}^3 \longrightarrow \mathbb{Z}^3$ be the homomorphism defined by the matrix

$$\begin{pmatrix} a & 6 & 6 \\ -3 & 6 & 0 \\ a & 3 & 3 \end{pmatrix} \quad \text{with } a \in \mathbb{Z}.$$

Find the values of $a$, if any exist, such that:

1. $\operatorname{Coker}\varphi$ is finite;
2. $\operatorname{Coker}\varphi$ is not cyclic.

**E. 10.56.** (→ p. 309) Let $M = \mathbb{Z}^3/N$, with $N$ the submodule generated by

$$m_1 = (0, a, b), \ m_2 = (3, 3, 0) \ \text{ and } \ m_3 = (3, -1, 0), \ \text{ with } a, b \in \mathbb{Z}.$$

Find all the values of $a$ and $b$ for which $M$ is finite, as well as those values for which $M$ is cyclic.

**E. 10.57.** (→ p. 309) Let $M = \mathbb{Z}^3/N$, with $N$ the submodule generated by

$$m_1 = (2, 4, -4), \ m_2 = (4, 12, -12) \ \text{ and } \ m_3 = (2, -4, -4).$$

Find the annihilator of $M$.

**E. 10.58.** (→ p. 309) Let $A$, $B$, $C \in M_3(\mathbb{Z})$ and let

$$D = \begin{pmatrix} A & C \\ 0 & B \end{pmatrix},$$

with $\det A = 28$ and $\det B = 7$.
Find all possible Smith forms of $D$ and provide an example of $A$, $B$, and $C$ for each case.

**E. 10.59.** (→ p. 310) Let $M$ be the Abelian group generated by elements $m_1$, $m_2$ and $m_3$ which satisfy the relations

$$3m_1 + m_3 = 0, \ 2m_1 - 2m_2 + m_3 = 0 \ \text{ and } \ m_1 + 4m_2 + 2m_3 = 0.$$

Find all the possible orders of an element of $M$.

**E. 10.60.** (→ p. 310) Consider $R \in M_n(\mathbb{Z})$, and let $M$ be the $\mathbb{Z}$-module $\mathbb{Z}^n/R\mathbb{Z}^n$.
Assume that, for all $m \in M$, there exists a prime $p_m \in \mathbb{Z}$ such that $p_m m = 0$. Prove that the rank of $R$ is $n$, and there exists a prime $p \in \mathbb{Z}$ such that $\det R = \pm p^k$, with $k \leq n$.

**E. 10.61.** (→ p. 310) Let $M$ be the $\mathbb{Z}$-module generated by elements $m_1$, $m_2$, and $m_3$ satisfying

$$\begin{cases} 2m_1 - 4m_2 - 2m_3 = 0 \\ 10m_1 - 6m_2 + 4m_3 = 0 \\ 6m_1 - 12m_2 + am_3 = 0. \end{cases}$$

1. Express $M$ as a direct sum of cyclic modules, as $a$ varies in $\mathbb{Z}$.
2. Find the values of $a$, if any exist, such that $\operatorname{Ann} M = 0$.

**E. 10.62.** (→ p. 311) Let $G_1 = \langle a, b, c, d \rangle_{\mathbb{Z}}$, where $a, b, c, d$ satisfy

$$\begin{cases} 2a + 2b + c + 3d = 0 \\ -2b + c + 3d = 0 \\ -4a + 4b - 3c - 15d = 0 \\ 6a + 4b + c + 9d = 0 \\ 12a + 4b + c + 21d = 0, \end{cases}$$

and $G_2(\alpha) = \operatorname{Coker} \varphi_\alpha$, where $\varphi_\alpha \colon \mathbb{Z}^3 \longrightarrow \mathbb{Z}^3$ is the $\mathbb{Z}$-linear map defined by

$$\varphi_\alpha(x, y, z) = (2x + 8y - 4z,\ \alpha x + 6y + \alpha z,\ -2x - 2y + 4z),$$

with $\alpha \in \mathbb{Z}$.
Find the values of $\alpha$, if any exist, such that $G_1$ and $G_2(\alpha)$ are isomorphic.

**E. 10.63.** (→ p. 312) Let $\alpha$, $\beta \in \mathbb{N}$, and let $R = R_{\alpha,\beta} \in M_6(\mathbb{R})$ with characteristic polynomial

$$p_R(x) = (x - 1)^\alpha (x - 2)^\beta (x^2 + 1).$$

Find all values of $\alpha$ and $\beta$, if any exist, such that there are exactly 4 possible Smith forms of the characteristic matrix $R - xI$.

**E. 10.64.** (→ p. 312) Let $M$ be the $\mathbb{Z}$-module generated by elements $m_1$, $m_2$, $m_3$, $m_4$ satisfying the relations

$$3m_1 = 0, \ am_1 + 3m_2 = 0, \ \text{ and } \ bm_2 + 3m_3 = 0,$$

with $a$, $b \in \mathbb{Z}$ such that $\gcd(a, b) = 1$.
Describe the torsion submodule $T(M)$ of $M$ as $a$ and $b$ vary.

**E. 10.65.** (→ p. 313) Consider an integer $a$, let $N$ be the submodule of $\mathbb{Z}^3$ generated by

$$m_1 = (2, 2, a),\ m_2 = (2, a, 0), \ \text{ and } \ m_3 = (0, 4, 2),$$

and let $M = \mathbb{Z}^3/N$.

1. Determine $M$ up to isomorphism, as $a$ varies.
2. Find all the values of $a$, if any exist, such that $\mathrm{Hom}_{\mathbb{Z}}(\mathbb{Z}/(7), M) \neq 0$.

**E. 10.66.** (→ p. 314) Let $M = \langle m_1, m_2, m_3\rangle_{\mathbb{Z}}$, where

$$2m_1 = m_2,\ m_1 = 3m_2,\ m_1 + m_2 = am_3, \ \text{ and } \ a \in \mathbb{Z}.$$

1. Construct, if possible, a non-trivial homomorphism $\varphi\colon \mathbb{Z}/(20) \longrightarrow M$ when $a = 3$.
2. Describe $\mathrm{Hom}_{\mathbb{Z}}(\mathbb{Z}/(20), M)$ as $a$ varies.

**E. 10.67.** (→ p. 314) Let $\psi\colon \mathbb{Z}^3 \longrightarrow M$ be a surjective $\mathbb{Z}$-module homomorphism such that $\operatorname{Ker}\psi = \langle m_1, m_2, m_3\rangle$, where

$$m_1 = (2, 4, 6),\ m_2 = (0, a, 2a), \ \text{ and } \ m_3 = (b, 4, 6),$$

with $a$, $b$ varying in $\mathbb{Z}$.
Find all the values of $a$ and $b$, if any exist, such that $M$ is simple.

**E. 10.68.** (→ p. 315) For each of the following cases, prove that the ring $A$ is a finitely generated $K[x]$-module, and express it as a direct sum of cyclic modules.

1. $A = K[x, y, z]/(x^3 - y^2 + z,\ x^2 - y^2)$;
2. $A = K[x, y, z]/(zy - 1,\ y^2 - z + x^2)$;
3. $A = K[x, y]/(x^2 - y^2,\ x^4 - x^3y + y)$.

**E. 10.69.** (→ p. 316) Let $A = K[x, y, z]$ and $I = (x^2 + y^2 - z,\ xy - 1)$.

1. Prove that $A/I$ is a finitely generated $K[z]$-module by finding a finite generating set.
2. Decompose $A/I$ as a direct sum of cyclic $K[z]$-modules.

**E. 10.70.** (→ p. 316) Let

$$I = (z^2 + xy,\ x^2y - y^2z + z^2,\ x^2 + xy + 2yz,\ x^2 - yz) \subset K[x, y, z].$$

1. Prove that $I$ is monomial.
2. Determine a generating set of the $K[y]$-module $M = K[x, y, z]/I$.
3. Is $M$ free?
4. Represent $M$ as the cokernel of a $K[y]$-module homomorphism, and decompose it as a direct sum of cyclic modules.

**E. 10.71.** (→ p. 317) Let $A = K[x, y]/(y^3 - xy^2 - y + x,\ x^2 - xy + x - y)$.

1. Prove that $A$ is finitely generated as a $K[x]$-module.
2. Represent $A$ as the cokernel of a $K[x]$-module homomorphism, and decompose it as a direct sum of cyclic modules.

# Chapter 11
# Tensor Product

**E. 11.1.** (→ p. 319) Let $A$ be a ring, and let $M$, $N$, and $P$ be $A$-modules. For all $b, b' \in \mathrm{Bil}(M, N; P)$ and $\alpha \in A$ define

$$(b + b')(m, n) = b(m, n) + b'(m, n) \quad \text{and} \quad (\alpha b)(m, n) = \alpha b(m, n),$$

for all $m \in M$, $n \in N$.
Show that $\mathrm{Bil}(M, N; P)$, equipped with the above operations, is an $A$-module.

**E. 11.2.** (→ p. 319) Compute $\mathbb{Z}/(a) \otimes \mathbb{Z}/(b)$ when $\gcd(a, b) = 1$.

**E. 11.3.** (→ p. 319) Let $I, J$ be ideals of a ring $A$.
Prove that

$$A/I \otimes_A A/J \simeq A/(I + J).$$

**E. 11.4.** (→ p. 320) Let $A$ be a ring, and let $M$, $N$ be free $A$-modules.
Prove that $M \otimes N$ is free.

**E. 11.5.** (→ p. 321) 1. Prove that $\mathbb{Q} \otimes_{\mathbb{Z}} \mathbb{Q} \simeq \mathbb{Q}$.

2. Consider $\mathbb{C}$ with the $\mathbb{R}$-module structure given by restriction of scalars via the inclusion homomorphism $\mathbb{R} \longrightarrow \mathbb{C}$.
   Prove that in $\mathbb{C} \otimes_{\mathbb{R}} \mathbb{C}$ there exist non-simple tensors.
3. Can the same line of reasoning as in the proof of part 1 be used to prove that $\mathbb{C} \otimes_{\mathbb{R}} \mathbb{C} \simeq \mathbb{C}$?

**E. 11.6.** (→ p. 322) Let $A$ be a ring, and let $f \colon M \longrightarrow M'$ and $g \colon N \longrightarrow N'$ be $A$-module homomorphisms.
Prove that

$$f \otimes g \colon M \otimes N \longrightarrow M' \otimes N' \quad \text{defined by} \quad (f \otimes g)(m \otimes n) = f(m) \otimes g(n)$$

is an $A$-module homomorphism.



A. Bandini et al., *Commutative Algebra through Exercises*, UNITEXT 159,
https://doi.org/10.1007/978-3-031-56910-4_11

**E. 11.7.** (→ p. 322) Let $A$ be a ring, and let $f\colon M \longrightarrow M'$, $f'\colon M' \longrightarrow M''$, $g\colon N \longrightarrow N'$, and $g'\colon N' \longrightarrow N''$ be $A$-module homomorphisms.
Prove that
$$(f' \circ f) \otimes (g' \circ g) = (f' \otimes g') \circ (f \otimes g).$$

**E. 11.8.** (→ p. 322) Prove that $\mathbb{Q} \otimes_{\mathbb{Z}} \mathbb{Z}/(n) = 0$ for every $n \in \mathbb{N}_+$.

**E. 11.9.** (→ p. 322) Let $N_1$ and $N_2$ be $A$-modules.
Prove:

1. $N_1$ and $N_2$ are projective $\Longleftrightarrow N_1 \oplus N_2$ is projective;
2. $N_1$ and $N_2$ are projective $\Rightarrow N_1 \otimes N_2$ is projective;
3. the converse of part 2 does not hold in general;
4. $N_1$ and $N_2$ are flat $\Longleftrightarrow N_1 \oplus N_2$ is flat;
5. $N_1$ and $N_2$ are flat $\Rightarrow N_1 \otimes N_2$ is flat;
6. the converse of part 5 does not hold in general.

**E. 11.10.** (→ p. 323) Let $(A, \mathfrak{m}, K)$ be a local ring, and let $M$ and $N$ be finitely generated $A$-modules.
Prove that
$$\mu(M \otimes_A N) = \mu(M)\mu(N),$$
where $\mu(M) = \dim_K M/\mathfrak{m}M$.

**E. 11.11.** (→ p. 324) Let $(A, \mathfrak{m}, K)$ be a local ring, and let $M$ and $N$ be finitely generated $A$-modules.
Prove that
$$M \otimes_A N = 0 \text{ implies } M = 0 \text{ or } N = 0.$$

**E. 11.12.** (→ p. 324) Let $p \in \mathbb{Z}$ be a prime, and let
$$M = \left\{ \tfrac{a}{p^n} \colon a \in \mathbb{Z},\ n \in \mathbb{N} \right\}.$$
Verify that $M$ is a $\mathbb{Z}$-module and prove that $M \otimes_{\mathbb{Z}} (M/\mathbb{Z}) = 0$.

**E. 11.13.** (→ p. 324) 1. Compute the dimension of the $\mathbb{Q}$-vector space
$$\mathbb{Q}[x]/(x) \otimes_{\mathbb{Q}[x]} \mathbb{Q}[x]/(x^2 + 1).$$

2. Let $\alpha = \sqrt[5]{3}$. Compute $\dim_{\mathbb{C}}(\mathbb{C} \otimes_{\mathbb{Q}} \mathbb{Q}[\alpha])$.

**E. 11.14.** (→ p. 324) Let $(A, \mathfrak{m}, K)$ be a local ring.
Prove that every finitely generated projective $A$-module is free.

**E. 11.15.** (→ p. 325) We say that a ring homomorphism $f\colon A \longrightarrow B$ is *flat* if $B$ is a flat $A$-module by restriction of scalars via $f$.
Let $I$, $J$ be ideals of $A$, and let $f\colon A \longrightarrow B$ be a flat ring homomorphism. Prove that

$$(I \cap J)B = IB \cap JB.$$

**E. 11.16.** (→ p. 325) Let $A$ be a ring and $a \in A$.
Prove that the following are equivalent:

1. $(a) = (a^2)$;
2. $(a)$ is a direct summand of $A$;
3. $A/(a)$ is a flat $A$-module.

**E. 11.17.** (→ p. 326) Let $M$ be an $A$-module such that $\mathfrak{m}M \neq M$ for all $\mathfrak{m} \in \operatorname{Max} A$.
Prove:

1. $M/IM \neq 0$ for every proper ideal $I$ of $A$;
2. if $M$ is a flat $A$-module, then $M \otimes N \neq 0$ for every $A$-module $N \neq 0$.

**E. 11.18.** (→ p. 326) Let $A$ be a ring, and let $a \in A \setminus \mathcal{D}(A)$.
Prove that if $N$ is a flat $A$-module, then $an \neq 0$ for all $n \in N \setminus \{0\}$.

**E. 11.19.** (→ p. 326) Let $A = K[x, y]$, $I = (x)$, and $J = (y)$.
Prove:

1. $I$, $J$, and $I \cap J$ are free $A$-modules;
2. $I + J$ is torsion-free, but not flat.

**E. 11.20.** (→ p. 326) Let $M$ and $N$ be free $A$-modules of finite rank.
Prove that

$$\operatorname{End}_A M \otimes \operatorname{End}_A N \simeq \operatorname{End}_A(M \otimes N).$$

**E. 11.21.** (→ p. 327) Consider $M = \mathbb{Z}/(15)$. Let $\varphi\colon \mathbb{Z}^2 \longrightarrow \mathbb{Z}^3$ be the homomorphism given by

$$\varphi(x, y) = (4x + 8y,\ 4x - 4y,\ 16x + 20y).$$

Compute $M \otimes_{\mathbb{Z}} T(\operatorname{Coker} \varphi)$.

**E. 11.22.** (→ p. 327) Let $a \in \mathbb{N}_+$, and let $M_a$ be the $\mathbb{Z}$-module generated by elements $m_1$, $m_2$, $m_3$ satisfying the relations

$$2m_1 - m_2 = 0,\ \ m_1 + m_2 + m_3 = 0\ \text{ and }\ m_1 + am_2 = 0.$$

Determine the values of $n \in \mathbb{N}$, if any exist, such that $M_a \otimes_{\mathbb{Z}} \mathbb{Z}/(n) \neq 0$, as $a$ varies.

# Chapter 12
# Localization

**E. 12.1.** ($\to$ p. 328) Let $A$ be a ring and $a \notin \mathcal{N}(A)$.
Prove that there exists a prime ideal $\mathfrak{p}$ of $A$ such that $a \notin \mathfrak{p}$.

**E. 12.2.** ($\to$ p. 328) Let $A$ be a ring, and let $S \subset A$ be a multiplicative subset.
Prove that $\sigma_S$ is an isomorphism if and only if $S \subseteq A^*$.

**E. 12.3.** ($\to$ p. 328) Let $A$ be a finite ring, and let $S \subset A$ be a multiplicative subset such that $\sigma_S$ is injective.
Prove that $S^{-1}A \simeq A$.

**E. 12.4.** ($\to$ p. 328) Describe the ring $S^{-1}A$ in the following situations:

1. $A = \mathbb{Z}$ and $S = A \setminus (p)$ with $p$ prime;
2. $A = \mathbb{Z}$ and $S = A \setminus \bigcup_{i=1}^{n} (p_i)$ with $p_i$ distinct primes;
3. $A = \mathbb{Z}/(12)$ and $S = \{\overline{2}^n : n \in \mathbb{N}\}$;
4. $A = \mathbb{Z}/(12)$ and $S = A \setminus (\overline{2})$;
5. $A = \mathbb{Z}/(12)$ and $S = A \setminus (\overline{3})$.

**E. 12.5.** ($\to$ p. 329) Let $A$ be a finite ring, and let $S \subset A$ be a multiplicative subset.

1. Prove that $\sigma_S$ is surjective.
2. Let $A = \mathbb{Z}/(24)$ and $S = \{\overline{2}^n : n \in \mathbb{N}\}$.
   Find $\operatorname{Ker} \sigma_S$ and describe $S^{-1}A$.

**E. 12.6.** ($\to$ p. 329) Let $A$ be a ring, and let $I \subset A$ be an ideal.

1. Let
$$S = 1 + I = \{1 + i \colon i \in I\}.$$
   Prove that $S$ is multiplicative and that $S^{-1}I \subseteq \mathcal{J}(S^{-1}A)$.


A. Bandini et al., *Commutative Algebra through Exercises*, UNITEXT 159,
https://doi.org/10.1007/978-3-031-56910-4_12

2. Let $A = \mathbb{Z}/(60)$ and $S = 1 + \overline{4}A$.
   Describe all ideals of $S^{-1}A$. Is $S^{-1}A$ a local ring?
3. Let again $A = \mathbb{Z}/(60)$.
   Is there an $\overline{m} \neq \overline{4}$ such that $T^{-1}A = (1 + \overline{m}A)^{-1}A$ is not local?

**E. 12.7.** (→ p. 330) Let $A$ and $B$ be rings with $B \neq 0$, and let $C$ be the ring $C = A \times B$.
Prove:

1. if $S = \{1\} \times B \subset C$, then $S^{-1}C \simeq A$;
2. if $T = \{(1,0), (1,1)\} \subset C$, then $T^{-1}C \simeq A$;
3. the relation defined on the elements of $C \times T$ as

$$(x, s) \sim (y, t) \text{ if and only if } xt = ys$$

   is not an equivalence relation.

**E. 12.8.** (→ p. 330) Let $A$ be a ring, and let $f \in A$.
Prove that

$$A_f \simeq A[x]/(1 - fx).$$

**E. 12.9.** (→ p. 331) Let $\mathbb{Z}\left[\frac{2}{3}\right] = \left\{p\left(\frac{2}{3}\right) : p(x) \in \mathbb{Z}[x]\right\} \simeq \mathbb{Z}[x]/(3x - 2)$.

1. Prove that $\mathbb{Z}\left[\frac{2}{3}\right] \simeq \mathbb{Z}_3$.
2. Find all rings $A$ such that $\mathbb{Z} \subsetneq A \subsetneq \mathbb{Q}$, and describe them as localizations of $\mathbb{Z}$.

**E. 12.10.** (→ p. 331) Let $I = (yz - y,\ xy + 2z^2,\ y - z) \subset \mathbb{Q}[x, y, z]$, and let $f = x^3z - y^2$, see **E.9.9**. Let also $J = I\mathbb{Q}[x, y, z]_{(x,y,z)}$.
Does the image of the polynomial $f$ in $\mathbb{Q}[x, y, z]_{(x,y,z)}$ belong to $J$?

**E. 12.11.** (→ p. 332) Let $I = (x^2 + 2y^2 - 3,\ x^2 + xy + y^2 - 3) \subset \mathbb{C}[x, y]$, see **E.9.13**. Let also $\mathfrak{p}_1 = (x - 1,\ y - 1)$ and $\mathfrak{p}_2 = (x,\ y)$.
Describe $I_{\mathfrak{p}_1} \cap \mathbb{C}[x, y]$ and $I_{\mathfrak{p}_2} \cap \mathbb{C}[x, y]$.

**E. 12.12.** (→ p. 332) Let $K$ be a field of characteristic different from 2, let $I = (x^2 + y^2,\ x^3y^3 + y^4) \subset K[x, y]$, and let $f = x^2 + 5x$, see **E.9.15**.
Verify whether the image of $f$ in $K[x, y]_{(x,y)}$ is an element of $\sqrt{I_{(x,y)}}$.

**E. 12.13.** (→ p. 332) Let $I = (x^2y + xz + yz,\ y^2z)$ be an ideal of $\mathbb{R}[x, y, z]$, and let $A = \mathbb{R}[x, y, z]/I$, see **E.9.18**. Let also $\mathfrak{p} = (\overline{x},\ \overline{z})A$.
Describe $A_{\mathfrak{p}}$.

**E. 12.14.** (→ p. 332) Let $A = K[x, y, z]$ and $I = (xz - y,\ yz - x) \subset A$, see **E.9.24**. Let also $S = A \setminus (x,\ y)$.
Describe $S^{-1}(A/I)$.

**E. 12.15.** (→ p. 332) Let $I = (y^2 - xz,\ x^2 - y^2,\ x^2 - yz) \subset \mathbb{Q}[x, y, z]$, and let $A = \mathbb{Q}[x, y, z]/I$, see **E.9.34**. Let also $S = A \setminus (\overline{x},\ \overline{y})A$.
Describe $S^{-1}A$.

**E. 12.16.** (→ p. 333) Let $p$ be a prime in $\mathbb{Z}$.
Prove that $\mathbb{Z}_p \otimes_{\mathbb{Z}} (\mathbb{Z}_p/\mathbb{Z}) = 0$.

**E. 12.17.** (→ p. 333) Let $A$ be a ring, and let $S,\ T \subseteq A$ be two multiplicative subsets.
Prove:

1. if $S \subseteq T$ and $T_1 = \sigma_S(T)$, then $T^{-1}A \simeq T_1^{-1}(S^{-1}A)$;
2. $\sigma_S(T)^{-1}(S^{-1}A) \simeq \sigma_T(S)^{-1}(T^{-1}A)$.

**E. 12.18.** (→ p. 334) Let $p$ be a prime in $\mathbb{Z}$.
Prove that

$$\mathbb{Z}_2 \otimes_{\mathbb{Z}} \mathbb{Z}_{(p)} \simeq \begin{cases} (\mathbb{Z}_2)_{(p)\mathbb{Z}_2} & \text{if } p \neq 2; \\ \mathbb{Q} & \text{if } p = 2. \end{cases}$$

**E. 12.19.** (→ p. 335) [**Total Quotient Ring**] Let $A$ be a ring, and consider the multiplicative subset $S = A \setminus \mathcal{D}(A)$.
We call the ring $S^{-1}A$ the *total quotient ring* or *total ring of fractions* of $A$, and we denote it by $Q(A)$.
Prove:

1. $S$ is the largest multiplicative subset such that $\sigma_S\colon A \longrightarrow S^{-1}A$ is injective;
2. each element of $Q(A)$ is either invertible or a zero-divisor;
3. if $A = A^* \sqcup \mathcal{D}(A)$, then $\sigma_S$ is an isomorphism;
4. if $A$ is a domain, then $Q(A)$ is the smallest field containing $A$.
   In this case $Q(A)$ is called the *quotient field* or *fraction field* of $A$.

**E. 12.20.** (→ p. 335) Let $\mathfrak{p} \subset A$ be a prime ideal of a ring $A$.
Prove that

$$Q(A/\mathfrak{p}) \simeq A_{\mathfrak{p}}/\mathfrak{p}A_{\mathfrak{p}}.$$

**E. 12.21.** (→ p. 336) Let $A$ be a domain with finitely many prime ideals, and let $Q(A)$ be its fraction field.
Prove that there exists an element $a \in A$ such that $Q(A) = A_a$.

**E. 12.22.** (→ p. 336) Let $B$ be a ring, and let $\mathfrak{p}_1, \ldots, \mathfrak{p}_n$ be prime ideals of $B$ such that $\mathfrak{p}_i \not\subset \mathfrak{p}_j$, when $i \neq j$.

1. Let $A = B/\mathfrak{p}_1 \cap \ldots \cap \mathfrak{p}_n$.
   Prove that
$$Q(A) \simeq \bigoplus_{i=1}^{n} Q(B/\mathfrak{p}_i).$$
2. Let $A = \mathbb{C}[x, y]/(xy)$ and $S = A \setminus \mathcal{D}(A)$.
   Describe $S^{-1}A$.
3. Let $A = \mathbb{C}[x, y]/(x^2 - y^3)$ and $S = A \setminus \mathcal{D}(A)$.
   Describe $S^{-1}A$.

**E. 12.23.** (→ p. 337) Let $K$ be a field, and let $B \subset K$ be a ring which is not a field.
Prove that $K$ is not finitely generated as a $B$-module.

**E. 12.24.** (→ p. 337) Let $A = K[x]_{(x)}$.
Prove:

1. $Q(A)/(x)Q(A)$ is a finitely generated $A$-module;
2. $Q(A)$ is not finitely generated as an $A$-module.

**E. 12.25.** (→ p. 337) Let $A$ be a ring, let $S \subset A$ be a multiplicative subset, and let $M$ be an $A$-module.

1. Prove that if $\operatorname{Ann} M \cap S \neq \emptyset$, then $S^{-1}M = 0$.
2. Prove that if $M$ is finitely generated, then the converse of part 1 also holds.
3. Provide an example of an $A$-module $M$ and a multiplicative subset $S$ such that $S^{-1}M = 0$ and $\operatorname{Ann} M \cap S = \emptyset$.

**E. 12.26.** (→ p. 338) Let $A$ be a ring, and let $\{f_h\}_{h \in H} \subset A$ be such that $(f_h : h \in H) = A$. Let also $M$ be an $A$-module, and consider $m \in M$.
Prove that if the image of $m$ is zero in $M_{f_h}$ for every $h \in H$, then $m = 0$.

**E. 12.27.** (→ p. 338) Let $M$ be an $A$-module.
We define the *support* of $M$ to be the set
$$\operatorname{Supp} M = \{\mathfrak{p} \in \operatorname{Spec} A : M_{\mathfrak{p}} \neq 0\}.$$

Prove:

1. if $M$ is finitely generated, then $\mathfrak{p} \in \operatorname{Supp} M$ if and only if $\mathfrak{p} \supseteq \operatorname{Ann} M$;
2. if $0 \longrightarrow M' \longrightarrow M \longrightarrow M'' \longrightarrow 0$ is an exact sequence, then
$$\operatorname{Supp} M = \operatorname{Supp} M' \cup \operatorname{Supp} M'';$$

3. if $M$ and $N$ are finitely generated, then

$$\operatorname{Supp}(M \otimes_A N) = \operatorname{Supp} M \cap \operatorname{Supp} N.$$

**E. 12.28.** (→ p. 338) Let $M = \mathbb{Z}/(10) \oplus \mathbb{Z}/(12)$.

1. Find all primes $p \in \mathbb{Z}$ such that $M_{(p)} \neq 0$.
2. Describe $M_{(3)}$.

**E. 12.29.** (→ p. 338) Let $A = \mathbb{Q}[x, y, z]$, and let $M$ be the $A$-module

$$A/(xyz - z^2,\, xy^2 - 4) \otimes_A A/(yz,\, x - y^2).$$

1. Compute the dimension of $M$ as a $\mathbb{Q}$-vector space.
2. Find the support of $M$.

**E. 12.30.** (→ p. 339) Let $A$ be a ring, and let $0 \neq I \subset A$ be a finitely generated ideal such that either $I_{\mathfrak{m}} = 0$ or $I_{\mathfrak{m}} = A_{\mathfrak{m}}$ for every maximal ideal $\mathfrak{m} \subset A$. Prove that $I$ is principal and generated by an idempotent element.

**E. 12.31.** (→ p. 339) Let

$$p(x) = (x-1)^2(x^2+2) \in \mathbb{Q}[x], \quad S = \{q(x) \in \mathbb{Q}[x] \colon (q(x), p(x)) = 1\},$$

and let $A = S^{-1}\mathbb{Q}[x]$.

1. Prove that $S$ is a multiplicative subset of $\mathbb{Q}[x]$.
2. Prove that $|\operatorname{Spec} A| = 3$.
3. Describe $A/\mathfrak{p}$ for any $\mathfrak{p} \in \operatorname{Spec} A$.
4. Describe $A/\mathfrak{p} \otimes_A A/\mathfrak{q}$ for any pair of ideals $\mathfrak{p} \neq \mathfrak{q}$ in $\operatorname{Spec} A$.

**E. 12.32.** (→ p. 340) Let $M$ be an $A$-module, and let $I \subset A$ be an ideal such that $M_{\mathfrak{m}} = 0$ for any maximal ideal $\mathfrak{m}$ containing $I$.
Prove that $M = IM$.

**E. 12.33.** (→ p. 340) Let $A$ be a ring, and let $\mathfrak{p} \in \operatorname{Min} A$.
Prove:

1. every element of $\mathfrak{p}A_{\mathfrak{p}}$ is nilpotent;
2. every element of $\mathfrak{p}$ is a zero-divisor in $A$;
3. if $A$ is reduced, then $A_{\mathfrak{p}}$ is a field.

**E. 12.34.** (→ p. 340) Let $S$ be a multiplicative subset of a ring $A$.
Prove that $\overline{S}$, the saturation of $S$, is a saturated multiplicative subset.

**E. 12.35.** ($\rightarrow$ p. 340) Let $A$ and $B$ be rings, let $T \subset B$ be a subset, and let $f\colon A \longrightarrow B$ be a ring homomorphism.
Prove:

1. if $T$ is multiplicative, then $f^{-1}(T)$ is multiplicative;
2. if $T$ is saturated, then $f^{-1}(T)$ is saturated;
3. if $f$ is surjective, then the converse of parts 1 and 2 also hold.

**E. 12.36.** ($\rightarrow$ p. 341) Let $A$ be a ring, let $S \subset A$ be a multiplicative subset, and let $\sigma_S\colon A \longrightarrow S^{-1}A$ be the localization homomorphism.
For any ideal $I \subset A$, we define the *saturation* of $I$ with respect to $S$ to be the set

$$I^S = \bigcup_{s \in S} I : (s) = \{a \in A\colon \text{ there exists } s \in S \text{ such that } as \in I\}.$$

We say that $I$ is *saturated with respect to* $S$ if $I = I^S$.
Prove:

1. $I \subseteq I^S$ and $I^S$ is an ideal;
2. $(0)^S = \operatorname{Ker} \sigma_S$;
3. if $J \subset A$ is an ideal, and $I \subseteq J$, then $I^S \subseteq J^S$;
4. $(I^S)^S = I^S$;
5. for any ideal $J$ of $A$, we have $(I^S J^S)^S = (IJ)^S$.

**E. 12.37.** ($\rightarrow$ p. 341) Let $A$ be a ring, and let $S \subset A$ be a multiplicative subset. Let also $M$ be an $A$-module, and let $\sigma_S\colon M \longrightarrow S^{-1}M$ be the localization homomorphism.
For any submodule $N \subseteq M$, we define the *saturation* of $N$ with respect to $S$ to be the set

$$N^S = \bigcup_{s \in S} N :_M s = \{m \in M\colon \text{ there exists } s \in S \text{ such that } sm \in N\}.$$

We say that $N$ is *saturated with respect to* $S$ if $N = N^S$.
Prove:

1. $N \subseteq N^S$ and $N^S$ is a submodule of $M$;
2. if $P$ is a submodule of $M$, and $N \subseteq P$, then $N^S \subseteq P^S$;
3. if $Q$ is a submodule of $S^{-1}M$, then $\sigma_S^{-1}(Q) = \sigma_S^{-1}(Q)^S$;
4. $N^S = \sigma_S^{-1}(S^{-1}N)$, and, in particular, $(0)^S = \operatorname{Ker} \sigma_S$;
5. $(N^S)^S = N^S$;
6. for any submodule $P \subset M$, we have $(N \cap P)^S = N^S \cap P^S$;
7. for any submodule $P \subset M$, we have $N^S + P^S \subseteq (N + P)^S$.

**E. 12.38.** (→ p. 342) Let

$$A = (\mathbb{Z}/(200))_{\overline{18}}, \quad B = (\mathbb{Z}/(200))_{\overline{6}},$$

$$C = \mathbb{Z}/(25) \otimes_{\mathbb{Z}} \mathbb{Z}/(40), \quad \text{and} \quad D = \mathbb{Z}_{(3)}[x]/(6x-1).$$

Determine which pairs of these rings are isomorphic.

**E. 12.39.** (→ p. 343) Let $A$ be a ring, and let $S \subset A$ be a multiplicative subset such that $S^{-1}A \neq 0$.
Prove that $S$ is maximal with respect to inclusion among multiplicative subsets $U \subset A$ such that $U^{-1}A \neq 0$ if and only if $A \setminus S$ is a minimal prime.

**E. 12.40.** (→ p. 343) Let $S = \{24^n\}_{n\in\mathbb{N}}$ and $T = \{4^n 6^m\}_{n,m\in\mathbb{N}}$.
Prove that $S^{-1}\mathbb{Z} = T^{-1}\mathbb{Z}$.

**E. 12.41.** (→ p. 343) Let $A$ be a ring, let $f$, $g \in A$, and let $I \subset A$ be an ideal.
Prove:

1. the saturation of $S = 1 + I$ is
$$\overline{S} = A \setminus \bigcup_{\mathfrak{p}\in\mathcal{V}(I)} \mathfrak{p},$$
where $\mathcal{V}(I) = \{\mathfrak{p} \in \operatorname{Spec} A \colon \mathfrak{p} \supseteq I\}$;
2. $\overline{S_f} \subseteq \overline{S_g}$ if and only if $\sqrt{(f)} \supseteq \sqrt{(g)}$.

**E. 12.42.** (→ p. 344) Let $K$ be a field of characteristic different from 2.
Let $A = K[x,y]/(x^2 - y^2)$, $\mathfrak{p} = (x+y)A$, and $\mathfrak{q} = (x,y)A$.
Describe:

1. the prime ideals of $A_{\mathfrak{q}}$.
2. $(A_{\mathfrak{q}})_{\mathfrak{p}A_{\mathfrak{q}}}$.

# Chapter 13
# Noetherian and Artinian Modules

**E. 13.1.** (→ p. 345) Let $A$ be a ring.
Prove:

1. if $A = \mathbb{Z}$, then $A$ satisfies a.c.c. but not d.c.c.;
2. if $A = \mathbb{R}[x]/(x^2 + 2)$, then $A$ satisfies both a.c.c. and d.c.c..

**E. 13.2.** (→ p. 345) Let $K$ be a field, and let $V$ be a $K$-vector space.
Prove that the following are equivalent:

1. $\dim_K V < \infty$;
2. $V$ is a Noetherian $K$-module;
3. $V$ is an Artinian $K$-module.

**E. 13.3.** (→ p. 345) Find an alternative proof of **T.7.2**.2 using the characterization of Noetherian modules given in **T.7.2**.1.

**E. 13.4.** (→ p. 345) Let $A$ be a Noetherian ring, and let $\varphi\colon A \longrightarrow A$ be a surjective homomorphism.

1. Prove that $\varphi$ is injective.
2. Prove that $\varphi(I \cap J) = \varphi(I) \cap \varphi(J)$ for all ideals $I$, $J \subset A$.
3. Do the previous statements hold when $A$ is not Noetherian?

**E. 13.5.** (→ p. 346) Let $A$ be a Noetherian ring, and let $I \subset A$ be an ideal such that $I = I^2$.
Prove that $I$ is principal, and generated by an idempotent.

**E. 13.6.** (→ p. 346) Let $N_1$ and $N_2$ be submodules of an $A$-module $M$.
Prove that if $M/N_1$ and $M/N_2$ are Noetherian, then $M/(N_1 \cap N_2)$ is also Noetherian.



A. Bandini et al., *Commutative Algebra through Exercises*, UNITEXT 159,
https://doi.org/10.1007/978-3-031-56910-4_13

**E. 13.7.** (→ p. 346) Let $A = K[x]/(fg^2)$, with $(f, g) = 1$.
Prove that an $A$-module $M$ is Noetherian if and only if $M/\overline{f}M$ and $M/\overline{g}^2M$ are Noetherian.

**E. 13.8.** (→ p. 347) Let $A$ be a domain such that for every non-zero ideal $I$, there exist an ideal $J \neq 0$ and an element $d \in A$ such that $IJ = (d)$.
Prove:

1. there exists a finitely generated ideal $\widetilde{J} = (g_1, \dots, g_k)$ such that $I\widetilde{J} = (d)$;
2. $A$ is Noetherian.

**E. 13.9.** (→ p. 347) Let $f\colon A \longrightarrow C$ and $g\colon B \longrightarrow C$ be surjective ring homomorphisms. Define $A \times_C B = \{(a, b) \in A \times B\colon f(a) = g(b)\}$.
Prove:

1. $A \times_C B$ is a subring of $A \times B$;
2. if $A$ and $B$ are Noetherian, then $A \times_C B$ is Noetherian.

**E. 13.10.** (→ p. 347) Let $A$ be a local ring with principal maximal ideal $\mathfrak{m} = (m)$.
Prove:

1. every $0 \neq a \in \mathfrak{m}$ admits a factorization $a = um^k$, where $u$ is invertible and $k$ is a positive integer, if and only if $\bigcap_{n \in \mathbb{N}} \mathfrak{m}^n = 0$;
2. if $\bigcap_{n \in \mathbb{N}} \mathfrak{m}^n = 0$ and $0 \neq I \subsetneq A$ is a proper ideal, then $I = \mathfrak{m}^h$ for some positive integer $h$;
3. if $A$ is Noetherian, then $A$ is a PIR.

**E. 13.11.** (→ p. 348) Let $A$ be a Noetherian ring.
Prove that, if every maximal ideal of $A$ is principal, then $\dim A \leq 1$.

**E. 13.12.** (→ p. 348) Let $M$ be a Noetherian $A$-module.
Prove that $A/\operatorname{Ann}_A M$ is Noetherian.

**E. 13.13.** (→ p. 348) Let $A$ be a Noetherian ring, and let $I$, $J$ be ideals of $A$ such that every prime ideal of $A$ contains either $I$ or $J$, but not both.
Prove:

1. $A = I + J$;
2. there exists $n \in \mathbb{N}_+$ such that $(IJ)^n = 0$.

**E. 13.14.** (→ p. 349) Let $(A, \mathfrak{m}, K)$ be a Noetherian local ring.
Prove:

1. if $I \subseteq A$ is an ideal and $\mu(I) > 1$, then $\mu(I^2) < \mu(I)^2$, where, for any ideal $J$ of $A$, $\mu(J) = \dim_K J/\mathfrak{m}J$;

2. if every ideal of $A$ is a flat $A$-module, then $A$ is a PID.

**E. 13.15.** (→ p. 349) Let $A$ be a ring. Given a prime ideal $\mathfrak{p} \in \operatorname{Spec} A$ we define its *height* $\operatorname{ht} \mathfrak{p}$ as the supremum of the lengths of strictly ascending chains of prime ideals contained in $\mathfrak{p}$:

$$\operatorname{ht} \mathfrak{p} = \sup\{l\colon\ \mathfrak{p}_0 \subsetneq \mathfrak{p}_1 \subsetneq \ldots \subsetneq \mathfrak{p}_l = \mathfrak{p},\ \ \mathfrak{p}_i \in \operatorname{Spec} A\}.$$

Given any ideal $I \subset A$, we define its *height* $\operatorname{ht} I$as the infimum of the heights of the prime ideals containing it:

$$\operatorname{ht} I = \inf\{\operatorname{ht} \mathfrak{p}\colon\ \mathfrak{p} \in \mathcal{V}(I)\}.$$

Let $A$ be a Noetherian ring, and let $I$, $J$ be ideals of $A$.
Prove:

1. $\sqrt{I} = \sqrt{J}$ if and only if $I$ and $J$ have the same minimal primes;
2. if $\sqrt{I} = \sqrt{J}$, then $\operatorname{ht} I = \operatorname{ht} J$ and $\dim A/I = \dim A/J$.

**E. 13.16.** (→ p. 350) Let $A$ be a Noetherian ring, and let $M$ be an $A$-module such that $0 \neq \operatorname{Ann} M$ is 0-dimensional.
Prove that $M$ contains a non-zero simple submodule.

**E. 13.17.** (→ p. 350) Let $A$ be a Noetherian ring and let $\mathfrak{p} \in \operatorname{Spec} A$.
Prove that $\mathfrak{p}$ is a minimal prime of $A$ if and only if there exist a non-nilpotent element $a \in A$ and $n \in \mathbb{N}$ such that $a\mathfrak{p}^n = 0$.

**E. 13.18.** (→ p. 350) Let $A$ be a Noetherian ring, and let $J \subset A$ be an ideal. Prove that if $J$ contains a 0-dimensional radical ideal, then either $J$ is 0-dimensional and radical or $J = (1)$.

**E. 13.19.** (→ p. 351) Find the minimal primes of the ideal

$$I = (x^2zt,\ yt^3,\ xyzt,\ x^5z^6) \subset K[x, y, z, t].$$

**E. 13.20.** (→ p. 351) Let $A = \mathbb{Q}[x, y, z, t]$ and

$$I = (x^2z - x^2t^3,\ x^2y^4t + x^2y^3 - x^3z,\ xt^2) \subset A.$$

1. Determine whether $I$ is a monomial ideal.
2. Find a primary decomposition of $I$ and determine its associated and minimal primes.
3. Find $\mathcal{N}(A/I)$ and $\mathcal{D}(A/I)$.

**E. 13.21.** (→ p. 351) Let $A = \mathbb{Q}[x]/(x^5 - 3x^2) \oplus \mathbb{Z}/(12)$.
Find:

1. $\mathcal{N}(A)$ and $\mathcal{D}(A)$;
2. the ideals $\mathfrak{p} \in \operatorname{Spec} A$ such that $A_\mathfrak{p}$ is a field, if any exist.

**E. 13.22.** (→ p. 352) Let $I = (3x^2 + 10y - 2,\ (x+y)^3,\ 45) \subset \mathbb{Z}[x,y]$, and let $A = \mathbb{Z}[x,y]/I$.
Find:

1. $\mathcal{D}(A)$;
2. the ideals $\mathfrak{p} \in \operatorname{Spec} A$ such that $A_\mathfrak{p}$ is not a domain, if any exist.

**E. 13.23.** (→ p. 353) Let $K = \overline{K}$, and let $I \subsetneq A = K[x_1, \dots, x_n]$ be an ideal.
Prove that the following are equivalent:

1. $\dim A/I = 0$;
2. $I \cap K[x_i] \neq 0$ for all $i = 1, \dots, n$.

**E. 13.24.** (→ p. 353) [**Computation of the Radical of a 0-dimensional Ideal**] Assume $K = \overline{K}$, and let $I \subsetneq K[x_1, \dots, x_n]$ be a 0-dimensional ideal. For all $i = 1, \dots, n$, let $0 \neq h_i \in I \cap K[x_i]$, and let $\sqrt{h_i}$ be its squarefree part. Prove that

$$\sqrt{I} = (I, \sqrt{h_1}, \dots, \sqrt{h_n}).$$

**E. 13.25.** (→ p. 354) Let $M$ be an $A$-module.
Define *the set of associated primes of* $M$ as

$$\operatorname{Ass} M = \{\mathfrak{p} \in \operatorname{Spec} A \colon \text{ there exists } m \in M \setminus \{0\} \text{ such that } \mathfrak{p} = \operatorname{Ann} m\}.$$

Prove:

1. the maximal elements of the set $\Sigma = \{\operatorname{Ann} m \colon 0 \neq m \in M\}$ are prime, and thus, they belong to $\operatorname{Ass} M$;
2. if $A$ is Noetherian and $M \neq 0$, then $\operatorname{Ass} M \neq \emptyset$.

**E. 13.26.** (→ p. 354) Let $M$ be an $A$-module, and let $\mathfrak{p} \in \operatorname{Spec} A$.
Prove that $\mathfrak{p} \in \operatorname{Ass} M$ if and only if there exists an injective homomorphism $A/\mathfrak{p} \longrightarrow M$, *i.e.*, if and only if $M$ contains a submodule isomorphic to $A/\mathfrak{p}$.

**E. 13.27.** (→ p. 354) Let $A$ be a ring.
Prove:

1. if $0 \longrightarrow N \xrightarrow{f} M \xrightarrow{g} P \longrightarrow 0$ is an exact sequence of $A$-modules, then

$$\operatorname{Ass} N \subseteq \operatorname{Ass} M \subseteq \operatorname{Ass} N \cup \operatorname{Ass} P;$$

2. if $A \neq 0$ is Noetherian and $M \neq 0$ is a Noetherian $A$-module, then there exist a chain of submodules $M_0 \subsetneq M_1 \subsetneq \ldots \subsetneq M_t = M$ and prime ideals $\mathfrak{p}_1, \ldots, \mathfrak{p}_t$, such that

$$M_i/M_{i-1} \simeq A/\mathfrak{p}_i \quad \text{for all } i;$$

3. [**Third Finiteness Theorem**] if $A$ is Noetherian ring and $M$ is a finitely generated $A$-module, then

$$\operatorname{Ass} M \quad \text{is finite.}$$

**E. 13.28.** ($\rightarrow$ p. 355) Let $M$ be an $A$-module, and let $I \subseteq \operatorname{Ann} M$.
Prove that $M$ is Artinian as an $A$-module if and only if it is Artinian as an $A/I$-module.

**E. 13.29.** ($\rightarrow$ p. 355) Let $M$ be an Artinian $A$-module, and let $\varphi\colon M \longrightarrow M$ be an injective homomorphism.
Prove that $\varphi$ is an isomorphism.

**E. 13.30.** ($\rightarrow$ p. 355) Let $A$ be an Artinian ring.
Prove:

1. $A = A^* \sqcup \mathcal{D}(A)$;
2. if $A$ is local and $S \subseteq A$ is a multiplicative subset, then the homomorphism $\sigma_S\colon A \longrightarrow S^{-1}A$ is surjective.

**E. 13.31.** ($\rightarrow$ p. 355) Let $M$ be an $A$-module, and let $\mathfrak{m}_1, \ldots, \mathfrak{m}_n$ be maximal ideals of $A$ not necessarily distinct and such that $\left(\prod_{i=1}^{n} \mathfrak{m}_i\right) M = 0$.
Prove that $M$ is Noetherian if and only if $M$ is Artinian.

# Chapter 14
# True or False?

For each of the exercises in this chapter, decide whether the statements are true or false. Support your answer with a proof or a counterexample.

**ToF. 14.1.** ($\rightarrow$ p. 357) Let $A$ be a ring, and let $I \subset A$ be an ideal.
An element $a \in A$ is not a zero-divisor in $A/I$ if and only if $I : (a) = I$.

**ToF. 14.2.** ($\rightarrow$ p. 357) Let $A$ be a domain, and let $Q$ be its quotient field.
Then,

$$Q[x] \otimes_{A[x]} Q[x] \simeq Q[x].$$

**ToF. 14.3.** ($\rightarrow$ p. 357) Let $A$ be a PID, let $B$ be a domain, and let $\varphi\colon A \longrightarrow B$ be a surjective homomorphism.
Then, either $B$ is a field or $\varphi$ is an isomorphism.

**ToF. 14.4.** ($\rightarrow$ p. 357) Let $I$, $J$, and $H \subset A$ be ideals.
Then:

1. $\sqrt{I + JH} = \sqrt{I + J} \cap \sqrt{I + H}$.
2. $\sqrt{I + \sqrt{J}} = \sqrt{I + J}$.
3. $\sqrt{I} + \sqrt{J} = \sqrt{I + J}$.

**ToF. 14.5.** ($\rightarrow$ p. 357) In any ring $A$, the sum of ideals is distributive with respect to intersection.

**ToF. 14.6.** ($\rightarrow$ p. 357) Let $A$ be a ring, and let $I \subset A$ be an ideal such that $\mathcal{N}(A/I) = (\bar{0})$.
Then, $I$ is prime.



A. Bandini et al., *Commutative Algebra through Exercises*, UNITEXT 159,
https://doi.org/10.1007/978-3-031-56910-4_14

**ToF. 14.7.** (→ p. 357) The $\mathbb{Q}[x]$-module $\mathbb{Q}[x]/(x^2-1) \otimes_{\mathbb{Q}[x]} \mathbb{Q}[x]/(x^2+1)$ is trivial.

**ToF. 14.8.** (→ p. 357) If $A[x]$ is Noetherian, then $A$ is Noetherian.

**ToF. 14.9.** (→ p. 357) Let $>$ be a monomial ordering on $K[X]$, and let $I \neq 0$ be an ideal.
If $\mathrm{Lt}_>(I)$ is prime, then $I$ is prime.

**ToF. 14.10.** (→ p. 357) Let $I$, $J \subset A$ be ideals.
Then, $\sqrt{I : J} \subseteq \sqrt{I} : \sqrt{J}$.

**ToF. 14.11.** (→ p. 358) Let $I$, $J \subset A$ be maximal ideals such that $I \cap J = (0)$.
Then, $A$ is Artinian.

**ToF. 14.12.** (→ p. 358) Let $A$ be a domain, and let $M$, $N$ be two $A$-modules.
Then, $T(M \otimes N) \simeq T(M) \otimes T(N)$.

**ToF. 14.13.** (→ p. 358) Let $f, g \in \mathbb{C}[x, y] \setminus \mathbb{C}$, and let $I = (f, g)$.
Then, $\mathbb{V}_{\mathbb{C}}(I)$ is infinite if and only if $\gcd(f, g) \neq 1$.

**ToF. 14.14.** (→ p. 358) Let $A$ be a ring, and let $I \subset A$ be an ideal.
Then, $A^n/IA^n$ is isomorphic to $(A/I)^n$.

**ToF. 14.15.** (→ p. 358) Let $p(x) \in K[x]$ be an irreducible polynomial.
Then, the ideal $(p(x), p(y))$ is prime in $K[x, y]$.

**ToF. 14.16.** (→ p. 358) Let $M$ be a projective $A$-module, and let $N \subseteq M$ be a submodule.
Then, $N$ is projective.

**ToF. 14.17.** (→ p. 358) Let $I \subset A$ be a proper ideal.
Then, $I$ is maximal if and only if, for every ideal $J$ of $A$, either $J \subseteq I$ or $I + J = A$.

**ToF. 14.18.** (→ p. 358) Let $\mathfrak{p} \subset A$ be a prime ideal such that $A/\mathfrak{p}$ is finite.
Then, $\mathfrak{p}$ is maximal.

**ToF. 14.19.** (→ p. 358) Let $A$ be a ring, and let $I$, $J \subset A$ be ideals.
Then:

1. $I + J = A$ if and only if $I^n + J^n = A$ for every $n \in \mathbb{N}$.
2. $\sqrt{I : J} \subseteq \sqrt{I} : J$.
3. $\sqrt{I : J} = \sqrt{I} : J$.
4. $I : J = I : \sqrt{J}$.

**ToF. 14.20.** (→ p. 359) Let $I = (x^2+1,\ y^2-1)$ and $J = (x^2+xy,\ y^2+xy+1)$ be ideals of $\mathbb{Q}[x,y]$.
Then, $\mathbb{Q}[x,y]/I \simeq \mathbb{Q}[x,y]/J$.

**ToF. 14.21.** (→ p. 359) Let $I \subset K[x,y]$ be an ideal such that $I \cap K[x] = 0$, and let $K[x,y] \longrightarrow K(x)[y]$ be the inclusion homomorphism.
Then, $I$ is prime if and only if $I^e$ is prime and $I^{ec} = I$.

**ToF. 14.22.** (→ p. 359) Let $f, g \in K[x,y]$.
Then, $\sqrt{(f,g)} = \sqrt{(f^2, g^3)}$.

**ToF. 14.23.** (→ p. 359) Let $(A, \mathfrak{m})$ be a local ring, and let $\pi\colon A \longrightarrow A/\mathfrak{m}$ be the canonical projection.
Then, $a \in A^*$ if and only if $\pi(a) \in (A/\mathfrak{m})^*$.

**ToF. 14.24.** (→ p. 359) Let $(A, \mathfrak{m})$ be a Noetherian local ring.
If the images of the elements $a_1, \ldots, a_n \in A$ in $A/\mathfrak{m}^2$ generate $\mathfrak{m}/\mathfrak{m}^2$ as an $A/\mathfrak{m}$-vector space, then $\mathfrak{m} = (a_1, \ldots, a_n)$.

**ToF. 14.25.** (→ p. 359) A subring of a Noetherian ring is Noetherian.

**ToF. 14.26.** (→ p. 360) Let $A = \mathbb{Z}/(18)$, and let $M_1$, $M_2$ be the $A$-modules $(\overline{2})$ and $(\overline{3})$, respectively.
Then:

1. $M_1$ is projective.
2. $M_2$ is projcctive.
3. $M_1$ is free.
4. $M_2$ is free.

**ToF. 14.27.** (→ p. 360) The sequence of $K[x]$-modules

$$0 \longrightarrow (x) \longrightarrow K[x] \longrightarrow K \longrightarrow 0$$

splits.

**ToF. 14.28.** (→ p. 360) Let $M$ be the $\mathbb{Z}$-module $\mathbb{Z}/(12) \otimes_{\mathbb{Z}} \mathbb{Z}/(30)$.
Then, $\operatorname{Supp} M = \{(2), (3)\}$.

**ToF. 14.29.** (→ p. 360) Let $A = K[x,y]/(xy)$.
Then, an element $a \notin \mathcal{D}(A)$ if and only if $a \in K$.

**ToF. 14.30.** (→ p. 360) Assume $K = \overline{K}$, and take $f, g \in K[X]$ with $f$ irreducible.
If $\mathbb{V}(f) \subseteq \mathbb{V}(g)$, then $f$ divides $g$.

**ToF. 14.31.** (→ p. 360) Let $P$, $Q$ be projective $A$-modules, and consider a surjective homomorphism $\varphi \in \operatorname{Hom}_A(P, Q)$.
Then, $\operatorname{Ker}\varphi$ is projective.

**ToF. 14.32.** (→ p. 360) Let $f \in \mathbb{Q}[x]$ be a polynomial such that $\gcd(f, f') = 1$.
Then, $\mathcal{N}(\mathbb{C} \otimes_{\mathbb{Q}} \mathbb{Q}[x]/(f)) \neq 0$.

**ToF. 14.33.** (→ p. 361) Let $S \subset A$ be a multiplicative subset, and let $M$ be a Noetherian $A$-module.
Then, $S^{-1}M$ is a Noetherian $A$-module.

**ToF. 14.34.** (→ p. 361) Let $\mathfrak{m}$, $\mathfrak{n} \subset A$ be distinct maximal ideals, and let $M$ be an $A$-module.
Then, $M/\mathfrak{m}M \otimes_A M/\mathfrak{n}M = 0$.

**ToF. 14.35.** (→ p. 361) Let $>$ be a monomial ordering on $K[X]$, and let $I$ be an ideal.
If $\mathrm{Lt}_>(I)$ is primary, then $I$ is primary.

**ToF. 14.36.** (→ p. 361) Let $\varphi\colon \mathbb{Q}[x]^3 \longrightarrow \mathbb{Q}[x]^3$ be the homomorphism defined by the matrix

$$\begin{pmatrix} x-1 & 0 & 0 \\ 0 & x & 0 \\ 0 & 0 & x(x-1)^3 \end{pmatrix}.$$

Then, $\dim_{\mathbb{Q}} \operatorname{Coker}\varphi = 6$.

**ToF. 14.37.** (→ p. 361) Let $A \neq 0$ be a ring.
Then, $A$ is a field if and only if every $A$-module $M$ is free.

**ToF. 14.38.** (→ p. 361) Let $A$ be a domain, and let $M \neq 0$ be an $A$-module.
If, for every prime $\mathfrak{p} \subset A$, the module $M_{\mathfrak{p}}$ is torsion-free, then $M$ is torsion-free.

**ToF. 14.39.** (→ p. 361) Let $f\colon A \longrightarrow B$ be a ring homomorphism.
If $M$ is a free $A$-module of rank $k$, then $M \otimes_A B$ is a free $B$-module of rank $k$.

**ToF. 14.40.** (→ p. 361) Let $A$ be a ring, let $I \subset A$ be an ideal, and let $f \in A \setminus \mathcal{N}(A)$.
Then, $\sqrt{I} = \sqrt{IA_f \cap A} \cap \sqrt{(I, f)}$.

**ToF. 14.41.** (→ p. 361) Every Artinian domain is a field.

**ToF. 14.42.** (→ p. 362) The Jacobson radical of a PID is always zero.

**ToF. 14.43.** (→ p. 362) The $\mathbb{Z}$-modules $\mathbb{Q} \otimes_{\mathbb{Z}} \mathbb{R}$ and $\mathbb{R}$ are isomorphic.

**ToF. 14.44.** (→ p. 362) Let $N$ and $N'$ be submodules of an $A$-module $M$.
If $N'_{\mathfrak{m}} \subseteq N_{\mathfrak{m}}$ for every maximal ideal $\mathfrak{m}$ of $A$, then $N' \subseteq N$.

**ToF. 14.45.** (→ p. 362) Let $A$ be a ring such that $A_{\mathfrak{m}}$ is a domain for each $\mathfrak{m} \in \operatorname{Max} A$.
Then, $A$ is a domain.

**ToF. 14.46.** (→ p. 362) Let $A = \mathbb{C}[t]$ and $M = A[x]/(x^2 - t)$.
Then, $M$ is a flat $A$-module.

**ToF. 14.47.** (→ p. 362) Let $A$ be a PID, and let $M$ be a torsion-free $A$-module.
Then, $M$ is free.

**ToF. 14.48.** (→ p. 362) Let $A = \mathbb{Z}$, let $S = \{3^n 5^m : m, n \in \mathbb{N}\}$, and let $T = \{15^n : n \in \mathbb{N}\}$.
Then, $S^{-1}\mathbb{Z} = T^{-1}\mathbb{Z}$.

**ToF. 14.49.** (→ p. 363) Let $M$, $N$ be finitely generated $A$-modules such that $M \otimes_A N = 0$.
Then, $\operatorname{Ann} M + \operatorname{Ann} N = A$.

**ToF. 14.50.** (→ p. 363) Let $(A, \mathfrak{m}, K)$ be a local ring, and let $M \neq 0$ be a finitely generated $A$-module.
Then, $\mathcal{V}(\operatorname{Ann}(M/\mathfrak{m}M)) = \{\mathfrak{m}\}$.

**ToF. 14.51.** (→ p. 363) Let $M$ be a flat $A$-module.
Then, $I \otimes_A M \simeq IM$ for any ideal $I \subset A$.

**ToF. 14.52.** (→ p. 363) Let $A$ be a Noetherian ring.
Then, every surjective endomorphism of $A$ is an isomorphism.

**ToF. 14.53.** (→ p. 363) Let $A$ be a ring, and let $S \subset A$ be a multiplicative subset such that $A \simeq S^{-1}A$.
Then, $S \subset A^*$.

**ToF. 14.54.** (→ p. 363) Let $A$ be a ring such that every submodule of a free $A$-module is free.
Then, $A$ is a PID.

**ToF. 14.55.** (→ p. 363) Let $A$ be a local ring.
Then, there exist a ring $B$ and a prime ideal $\mathfrak{p}$ of $B$ such that $A \simeq B_{\mathfrak{p}}$.

**ToF. 14.56.** (→ p. 363) The polynomial $p(x) = \overline{30}x^5 + \overline{60}x^3 + \overline{90}x + \overline{7}$ is invertible in $\mathbb{Z}/(540)[x]$.

**ToF. 14.57.** (→ p. 363) In the ring $\mathbb{Z}_{(2)}[x]$ every maximal ideal has at least two generators.

**ToF. 14.58.** (→ p. 364) The $\mathbb{Z}$-module $(\mathbb{Z}/(15) \oplus \mathbb{Z}/(18))_{(3)}$ is cyclic.

**ToF. 14.59.** (→ p. 364) Let $a \in \mathbb{Z}$, and let $M$ be a $\mathbb{Z}$-module generated by elements $m_1, m_2, m_3$ which satisfy the relations

$$2m_1 - m_2 = 0, \quad m_1 + m_2 + m_3 = 0, \quad m_1 + am_2 = 0.$$

Then, for every $a > 0$, there exists an integer $n$ such that $M \otimes_{\mathbb{Z}} \mathbb{Z}/(n) \neq 0$.

**ToF. 14.60.** (→ p. 364) Let $I \subset \mathfrak{p} \subset A$ where $I$ is an ideal and $\mathfrak{p} \in \operatorname{Spec} A$. If $I_{\mathfrak{p}}$ is primary, then $I$ is primary.

**ToF. 14.61.** (→ p. 364) Let $f(x, y) \in K[x, y]$ be a polynomial of total degree $n$. Let also $\mathcal{C} = \mathbb{V}(f)$ be the plane curve of $K^2$ defined by $f$, and let $\ell$ be a line not contained in $\mathcal{C}$.
Then, $|\mathcal{C} \cap \ell| \leq n$.

**ToF. 14.62.** (→ p. 364) Let $A$ be a domain, and let $M$ be a flat $A$-module. Then, $M$ is torsion-free.

**ToF. 14.63.** (→ p. 364) Let $A$ be a PID, let $Q(A)$ be its fraction field, and let $M \simeq A^n \oplus \bigoplus_{i=1}^{k} A/(a_i)$ with $a_i \neq 0$ for all $i$.
Then, $\dim_{Q(A)} Q(A) \otimes_A M = n$.

**ToF. 14.64.** (→ p. 365) Let $A$ be a ring, and let $M$ be an $A$-module.
If $\mathcal{J}(A)$ is finitely generated, and $\mathcal{J}(A)M = M$, then $M = 0$.

**ToF. 14.65.** (→ p. 365) Let $I \subset \mathbb{Z}[x]$ be an ideal.
If $\sqrt{I} = (f)$ with $f$ irreducible modulo $p$ for every $p$ prime in $\mathbb{Z}$, then $I$ is primary.

**ToF. 14.66.** (→ p. 365) A direct summand of a finitely generated module is finitely generated.

**ToF. 14.67.** (→ p. 365) Let $I = (y^2 - z^2, x^3y^3 - yz) \subseteq \mathbb{Q}[x, y, z]$.
Then, $f = x^3y^2 + x^3yz + y + z \in \sqrt{I}$.

**ToF. 14.68.** (→ p. 365) Let $I$ be an ideal of a ring $A$ such that $A/I$ is a flat $A$-module.
Then, $I = I^2$.

**ToF. 14.69.** (→ p. 365) Let $A$ be a domain, and let

$$0 \longrightarrow M \xrightarrow{f} N \xrightarrow{g} L \longrightarrow 0$$

be an exact sequence of $A$-modules.
Then:

1. $\operatorname{Ann} N \subseteq \operatorname{Ann} M$, $\operatorname{Ann} N \subseteq \operatorname{Ann} L$ and $\operatorname{Ann} M \cdot \operatorname{Ann} L \subseteq \operatorname{Ann} N$.

2. the sequence $T(M) \xrightarrow{f_{|T(M)}} T(N) \xrightarrow{g_{|T(N)}} T(L)$ is well-defined.
3. the sequence $T(M) \xrightarrow{f_{|T(M)}} T(N) \xrightarrow{g_{|T(N)}} T(L) \longrightarrow 0$ is exact.
4. the sequence $0 \longrightarrow T(M) \xrightarrow{f_{|T(M)}} T(N) \xrightarrow{g_{|T(N)}} T(L)$ is exact.

**ToF. 14.70.** (→ p. 366) Let $I = \{p(x) \in \mathbb{Z}[x] \colon\ p(5) \text{ is even}\}$.
Then, $\mathbb{Z}[x]/I$ is a field.

**ToF. 14.71.** (→ p. 366) There is a unique ring homomorphism $\varphi\colon \mathbb{C} \longrightarrow \mathbb{R}$.

**ToF. 14.72.** (→ p. 366) Let $A$ be a ring, let $\mathfrak{p} \subset A$ be a prime, and let $M$ be a finitely generated $A$-module.
If the $A/\mathfrak{p}$-module $M/\mathfrak{p}M$ is trivial, then $M_\mathfrak{p} = 0$.

**ToF. 14.73.** (→ p. 366) Let $I$ be an irreducible and radical ideal in a ring $A$.
Then, $I$ is prime.

**ToF. 14.74.** (→ p. 366) Let $I_1 \supset I_2 \supset I_3 \supset \ldots$ be a chain of prime ideals in a ring $A$.
Then, $I = \bigcap_j I_j$ is a prime ideal.

**ToF. 14.75.** (→ p. 366) The ring $K[[x_1, \ldots, x_n]]$ is local.

**ToF. 14.76.** (→ p. 367) Let $A$ be a Noetherian ring of finite Krull dimension, and let $\{\mathfrak{p}_1, \ldots, \mathfrak{p}_k\}$ be the set of minimal prime ideals of $A$.
Then, for any $i = 1, \ldots, k$,

$$\dim A = \dim A_{\mathfrak{p}_i} + \max_i\{\dim A/\mathfrak{p}_i\}.$$

**ToF. 14.77.** (→ p. 367) Let $A$ be a domain such that $IJ = I \cap J$ for any pair of ideals $I$, $J$ of $A$.
Then, $A$ is a field.

**ToF. 14.78.** (→ p. 367) Let $A$ be a PID.
Then, a finitely generated $A$-module is flat if and only if it is free.

**ToF. 14.79.** (→ p. 367) Let $A = K[x,y]/(xy)$, $\mathfrak{p} = (x)$, and $\mathfrak{q} = (x,\, y)$.
Then, there exists an injective homomorphism from $A_\mathfrak{q}$ to $A_\mathfrak{p}$.

**ToF. 14.80.** (→ p. 367) Every localization of a projective module is projective.

**ToF. 14.81.** (→ p. 368) Let $f$, $g \in \mathbb{Q}[x,y]$ be irreducible polynomials of total degree $n \geq 1$.
Then, $\dim_\mathbb{Q} \left(\mathbb{Q}[x,y]/(f) \otimes_{\mathbb{Q}[x,y]} \mathbb{Q}[x,y]/(g)\right) = n$.

**ToF. 14.82.** (→ p. 368) Let $f\colon K[x] \longrightarrow K[x,y]/(xy-1)$ be the homomorphism defined by the composition of the inclusion $K[x] \longrightarrow K[x,y]$ with the canonical projection onto $K[x,y]/(xy-1)$.
Then, there exists a prime ideal $\mathfrak{p} \subset K[x,y]/(xy-1)$ whose contraction via $f$ is equal to $(x)$.

**ToF. 14.83.** (→ p. 368) Let $I$, $J \subset A$ be ideals.
Then, $I \subset J$ if and only if $I_{\mathfrak{m}} \subset J_{\mathfrak{m}}$ for all $\mathfrak{m} \in \operatorname{Max} A$.

**ToF. 14.84.** (→ p. 368) Let $A$ be a local ring, and let $M$ be a finitely generated $A$-module.
Then, $M$ is projective if and only if it is free.

**ToF. 14.85.** (→ p. 368) Let $n \in \mathbb{Z} \setminus \{0\}$.
Then, $\operatorname{Hom}_{\mathbb{Z}}(\mathbb{Z}/(n), \mathbb{Q}/\mathbb{Z}) \simeq \mathbb{Z}/(n)$.

**ToF. 14.86.** (→ p. 368) Let $A = \mathbb{Q}[x]$, let $\varphi : A^3 \longrightarrow A^3$ be the homomorphism defined by

$$\varphi(f,g,h) = ((x-1)f, (x^3-1)g, (x^2-1)h),$$

and let $M = \operatorname{Coker} \varphi$.
Then, $M \simeq \mathbb{Q}^6$ as an $A$-module.

**ToF. 14.87.** (→ p. 368) Let $S$ be a subset of $K^n$, and let

$$\mathbb{I}(S) = (f \in K[x_1, \ldots, x_n]\colon f(s) = 0 \text{ for all } s \in S).$$

Then, $\mathbb{V}(\mathbb{I}(S)) = S$.

**ToF. 14.88.** (→ p. 368) Let $A$ be a ring with $\dim A = n$.
Then, for any multiplicative subset $S \subseteq A$, we have $\dim S^{-1}A \leq n$, and there exists at least a multiplicative subset $T \subseteq A$ such that $\dim T^{-1}A = n$.

**ToF. 14.89.** (→ p. 369) Let $K$ be a field, and let $\varphi\colon \mathbb{Z}[x] \longrightarrow K$ be a surjective ring homomorphism.
Then, $K$ is a finite field.

**ToF. 14.90.** (→ p. 369) Let $A$ be a ring, and let $P$ be a finitely generated projective $A$-module.
Then, $\operatorname{Hom}_A(P, A)$ is projective.

**ToF. 14.91.** (→ p. 369) Direct product and tensor product commute.

**ToF. 14.92.** (→ p. 369) Let $S = \mathbb{Z} \setminus \{0\}$.
Then,

$$S^{-1}\left(\prod_{n\in\mathbb{N}_+} \mathbb{Z}/(n)\right) \simeq \prod_{n\in\mathbb{N}_+} S^{-1}(\mathbb{Z}/(n)).$$

# Chapter 15
# Review Exercises

**E. 15.1.** (→ p. 370) Let $A = \mathbb{C}[x_1, \ldots, x_n]$, with $n \geq 2$, be equipped with a monomial ordering $>$, and let $m_1, m_2 \in A \setminus \mathbb{C}$ be monomials such that

$$m_1^k < m_2 \quad \text{for all } k \in \mathbb{N}.$$

Prove that there exist indices $i \neq j$ such that $x_i^k < x_j$ for all $k \in \mathbb{N}$.

**E. 15.2.** (→ p. 370) Let $A = K[x, y, z]$, and let $B = K[t]$.

1. Prove that the map defined as

$$x \mapsto t, \quad y \mapsto t^2, \quad z \mapsto t^3$$

induces a surjective homomorphism $\varphi\colon A \longrightarrow B$.

2. Compute $\operatorname{Ker} \varphi$.

**E. 15.3.** (→ p. 371) Let $M$, $N$, and $L$ be $A$-modules.
Prove:

1. $\operatorname{Hom}(A^n, M) \otimes N \simeq \operatorname{Hom}(A^n, M \otimes N)$ for all $n \in \mathbb{N}$;
2. if $N$ is flat and $L$ is finitely generated, then there exists an injective homomorphism

$$\operatorname{Hom}(L, M) \otimes N \longrightarrow \operatorname{Hom}(L, M \otimes N).$$

**E. 15.4.** (→ p. 372) Let $A$ be a ring such that every prime ideal is finitely generated.
Prove that every ideal of $A$ is finitely generated.



A. Bandini et al., *Commutative Algebra through Exercises*, UNITEXT 159,
https://doi.org/10.1007/978-3-031-56910-4_15

**E. 15.5.** (→ p. 373) Let $A = \mathbb{Q}[x, y, z]$, and let $I = (x^2 - x^2z,\ xz + xyz) \subset A$.

1. Prove that $\dim A/I \geq 2$.
2. Determine whether the escalier of $I$ with respect to the lex order with $x > y > z$ is equal to the escalier of $I$ with respect to the degrevlex order with $x > y > z$.
3. Compute $I \cap \mathbb{Q}[y, z]$ and $I \cap \mathbb{Q}[x, y]$.
4. Find the irreducible components of $\mathbb{V}(I)$.
5. Is $(A/I)_\mathfrak{p} \neq 0$ for all $\mathfrak{p} \in \operatorname{Min} I$?
   Is it true that $(A/I)_\mathfrak{p} \neq 0$ if and only if $\mathfrak{p} \in \operatorname{Ass} I$?
6. Determine the primary components of $I$ associated to its minimal primes.

**E. 15.6.** (→ p. 374) Let $I_1$ and $I_2$ be ideals of a Noetherian ring $A$.
Prove that there exist $t \in \mathbb{N}$ and an ideal $J$ such that

$$I_1^t \subseteq J \quad \text{and} \quad I_1 I_2 = J \cap I_2.$$

**E. 15.7.** (→ p. 374) Let $0 \longrightarrow M_1 \xrightarrow{\varphi} M_2 \xrightarrow{\psi} M_3 \longrightarrow 0$ be an exact sequence of $A$-modules, where $M_3$ is flat.
Prove that, for every $A$-module $N$, the sequence

$$0 \longrightarrow N \otimes M_1 \xrightarrow{\mathrm{id}_N \otimes \varphi} N \otimes M_2 \xrightarrow{\mathrm{id}_N \otimes \psi} N \otimes M_3 \longrightarrow 0$$

is exact.

**E. 15.8.** (→ p. 375) Let $n$ be a positive integer, and let $\mathfrak{p}$ be a prime ideal of a ring $A$. Define $\mathfrak{p}^{(n)} = \mathfrak{p}^n A_\mathfrak{p} \cap A$ to be the *n-th symbolic power* of $\mathfrak{p}$.

1. Prove that, if $\mathfrak{p}^n$ is primary, then $\mathfrak{p}^{(n)} = \mathfrak{p}^n$.
2. If $\mathfrak{p}^n$ is not primary, does the equality $\mathfrak{p}^{(n)} = \mathfrak{p}^n$ still hold?

**E. 15.9.** (→ p. 375) Let $A = \mathbb{C}[x]$, and let $M = \langle m_1, \ldots, m_4 \rangle_A$ be an $A$-module such that

$$\begin{aligned}
&(x^2 - 1)m_1 + (3x + 3)m_2 + (3x + 3)m_3 + (3x + 3)m_4 = 0,\\
&(x^2 + x)m_2 + (x + 1)m_3 + (x + 1)m_4 = 0,\\
&(x^2 - 1)m_1 + (x + 1)m_2 + (x + 1)m_3 + (x + 1)m_4 = 0,\\
&(x^2 - 1)m_1 + (2x + 2)m_2 + (2x + 2)m_3 + (2x + 2)m_4 = 0.
\end{aligned}$$

1. Compute the free part of $M$, $T(M)$, and $\operatorname{Ann}_A(M)$.
2. Compute $M \otimes_A A/(x - 1)$ and $M \otimes_A A/(x - i)$.
3. Find, if possible, a non-zero $A$-module $N$ such that $\operatorname{Hom}_A(M, N) \simeq N$.

**E. 15.10.** (→ p. 376) Let $A$ be a PIR, let $\mathfrak{p}_1, \mathfrak{p}_2 \in \operatorname{Spec} A$, and let $\mathfrak{q}$ be a primary ideal.
Prove:

1. if $\mathfrak{q} \subseteq \mathfrak{p}_1$ and $\mathfrak{p}_2 \subsetneq \mathfrak{p}_1$, then $\mathfrak{p}_2 \subseteq \mathfrak{q}$;
2. if $\mathfrak{p}_2 = \sqrt{\mathfrak{q}} \subsetneq \mathfrak{p}_1$, then $\mathfrak{p}_2 = \mathfrak{q}$;
3. if $\mathfrak{p}_2 \subsetneq \mathfrak{p}_1$, then $\mathfrak{p}_2$ is the intersection of all primary ideals contained in $\mathfrak{p}_1$;
4. if $\mathfrak{p}_1$ and $\mathfrak{p}_2$ are not comaximal, then one of them is contained in the other.

**E. 15.11.** (→ p. 376) Let $I = (x^2y - y^2,\, x^3 - xy) \subset \mathbb{R}[x, y]$.

1. Prove that $\{g_1 = x^2y - y^2,\, g_2 = x^3 - xy\}$ is a Gröbner basis of $I$ with respect to all possible lexicographic monomial orderings.
2. Decompose $\sqrt{I}$ as an intersection of prime ideals and verify whether $I$ is radical.
3. Let
$$J = (xz^2 + z^2,\, yz^2 + z^2, I) \subseteq \mathbb{R}[x, y, z].$$
Is it true that for every $\alpha \in \mathbb{V}_{\mathbb{R}}(I)$, there exists $b \in \mathbb{R}$ such that $(\alpha, b)$ belongs to $\mathbb{V}_{\mathbb{R}}(J)$?

**E. 15.12.** (→ p. 377) Let $A = \mathbb{Q}[x, y]$, and let
$$I = (x^2 - y^2,\, x^4 - x^3y + y,\, xy + y^2) \subset A.$$

1. Decompose the radical of $I$ as an intersection of prime ideals.
2. Decompose $I$ as an intersection of primary ideals.
3. Compute $\mathcal{D}(A/I)$ and $\mathcal{N}(A/I)$.
4. Let $\mathfrak{p} = (x, y)$ and $\mathfrak{q} = (y)$.
Compute $I_{\mathfrak{p}} \cap A$ and $I_{\mathfrak{q}} \cap A$.
5. Describe the rings $(A/I)_{\mathfrak{p}}$ and $(A/I)_{\mathfrak{q}}$.

**E. 15.13.** (→ p. 378) Let $\varphi_{ab}\colon \mathbb{Q}[x]^3 \to \mathbb{Q}[x]^3$, with $a, b \in \mathbb{Q}$, be the homomorphism defined by the matrix
$$B_{ab} = \begin{pmatrix} x - a & 0 & 0 \\ 0 & 1 - x & x - a \\ b & a - x^2 & 0 \end{pmatrix}.$$

1. Find all values of $a, b \in \mathbb{Q}$ such that $\operatorname{Coker} \varphi_{ab}$ is cyclic.
2. Compute $\operatorname{Coker} \varphi_{ab}$ for all values of $a$ and $b$ such that it is not cyclic.

**E. 15.14.** (→ p. 379) Let $I = (xy,\, x - yz)$, $\mathfrak{q}_1 = (x,\, z)$, and $\mathfrak{q}_2 = (y^2,\, x - yz)$ be ideals of $K[x, y, z]$.
Prove that $I = \mathfrak{q}_1 \cap \mathfrak{q}_2$ is a minimal primary decomposition of $I$.

**E. 15.15.** (→ p. 380) Let $A$ be a ring such that:

i) $A_{\mathfrak{m}}$ is Noetherian for all $\mathfrak{m} \in \operatorname{Max} A$;
ii) the set $\mathcal{M}_x = \{\mathfrak{m} \in \operatorname{Max} A \colon x \in \mathfrak{m}\}$ is finite for all $x \in A \setminus \{0\}$.

Prove that $A$ is Noetherian.

**E. 15.16.** (→ p. 380) Let $I$ be an ideal of a ring $A$.
Define, for all $n \in \mathbb{N}$,

$$I^{[n]} = (a^n \colon\ a \in I).$$

1. Prove that if $A = \mathbb{Q}[x, y]$, then $I^2 = I^{[2]}$.
2. Find an example of a ring $B$ and an ideal $I$ such that $I^2 \neq I^{[2]}$.
3. Prove that if $A = \mathbb{Q}[x, y]$ and $I = (x,\ y)$, then $I^n = I^{[n]}$ for all $n \in \mathbb{N}$.

**E. 15.17.** (→ p. 381) Let $I$ and $J$ be ideals of a ring $A$.
Define

$$I : J^{\infty} = \bigcup_{n \in \mathbb{N}} (I : J^n).$$

Prove:

1. $I : J^{\infty}$ is an ideal of $A$;
2. if $A$ is Noetherian and $a, b \in A$, then $I : (b)^{\infty} = I : (a, b)^{\infty}$ if and only if every associated prime of $I$ containing $b$ also contains $a$.

**E. 15.18.** (→ p. 381) Let $I = (x^2 + xy + y^2,\ y^3 - 1) \subset K[x, y]$, and let $\mathbb{V}_K(I) \subset K^2$ be the affine variety associated to $I$.

1. Find $\mathbb{V}_{\mathbb{Q}}(I)$ and $\mathbb{V}_{\mathbb{C}}(I)$.
2. Compute $\mathcal{D}(\mathbb{Q}[x, y]/I)$ and $\mathcal{N}(\mathbb{Q}[x, y]/I)$.
3. Describe the set of prime ideals of $\mathbb{Q}[x, y]/I$.
4. Find, if any exists, a prime ideal $\mathfrak{p} \subset \mathbb{Q}[x, y]/I$ such that $(\mathbb{Q}[x, y]/I)_{\mathfrak{p}}$ is a field.

**E. 15.19.** (→ p. 382) Let $A$ be a domain, and let $K$ be its fraction field.
For any ideal $I$ of $A$, we define

$$I^{-1} = \{x \in K \colon xI \subseteq A\},$$

and we say that $I$ is *invertible* if $I\,I^{-1} = A$.
Prove that, if $I$ is invertible, then:

1. $I$ is finitely generated;
2. $I$ is a projective $A$-module.

**E. 15.20.** (→ p. 382) Let $I \subset \mathbb{C}[x_1, \ldots, x_n]$ be a 0-dimensional radical ideal such that:

i) $I \cap \mathbb{C}[x_n] = (p_n = x_n^m + a_{m-1}x_n^{m-1} + \ldots + a_0)$;
ii) $|\mathbb{V}(I)| = m$.

Prove that there exist $p_1, \ldots, p_{n-1} \in \mathbb{C}[x_n]$, with $\deg p_i < m$ for all $i$, such that the reduced Gröbner basis of $I$ with respect to the lex order with $x_1 > \ldots > x_n$ is

$$\{x_1 - p_1,\ x_2 - p_2, \ldots,\ x_{n-1} - p_{n-1},\ p_n\}.$$

**E. 15.21.** (→ p. 383) [**Schanuel's Lemma**] Let

$$0 \longrightarrow N \xrightarrow{\varphi} P \xrightarrow{\psi} M \longrightarrow 0,$$

$$0 \longrightarrow N' \xrightarrow{\varphi'} P' \xrightarrow{\psi'} M \longrightarrow 0$$

be two exact sequences of $A$-modules, with $P$ projective.
Prove:

1. there exist homomorphisms $f\colon P \longrightarrow P'$ and $g\colon N \longrightarrow N'$ such that $\operatorname{Ker} f \simeq \operatorname{Ker} g$ and $\operatorname{Coker} f \simeq \operatorname{Coker} g$;
2. if $P'$ is also projective, then $N \oplus P' \sim N' \oplus P$.

# Part III
# Proofs and Solutions

# Chapter 16
# Proofs of Theoretical Results

## 16.1 Chapter 1

**Proof T. 1.1.** We prove that if $a \notin \mathcal{D}(A)$, then $a \in A^*$.
Consider the map $m_a \colon A \longrightarrow A$ that represents the multiplication by $a$, defined by $m_a(b) = ab$. The equality $m_a(b) = m_a(c)$ yields $a(b-c) = 0$. Since $a \notin \mathcal{D}(A)$, it follows that $b - c = 0$. Therefore, $m_a$ is injective.
Since $A$ is finite, the map $m_a$ is also surjective, and there exists $a' \in A$ such that $1 = m_a(a') = aa'$, that is, $a \in A^*$. □

**Proof T. 1.3.** Clearly, $0 \in I : J$. Let $a, b \in I : J$. Then, for any $j \in J$ we have $aj, bj \in I$. Since $I$ is an ideal, $(a-b)j = aj - bj \in I$, that is, $a - b \in I : J$. Similarly, if $aj \in I$ for every $j \in J$, then, for every $c \in A$, we have $(ca)j = c(aj) \in I$. Therefore, $ca \in I : J$.

Obviously, $\sqrt{A} = A$. Now assume $I \subsetneq A$. Thus, $a^n \in I$ implies $n > 0$. It is clear that $0 \in \sqrt{I}$. If $a \in \sqrt{I}$, then there exists $n \in \mathbb{N}_+$ such that $a^n \in I$. Therefore, for any $c \in A$, we have $(ca)^n \in I$, that is, $ca \in \sqrt{I}$.
Let $a, b \in \sqrt{I}$. Then, there exist $n, m \in \mathbb{N}_+$ such that $a^n, b^m \in I$.
Consider

$$(a+b)^{n+m-1} = \sum_{k=0}^{n+m-1} \binom{n+m-1}{k} a^k b^{n+m-1-k}$$

$$= b^m \sum_{k=0}^{n-1} \binom{n+m-1}{k} a^k b^{n-1-k} + a^n \sum_{k=n}^{n+m-1} \binom{n+m-1}{k} a^{k-n} b^{n+m-1-k},$$

and note that both the first and the second addend belong to $I$, therefore $(a+b)^{n+m-1} \in I$, that is, $a + b \in \sqrt{I}$. □

**Proof T. 1.4.** We proceed by induction on the number $n$ of ideals. The case $n = 2$ has already been shown.
Let $I = \bigcap_{i=1}^{n-1} I_i$. By the inductive hypothesis, $I = \prod_{i=1}^{n-1} I_i$.
Since $I_n + I_j = (1)$ for every $j < n$, there exist elements $a_j \in I_n$ and $b_j \in I_j$ such that $a_j + b_j = 1$.



A. Bandini et al., *Commutative Algebra through Exercises*, UNITEXT 159,
https://doi.org/10.1007/978-3-031-56910-4_16

Thus,

$$1=\prod_{j=1}^{n-1}(a_j+b_j)=a+b_1\cdots b_{n-1}\in I_n+I.$$

Since $I$ and $I_n$ are comaximal, we have

$$\prod_{i=1}^{n} I_i=\prod_{i=1}^{n-1} I_i\cdot I_n=I\,I_n=I\cap I_n=\bigcap_{i=1}^{n} I_i.$$ □

**Proof T. 1.5.** Recall that, by definition, the sum of two ideals $I+J$ is the smallest ideal containing both $I$ and $J$.

1. Consider $i\in I$, $j\in J$, and $h\in H$. Then, we have

$$(i+j)h=ih+jh\in IH+JH.$$

As the ideal $(I+J)H$ is generated by elements of the form $(i+j)h$, it follows that $(I+J)H\subseteq IH+JH$.

To prove the opposite inclusion, note that

$$IH\subseteq(I+J)H \text{ and } JH\subseteq(I+J)H.$$

Therefore, we also have $IH+JH\subseteq(I+J)H$.

2. Using part 1 we obtain that

$$(I+J)(I\cap J)=I(I\cap J)+J(I\cap J)\subseteq IJ+JI=IJ.$$

3. Since $J,\ H\subseteq J+H$, we have that both $I\cap J$ and $I\cap H$ are contained in $I\cap(J+H)$.
Therefore, $(I\cap J)+(I\cap H)\subseteq I\cap(J+H)$.

4. Due to part 3, it is sufficient to prove that if $I\supseteq J$, then

$$I\cap(J+H)\subseteq J+(I\cap H).$$

Write $i\in I\cap(J+H)$ as $i=j+h$, with $i\in I$, $j\in J$, and $h\in H$.
Then, $h=i-j\in I\cap H$, and thus, $i=j+h\in J+I\cap H$.

5. Since $JH$ is contained in both $J$ and $H$, the proof is immediate. □

**Proof T. 1.6.** Parts 1, 2, and 4 immediately follow from the definition.

3. We will prove that

$$\sqrt{IJ}\subseteq\sqrt{I\cap J}\subseteq\sqrt{I}\cap\sqrt{J}\subseteq\sqrt{IJ}.$$

The first inclusion can be deduced from part 1 since $IJ\subseteq I\cap J$.
The second inclusion is obvious.

To prove the third inclusion, let $a \in \sqrt{I} \cap \sqrt{J}$. Then, there exist $m, n \in \mathbb{N}$ such that $a^m \in I$ and $a^n \in J$. Therefore, $a^{m+n} = a^m a^n \in IJ$, and the proof is complete.

5. It is evident that $1 = 1^n \in I$ if and only if $1 \in \sqrt{I}$.

6. From part 1, we have $I + J \subseteq \sqrt{I} + \sqrt{J}$. We obtain the inclusion $\subseteq$ by applying part 1 again.

To prove the opposite inclusion, note that if $a \in \sqrt{\sqrt{I} + \sqrt{J}}$, then there exist $i \in \sqrt{I}$ and $j \in \sqrt{J}$ such that $a^t = i + j$ for some positive integer $t$. Let $n, m \in \mathbb{N}$ be such that $i^n \in I$ and $j^m \in J$. Then,

$$a^{t(n+m-1)} = \sum_{k=0}^{n-1} \binom{n+m-1}{k} i^k j^{m+n-1-k} + \sum_{k=n}^{n+m-1} \binom{n+m-1}{k} i^k j^{n+m-1-k}$$

belongs to $J + I$, and thus, $a \in \sqrt{I + J}$.

7. The inclusion $\subseteq$ follows from **T.1.5**.5 and from part 3.

To prove the opposite inclusion, let $a \in \sqrt{I + J} \cap \sqrt{I + H}$. Then, $a^n = i_1 + j$ and $a^m = i_2 + h$ for some $n, m \in \mathbb{N}$, $i_1, i_2 \in I$, $j \in J$, and $h \in H$. Hence,

$$a^{n+m} = i_1 i_2 + i_1 h + i_2 j + jh \in I + JH,$$

and $a \in \sqrt{I + JH}$. □

**Proof T. 1.7.** 1. If $ab \in \sqrt{I}$, then there exists $k \in \mathbb{N}$ such that $(ab)^k \in I$. Since $I$ is primary, either $a^k \in I$ or there exists $n \in \mathbb{N}$ such that $b^{kn} \in I$, *i.e.*, either $a \in \sqrt{I}$ or $b \in \sqrt{I}$.

2. We will prove that if $ab \in I$ and $b \notin \mathfrak{m}$, then $a \in I$.
Since $\mathfrak{m}$ is a maximal ideal, $(b) + \mathfrak{m} = (1)$ and there exist $c \in A$ and $m \in \mathfrak{m}$ such that $1 = cb + m$. Let $n \geq 1$ be an integer such that $m^n \in I$. Then, $1 = 1^n = (cb + m)^n \in (b) + I$.
By multiplying by $a$ we obtain $a \in (ab) + I \subseteq I$, as required. □

**Proof T. 1.8.** We have to verify that if $\overline{a} = \overline{c}$ and $\overline{b} = \overline{d}$ in $A/I$, then $\overline{ab} = \overline{cd}$, that is, if $a - c, b - d \in I$, then $ab - cd \in I$.
The element

$$ab - cd = ab - ad - cd + ad = a(b - d) + d(a - c)$$

is in $I$, because both its addends are elements of $I$. It is easy to verify that this product defines a commutative ring structure on $A/I$. □

**Proof T. 1.9.** 1. To prove the first theorem it is sufficient to restrict the range of $f$ to $\operatorname{Im} f$. In this way the induced map $\overline{f}\colon A/\operatorname{Ker} f \longrightarrow \operatorname{Im} f$ is a homomorphism that is injective and surjective by construction.

2. Let $\overline{j_1}, \overline{j_2} \in J/I = \{\overline{j} : j \in J\}$ and $\overline{a} \in A/I$. We have $\overline{j_1} + \overline{j_2} = \overline{j_1 + j_2}$, with $j_1 + j_2 \in J$, and $\overline{a}\overline{j_1} = \overline{aj_1}$, with $aj_1 \in J$. Thus, $J/I$ is an ideal of $A/I$, and we can define the quotient ring $(A/I)/(J/I)$.

Consider the map $\pi\colon A/I \longrightarrow A/J$, where $\overline{a} \mapsto \overline{\overline{a}}$. This map is well-defined because if $\overline{a} = \overline{b}$, then $a - b \in I \subseteq J$, therefore $\overline{\overline{a}} = \overline{\overline{b}}$.
Clearly, $\pi$ is a surjective homomorphism and $J/I \subseteq \operatorname{Ker}\pi$.
Let $\overline{a} \in \operatorname{Ker}\pi$, that is, $0 = \pi(\overline{a}) = \overline{\overline{a}}$. This means that $a \in J$, thus $J/I = \operatorname{Ker}\pi$. The claim follows from part 1.

3. Parts a, b and c are easy to verify and only needed in order to define the last isomorphism.
Consider the map

$$f\colon B \longrightarrow B + I \longrightarrow (B + I)/I,$$

defined as the composition of the inclusion map of $B$ in $B + I$ and of the projection onto the quotient. The homomorphism $f$ is surjective because each element $\overline{b + i}$ is the image of $b \in B$. Moreover, if $b \in B$ is such that $\overline{b} = 0$, then $b \in I$ and $b \in B \cap I$.
Thus, the kernel of $f$ is $B \cap I$, and the claim follows again from part 1. □

**Proof T. 1.10.** Let $J$ be an ideal of $A$ containing $I$. Since the projection is surjective, $\pi(J)$ is an ideal of $A/I$, see **T.1.18**.1.

Conversely, for any ideal $H$ in $A/I$ its preimage is an ideal of $A$ containing $I$, because $\pi(I) = \overline{0} \in H$.
From the definition of maximal ideals it immediately follows that the correspondence is one-to-one on maximal ideals.
The preimage of a prime ideal is always a prime ideal, see **T.1.17**.7.

Conversely, suppose $\mathfrak{p}$ is a prime ideal of $A$ containing $I$ and take $\overline{ab} \in \pi(\mathfrak{p})$. Then, there exists $c \in \mathfrak{p}$ such that $\overline{c} = \overline{ab}$. Hence, $c - ab \in I \subseteq \mathfrak{p}$, and thus, $ab \in \mathfrak{p}$ and the claim follows from the primality of $\mathfrak{p}$. □

**Proof T. 1.11.** 1. The quotient ring $A/I$ is non-trivial if and only if $\overline{1} \neq \overline{0}$ in $A/I$, *i.e.*, if and only if $1 \notin I$, which is exactly when $I$ is a proper ideal.

2. The quotient ring $A/I$ is a field if and only if its only ideals are $(\overline{0})$ and $(\overline{1})$. By **T.1.10**, the only ideals containing $I$ are $A$ and $I$ itself, *i.e.*, $I$ is maximal.

3. The quotient ring $A/I$ is a domain if and only if for each pair of elements $\overline{a}, \overline{b} \neq \overline{0}$ of $A/I$ the product $\overline{ab} \neq \overline{0}$. Thus, $A/I$ is a domain if and only if for every pair of elements $a, b \in A \setminus I$ we have $ab \notin I$, that is, when $I$ is a prime ideal of $A$.

4. The quotient ring $A/I$ is reduced if and only if $\mathcal{N}(A/I) = (\overline{0})$, that is, if $\overline{a}^k = \overline{0}$ implies $\overline{a} = \overline{0}$. Thus, $A/I$ is reduced if and only if $a^k \in I$ implies $a \in I$, that is, if and only if $\sqrt{I} \subseteq I$, *i.e.*, if and only if $I$ is radical.

5. We assume $\mathcal{D}(A/I) \subseteq \mathcal{N}(A/I)$ and prove that $I$ is a primary ideal of $A$. We show that if $a, b \in A$ are such that $a \notin I$ and $ab \in I$, then $b^n \in I$ for some positive integer $n$. Indeed, $\overline{a} \neq \overline{0}$ and $\overline{a}\overline{b} = \overline{0}$ imply $\overline{b} \in \mathcal{D}(A/I) \subseteq \mathcal{N}(A/I)$, that is, $\overline{b^n} = \overline{b}^n = \overline{0}$ for some $n$. Therefore, $b \in \sqrt{I}$.

Conversely, reversing the implications, we can prove that if $I$ is primary, then $\mathcal{D}(A/I) \subseteq \mathcal{N}(A/I)$. Since $\mathcal{N}(A/I) \subseteq \mathcal{D}(A/I))$ always holds, the conclusion follows.

6. If $A/I$ is a field, then $A/I$ is a domain.

7. If $A/I$ is a domain, then $\mathcal{N}(A/I) = \mathcal{D}(A/I) = \{\bar{0}\}$. In this case $A/I$ is reduced and the conclusion follows from part 4.

8. If $I$ is prime, then $\mathcal{N}(A/I) = \mathcal{D}(A/I) = \{\bar{0}\}$. Therefore, the claim follows from part 5. □

**Proof T. 1.13.** Let $I \subset A$ be a prime ideal, and let $I = J \cap H$ be a decomposition of $I$. Then, either $I = J$ or $I = H$ by **T.1.12**.2, *i.e.*, the given decomposition is trivial. □

**Proof T. 1.15.** Let $a \in A$ be such that $a \notin \mathcal{J}(A)$.
Then, there exists a maximal ideal $\mathfrak{m}$ such that $a \notin \mathfrak{m}$, and this implies that $(a, \mathfrak{m}) = (1)$. Therefore, there exists $b \in A$ such that the element $1 - ab \in \mathfrak{m}$ is not invertible.

Conversely, let $a \in \mathcal{J}(A)$ and assume, by contradiction, that there exists $b \in A$ such that $1-ab$ is not invertible. Then, by **T.1.2**.3, $1-ab \in \mathfrak{m}$, for some $\mathfrak{m} \in \operatorname{Max} A$. Since $ab \in \mathfrak{m}$, this implies $1 \in \mathfrak{m}$, which is a contradiction. □

**Proof T. 1.16.** 1. For each proper ideal $I \subsetneq A$ we have that $I \subseteq A \setminus A^*$. Therefore, $I \subseteq \mathfrak{m}$ and $\mathfrak{m}$ is the only maximal ideal of $A$.

2. If we show that every element $a \notin \mathfrak{m}$ is invertible, then the claim follows from part 1. Assume $a \notin \mathfrak{m}$. Then, since $\mathfrak{m}$ is maximal, $(a, \mathfrak{m}) = (1)$, *i.e.*, there exist $b \in A$ and $m \in \mathfrak{m}$ such that $ba + m = 1$.
Therefore, by hypothesis, $ba = 1 - m \in A^*$ and it follows that $a \in A^*$. □

**Proof T. 1.17.** Parts 1 and 2 immediately follow from the definitions.

3. If $a \in I$, then $f(a) \in f(I) \subseteq I^e$. Thus, $a \in I^{ec}$.
If we consider the inclusion homomorphism $\mathbb{Z} \longrightarrow \mathbb{Q}$ and the ideal $I = (n)$ with $n \neq 0, \pm 1$, then $I^e = \mathbb{Q}$ and $I^{ec} = \mathbb{Z} \supsetneq I$.

4. Since by definition $f(J^c) \subseteq J$, it follows that $J^{ce} = (f(J^c)) \subseteq J$.
If we consider the inclusion homomorphism $K \longrightarrow K[x]$ and the ideal $J = (x)$, then $J^c = (0)$ and $J^{ce} = (0) \subsetneq J$.

5. By parts 3 and 1, $I \subseteq I^{ec}$ implies $I^e \subseteq I^{ece}$.

To prove the opposite inclusion, we observe that $I^{ec} = f^{-1}(I^e)$. Therefore, we have $f(I^{ec}) = f(f^{-1}(I^e)) \subseteq I^e$, and finally $I^{ece} \subseteq I^e$.

6. By parts 4 and 2, $J^{ce} \subseteq J$ implies $J^{cec} \subseteq J^c$.

To prove the opposite inclusion, we have $f(J^c) \subseteq J^{ce}$, which immediately implies $J^c \subseteq J^{cec}$.

7. If $ab \in J^c$, then $f(a)f(b) = f(ab) \in J$. Since $J$ is prime, either $f(a) \in J$ or $f(b) \in J$. Then, either $a \in J^c$ or $b \in J^c$.

8. If $ab \in J^c$, then $f(a)f(b) = f(ab) \in J$. Since $J$ is primary, either $f(a) \in J$ or $f(b^n) = f(b)^n \in J$ for some $n$. Consequently, either $a \in J^c$ or $b^n \in J^c$.

9. If $a \in \sqrt{J^c}$, then $f(a)^n = f(a^n) \in J$ for some $n$. Therefore, $f(a) \in \sqrt{J} = J$ and $a \in J^c$. Since the other inclusion always holds, this shows that $J^c$ is radical.

Finally, the inclusion homomorphism $\mathbb{Z} \longrightarrow \mathbb{Q}$ and a non-zero prime ideal $(p)$ of $\mathbb{Z}$ provide a counterexample to parts 7 and 8 for extension of ideals. The inclusion homomorphism $\mathbb{Z} \longrightarrow \mathbb{Z}[i]$ and the ideal $(2)$ show that the extension of a radical ideal is not necessarily radical, because $(2)^e = (1+i)^2$. □

**Proof T. 1.18.** 1. Clearly, $f(I) \subseteq I^e$.

To prove the opposite inclusion, let $b \in I^e = (f(I))$. Then, there exist $b_j \in B$ and $i_j \in I$ such that $b = \sum_j b_j f(i_j)$. Since $f$ is surjective, there exists $a_j \in A$ such that $b_j = f(a_j)$ for every $j$.
Therefore,

$$b = \sum_j b_j f(i_j) = \sum_j f(a_j)f(i_j) = \sum_j f(a_j i_j) = f(\sum_j a_j i_j) \in f(I),$$

because $f$ is a ring homomorphism.

Alternatively, one can directly prove that $f(I)$ is an ideal of $B$.

2. We have $I^{ec} = f^{-1}(I^e)$. By part 1, we also have $I^e = f(I)$. Therefore, if $a \in I^{ec}$, then $f(a) = f(b)$ for some $b \in I$. This implies that $a - b \in \operatorname{Ker} f \subseteq I$, and then, $a \in I$.

Since $f$ is surjective, it follows that $J^{ce} = f(J^c) = f(f^{-1}(J)) = J$.

3. We remark that the homomorphism

$$A \xrightarrow{f} B \xrightarrow{\pi} B/I^e$$

is surjective and that $I$ is its kernel, since $\operatorname{Ker} f \subseteq I$. By the homomorphism theorem **T.1.9**.1, we have $A/I \simeq B/I^e$.
Similarly, it can be shown that $A/J^c \simeq B/J$.
All the statements easily follow from **T.1.11**.3, 4 and 5. □

**Proof T. 1.21.** 1. If the ideals are pairwise comaximal, then $f$ is surjective by **T.1.20**.

Conversely, if $f$ is surjective, then for every $i$ we can choose elements $a_i \in A$ such that $f(a_i) = (0, \ldots, 0, 1, 0, \ldots, 0)$, where the 1 is in the $i$-th position. In this way, $a_i \equiv 1 \mod I_i$ and $a_i \equiv 0 \mod I_j$ for each $j \neq i$. For every $j \neq i$, we have $1 = (1 - a_i) + a_i \in I_i + I_j$, as desired.

2. Obviously, $\operatorname{Ker} f = \bigcap_{i=1}^{n} I_i$. □

**Proof T. 1.22.** 1. For every $a \in A$, it holds that $(a) \subseteq (1)$.
An element $a$ is invertible if and only if there exists $b \in A$ such that $ab = 1$, *i.e.*, if and only if $(1) \subseteq (a)$.

2. The element $a$ is a divisor of $b$ if and only if there exists $c \in A$ such that $ac = b$, *i.e.*, if and only if $(b) \subseteq (a)$.

3. Two elements $a$ and $b$ are associate if and only if there exists $c \in A^*$ such that $ac = b$ and $a = bc^{-1}$, *i.e.*, if and only if $(b) \subseteq (a)$ and $(a) \subseteq (b)$.

4. If $a$ is a proper divisor of $b \neq 0$, then it is impossible for $(a) = (b)$ to hold. Indeed, if that were the case, then we would have $b = ac = bdc$ for some $c, d \in A$. However, since $A$ is a domain, $b(1 - dc) = 0$ would imply that $c$ is invertible, which contradicts the given hypothesis.
Since $a$ is not invertible, we have $(b) \subsetneq (a) \subsetneq (1)$.

5. Let $a \notin A^*$. Then, by part 2, the element $a$ being prime, *i.e.*, $a \mid bc$ if and only if either $a \mid b$ or $a \mid c$, is equivalent to the condition: $(bc) \subseteq (a)$ if and only if either $(b) \subseteq (a)$ or $(c) \subseteq (a)$. The latter condition is equivalent to the ideal $(a)$ being prime, *i.e.*, $bc \in (a)$ if and only if either $b \in (a)$ or $c \in (a)$. □

**Proof T. 1.23.** 1. Suppose $a$ is not irreducible. Then, there exist $b, c \notin A^*$ such that $a = bc$. Therefore, neither $b$ nor $c$ are elements of $(a)$, because if, for example, $b = ab_1$ for some $b_1 \in A$, then $a = bc = ab_1c$. This implies that $c$ is invertible, which is a contradiction. Thus, we have found $b, c \notin (a)$ such that $bc \in (a)$, that is, $(a)$ is not prime.
This contradicts the hypothesis by **T.1.22**.5.

2. Suppose $(a)$ is not prime. Thus, there exist $b, c \in A \setminus (a)$ such that $bc \in (a)$. Since $A$ is a PID, we have $(a) \subsetneq (a, b) = (d)$ for some $d$. Hence, $a = da_1$ with $a_1 \notin A^*$, otherwise $b \in (d) = (a)$.
To complete the proof, it suffices to show that $d \notin A^*$, because then we would have a non-trivial factorization $da_1$ of $a$, which is a contradiction.
If $d \in A^*$, that is, $(1) = (d) = (a, b)$, then there exist $\alpha, \beta \in A$ such that $1 = \alpha a + \beta b$, from which it follows that $c = c \cdot 1 \in (a) + (bc) \subseteq (a)$. This is impossible. □

**Proof T. 1.24.** It can be easily verified that the union $I$ of the ideals of $\mathcal{C}$ is an ideal. Furthermore, since $A$ is a PID, $I$ is principal. Let $I = (a)$. Then, there exists $h_0$ such that $I_{h_0} \subseteq I = (a) \subseteq I_{h_0}$. □

**Proof T. 1.26.** Let $A$ be a ring. By **T.1.23**.2, in a PID every irreducible element is also prime. Therefore, by **T.1.25**, it is sufficient to prove that every non-zero and non-invertible element of $A$ can be expressed as the product of a finite number of irreducible elements.
Assume, by contradiction, that $a_0 \in A \setminus \{A^* \cup \{0\}\}$ is an element which cannot be written as a finite product of irreducible elements. This means that $a_0$ is not irreducible and generates a proper ideal. Thus, there exist $a_1, b_1 \in A \setminus A^*$ such that $a_0 = a_1 b_1$. If both $a_1$ and $b_1$ could be written as a finite product of irreducible elements, then $a_0$ would have a similar factorization. Hence, we can assume that $a_1$ has no such factorization, and, in particular, it is not irreducible. Iterating this process, we can construct an ascending chain of ideals $(a_0) \subsetneq (a_1) \subsetneq \ldots$ of $A$ that is not stationary.
This contradicts **T.1.24**. □

**Proof T. 1.27.** Let $I$ be an ideal of $A$. If $I = (0)$, we are done.
Otherwise, $I \neq 0$, and let $\delta$ be the degree function which makes $A$ Euclidean. Then, $\delta(I \setminus \{0\})$ is a non-empty subset of $\mathbb{N}$, and thus, it has a minimum.
Choose $a \in I \setminus \{0\}$ such that $\delta(a)$ is minimal. We claim that $I = (a)$, *i.e.*, $A$ is a PID, hence a UFD, by **T.1.26**.

Let $b \in I$. By hypothesis, there exist $q$, $r \in A$ such that $b = qa + r$ with either $r = 0$ or $\delta(r) < \delta(a)$. Since $r = b - qa \in I$, the second case cannot occur because $\delta(a)$ is minimal. Therefore, $r = 0$, as desired. □

**Proof T. 1.28.** Suppose $a$ is irreducible, and let $I$, $J \subset A$ be ideals such that $(a) = I \cap J$. Then, $(a) \subseteq I$ and $(a) \subseteq J$.
If, by contradiction, $(a) \neq I$ and $(a) \neq J$, then there exist $b \in I \setminus (a)$ and $c \in J \setminus (a)$. Since $bc \in I \cap J = (a)$, we have that $a \nmid b$ and $a \nmid c$, but $a \mid bc$. This is impossible because $a$ is irreducible, hence prime by **T.1.25**. □

## 16.2 Chapter 2

**Proof T. 2.1.** Since $I$ is an ideal, if $X^{\mathbf{a}} \in I$ for every $\mathbf{a}$ such that $c_{\mathbf{a}} \neq 0$, then $f \in I$.

Conversely, suppose $f \in I$.
If we denote the generators of $I$ by $X^{\mathbf{b}}$, then

$$f = \sum_{\mathbf{a}} c_{\mathbf{a}} X^{\mathbf{a}} = \sum_{\mathbf{b}} p_{\mathbf{b}}(X) X^{\mathbf{b}} = \sum_{\mathbf{b}} \Big( \sum_{\mathbf{d}} c_{\mathbf{b},\mathbf{d}} X^{\mathbf{d}} \Big) X^{\mathbf{b}} = \sum_{\mathbf{b},\mathbf{d}} c_{\mathbf{b},\mathbf{d}} X^{\mathbf{b}+\mathbf{d}}$$

for some $\mathbf{b}, \mathbf{d} \in \mathbb{N}^n$, $c_{\mathbf{b},\mathbf{d}} \in K$, and $p_{\mathbf{b}}(X) = \sum_{\mathbf{d}} c_{\mathbf{b},\mathbf{d}} X^{\mathbf{d}} \in A$.
Since $\mathrm{Mon}(A)$ is a basis of the $K$-vector space $A$, it follows that each $X^{\mathbf{a}}$ with $c_{\mathbf{a}} \neq 0$ can be written as $X^{\mathbf{b}+\mathbf{d}}$.
Therefore, $X^{\mathbf{a}}$ belongs to the monomial ideal $I$. □

**Proof T. 2.2.** Let $S$ be a set of monomial generators of $I$, and let

$$E = \{\mathbf{a} + \mathbb{N}^n \colon X^{\mathbf{a}} \in S\}.$$

Clearly, $E \neq \emptyset$, and by construction, it is an $\mathcal{E}$-subset with the required property.

By definition, for every boundary $F$ of $E$, we have $(X^{\mathbf{a}} \colon \mathbf{a} \in F) \subseteq I$.
Let $\mathbf{b} \in E$. There exist $\mathbf{a} \in F$ and $\mathbf{c} \in \mathbb{N}^n$ such that $\mathbf{b} = \mathbf{a} + \mathbf{c}$. Therefore, $X^{\mathbf{b}} = X^{\mathbf{a}} X^{\mathbf{c}} \in (X^{\mathbf{a}} \colon \mathbf{a} \in F)$. □

**Proof T. 2.4.** Let $F$ be a minimal boundary of $E$. By Dickson's Lemma, there exists a finite boundary $F'$ of $E$, and each $\mathbf{a} \in F$ can be written as $\mathbf{a} = \mathbf{b} + \mathbf{c}$, with $\mathbf{b} \in F'$ and $\mathbf{c} \in \mathbb{N}^n$. Moreover, there exist $\mathbf{a}_1 \in F$ and $\mathbf{c}_1 \in \mathbb{N}^n$ such that $\mathbf{b} = \mathbf{a}_1 + \mathbf{c}_1$.
By the minimality of $F$, $\mathbf{a} = \mathbf{b} = \mathbf{a}_1$, and therefore, $F \subseteq F'$. □

**Proof T. 2.5.** Let $F = \{\mathbf{a}_1, \dots, \mathbf{a}_s\}$ and $F' = \{\mathbf{b}_1, \dots, \mathbf{b}_t\}$ be two minimal boundaries of an $\mathcal{E}$-subset $E$.
Then,

$$E = \bigcup_{i=1}^{s} (\mathbf{a}_i + \mathbb{N}^n) = \bigcup_{j=1}^{t} (\mathbf{b}_j + \mathbb{N}^n).$$

Since $\mathbf{a}_i \in E$, for every $i$ there exists $\mathbf{b}_j \in F'$ such that $\mathbf{a}_i \in \mathbf{b}_j + \mathbb{N}^n$.
Define

$$\eta \colon \{1, \dots, s\} \longrightarrow \{1, \dots, t\} \ \text{ by } \ \eta(i) = \min\{j \colon \mathbf{a}_i \in \mathbf{b}_j + \mathbb{N}^n\}.$$

This map is surjective, because otherwise

$$E = \bigcup_{i=1}^{s} (\mathbf{a}_i + \mathbb{N}^n) \subseteq \bigcup_{i=1}^{s} (\mathbf{b}_{\eta(i)} + \mathbb{N}^n) \subsetneq \bigcup_{j=1}^{t} (\mathbf{b}_j + \mathbb{N}^n) = E,$$

which is not possible. Therefore, $s \geq t$.
In the same way, exchanging the roles of $F$ and $F'$, we can define a surjective map $\nu\colon \{1,\ldots,t\} \longrightarrow \{1,\ldots,s\}$, and obtain $t \geq s$. Thus, $s = t$, and it follows that $\eta$ and $\nu$ are permutations.
Finally, $\mathbf{a}_i \in \mathbf{b}_{\eta(i)} + \mathbb{N}^n \subseteq \mathbf{a}_{\nu(\eta(i))} + \mathbb{N}^n$, and hence, the minimality of $F$ implies $\mathbf{a}_i = \mathbf{a}_{\nu(\eta(i))} = \mathbf{b}_{\eta(i)}$. □

**Proof T. 2.8.** We have to prove that the Division Algorithm terminates, and that the polynomials $u_1, \ldots, u_s$, and $r$ satisfy the required properties. Using the notation from the algorithm, we will first prove that the following relation holds at every step

$$f = u_1 f_1 + \ldots + u_s f_s + p + r.$$

The equation is clearly true at the first step.

Assume it holds at the $(n-1)$-th step. When executing the $n$-th step, the algorithm proceeds in one of the following two ways:

a) if $\mathrm{lt}(f_i) \mid \mathrm{lt}(p)$, then

$$u_i f_i + p = \Big(u_i + \frac{\mathrm{lt}(p)}{\mathrm{lt}(f_i)}\Big) f_i + p - \frac{\mathrm{lt}(p)}{\mathrm{lt}(f_i)} f_i.$$

Therefore $u_i f_i + p$ does not change;

b) if the remainder is updated, *i.e.*, if $\mathrm{lt}(f_i) \nmid \mathrm{lt}(p)$ for every $i$, then

$$p + r = (p - \mathrm{lt}(p)) + (r + \mathrm{lt}(p)).$$

Therefore $p + r$ does not change.

In both cases the relation $f = u_1 f_1 + \ldots + u_s f_s + p + r$ continues to hold.

To prove that the algorithm terminates, we recall that $>$ is a well-order, and, at each step, in both cases, $\mathrm{lt}(p)$ is replaced by the leading term of either $p - \mathrm{lt}(p)$ or $p - \frac{\mathrm{lt}(p)}{\mathrm{lt}(f_i)} f_i$, which are polynomials of multidegree strictly smaller than the multidegree of $p$. Therefore, this process terminates with $p = 0$, and the relation $f = u_1 f_1 + \ldots + u_s f_s + p + r$ provides the equation i).

To verify ii), note that the algorithm adds to $r$ only monomials not divisible by any of the $\mathrm{lt}(f_i)$ for all $i$. Therefore, by construction, $r$ is reduced with respect to $F$.

To verify iii), note that $\mathrm{Deg}(p) \leq \mathrm{Deg}(f)$. Moreover, when $u_i$ is modified, we add to it a term of the form $\frac{\mathrm{lt}(p)}{\mathrm{lt}(f_i)}$, where $\frac{\mathrm{lt}(p)}{\mathrm{lt}(f_i)} f_i$ cancels the current $\mathrm{lt}(p)$. Hence, $\mathrm{Deg}(u_i f_i) \leq \mathrm{Deg}(p) \leq \mathrm{Deg}(f)$. □

**Proof T. 2.9.** By the Division Theorem, for any polynomial $0 \neq f \in A$, dividing by $G$, we obtain $f = \sum_{i=1}^{t} u_i g_i + r$, where the remainder $r$ is reduced with respect to $G$.

$1 \Rightarrow 2$. If $f \xrightarrow{G}_* 0$, then we have $f = \sum_{i=1}^{t} u_i g_i$. Thus, $f \in (G) \subseteq I$.

Conversely, assume $f \in I$. In this case, $r = \sum r_{\mathbf{a}} X^{\mathbf{a}} \in I$, and if $r \neq 0$, then $\operatorname{lt}(r) \in \operatorname{Lt}(G)$, which contradicts the fact that $r$ is reduced with respect to $G$. Therefore, $r = 0$.

$2 \Rightarrow 3$. It is evident that if $f = \sum_{i=1}^{t} u_i g_i$, then $f \in I$.

Conversely, let $f \in I$. By hypothesis, $f \xrightarrow{G}_* 0$, and therefore, by the Division Algorithm, $f$ is a combination of the $g_i$ with coefficients $u_i \in A$. The equality $\operatorname{lm}(f) = \max_{u_i \neq 0}\{\operatorname{lm}(u_i)\operatorname{lm}(g_i)\}$ follows from (2.1), together with the fact that $r = 0$.

$3 \Rightarrow 1$. We have to prove that $\operatorname{Lt}(I) = \operatorname{Lt}(G)$.
Since $\operatorname{Lt}(G) \subseteq \operatorname{Lt}(I)$, it suffices to prove that, for every $0 \neq f \in I$, we have $\operatorname{lm}(f) \in (\operatorname{lm}(g_1), \ldots, \operatorname{lm}(g_t))$.
By hypothesis, there exist $u_1, \ldots, u_t$ such that

$$f = \sum_{i=1}^{t} u_i g_i \quad \text{with} \quad \operatorname{lm}(f) = \max_{u_i \neq 0}\{\operatorname{lm}(u_i)\operatorname{lm}(g_i)\} \in (\operatorname{lm}(g_1), \ldots, \operatorname{lm}(g_t)),$$

and the conclusion follows. □

**Proof T. 2.10.** It is an immediate consequence of **T.2.9**.
If $f \in I$, then $f \xrightarrow{G}_* 0$, that is, $f = \sum_{i=1}^{t} u_i g_i$ for some $u_i \in A$. Therefore, $I \subseteq (g_1, \ldots, g_t)$. □

**Proof T. 2.11.** Let $f \in A \setminus \{0\}$. By the Division Algorithm, we can write

$$f = \sum_{i=1}^{t} u_i g_i + r \quad \text{with} \quad r = \sum_{\mathbf{a}} r_{\mathbf{a}} X^{\mathbf{a}}.$$

If $r_{\mathbf{a}} \neq 0$, then $r_{\mathbf{a}} X^{\mathbf{a}} \notin \operatorname{Lt}(G) = \operatorname{Lt}(I)$. If $r'$ is another remainder, then $r - r' \in I$. Since none of its monomials belong to $\operatorname{Lt}(I)$, we have $r - r' = 0$. □

**Proof T. 2.12.** Note that $r = f - h$ and $r' = f - h'$ for some $h, h' \in I$. Therefore, $r - r' \in I$, and if $r - r' \neq 0$, then $\operatorname{lt}(r - r') \in \operatorname{Lt}(I)$.
Under this assumption, there would exist $g_i \in G$ and $g'_j \in G'$ such that $\operatorname{lt}(g_i)$ and $\operatorname{lt}(g'_j)$ divide $\operatorname{lt}(r - r')$, which is either a term of $r$ or of $r'$. In either case, we obtain a contradiction, because $r$ is reduced modulo $G$, $r'$ is reduced modulo $G'$, and $\operatorname{Lt}(G) = \operatorname{Lt}(G')$. □

**Proof T. 2.16.** We write $f = \operatorname{lt}(f) + f_1$ and $g = \operatorname{lt}(g) + g_1$.
From $\gcd(\operatorname{lt}(f), \operatorname{lt}(g)) = 1$ it follows that

$$S(f, g) = \operatorname{lt}(g) f - \operatorname{lt}(f) g = (g - g_1) f - (f - f_1) g = f_1 g - g_1 f.$$

If either, or both, $f_1 = 0$ or $g_1 = 0$, then $S(f, g)$ clearly reduces to zero, using either $f$ or $g$.
Otherwise, $f_1 g_1 \neq 0$ and we have $\operatorname{Deg}(f_1) < \operatorname{Deg}(f)$ and $\operatorname{Deg}(g_1) < \operatorname{Deg}(g)$.
Note that for any non-zero polynomials $p$ and $q$, such that $\operatorname{Deg}(p) < \operatorname{Deg}(g)$ and $\operatorname{Deg}(q) < \operatorname{Deg}(f)$, we have $\operatorname{lt}(pf) \neq \operatorname{lt}(qg)$.

In fact, if $\mathrm{lt}(pf) = \mathrm{lt}(qg)$, then $\mathrm{lt}(f) \mid \mathrm{lt}(q)$ and $\mathrm{lt}(g) \mid \mathrm{lt}(p)$, which violates the degree constraints.
Thus, no cancellation occurs between the leading terms of $g_1 f$ and $f_1 g$, and then, either $\mathrm{lt}(S(f,g)) = \mathrm{lt}(f_1 g)$ or $\mathrm{lt}(S(f,g)) = \mathrm{lt}(-g_1 f)$.
Assume for example $\mathrm{lt}(S(f,g)) = \mathrm{lt}(g_1 f)$. Then, we can perform one step of the reduction by subtracting $\mathrm{lt}(g_1)f$. This produces a new polynomial of the form $f_1 g - g_2 f$, where $g_2 = g_1 - \mathrm{lt}(g_1)$. In this expression $f_1$ and $g_2$ again satisfy our degree constraints, hence we can repeat the argument reducing to 0 both $g_2 f$ and $f_1 g$, using only $f$ and $g$. □

**Proof T. 2.17.** Since the basis $G$ is minimal, $\mathrm{lt}(g_i) \nmid \mathrm{lt}(g_j)$ for all $i \neq j$. Hence, $\mathrm{lt}(g_i) = \mathrm{lt}(g_i')$ for all $i = 1, \ldots, t$.
Therefore, $G'$ is a minimal Gröbner basis of $I$.
Moreover, since at each step the reduction modulo $G_i'$ is performed using $\mathrm{lt}(g_1'), \ldots, \mathrm{lt}(g_{i-1}'), \mathrm{lt}(g_{i+1}), \ldots, \mathrm{lt}(g_t)$, and since $\mathrm{lt}(g_j) = \mathrm{lt}(g_j')$ for each $j$, it follows that $G'$ is reduced. □

**Proof T. 2.18.** Since both $G$ and $G'$ are minimal, without loss of generality, we can assume $\mathrm{lm}(g_i) = \mathrm{lm}(g_i')$ for all $i = 1, \ldots, t$, see **T.2.5**.
For any $i$ consider the polynomial $g_i - g_i' \in I$.
If $g_i - g_i' \neq 0$, then there exists $g_j$ such that $\mathrm{lm}(g_j) \mid \mathrm{lm}(g_i - g_i')$. Since $\mathrm{lm}(g_i - g_i') < \mathrm{lm}(g_i)$, we have that $i \neq j$. Thus, $\mathrm{lm}(g_j) = \mathrm{lm}(g_j')$ divides a term of $g_i - g_i'$, *i.e.*, a term of $g_i$ or of $g_i'$. This is not possible because $G$ and $G'$ are reduced. □

**Proof T. 2.21.** Let $G = \{g_1, \ldots, g_t\}$ be a Gröbner basis of $I$ with respect to $>$. To show that every element $j \in J$ is in $I$, we prove that $j \xrightarrow{G}_* 0$.
Let $r$ be the remainder of $j$, then

$$r = j - \sum_{i=1}^{t} u_i g_i \in J + I \subseteq J.$$

If $r \neq 0$, then none of its terms is in $\mathrm{Lt}(G) = \mathrm{Lt}(I) = \mathrm{Lt}(J)$, because $r$ is reduced. Since $r \in J$, we have that $\mathrm{lt}(r) \in \mathrm{Lt}(J)$, and hence, $r = 0$. □

**Proof T. 2.23.** 1. Each element of $A/I$ can be represented by the class $\overline{r}$, where $r = \overline{f}^G$ for some $f \in A$. Since $r$ is reduced with respect to $G$, the element $\overline{r}$ is a $K$-linear combination of monomials which are not divisible by $\mathrm{lt}(g_i)$ for all $i = 1, \ldots, t$. Thus, $\mathcal{B}$ is a set of generators.
To prove linear independence over $K$, assume, by contradiction, that there exists $m_1, \ldots, m_k$ in $\mathcal{B}$ such that $\overline{m_1} = \sum_{i=2}^k a_i \overline{m_i}$ for some $a_i \in K$. Then, there exists $g \in I$ such that

$$m_1 = g + \sum_{i=2}^{k} a_i m_i.$$

Therefore, in $A$ we have $m_1 \xrightarrow{G}_* m_1$ and $m_1 \xrightarrow{G}_* \sum_{i=2}^{k} a_i m_i$, because $m_i \in \mathcal{B}$. Since the $m_i$ are obviously linearly independent over $K$, this contradicts the uniqueness of the remainder.

2. The proof of the second statement is immediate. □

**Proof T. 2.24.** There exists $g \in A$ such that $fg \equiv 1 \mod I$ if and only if $fg - 1 \in I$, *i.e.*, if and only if $1 \in (I, f)$. □

**Proof T. 2.26.** 1. Let $f \in I \cap J$, then $f = tf + (1-t)f \in (tI, (1-t)J)$.

To prove the opposite inclusion, note that if $h_1, \ldots, h_u$ are generators of an ideal $H$, then the elements $th_1, \ldots, th_u$ and $(1-t)h_1, \ldots, (1-t)h_u$ generate $tH$ and $(1-t)H$, respectively.

Therefore, if $f(X) = g(t, X) + h(t, X) \in (tI, (1-t)J) \cap A$ for some $g(t, X) \in tI$ and $h(t, X) \in (1-t)J$, then

$$f(X) = g(1, X) = h(0, X) \in I \cap J$$

by the previous observation.

2. By **E.8.21**.5, we have $I : J = \bigcap_{i=1}^{s} I : (f_i)$. Moreover, for each $f \in A$, we have $I : (f) = \frac{1}{f}(I \cap (f))$, see **E.8.22**. □

**Proof T. 2.27.** The statement is obvious for $f = 0$, thus we assume $f \neq 0$. If $f \in \sqrt{I}$, then there exists $m \in \mathbb{N}$ such that $f^m \in I$.

Thus, from

$$1 = t^m f^m + (1 - t^m f^m) = t^m f^m + (1 - tf) \sum_{i=0}^{m-1} t^i f^i,$$

it follows that $1 \in (I, 1 - tf)$.

Conversely, if $I = (f_1, \ldots, f_k)$ and $(I, 1 - tf) = (1)$, then

$$1 = \sum_{i=1}^{k} h_i(x_1, \ldots, x_n, t) f_i + h(x_1, \ldots, x_n, t)(1 - tf).$$

Since the left-hand side of the equality does not depend on $t$, we can evaluate at $t = \frac{1}{f}$ obtaining

$$1 = \sum_{i=1}^{k} h_i(x_1, \ldots, x_n, 1/f) f_i \in K(x_1, \ldots, x_n).$$

Thus, clearing denominators, there exist polynomials $g_i = g_i(x_1, \ldots, x_n) \in A$ and a positive integer $m$ such that $f^m = \sum_{i=1}^{k} g_i f_i \in I$. □

## 16.3 Chapter 3

**Proof T. 3.1.** 1. If $\alpha \in \mathbb{V}(J)$, then $f(\alpha) = 0$ for each $f \in J$. Hence, by hypothesis, $f(\alpha) = 0$ for each $f \in I$, that is, $\alpha \in \mathbb{V}(I)$.

2. The statement is a direct consequence of the definition of $\mathbb{I}(\mathbb{V}(I))$.

3. It is evident that $\mathbb{V}(I) \subseteq \mathbb{V}(\mathbb{I}(\mathbb{V}(I)))$.

To prove the opposite inclusion, we observe that $I \subseteq \mathbb{I}(\mathbb{V}(I))$, by part 2. Therefore $\mathbb{V}(\mathbb{I}(\mathbb{V}(I))) \subseteq \mathbb{V}(I)$ by part 1.

4. We first show that if $V \subseteq W$, then $\mathbb{I}(W) \subseteq \mathbb{I}(V)$.
If $f \in \mathbb{I}(W)$, then $f(\alpha) = 0$ for all $\alpha \in W$. Therefore, $f(\alpha) = 0$ for all $\alpha \in V$, *i.e.*, $f \in \mathbb{I}(V)$.

Conversely, assume $\mathbb{I}(W) \subseteq \mathbb{I}(V)$. If $\alpha \in V$, then for every $f \in \mathbb{I}(V)$, we have $f(\alpha) = 0$. Therefore, $f(\alpha) = 0$ for all $f \in \mathbb{I}(W)$, that is, $\alpha \in \mathbb{V}(\mathbb{I}(W)) = W$, where the last equality follows from part 3.

5. Since $I, J \subseteq I + J$, we have $\mathbb{V}(I + J) \subseteq \mathbb{V}(I) \cap \mathbb{V}(J)$.

To prove the opposite inclusion, consider $\alpha \in \mathbb{V}(I) \cap \mathbb{V}(J)$. For each $f = i + j \in I + J$, we have $f(\alpha) = i(\alpha) + j(\alpha) = 0$. In other words, $\alpha \in \mathbb{V}(I + J)$.

6. Since $IJ \subseteq I$ and $IJ \subseteq J$, we have $\mathbb{V}(I) \cup \mathbb{V}(J) \subseteq \mathbb{V}(IJ)$.

To prove the opposite inclusion, consider $\alpha \in \mathbb{V}(IJ)$. Then, $fg(\alpha) = 0$ for every $f \in I$ and every $g \in J$. If $\alpha \notin \mathbb{V}(I)$, then there exists $f \in I$ such that $f(\alpha) \neq 0$. Therefore, $\alpha \in \mathbb{V}(J)$ since $f(\alpha)g(\alpha) = fg(\alpha) = 0$.

7. Since $I \cap J \subseteq I$ and $I \cap J \subseteq J$, we have $\mathbb{V}(I) \cup \mathbb{V}(J) \subseteq \mathbb{V}(I \cap J)$.

To prove the opposite inclusion, we use the fact that $IJ \subseteq I \cap J$. From part 1 and part 6, we obtain that $\mathbb{V}(I \cap J) \subseteq \mathbb{V}(IJ) = \mathbb{V}(I) \cup \mathbb{V}(J)$.

8. Since $I \subseteq \sqrt{I}$, we have $\mathbb{V}(I) \supseteq \mathbb{V}(\sqrt{I})$.

To prove the opposite inclusion, let $f \in \sqrt{I}$. Then, there exists $m \in \mathbb{N}$ such that $f^m \in I$. Hence, for every $\alpha \in \mathbb{V}(I)$, we have $f^m(\alpha) = 0$. Therefore, we have $f(\alpha) = 0$, which implies that $\alpha \in \mathbb{V}(\sqrt{I})$. □

**Proof T. 3.2.** Let $\{V_h\}_{h \in H}$ be a descending chain of affine varieties of $K^n$. By **T.3.1**.4, we obtain an ascending chain of ideals $\{\mathbb{I}(V_h)\}_{h \in H}$ of $A$.
Since $A$ is Noetherian by Hilbert's Basis Theorem, this chain is stationary. Hence, the chain $\{\mathbb{V}(\mathbb{I}(V_h))\}_{h \in H}$ is also stationary, and the conclusion follows from **T.3.1**.3. □

**Proof T. 3.4.** 1. Consider the set

$$\Sigma = \{V \colon V \text{ affine variety which is not a finite union of irreducible varieties}\}$$

ordered by $\supseteq$. Assume, by contradiction, that $\Sigma \neq \emptyset$.
By **T.3.2**, every descending chain in $\Sigma$ has a maximal element with respect to $\supseteq$. Therefore, there exists a minimal element $W$ in $\Sigma$ which is not irreducible. Hence, we can write $W = W_1 \cup W_2$ where $W_1, W_2 \subsetneq W$ are proper subvarieties.

By the minimality of $W$, there exist integers $r$ and $s$, as well as irreducible subvarieties $W_{1,j} \subset W_1$, for $j = 1, \ldots, r$, and $W_{2,h} \subset W_2$, for $h = 1, \ldots, s$, such that

$$W = W_1 \cup W_2 = \bigcup_{j=1}^{r} W_{1,j} \cup \bigcup_{h=1}^{s} W_{2,h}.$$

Therefore, $W$ can be expressed as a finite union of irreducible varieties, which contradicts the assumption that $W$ is an element of $\Sigma$.

2. Assume that $V = \bigcup_{i=1}^{r} V_i = \bigcup_{j=1}^{s} W_j$ are two minimal decompositions of $V$ into irreducible varieties.
Then,

$$V_1 = V_1 \cap V = \bigcup_{j=1}^{s} (V_1 \cap W_j).$$

Since $V_1$ is irreducible, $V_1 = V_1 \cap W_{j_0}$ for some $j_0 \in \{1, \ldots, s\}$. In the same way, we obtain that there exists $i_0 \in \{1, \ldots, r\}$ such that $W_{j_0} = W_{j_0} \cap V_{i_0}$. Therefore, $V_1 \subseteq W_{j_0} \subseteq V_{i_0}$. Since the first decomposition is minimal, we conclude that $V_1 = W_{j_0}$.
By iterating this process for all the $V_i$ and using the minimality of the second decomposition, we obtain $r = s$. Hence, the two decompositions are equal, up to a permutation of the indices. □

**Proof T. 3.5.** Consider $f \in A$. By dividing $f$ by the polynomials $x_i - a_i$, we can express $f$ as

$$f(x_1, \ldots, x_n) = \sum_i h_i(x_i - a_i) + r \quad \text{for some} \quad h_i \in A \quad \text{and } r \in K.$$

Since the polynomials $x_i - a_i$ are a Gröbner basis for the ideal they generate, see **T.2.16**, we have that $f \in \mathbb{I}(\{\alpha\})$ if and only if $f(\alpha) = r = 0$, *i.e.*, if and only if $f \in (x_1 - a_1, \ldots, x_n - a_n)$.

Clearly, $A/\mathbb{I}(\{\alpha\}) \simeq K$, and thus, $\mathbb{I}(\{\alpha\})$ is maximal. □

**Proof T. 3.9.** 1. Since $\alpha_1, \ldots, \alpha_m \in \overline{K}$ are the roots of $f$, we can write

$$f = a_m \prod_{i=1}^{m} (x - \alpha_i)$$

and, by **T.3.6**.3, $\operatorname{Res}(f, g) = a_m^{\ell} \operatorname{Res}(\hat{f}, g)$, where $\hat{f} = \frac{f}{a_m}$.
By applying $m$ times **T.3.8**.3 to $\hat{f}$ and $g$, and evaluating the resulting expression at $y_1 = \alpha_1, \ldots, y_m = \alpha_m$, we obtain

$$\operatorname{Res}(f, g) = a_m^{\ell} \prod_{i=1}^{m} g(\alpha_i) \operatorname{Res}(1, g(x)) = a_m^{\ell} \prod_{i=1}^{m} g(\alpha_i).$$

From this relation and from **T.3.6**.2, we obtain

$$\operatorname{Res}(f,g) = (-1)^{m\ell} \operatorname{Res}(g,f) = (-1)^{m\ell} b_\ell^m \prod_{j=1}^{\ell} f(\beta_j).$$

2. It is a consequence of part 1. Since $f = a_m \prod_{i=1}^m (x - \alpha_i)$, we have

$$\prod_{j=1}^{\ell} f(\beta_j) = a_m^\ell \prod_{j=1}^{\ell}\prod_{i=1}^{m} (\beta_j - \alpha_i) = (-1)^{m\ell} a_m^\ell \prod_{j=1}^{\ell}\prod_{i=1}^{m} (\alpha_i - \beta_j).$$

3. It immediately follows from part 1.
4. If $\alpha \in \overline{K}$ is a common root of $f$ and $g$, then the minimal polynomial of $\alpha$ in $K[x]$ divides both $f$ and $g$. □

**Proof T. 3.10.** Let $h = h(x_2, \ldots, x_n) = \operatorname{Res}_{x_1}(f,g)$.
Then,

$$h(\beta) = \det \begin{pmatrix} c_m(\beta) & \cdots & \cdots & c_0(\beta) & 0 & \cdots & 0 \\ 0 & c_m(\beta) & \cdots & \cdots & c_0(\beta) & & \vdots \\ \vdots & & \ddots & & & \ddots & 0 \\ 0 & \cdots & 0 & c_m(\beta) & \cdots & \cdots & c_0(\beta) \\ d_\ell(\beta) & \cdots & \cdots & d_0(\beta) & 0 & \cdots & 0 \\ 0 & d_\ell(\beta) & & & \ddots & & \vdots \\ \vdots & & \ddots & & & \ddots & 0 \\ 0 & \cdots & 0 & d_\ell(\beta) & \cdots & \cdots & d_0(\beta) \end{pmatrix}.$$

If $d_\ell(\beta) \neq 0$, *i.e.*, if $r = 0$, then this determinant is equal to the determinant of the Sylvester matrix of $f(x_1, \beta)$ and $g(x_1, \beta)$, and thus the result follows.

Otherwise, if $d_\ell(\beta) = 0$, by expanding the determinant with respect to the first column, whose unique non-zero element is $c_m(\beta)$, we obtain

$$h(\beta) = c_m(\beta) \det \begin{pmatrix} c_m(\beta) & \cdots & c_0(\beta) & \cdots & 0 \\ \vdots & \ddots & & \ddots & \vdots \\ 0 & \cdots & c_m(\beta) & \cdots & c_0(\beta) \\ d_{\ell-1}(\beta) & \cdots & d_0(\beta) & \cdots & 0 \\ \vdots & \ddots & & \ddots & \vdots \\ 0 & \cdots & d_{\ell-1}(\beta) & \cdots & d_0(\beta) \end{pmatrix} \begin{matrix} \left.\vphantom{\begin{matrix} a \\ b \\ c \end{matrix}}\right\} \ell - 1 \\ \left.\vphantom{\begin{matrix} a \\ b \\ c \end{matrix}}\right\} m \end{matrix}.$$

If $d_{\ell-1}(\beta) \neq 0$, *i.e.*, if $r = 1$, then this determinant is equal to the determinant of the Sylvester matrix of $f(x_1, \beta)$ and $g(x_1, \beta)$, and thus the result follows. If $d_{\ell-1}(\beta) = 0$, we can repeat this process until we find a non-zero coefficient $d_{\ell-r}(\beta) \neq 0$, which exists since $g(x_1, \beta) \neq 0$. □

**Proof T. 3.12.** We write $f = f_N + f_{N-1} + \ldots + f_0$, with $f_i$ homogenous of total degree $i$.

For a monomial $X^{\mathbf{a}}$ of degree $N$, its image under $\varphi$ is

$$y_1^{a_1}(y_2+\alpha_2 y_1)^{a_2}\cdots(y_n+\alpha_n y_1)^{a_n}.$$

Hence,

$$\varphi(f)=f(y_1,y_2+\alpha_2 y_1,\ldots,y_n+\alpha_n y_1)=f_N(1,\alpha_2,\ldots,\alpha_n)y_1^N+f',$$

with $\deg_{y_1} f'<N$.
Since

$$0\neq f_N(x_1,\ldots,x_n)=x_1^N f_N\left(1,\frac{x_2}{x_1},\ldots,\frac{x_n}{x_1}\right),$$

and $K$ is infinite, there exist elements $\alpha_2,\ldots,\alpha_n\in K$ such that

$$c=f_N(1,\alpha_2,\ldots,\alpha_n)\in K^*.$$

□

**Proof T. 3.15.** It is always true that an ideal of the form $(x_1-a_1,\ldots,x_n-a_n)$ is maximal, as shown in **T.3.5**.

The converse is only true when $K=\overline{K}$. Let $\mathfrak{m}\subset A$ be a maximal ideal. By the weak form of the Nullstellensatz, we know that $\mathbb{V}(\mathfrak{m})\neq\emptyset$, which implies that there exists $\alpha\in\mathbb{V}(\mathfrak{m})$.
Therefore,

$$\mathfrak{m}_\alpha=\mathbb{I}(\{\alpha\})\supseteq\mathbb{I}(\mathbb{V}(\mathfrak{m}))=\sqrt{\mathfrak{m}}=\mathfrak{m},$$

where the penultimate equality is given by the strong form of the Nullstellensatz. The claim follows from the maximality of $\mathfrak{m}$. □

**Proof T. 3.16.** According to **T.3.1**.8, we have $\mathbb{V}(I)=\mathbb{V}(\sqrt{I})$.
Let

$$\mathbb{V}(I)=V_1\cup\ldots\cup V_r$$

be the minimal decomposition of $\mathbb{V}(I)$ into irreducible subvarieties, see **T.3.4**. Taking the associated ideals and applying the strong form of the Nullstellensatz, we obtain

$$\sqrt{I}=\mathbb{I}(\mathbb{V}(I))=\mathbb{I}(V_1\cup\ldots\cup V_r)=\mathbb{I}(V_1)\cap\ldots\cap\mathbb{I}(V_r).$$

By **T.3.3**, this shows that $\sqrt{I}$ is a finite intersection of prime ideals. Since the decomposition of the variety is minimal, this intersection is also minimal, *i.e.*, there are no inclusions between the primes being intersected, see **T.1.12**.2.

Finally, recalling that an ideal belongs to $\operatorname{Min} I$ if and only if it is in $\operatorname{Min}\sqrt{I}$, from **T.1.14** and **T.1.12**.2 we deduce that these primes are exactly the elements of $\operatorname{Min} I$. □

**Proof T. 3.18.** If $K=\overline{K}$, the variety $\mathbb{V}(I)=\{\alpha_1,\ldots,\alpha_s\}$ is a finite set, and $I$ is radical, then

$$I = \sqrt{I} = \mathbb{I}(\mathbb{V}(I)) = \mathbb{I}(\{\alpha_1, \dots, \alpha_s\}) = \bigcap_{i=1}^{s} \mathbb{I}(\{\alpha_i\}) = \bigcap_{i=1}^{s} \mathfrak{m}_{\alpha_i}.$$

Therefore, from **T.3.15** it follows that $I$ is 0-dimensional.
Since the ideals $\mathfrak{m}_{\alpha_i}$ are distinct and comaximal, we have

$$A/I = A/\bigcap_{i=1}^{s} \mathfrak{m}_{\alpha_i} = A/\mathfrak{m}_{\alpha_1} \cdots \mathfrak{m}_{\alpha_s} \simeq \prod_{i=1}^{s} A/\mathfrak{m}_{\alpha_i} \simeq K^s,$$

by the Chinese Remainder Theorem. □

**Proof T. 3.19.** The first equality is obvious, the second one was proved in **T.3.18**, and the last inequality follows from **T.3.17**.
To prove that $\dim_K(A/\sqrt{I}) \leq \dim_K(A/I)$, first fix a monomial ordering $>$. By **T.2.23**, we have

$$\dim_K(A/\sqrt{I}) = \dim_K(A/\mathrm{Lt}(\sqrt{I})) \leq \dim_K(A/\mathrm{Lt}(I)) = \dim_K(A/I),$$

where the inequality follows from $\mathrm{Lt}(I) \subseteq \mathrm{Lt}(\sqrt{I})$. Moreover, by **T.2.21**, equality holds if and only if $I$ is radical.

Regarding the Krull dimension, note that $\dim(A/I) = \dim(A/\sqrt{I})$ holds in general by **T.1.14**. Therefore, $I$ is 0-dimensional if and only if $\sqrt{I}$ is 0-dimensional, and this follows from **T.3.18**. □

**Proof T. 3.20.** The whole space $K^n = \mathbb{V}((0))$ and the empty set $\emptyset = \mathbb{V}((1))$ are closed sets. Furthermore, it can be easily verified that the finite union and the intersection of any family of closed sets are closed sets as well. □

**Proof T. 3.21.** The definition implies that the Zariski closure $\overline{S}$ is an affine variety containing $S$.
If $W \supseteq S$ is a variety containing $S$, then $\mathbb{I}(W) \subseteq \mathbb{I}(S)$. Hence,

$$W = \mathbb{V}(\mathbb{I}(W)) \supseteq \mathbb{V}(\mathbb{I}(S)) = \overline{S}.$$

Therefore, $\overline{S}$ is the intersection of all affine varieties containing $S$. □

**Proof T. 3.24.** 1. It is clear from the definitions. Therefore, $\mathcal{X}$ and $\emptyset$ are closed sets.

2. Obvious.

3. Since $(E)$ is the smallest ideal containing $E$, every prime containing $E$ also contains $(E)$.
Thus,

$$\sqrt{(E)} = \bigcap_{(E)\subset\mathfrak{p}\in\mathcal{X}} \mathfrak{p}.$$

Therefore, $\mathcal{V}(E) = \mathcal{V}((E)) = \mathcal{V}(\sqrt{(E)})$ for every subset $E$ of $A$.

4. From $IJ \subseteq I \cap J \subseteq I,\ J$, it follows that

$$\mathcal{V}(IJ) \supseteq \mathcal{V}(I \cap J) \supseteq \mathcal{V}(I) \cup \mathcal{V}(J).$$

To prove the opposite inclusions, recall that if $I \cap J \subseteq \mathfrak{p}$, then either $I \subseteq \mathfrak{p}$ or $J \subseteq \mathfrak{p}$, see **T.1.12**.2. Moreover, if $IJ \subseteq \mathfrak{p}$, then either $I \subseteq \mathfrak{p}$ or $J \subseteq \mathfrak{p}$, by the definition of prime ideal.
Hence,

$$\mathcal{V}(I) \cup \mathcal{V}(J) \supseteq \mathcal{V}(I \cap J) \text{ and } \mathcal{V}(I) \cup \mathcal{V}(J) \supseteq \mathcal{V}(IJ).$$

It immediately follows that the finite union of closed sets is itself a closed set.
5. Note that $\mathfrak{p} \supseteq \bigcup_{\alpha \in \Lambda} I_\alpha$ if and only if $\mathfrak{p} \supseteq I_\alpha$ for all $\alpha$, *i.e.*, if and only if $\mathfrak{p} \in \bigcap_{\alpha \in \Lambda} \mathcal{V}(I_\alpha)$. □

**Proof T. 3.25.** We have to show that every open set $\mathcal{X} \setminus \mathcal{V}(I)$ can be expressed as a union of $\mathcal{X}_f$ for some $f \in A$. From **T.3.24**.5, it follows that $\mathcal{V}(I) = \bigcap_{f \in I} \mathcal{V}(\{f\})$.
Thus,

$$\mathcal{X} \setminus \mathcal{V}(I) = \mathcal{X} \setminus \Big(\bigcap_{f \in I} \mathcal{V}(\{f\})\Big) = \bigcup_{f \in I} (\mathcal{X} \setminus \mathcal{V}(\{f\})) = \bigcup_{f \in I} \mathcal{X}_f. \qquad \square$$

**Proof T. 3.26.** A topological space $\mathcal{X}$ is compact if and only if every open cover of $\mathcal{X}$ has a finite subcover.
Let

$$\mathcal{X} = \bigcup_{\alpha \in \Lambda} \mathcal{X}_{f_\alpha} = \bigcup_{\alpha \in \Lambda} (\mathcal{X} \setminus \mathcal{V}(\{f_\alpha\})) = \mathcal{X} \setminus \bigcap_{\alpha \in \Lambda} \mathcal{V}(\{f_\alpha\}) = \mathcal{X} \setminus \mathcal{V}(\bigcup_{\alpha \in \Lambda} \{f_\alpha\}).$$

Then, $\mathcal{V}(\bigcup_{\alpha \in \Lambda} \{f_\alpha\}) = \emptyset$, *i.e.*, $\{f_\alpha\}_{\alpha \in \Lambda}$ generates $A$.
Therefore, there exist a finite subset $\Lambda' \subseteq \Lambda$ and elements $g_\lambda \in A$ such that $1 = \sum_{\lambda \in \Lambda'} g_\lambda f_\lambda$. As a result, we have $\mathcal{V}(\bigcup_{\lambda \in \Lambda'} \{f_\lambda\}) = \emptyset$, which implies that

$$\mathcal{X} = \bigcup_{\lambda \in \Lambda'} \mathcal{X}_{f_\lambda},$$

*i.e.*, $\{\mathcal{X}_{f_\lambda} : \lambda \in \Lambda'\}$ is a finite subcover. □

**Proof T. 3.27.** Since $\overline{\mathcal{Y}}$ is closed, we have $\overline{\mathcal{Y}} = \mathcal{V}(I)$, where $I$ is an ideal of $A$. We also have $\mathcal{Y} \subseteq \mathcal{V}(\bigcap_{\mathfrak{p} \in \mathcal{Y}} \mathfrak{p})$, because if $\mathfrak{q} \in \mathcal{Y}$, then $\mathfrak{q} \supseteq \bigcap_{\mathfrak{p} \in \mathcal{Y}} \mathfrak{p}$, and therefore $\mathfrak{q} \in \mathcal{V}(\bigcap_{\mathfrak{p} \in \mathcal{Y}} \mathfrak{p})$.
Moreover, $\mathcal{V}(I) \supseteq \mathcal{V}(\bigcap_{\mathfrak{p} \in \mathcal{Y}} \mathfrak{p})$. To prove this, it is enough to show that $I \subseteq \bigcap_{\mathfrak{p} \in \mathcal{Y}} \mathfrak{p}$, which is true because $\mathfrak{p} \in \mathcal{Y} \subseteq \mathcal{V}(I)$ implies $\mathfrak{p} \supseteq I$.

Therefore,

$$\mathcal{Y} \subseteq \mathcal{V}(\bigcap_{\mathfrak{p}\in\mathcal{Y}} \mathfrak{p}) \subseteq \mathcal{V}(I) = \overline{\mathcal{Y}},$$

and we conclude by taking the closure. □

**Proof T. 3.28.** By **T.3.27**, if $\mathfrak{p}$ is a prime ideal, then $\overline{\{\mathfrak{p}\}} = \mathcal{V}(\mathfrak{p})$. Therefore, $\overline{\{\mathfrak{p}\}} = \{\mathfrak{p}\}$ if and only if $\mathfrak{p}$ is maximal. Since $\dim A > 0$, there exist prime ideals which are not maximal. Therefore, not all the points of $\mathcal{X}$ are closed. Hence, $\mathcal{X}$ is not $T_1$.
Consequently, $\mathcal{X}$ cannot be a $T_2$ space.

To prove that $\mathcal{X}$ is a $T_0$ space, consider two distinct points $\mathfrak{p}_1$ and $\mathfrak{p}_2$ of $\mathcal{X}$. Then, there exists $f \in \mathfrak{p}_1 \setminus \mathfrak{p}_2$, which implies that $\mathcal{X}_f = \mathcal{X} \setminus \mathcal{V}(f)$ is an open set containing $\mathfrak{p}_2$ and not $\mathfrak{p}_1$. □

**Proof T. 3.29.** The contraction of a prime ideal is a prime ideal, therefore the map $\phi^*$ is well-defined. We will prove that the preimage of an open set of the basis of $\mathcal{X}$ is open in $\mathcal{Y}$.
It is sufficient to show that for any $f \in A$, we have $(\phi^*)^{-1}(\mathcal{X}_f) = \mathcal{Y}_{\phi(f)}$. Indeed, we have

$$\begin{aligned}(\phi^*)^{-1}(\mathcal{X}_f) &= \{\mathfrak{q} \in \mathcal{Y} \colon \phi^*(\mathfrak{q}) \in \mathcal{X}_f\} = \{\mathfrak{q} \in \mathcal{Y} \colon f \notin \phi^*(\mathfrak{q})\} \\ &= \{\mathfrak{q} \in \mathcal{Y} \colon f \notin \phi^{-1}(\mathfrak{q})\} = \{\mathfrak{q} \in \mathcal{Y} \colon \phi(f) \notin \mathfrak{q}\} = \mathcal{Y}_{\phi(f)}.\end{aligned}$$ □

**Proof T. 3.30.** By the previous result, the map $\pi^* \colon \operatorname{Spec}(A/\mathcal{N}(A)) \longrightarrow \mathcal{X}$, induced by the projection $\pi \colon A \longrightarrow A/\mathcal{N}(A)$, is continuous.
Moreover, $\pi^*$ is bijective, because $\pi$ establishes a one-to-one correspondence between prime ideals of $A/\mathcal{N}(A)$ and primes of $A$ containing $\mathcal{N}(A)$, *i.e.*, all prime ideals of $A$. Since for any ideal $I$ of $A/\mathcal{N}(A)$ we have

$$\begin{aligned}\pi^*(\mathcal{V}(I)) &= \pi^*\left(\{\mathfrak{p} \in \operatorname{Spec}(A/\mathcal{N}(A)) \colon I \subseteq \mathfrak{p}\}\right) \\ &= \left\{\pi^{-1}(\mathfrak{p}) \colon \mathfrak{p} \in \operatorname{Spec}(A/\mathcal{N}(A)) \text{ and } I \subseteq \mathfrak{p}\right\} \\ &= \{\mathfrak{q} \in \mathcal{X} \colon I \subseteq \pi(\mathfrak{q})\} = \left\{\mathfrak{q} \in \mathcal{X} \colon \pi^{-1}(I) \subseteq \mathfrak{q}\right\} = \mathcal{V}(\pi^{-1}(I)),\end{aligned}$$

the map $\pi^*$ is closed. Therefore, it is a homeomorphism. □

## 16.4 Chapter 4

**Proof T. 4.1.** We only need to verify that the product is well-defined, since the linearity conditions are obviously verified. We prove that for every $a$, $b \in A$ such that $\overline{a} = \overline{b} \in A/I$, we have $am = bm \in M$ for any $m \in M$.
This holds if and only if $(a - b)m = 0$, *i.e.*, if and only if $a - b \in \operatorname{Ann} M$. Since, by hypothesis, $a - b \in I \subseteq \operatorname{Ann} M$, the conclusion follows. □

**Proof T. 4.2.** It is straightforward to prove that $\operatorname{Hom}_A(M, N)$, with the defined operations, is an $A$-module (verify!).
Let $\varphi\colon M \longrightarrow \operatorname{Hom}_A(A, M)$ be defined by

$$\varphi(m) = f_m, \quad \text{where} \quad f_m(a) = am.$$

This map is $A$-linear, since it is easy to verify that for any $\alpha$, $\beta \in A$

$$\varphi(\alpha m_1 + \beta m_2) = f_{\alpha m_1 + \beta m_2} = \alpha f_{m_1} + \beta f_{m_2}.$$

It is injective because $\varphi(m) = 0$ if and only if $f_m = 0$, *i.e.*, if and only if $am = 0$ for all $a \in A$. In particular, for $a = 1$, it follows that $m = 0$.
It is surjective since every $A$-module homomorphism from $A$ to $M$ is completely determined by the image of 1 due to $A$-linearity. Consequently, if $f \in \operatorname{Hom}_A(A, M)$ is such that $f(1) = m$, then $f = f_m = \varphi(m)$. □

**Proof T. 4.3.** The proof is similar to the one of the theorems for rings.

1. It is easy to prove that the map

$$\overline{f}\colon M/\operatorname{Ker} f \longrightarrow \operatorname{Im} f \ \text{ given by } \ \overline{f}(\overline{m}) = f(m)$$

is well-defined and is an isomorphism.

2. It is easy to see that $N/P$ is a submodule of $M/P$.
The map $f\colon M/P \longrightarrow M/N$ given by $f(\overline{m}) = \overline{\overline{m}}$ is well-defined. Indeed, if $\overline{m} = \overline{n}$, then $m - n \in P \subseteq N$ and $\overline{\overline{m}} = \overline{\overline{n}}$.
Moreover, it is easy to prove that $f$ is a surjective homomorphism. Finally, $\overline{m} \in \operatorname{Ker} f$ if and only if $\overline{\overline{m}} = 0$, *i.e.*, if and only if $m \in N$, and $\overline{m} \in N/P$.

3. Consider the homomorphism

$$\pi \circ i\colon N \xrightarrow{\ i\ } N + P \xrightarrow{\ \pi\ } (N + P)/P,$$

where $i$ is the inclusion homomorphism of $N$ into $N+P$ and $\pi$ is the canonical projection on the quotient. It is evident that $\pi \circ i$ is surjective.
Since for any $n \in N$, we have $(\pi \circ i)(n) = \overline{n} = 0$ if and only if $n \in N \cap P$, by part 1, we can conclude that $N/(N \cap P) \simeq (N + P)/P$. □

**Proof T. 4.4.** Assume that $M$ is free. Since $S$ is a generating set of $M$, every element of $M$ can be written as a linear combination of elements of $S$.

Assume we have two expressions for the same element, $\sum_i a_i s_i = \sum_j b_j s_j$. Reordering the indices if necessary, we obtain $\sum_i (a_i - b_i)s_i = 0$. Since $S$ is free, this implies $a_i = b_i$ for every $i$, that is, the expression is unique.

Conversely, a zero linear combination of elements of $S$ is a way of writing 0 as a linear combination of elements of $S$. The uniqueness of the expression implies that the coefficients must be zero. □

**Proof T. 4.8.** Let $\{m_h\}_{h\in H}$ be a generating set of $M$. Consider the free module $A^H$ and the homomorphism defined by $f(e_h) = m_h$, where $\{e_h\}_{h\in H}$ is the canonical basis of $A^H$.
Such $f$ is clearly surjective, hence $M \simeq A^H/\operatorname{Ker} f$. □

**Proof T. 4.9.** In the proof of **T.4.7** it has been shown that if $M$ is a free module with basis $S$, then $M \simeq A^S$. □

**Proof T. 4.12.** Let $N = \langle n_1, \dots, n_k\rangle \subseteq M$, and consider the homomorphism

$$N \xrightarrow{i} M \xrightarrow{\pi} M/\mathfrak{m}M,$$

where $i$ is the inclusion of $N$ into $M$. By hypothesis, $\pi \circ i$ is surjective. We claim that $M = N + \mathfrak{m}M$. Clearly, $N + \mathfrak{m}M \subseteq M$.

To prove the opposite inclusion, note that for any $m \in M$ there exists $n \in N$ such that $\overline{m} = \overline{n}$. Hence, $m - n \in \mathfrak{m}M$ and $m = n + (m-n) \in N + \mathfrak{m}M$. Since the ring is local, we have $\mathfrak{m} = \mathcal{J}(A)$. Thus, $M = N$ by Nakayama's Lemma 3. □

**Proof T. 4.15.** Let $S = \{m_1, \dots, m_r\}$ and $\{n_1, \dots, n_r\}$ be a basis and a set of generators of $M$, respectively.
The map $f\colon S \longrightarrow M$ defined by $m_i \mapsto n_i$ for any $i = 1, \dots, r$ induces a surjective endomorphism $\tilde{f}$ of $M$ by **T.4.7**. Therefore, by **T.4.14**, $\tilde{f}$ is an isomorphism.
Hence, $\{n_1, \dots, n_r\}$ is a free set. □

**Proof T. 4.17.** 1. Since the Hom sequence is exact for any $A$-module $N$, we can choose appropriate modules $N$ to prove the exactness of the initial sequence.

At $M_2$ We choose $N = \operatorname{Coker} g = M_2/\operatorname{Im} g$ and prove that $N = 0$.
Consider the diagram

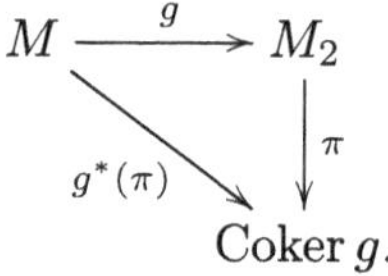


To prove that $N = 0$, we show that the projection homomorphism $\pi$ is the zero homomorphism. By construction, $g^*(\pi) = 0$, and the conclusion follows from the assumption that $g^*$ is injective, which holds for any $A$-module $N$.

At $M$ Since $(g \circ f)^* = f^* \circ g^* = 0$, it immediately follows that $\operatorname{Im} f \subseteq \operatorname{Ker} g$. Indeed, by setting $N = M_2$, we obtain

$$0 = (g \circ f)^*(\mathrm{id}_{M_2}) = \mathrm{id}_{M_2} \circ g \circ f = g \circ f.$$

To prove the opposite inclusion, we choose $N = \operatorname{Coker} f = M/\operatorname{Im} f$, and consider the diagram

$$\begin{array}{ccc} M_1 & \xrightarrow{f} & M \\ & \searrow^{f^*(\pi)} & \downarrow \pi \\ & & \operatorname{Coker} f. \end{array}$$

By construction, $\operatorname{Im} f = \operatorname{Ker} \pi$ and $f^*(\pi) = 0$. Therefore, $\pi \in \operatorname{Ker} f^* = \operatorname{Im} g^*$, and there exists $\varphi \in \operatorname{Hom}(M_2, N)$ such that $\varphi \circ g = g^*(\varphi) = \pi$.
This implies $\operatorname{Im} f = \operatorname{Ker} \pi \supseteq \operatorname{Ker} g$.

2. As in the previous part, we prove that the initial sequence is exact by choosing appropriate modules $M$ for the Hom sequence.

**At $N_1$** We choose $M = \operatorname{Ker} f$.
Consider the inclusion $i\colon \operatorname{Ker} f \longrightarrow N_1$ and the diagram

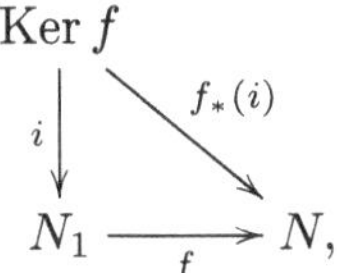


where, by construction, $f_*(i) = 0$.
Since, by hypothesis, $f_*$ is injective for any $M$, we obtain $i = 0$, that is, $f$ is injective.

**At $N$** We show both inclusions $\operatorname{Im} f \subseteq \operatorname{Ker} g$ and $\operatorname{Im} f \supseteq \operatorname{Ker} g$.
The first follows from $(g \circ f)_* = g_* \circ f_* = 0$. Indeed, taking $M = N_1$ and $\mathrm{id}_{N_1} \in \operatorname{Hom}_A(M, N_1)$, we have $g \circ f = 0$, that is, $\operatorname{Im} f \subseteq \operatorname{Ker} g$.

To prove the opposite inclusion, we choose $M = \operatorname{Ker} g$ and consider the diagram

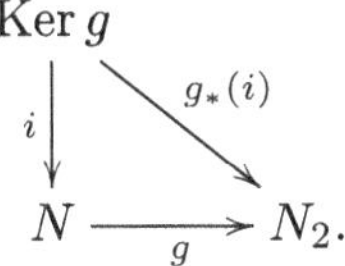


By construction, $g_*(i) = 0$, that is, $i \in \operatorname{Ker} g_* = \operatorname{Im} f_*$.
Hence, there exists $\varphi \in \operatorname{Hom}_A(\operatorname{Ker} g, N_1)$ such that $f \circ \varphi = f_*(\varphi) = i$.
Since $i$ is the inclusion homomorphism of $\operatorname{Ker} g$ into $N$, the last equality yields $\operatorname{Im} f \supseteq \operatorname{Im} i = \operatorname{Ker} g$, as desired.

Alternatively, we can choose $M = A$ and observe that, by **T.4.2**, in the diagram

$$\begin{array}{ccccccc}
0 & \longrightarrow & \operatorname{Hom}_A(A, N_1) & \xrightarrow{f_*} & \operatorname{Hom}_A(A, N) & \xrightarrow{g_*} & \operatorname{Hom}_A(A, N_2) \\
 & & \downarrow \simeq & & \downarrow \simeq & & \downarrow \simeq \\
0 & \longrightarrow & N_1 & \xrightarrow{f} & N & \xrightarrow{g} & N_2
\end{array}$$

the vertical arrows, which are all defined by $\varphi \mapsto \varphi(1_A)$, are isomorphisms and all the squares are commutative. □

**Proof T. 4.20.** If $\alpha$ and $\beta$ are isomorphisms, then

$$\operatorname{Ker} \alpha = \operatorname{Ker} \beta = \operatorname{Coker} \alpha = \operatorname{Coker} \beta = 0.$$

The Snake Lemma provides the exact sequence

$$0 \longrightarrow 0 \longrightarrow 0 \longrightarrow \operatorname{Ker} \gamma \longrightarrow 0 \longrightarrow 0 \longrightarrow \operatorname{Coker} \gamma \longrightarrow 0.$$

Therefore, $\operatorname{Ker} \gamma = \operatorname{Coker} \gamma = 0$, *i.e.*, $\gamma$ is also an isomorphism.

The proof of the remaining two cases is essentially identical. □

**Proof T. 4.21.** Let $P$ be a free $A$-module, and let $\mathcal{B}$ be a basis of $P$. Consider the diagram

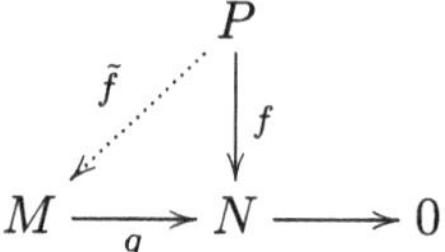

with $g$ surjective.

For any $b \in \mathcal{B} \subseteq P$ there exists $m_b \in M$ such that $g(m_b) = f(b)$. We define $\hat{f}$ on the elements of $\mathcal{B}$ by $\hat{f}(b) = m_b$ for any $b \in \mathcal{B}$. According to **T.4.7**, the map $\hat{f}$ lifts to a unique homomorphism $\tilde{f}$ such that $g \circ \tilde{f} = f$, making $P$ projective. □

**Proof T. 4.24.** If $M$ is projective, then, according to **T.4.22**, it is a direct summand of a free module. Thus, $M$ is isomorphic to a submodule of a free module, which is free by **T.4.23**.1. Hence, $M$ is free.

The converse always holds in general, see **T.4.21**. □

**Proof T. 4.25.** It is sufficient to prove that, for any invertible matrix $R$, the equality $\Delta_i(RX) = \Delta_i(X)$ holds. Since this implies that for any invertible matrix $S$

$$\Delta_i(XS) = \Delta_i((XS)^{\mathrm{t}}) = \Delta_i(S^{\mathrm{t}} X^{\mathrm{t}}) = \Delta_i(X^{\mathrm{t}}) = \Delta_i(X),$$

the conclusion will follow.

We begin by observing that the rows of $RX$ are linear combination of the rows of $X$. Since the determinant is multilinear, the determinants of the $i \times i$ submatrices of $RX$ are linear combination of the determinants of the $i \times i$ submatrices of $X$. Consequently, $\Delta_i(RX) \subseteq \Delta_i(X)$.

To prove the opposite inclusion, observe that $R$ is invertible, and thus

$$\Delta_i(RX) \supseteq \Delta_i(R^{-1}RX) = \Delta_i(X). \qquad \square$$

**Proof T. 4.28.** By **T.4.25**, since $D$ is diagonal and $d_{11} \mid d_{22} \mid \ldots \mid d_{tt}$, we have

$$\Delta_i(X) = \Delta_i(D) = (d_{11} \cdots d_{ii}).$$

Both statements immediately follow from this observation. $\square$

**Proof T. 4.30.** 1. Clearly, $\{\overline{v_1}, \ldots, \overline{v_r}\}$ is a generating set of $L/N$.
Thus, $L/N = \sum_i \langle \overline{v_i} \rangle$, and we have to show that this sum is a direct sum.
Suppose $a\overline{v_i} \in \sum_{j \neq i} \langle \overline{v_j} \rangle$, for some $a \in A$ and $i \in \{1, \ldots, r\}$.
Then, there exist $a_j$, $b_k \in A$ such that

$$av_i - \sum_{i \neq j} a_j v_j = \sum_{k=1}^{s} b_k d_k v_k \in N.$$

This implies that $a = 0$ if $i > s$ and $a = b_i d_i$, otherwise.
In both cases, $a\overline{v_i} = 0$, which shows that the sum is direct.

2. Note that $1 \mapsto \overline{v_i}$ induces a surjective homomorphism $A \longrightarrow \langle \overline{v_i} \rangle$ with kernel $\operatorname{Ann} \overline{v_i} = 0 : \langle \overline{v_i} \rangle$.
Now, let $a \in A$ be such that $a\overline{v_i} = 0$. If $i \leq s$, this relation holds if and only if $av_i = \sum_{j=1}^{s} b_j d_j v_j$, i.e., if and only if $a \in (d_i)$.
On the other hand, if $i > s$, this relation holds if and only if $a = 0$. Therefore, $\langle \overline{v_i} \rangle \simeq A/\operatorname{Ann} \overline{v_i}$, which in turn is isomorphic to $A/(d_i)$ if $i \leq s$, and to $A$ otherwise.
The claim follows from part 1. $\square$

**Proof T. 4.31.** Let $f\colon A^r \longrightarrow M$ be the $A$-module homomorphism defined by $f(e_i) = m_i$. Since $\operatorname{Ker} f$ is free, say of rank $s$, there exist a basis $\{v_1, \ldots, v_r\}$ of $A^r$ and scalars $d_1, \ldots, d_s$ such that $\{d_1 v_1, \ldots, d_s v_s\}$ is a basis of $\operatorname{Ker} f$, see also **T.4.29**.
By letting $I_i = 0 : \langle \overline{v_i} \rangle$ for all $i$, we have $I_i = (d_i)$ for all $i \leq s$ and $I_{s+1} = \ldots = I_r = 0$. Since $d_1 \mid d_2 \mid \ldots \mid d_s$, the ideals $I_i$ verify the required containment relations.
The conclusion follows from **T.4.30**, since $M \simeq A^r/\operatorname{Ker} f$. $\square$

**Proof T. 4.33.** 1. This is a particular instance of **T.4.23**.2.

2. By the structure theorem **T.4.31**, there exist principal ideals

$$I_1 = (d_1) \supseteq I_2 = (d_2) \supseteq \ldots \supseteq I_r = (d_r)$$

such that $M \simeq \bigoplus_{i=1}^{r} A/I_i$.
If $I_i = 0$ for all $i$, then $M \simeq A^r$ and $T(M) = 0$. Otherwise, let $s$ be such that $I_h \neq 0$ for $h \leq s$ and $I_i = 0$ for $i > s$.
Then, it is enough to note that

$$T(M) = \bigoplus_{i=1}^{s} A/I_i \quad \text{and} \quad M \simeq T(M) \oplus A^k,$$

where $k = r - s$.

3. As shown in part 2, we have $T(M) = \bigoplus_{i=1}^{s} A/I_i$.
As a result,

$$0 : T(M) = I_s = (d_s),$$

and the conclusion follows. □

**Proof T. 4.34.** By the structure theorem **T.4.31**, there exist principal ideals

$$I_1 = (d_1) \supseteq I_2 = (d_2) \supseteq \ldots \supseteq I_r = (d_r)$$

such that $M \simeq \bigoplus_{i=1}^{r} A/I_i$.
Since $M = M_{[p]}$, every $I_i$ contains a power of $p$. If $p^n \in I_i$, then $p^n = bd_i$. This implies that $I_i = (d_i) = (p^{k_i})$, as desired.

The inequalities among the exponents $k_i$ are a consequence of the inclusions of the ideals $I_i$. □

**Proof T. 4.35.** Recall that primary ideals in a PID are generated by powers of prime elements, see **E.8.64**.
By the structure theorem **T.4.31**, there exist principal ideals $I_1 = (d_1) \supseteq I_2 = (d_2) \supseteq \ldots \supseteq I_r = (d_r)$ such that $M \simeq \bigoplus_{i=1}^{r} A/I_i$.
Let $d_1, \ldots, d_s \neq 0$, $d_{s+1} = \ldots = d_r = 0$, and let

$$d_i = \prod_{j=1}^{h_i} p_{ij}^{e_{ij}}$$

be the decomposition of $d_i$ as a product of distinct primes, for $i = 1, \ldots, s$.
Thus, we have

$$M \simeq \bigoplus_{i=1}^{s} A/I_i \oplus A^{r-s}.$$

Moreover, by the Chinese Remainder Theorem, $A/I_i \simeq \bigoplus_{j=1}^{h_i} A/(p_{ij}^{e_{ij}})$.
Therefore,

$$M \simeq \bigoplus_{i=1}^{s} \bigoplus_{j=1}^{h_i} A/(p_{ij}^{e_{ij}}) \oplus A^{r-s},$$

where the ideals $(p_{ij}^{e_{ij}})$ are primary, as desired. □

**Proof T. 4.37.** 1. By **T.4.36**.5, $\{v_i, \ldots, \varphi^{d_i-1}(v_i)\}$ is a basis of $\langle v_i \rangle$ as a $K$-vector space. Since $V$ is the direct sum over $K[x]$, and thus, also over $K$, of the subspaces $\langle v_1 \rangle, \ldots, \langle v_s \rangle$, the conclusion follows.

2. It is a straightforward application of part 1. □

## 16.5 Chapter 5

**Proof T. 5.3.** 1. Note that $0 \otimes n = (0+0) \otimes n = 0 \otimes n + 0 \otimes n$.

2. It immediately follows from the definitions.

3. Saying that $L$ generates $M \otimes N$ is equivalent to saying that $M \otimes N/\langle L \rangle = 0$. Set $L = \mathcal{G}_1 \otimes \mathcal{G}_2$, and consider the diagram

$$\begin{array}{ccc} M \times N & \xrightarrow{\ 0\ } & M \otimes N/\langle L \rangle \\ {\scriptstyle \tau}\big\downarrow & \overset{\varphi}{\nearrow} & \\ M \otimes N. & & \end{array}$$

Obviously, $\varphi = 0$ makes the diagram commutative.
Moreover, let $(m, n) \in M \times N$, where $m = \sum_{i=1}^{h} a_i m_i$, with $m_i \in \mathcal{G}_1$ and $a_i \in A$ for all $i = 1, \ldots, h$, and $n = \sum_{j=1}^{k} b_j n_j$ with $n_j \in \mathcal{G}_2$ and $b_j \in A$ for all $j = 1, \ldots, k$.
Then, letting $\pi$ be the projection $M \otimes N \longrightarrow M \otimes N/\langle L \rangle$, we have

$$\begin{aligned} \pi(\tau(m,n)) = \pi(m \otimes n) &= \pi\left(\sum_{i=1}^{h} a_i m_i \otimes \sum_{j=1}^{k} b_j n_j\right) \\ &= \sum_{i,j} a_i b_j \pi(m_i \otimes n_j) = 0, \end{aligned}$$

where we used the fact that $\pi$ is a homomorphism, that $\otimes$ is bilinear, and that $m_i \in \mathcal{G}_1$, $n_j \in \mathcal{G}_2$. Hence, $\varphi = \pi$ makes the diagram commute as well. The uniqueness in the universal property yields $\pi = 0$, that is, $M \otimes N = \langle L \rangle$.

4. It follows immediately from part 3. □

**Proof T. 5.4.** 1. The map $f\colon A \times M \longrightarrow M$ defined by $f(a, m) = am$ is $A$-bilinear. By the universal property, there exists a unique homomorphism $\tilde{f}\colon A \otimes M \longrightarrow M$ such that $\tilde{f}(a \otimes m) = am$, which is clearly surjective.
Now, let $g\colon M \longrightarrow A \otimes M$ be the homomorphism defined by $g(m) = 1 \otimes m$. Then, for any simple tensor $a \otimes m$, we have

$$g \circ \tilde{f}(a \otimes m) = g(am) = 1 \otimes am = a \otimes m.$$

Therefore, $\tilde{f}$ is injective as well.

2. The map $f\colon N \times M \longrightarrow M \otimes N$ given by $f(n, m) = m \otimes n$ is bilinear, and therefore, induces a homomorphism $\tilde{f}\colon N \otimes M \longrightarrow M \otimes N$ defined as $\tilde{f}(n \otimes m) = m \otimes n$.
In an similar way, we can find a homomorphism $\tilde{g}\colon M \otimes N \longrightarrow N \otimes M$ such that $\tilde{g}(m \otimes n) = n \otimes m$.
These homomorphisms are mutually inverse, showing that the two modules are isomorphic.

3. If $m \in M$ is fixed, then the map

$$f_m : N \times P \longrightarrow (M \otimes N) \otimes P$$

$$f_m(n,p) = (m \otimes n) \otimes p$$

is bilinear, and thus induces a homomorphism $\widetilde{f_m} : N \otimes P \longrightarrow (M \otimes N) \otimes P$. We define the bilinear map

$$g \colon M \times (N \otimes P) \longrightarrow (M \otimes N) \otimes P$$

$$g(m, n \otimes p) = \widetilde{f_m}(n \otimes p) = (m \otimes n) \otimes p.$$

Since $g$ is bilinear, it induces a homomorphism

$$\tilde{g} \colon M \otimes (N \otimes P) \longrightarrow (M \otimes N) \otimes P$$

$$\tilde{g}(m \otimes (n \otimes p)) = (m \otimes n) \otimes p.$$

Similarly, we can construct a homomorphism

$$\tilde{h} \colon (M \otimes N) \otimes P \longrightarrow M \otimes (N \otimes P)$$

$$\tilde{h}((m \otimes n) \otimes p) = m \otimes (n \otimes p),$$

which is the inverse of $\tilde{g}$.

4. The $A$-module homomorphism

$$f \colon (M \oplus N) \times P \longrightarrow (M \otimes P) \oplus (N \otimes P)$$

$$f((m,n),p) = (m \otimes p, n \otimes p)$$

is bilinear, hence it induces a homomorphism

$$\tilde{f} \colon (M \oplus N) \otimes P \longrightarrow (M \otimes P) \oplus (N \otimes P)$$

$$\tilde{f}((m,n) \otimes p) = (m \otimes p, n \otimes p).$$

Using the bilinear maps

$$g \colon M \times P \longrightarrow (M \oplus N) \otimes P, \quad g(m,p) \mapsto (m,0) \otimes p$$

and

$$h \colon N \times P \longrightarrow (M \oplus N) \otimes P, \quad h(n,p) \mapsto (0,n) \otimes p,$$

we obtain homomorphisms

$$\tilde{g} \colon M \otimes P \longrightarrow (M \oplus N) \otimes P \text{ and } \tilde{h} \colon N \otimes P \longrightarrow (M \oplus N) \otimes P.$$

By the universal property of direct sum **T.4.6**.1, we have a homomorphism

$$(M \otimes P) \oplus (N \otimes P) \longrightarrow (M \oplus N) \otimes P$$

$$(m \otimes p, n \otimes q) \mapsto \tilde{g}(m \otimes p) + \tilde{h}(n \otimes q),$$

which is the inverse of $\tilde{f}$.

5. The map $f\colon A/I \times M \longrightarrow M/IM$ defined by $(\overline{a}, m) \mapsto \overline{am}$ is well-defined and bilinear. Therefore, it induces a homomorphism $\tilde{f}\colon A/I \otimes M \longrightarrow M/IM$ such that $\tilde{f}(\overline{a} \otimes m) = \overline{am}$.
We define a homomorphism $M/IM \longrightarrow A/I \otimes M$ by $\overline{m} \mapsto \overline{1} \otimes m$.
This homomorphism is well-defined because if $\overline{m} = \overline{n}$, then $m - n \in IM$, which implies that $m - n = \sum_i a_i m_i$ for some $a_i \in I$ and $m_i \in M$.
We obtain

$$\overline{1} \otimes m - \overline{1} \otimes n = \overline{1} \otimes (m - n) = \overline{1} \otimes \sum_i a_i m_i = \sum_i \overline{a_i} \otimes m_i = 0.$$

It is easy to verify that the two homomorphisms are inverses of each other.

6. By induction on $m$, we can prove that $A^m \otimes A^n \simeq A^{mn}$ for any $n \in \mathbb{N}$.
The case $m = 1$ follows from part 1.
For the general case we use part 4. We can assume that $M \simeq A^m$ and $N \simeq A^n$.
Then, we have

$$\begin{aligned} M \otimes N = A^m \otimes A^n &= (A^{m-1} \oplus A) \otimes A^n \\ &\simeq (A^{m-1} \otimes A^n) \oplus (A \otimes A^n) \\ &\simeq A^{(m-1)n} \oplus A^n \simeq A^{mn}. \end{aligned}$$

□

**Proof T. 5.5.** We already proved the second isomorphism in **T.5.1**.
To prove the first isomorphism, note that for any $f \in \mathrm{Bil}(M, N; P)$ there exists a unique homomorphism $\tilde{f}$ such that $\tilde{f}(m \otimes n) = f(m, n)$.
Define $\Phi\colon \mathrm{Bil}(M, N; P) \longrightarrow \mathrm{Hom}_A(M \otimes_A N, P)$ by $\Phi(f) = \tilde{f}$.
Assume $\tilde{f} = \Phi(f) = \Phi(g) = \tilde{g}$, then

$$f(m, n) = \tilde{f}(m \otimes n) = \tilde{g}(m \otimes n) = g(m, n)$$

for any $m \in M$ and $n \in N$, *i.e.*, $\Phi$ is injective.
The homomorphism $\Phi$ is also surjective. Indeed, take $f \in \mathrm{Hom}_A(M \otimes_A N, P)$ and define $\underline{f}\colon M \times N \longrightarrow P$ by $\underline{f}(m, n) = f(m \otimes n)$.
Then, $\underline{f}$ is bilinear, hence there exists $\widetilde{\underline{f}}$ such that

$$\widetilde{\underline{f}}(m \otimes n) = \underline{f}(m, n) = f(m \otimes n)$$

for any $m \in M$ and $n \in N$.
Therefore, $f = \widetilde{\underline{f}} = \Phi(\underline{f})$. □

**Proof T. 5.7.** It immediately follows from **T.5.4**.1 by setting $N = A$. □

## 16.6 Chapter 6

**Proof T. 6.1.** It is evident that this relation is reflexive and symmetric. To show transitivity, we assume $(a, s) \sim (b, t)$ and $(b, t) \sim (c, r)$. Then, there exist $u, v \in S$ such that $u(at - bs) = 0$ and $v(br - ct) = 0$.
Consequently,

$$vru(at - bs) = 0 \text{ and } vus(br - ct) = 0.$$

By adding these equalities, we obtain $vruat - vusct = 0$, *i.e.*, $vut(ar - cs) = 0$. Hence, $(a, s) \sim (c, r)$ because $vut \in S$. □

**Proof T. 6.2.** It is sufficient to prove that the operations are well-defined, since the existence of the neutral elements for sum and product, associativity, distributivity, and commutativity immediately follow from the corresponding properties of the operations in $A$.
Assume that $\frac{a}{s} = \frac{a'}{s'}$ and $\frac{b}{t} = \frac{b'}{t'}$.
We have to prove that

$$\frac{a}{s} + \frac{b}{t} = \frac{at + bs}{st} = \frac{a't' + b's'}{s't'} = \frac{a'}{s'} + \frac{b'}{t'}.$$

By hypothesis, there exist $u$, $v \in S$ such that

$$u(as' - a's) = 0 \text{ and } v(bt' - b't) = 0.$$

These yield $uvtt'(as' - a's) = 0$ and $uvss'(bt' - b't) = 0$, respectively.
Adding these equations, we obtain the desired equality.

A similar computation shows that the product is also well-defined. □

**Proof T. 6.3.** 1. By definition, an element $a \neq 0$ is such that $\sigma(a) = \frac{a}{1} = 0$ if and only if there exists $u \in S$ such that $ua = 0$.
This condition is equivalent to $u \in S \cap \mathcal{D}(A)$.
2. We have that $S^{-1}A = 0$ if and only if $\frac{0}{1} = \frac{1}{1}$, *i.e.*, by definition, if and only if $0 \in S$. Since $S$ is a multiplicative subset, this occurs if and only if $S$ contains a nilpotent element. □

**Proof T. 6.5.** Since $\tilde{g}(\frac{a}{s}) = g(a)g(s)^{-1}$, condition ii) immediately implies that $\tilde{g}$ is surjective.

To prove that $\tilde{g}$ is injective, let $\frac{a}{s} \in S^{-1}A$. If $\tilde{g}(\frac{a}{s}) = 0$, then $g(a) = 0$. By condition i), there exists $t \in S$ such that $ta = 0$. Hence, $\frac{a}{s} = 0$ in $S^{-1}A$. □

**Proof T. 6.6.** Let $\mathfrak{m} = \mathfrak{p}A_{\mathfrak{p}}$. Since for any $\frac{a}{s}$, $\frac{b}{t} \in \mathfrak{m}$, and any $\frac{\alpha}{\beta} \in A_{\mathfrak{p}}$, we have

$$\frac{a}{s} - \frac{b}{t} = \frac{at - bs}{st} \in \mathfrak{m} \text{ and } \frac{\alpha a}{\beta s} \in \mathfrak{m},$$

$\mathfrak{m}$ is an ideal of $A_{\mathfrak{p}}$.
Furthermore, $\mathfrak{m}$ is a proper ideal, since if $\frac{a}{s} = \frac{1}{1}$ for some $a \in \mathfrak{p}$, then there exists $u \in A \setminus \mathfrak{p}$ such that $ua = us \in A \setminus \mathfrak{p}$, which is a contradiction.

Finally, if $\frac{a}{s} \notin \mathfrak{m}$, then $a \in S$ and $\frac{s}{a} \in A_{\mathfrak{p}}$, which implies that $\frac{a}{s}$ is invertible. By **T.1.16**.1, $A_{\mathfrak{p}}$ is local with maximal ideal $\mathfrak{m}$. □

**Proof T. 6.7.** 1a. It is clear that $S^{-1}I$ is a subset of $I^e$.

To prove the opposite inclusion, let $i \in I^e$.
Then,

$$i = \sum_{h=1}^{k} \frac{a_h}{s_h}\frac{i_h}{1} \quad \text{for some } \frac{a_1}{s_1}, \ldots, \frac{a_k}{s_k} \in S^{-1}A \text{ and } i_1, \ldots, i_k \in I.$$

We can also write

$$i = \frac{\sum_{h=1}^{k} b_h i_h}{s_1 \cdots s_k}, \quad \text{for some } b_1, \ldots, b_k \in A.$$

Then, $i \in S^{-1}I$, since its numerator is in $I$ and its denominator is in $S$.

1b. If $s \in I \cap S$, then $1 = \frac{s}{s} \in S^{-1}I$. Therefore, $S^{-1}I = S^{-1}A$.

Conversely, if $I^e = S^{-1}I = S^{-1}A$, then $\frac{1}{1} \in I^e$, and there exist $a \in I$ and $s \in S$ such that $\frac{a}{s} = \frac{1}{1}$.
Therefore, there exists $t \in S$ such that $ta = ts \in I \cap S$.

1c. If $a \in \bigcup_{s \in S} I : (s)$, then there exists $s \in S$ such that $as \in I$.
Since $\frac{a}{1} = \frac{as}{s}$, we have $\frac{a}{1} \in S^{-1}I = I^e$, and thus, $a \in I^{ec}$.

To prove the opposite inclusion, let $b \in I^{ec}$.
Then, there exist $a \in I$ and $s \in S$ such that $\frac{b}{1} = \frac{a}{s} \in S^{-1}I = I^e$. Thus, there exists $t \in S$ such that $tsb = ta \in I$, which implies $b \in I : (ts)$, with $ts \in S$.

2. It suffices to prove that $J \subseteq J^{ce}$, since the inclusion $J \supseteq J^{ce}$ always holds. Let $\frac{a}{s} \in J$. Then, $\frac{a}{1} \in J$, which implies $a \in J^c$, and thus, $\frac{a}{s} = \frac{1}{s}\frac{a}{1} \in J^{ce}$. □

**Proof T. 6.8.** The inclusion $\mathfrak{p} \subseteq \mathfrak{p}^{ec}$ always holds.

To prove the opposite inclusion, let $a \in \mathfrak{p}^{ec}$. By **T.6.7**.1c, there exists $s \in S$ such that $a \in \mathfrak{p} : (s)$. Since $\mathfrak{p} \cap S = \emptyset$, we have $a \in \mathfrak{p}$, hence $\mathfrak{p}^{ec} \subseteq \mathfrak{p}$.

Now, we prove that $\mathfrak{p}^e$ is a prime ideal of $S^{-1}A$. Clearly, $\mathfrak{p}^e \subsetneq S^{-1}A$, by the hypothesis and **T.6.7**.1b.
Let $\frac{a}{s}, \frac{b}{t} \in S^{-1}A$ be such that $\frac{a}{s}\frac{b}{t} \in \mathfrak{p}^e$. Then, by **T.6.7**.1a, there exist $p \in \mathfrak{p}$ and $u \in S$ such that $\frac{ab}{st} = \frac{p}{u}$. Also, there exists $v \in S$ such that $vuab = vstp \in \mathfrak{p}$. Since $vu \in S$ and $\mathfrak{p} \cap S = \emptyset$ by hypothesis, we have $ab \in \mathfrak{p}$. Thus, either $\frac{a}{s} \in \mathfrak{p}^e$ or $\frac{b}{t} \in \mathfrak{p}^e$. □

**Proof T. 6.9.** Parts 1 and 2 immediately follow from the properties of extension of ideals, see **E.8.46**.1 and 2.

3. It is clear that $S^{-1}(I \cap J) \subseteq S^{-1}I \cap S^{-1}J$, see **E.8.46**.3.

To prove the opposite inclusion, let $\alpha = \frac{i}{s} = \frac{j}{t} \in S^{-1}I \cap S^{-1}J$ with $i \in I$, $j \in J$, and $s, t \in S$. Then, there exists $u \in S$ such that $u(ti - sj) = 0$.
Thus, $uti = usj \in I \cap J$, and we have $\alpha = \frac{uti}{uts} \in S^{-1}(I \cap J)$.

4. Let $\alpha \in S^{-1}\sqrt{I}$. Then, there exist $i \in \sqrt{I}$ and $s \in S$ such that $\alpha = \frac{i}{s}$.

If $i^n \in I$, then $\alpha^n = \frac{i^n}{s^n} \in S^{-1}I$. Therefore, $\alpha \in \sqrt{S^{-1}I}$.

To show the opposite inclusion, let $\beta = \frac{a}{t} \in \sqrt{S^{-1}I}$. Then, there exists $n$ such that $\beta^n = \frac{a^n}{t^n} \in S^{-1}I$, and hence $\beta^n = \frac{i}{s}$, for some $i \in I$ and $s \in S$. Thus, there exists $u \in S$ such that $usa^n = ut^n i \in I$. It follows that $usa \in \sqrt{I}$, and finally, $\beta = \frac{usa}{ust} \in S^{-1}\sqrt{I}$. □

**Proof T. 6.10.** Note that $\mu_s \circ \mu_t = \mu_{st}$, hence the homomorphisms $\mu_s$ and $\mu_t$ commute for any $s$, $t \in S$. Moreover, it is clear that, if they are invertible, their inverses also commute.

Assume that $N$ has an $S^{-1}A$-module structure compatible with the $A$-module structure. Let $s \in S$.
Then, $\mu_s = \mu_{\frac{s}{1}}$, and hence

$$\mu_s \circ \mu_{\frac{1}{s}} = \mu_{\frac{1}{s}} \circ \mu_s = \mu_1 = \mathrm{id}_N.$$

Thus, $\mu_s$ is bijective for any $s \in S$.

Conversely, let $N$ be an $A$-module. In order to define an outer product by elements of $S^{-1}A$ on $N$, compatible with the $A$-module structure, we must have

$$\frac{a}{1}n = an \quad \text{and} \quad \frac{1}{s}n = \left(\frac{s}{1}\right)^{-1} n.$$

Thus, we define $\frac{1}{s}n$ as the unique element $n' \in N$ such that $\mu_s(n') = sn' = n$. In this way we obtain

$$\frac{a}{s}n = an' = a\mu_s^{-1}(n).$$

To verify that this product is well-defined we show that it is independent from the representative of $\frac{a}{s}$. Let $\frac{a}{s} = \frac{b}{t}$, and let $u \in S$ be such that $u(at - bs) = 0$. Then, $u(at - bs)n = 0$ for any $n \in N$. Since $\mu_u$ is injective, we have $atn = bsn$, that is, $a\mu_t(n) = b\mu_s(n)$ for any $n \in N$.
It follows that

$$\frac{a}{s}n = a\mu_s^{-1}(n) = a\mu_t\mu_t^{-1}\mu_s^{-1}(n) = b\mu_s\mu_t^{-1}\mu_s^{-1}(n) = b\mu_s\mu_s^{-1}\mu_t^{-1}(n) = \frac{b}{t}n.$$

It is straightforward to verify that this product defines a unique $S^{-1}A$-module structure on $N$, which is compatible with the $A$-module structure of $N$. □

**Proof T. 6.11.** Note that if $\tilde{f}$ exists, then it is unique.
Indeed, for any $m \in M$ and $s \in S$ we have

$$\tilde{f}\left(\frac{m}{s}\right) = \tilde{f}\left(\frac{1}{s}\sigma(m)\right) = \frac{1}{s}\tilde{f}(\sigma(m)) = \frac{1}{s}f(m) = \mu_s^{-1}(f(m)),$$

where the last equality is guaranteed by **T.6.10**.

To verify that $\tilde{f}$ is a well-defined $S^{-1}A$-module homomorphism, it is sufficient to follow the same approach as in the proof of **T.6.4**. □

**Proof T. 6.12.** Since $S^{-1}g \circ S^{-1}f = S^{-1}(g \circ f) = 0$, we immediately have $\operatorname{Im} S^{-1}f \subseteq \operatorname{Ker} S^{-1}g$.

To prove the opposite inclusion, let $\frac{n}{s} \in \operatorname{Ker} S^{-1}g$.
Then, $\frac{g(n)}{s} = \frac{0}{1}$, and there exists $t \in S$ such that $g(tn) = tg(n) = 0$. Therefore, $tn \in \operatorname{Ker} g = \operatorname{Im} f$, and there exists $m \in M$ such that $f(m) = tn$.
Hence, in $S^{-1}N$, we have

$$\frac{n}{s} = \frac{tn}{ts} = \frac{f(m)}{ts} = S^{-1}f\left(\frac{m}{ts}\right) \in \operatorname{Im} S^{-1}f. \qquad \square$$

**Proof T. 6.14.** 1. Let $f\colon S^{-1}A \times M \longrightarrow S^{-1}M$ be defined by $f\left(\frac{a}{s}, m\right) = \frac{am}{s}$. This map is well-defined and $A$-bilinear. Therefore, using the universal property of tensor product, there exists a unique $A$-module homomorphism

$$\tilde{f}\colon S^{-1}A \otimes_A M \longrightarrow S^{-1}M, \text{ defined by } \tilde{f}\left(\frac{a}{s} \otimes_A m\right) = \frac{am}{s}.$$

Since $f$ is surjective, $\tilde{f}$ is also surjective.

Note that if $\alpha \in S^{-1}A \otimes_A M$, then $\alpha = \sum_{i=1}^{k} \frac{a_i}{s_i} \otimes m_i$ for some $a_i \in A$, $s_i \in S$, and $m_i \in M$. We set $s = \prod_{i=1}^{k} s_i$, $t_i = \frac{s}{s_i}$, and $n = \sum_{i=1}^{k} t_i a_i m_i$. Then, we can write

$$\alpha = \sum_{i=1}^{k} \frac{t_i a_i}{t_i s_i} \otimes m_i = \frac{1}{s} \otimes \sum_{i=1}^{k} t_i a_i m_i = \frac{1}{s} \otimes n.$$

Now, we prove that $\tilde{f}$ is injective. Let $\alpha \in \operatorname{Ker} \tilde{f}$.
Then,

$$0 = \tilde{f}(\alpha) = \tilde{f}\left(\frac{1}{s} \otimes n\right) = \frac{n}{s},$$

and there exists $u \in S$ such that $un = 0$.
It follows that

$$\alpha = \frac{1}{s} \otimes n = \frac{u}{us} \otimes n = \frac{1}{us} \otimes un = 0.$$

2. By part 1 and the properties of tensor product **T.5.4** and **T.5.8**, we obtain

$$\begin{aligned} S^{-1}M \otimes_{S^{-1}A} S^{-1}N &\simeq S^{-1}M \otimes_{S^{-1}A} (S^{-1}A \otimes_A N) \\ &\simeq (S^{-1}M \otimes_{S^{-1}A} S^{-1}A) \otimes_A N \simeq S^{-1}M \otimes_A N \\ &\simeq (S^{-1}A \otimes_A M) \otimes_A N \simeq S^{-1}A \otimes_A (M \otimes_A N) \\ &\simeq S^{-1}(M \otimes_A N). \end{aligned} \qquad \square$$

**Proof T. 6.15.** The implications $1 \Rightarrow 2 \Rightarrow 3$ are straightforward, thus it is sufficient to prove that $3 \Rightarrow 1$.
Assume, by contradiction, that $M \neq 0$. Then, there exists $0 \neq m \in M$ such that $\operatorname{Ann} m \subsetneq A$. This implies that there exists a maximal ideal $\mathfrak{m}$ containing

$\operatorname{Ann} m$. By hypothesis, $M_{\mathfrak{m}} = 0$. Thus, $\frac{m}{1} = \frac{0}{1}$, and there exists $u \in A \setminus \mathfrak{m}$ such that $um = 0$.
This is not possible, because $\operatorname{Ann} m \subseteq \mathfrak{m}$. □

**Proof T. 6.17.** 1. The conclusion immediately follows from the equality $S^{-1}\mathcal{N}(A) = \mathcal{N}(S^{-1}A)$, see **T.6.9**.4, and from **T.6.15**.

2. Since $S^{-1}$ is an exact functor, we have that $S^{-1}A$ is flat by **T.6.14**.1. Since tensor product of flat modules is flat, see **E.11.9**.5, the flatness of $M$ yields the flatness of $M_{\mathfrak{p}}$ for any prime $\mathfrak{p}$, again by **T.6.14**.1.

Conversely, let $f: N \longrightarrow N'$ be an injective $A$-module homomorphism. We need to show that $\mathrm{id}_M \otimes f: M \otimes N \longrightarrow M \otimes N'$ is also injective. Using the fact that $(M \otimes_A N)_{\mathfrak{p}} \simeq M_{\mathfrak{p}} \otimes_{A_{\mathfrak{p}}} N_{\mathfrak{p}}$, see **T.6.14**.2, we obtain

$$(\mathrm{id}_M \otimes f)_{\mathfrak{p}} = \mathrm{id}_{M_{\mathfrak{p}}} \otimes f_{\mathfrak{p}}.$$

From the flatness of $M_{\mathfrak{p}}$, it follows that

$$0 = \operatorname{Ker}(\mathrm{id}_{M_{\mathfrak{p}}} \otimes f_{\mathfrak{p}}) = (\operatorname{Ker}(\mathrm{id}_M \otimes f))_{\mathfrak{p}}$$

for any prime $\mathfrak{p}$.
Hence, by **T.6.15**, $\operatorname{Ker}(\mathrm{id}_M \otimes f) = 0$, and $M$ is flat. □

**Proof T. 6.18.** If $S$ is any subset of $A$ and $P = \{\mathfrak{p} \in \operatorname{Spec} A \colon \mathfrak{p} \cap S = \emptyset\}$, then it is clear that $S \subseteq A \setminus \bigcup_{\mathfrak{p} \in P} \mathfrak{p}$.

Now, we suppose that $S$ is a saturated multiplicative subset, and prove the opposite inclusion.
If $S = A$ we have nothing to prove. Otherwise, take $a \notin S$.
We will prove that there exists at least one prime $\mathfrak{p} \in P$ such that $a \in \mathfrak{p}$. Since $a \notin S$ and $S$ is saturated, the element $\frac{a}{1}$ is not invertible in $S^{-1}A$, hence it is contained in a maximal ideal $\mathfrak{m} \subset S^{-1}A$. By **T.6.8**, there exists a prime $\mathfrak{p} \subset A$ such that $\mathfrak{p} \cap S = \emptyset$ and $\mathfrak{m} = S^{-1}\mathfrak{p}$. Thus, $\frac{a}{1} = \frac{b}{t}$ where $b \in \mathfrak{p}$ and $t \in S$, and there exists $u \in S$ such that $uta = ub \in \mathfrak{p}$. It follows that $a \in \mathfrak{p}$.

Conversely, as previously noted, the complement of a union of prime ideals is a multiplicative subset. Furthermore, if $st \in S$, then $st \notin \mathfrak{p}$ for all $\mathfrak{p} \in P$. Hence, $s, t \in S$, and thus, $S$ is saturated. □

**Proof T. 6.20.** If $S^{-1}A = T^{-1}A$, then they have the same invertible elements. Moreover, $\sigma_S = \sigma_T$, and by **T.6.19**.5, we have

$$\overline{S} = {\sigma_S}^{-1}((S^{-1}A)^*) = {\sigma_T}^{-1}((T^{-1}A)^*) = \overline{T}.$$

Conversely, if $\overline{S} = \overline{T}$, then, by **T.6.19**.6, we have

$$S^{-1}A = \overline{S}^{-1}A = \overline{T}^{-1}A = T^{-1}A. \qquad \square$$

## 16.7 Chapter 7

**Proof T. 7.1.** 1 $\Rightarrow$ 2. Let $S$ be a non-empty subset of $\Sigma$, and let $s_1 \in S$. Assume that there are no maximal elements in $S$. Then, there exists $s_2 \in S$ such that $s_1 \leq s_2$. By repeating this argument, we can construct an infinite strictly ascending chain of $\Sigma$, leading to a contradiction.

2 $\Rightarrow$ 1. Let $\{s_\alpha\}_{\alpha\in\Lambda}$ be an ascending chain of elements of $\Sigma$. Since $\{s_\alpha\}_{\alpha\in\Lambda}$ is a non-empty subset of $\Sigma$, it has a maximal element $s_{\alpha_0}$, which satisfies $s_\alpha = s_{\alpha_0}$ for all $\alpha \geq \alpha_0$. □

**Proof T. 7.3.** The proof is essentially the same as the one of **T.7.2**.2. □

**Proof T. 7.5.** According to **T.7.2**.1, the ideal $\sqrt{I}$ is finitely generated, say by $f_1, \ldots, f_r$. Therefore, there exist $k_1, \ldots, k_r \in \mathbb{N}$ such that $f_i^{k_i} \in I$ for all $i = 1, \ldots, r$. Any integer $k \geq \sum_{i=1}^r k_i$ proves our claim. □

**Proof T. 7.7.** 1. This is evident since $1 \in \mathfrak{q} : (a)$.

2. Recall that $\mathfrak{q} : (a) \supseteq \mathfrak{q}$. We first prove $\sqrt{\mathfrak{q} : (a)} = \mathfrak{p}$, and then, that $\mathfrak{q} : (a)$ is primary.
Let $b \in \mathfrak{q} : (a)$. Then, $ab \in \mathfrak{q}$. Since $a \notin \mathfrak{q}$ and $\mathfrak{q}$ is primary, it follows that $b \in \sqrt{\mathfrak{q}} = \mathfrak{p}$. Thus, we have $\mathfrak{q} \subseteq \mathfrak{q} : (a) \subseteq \mathfrak{p}$.
Taking radicals, we obtain

$$\mathfrak{p} = \sqrt{\mathfrak{q}} \subseteq \sqrt{\mathfrak{q} : (a)} \subseteq \mathfrak{p}.$$

Now, let $bc \in \mathfrak{q} : (a)$, with $b \notin \mathfrak{p}$.
Since $\mathfrak{q}$ is primary, $bca \in \mathfrak{q}$ implies $ca \in \mathfrak{q}$. Thus, $c \in \mathfrak{q} : (a)$.

3. Let $b \in \mathfrak{q} : (a)$. Then, $ab \in \mathfrak{q}$, and, by assumption, $a \notin \mathfrak{p}$. Thus, $b \in \mathfrak{q}$, since $\mathfrak{q}$ is primary. Therefore, $\mathfrak{q} : (a) \subseteq \mathfrak{q}$. □

**Proof T. 7.9.** Since $\mathfrak{q}_i$ is $\mathfrak{p}_i$-primary, by **T.7.5**, there exists a positive integer $h$ such that $\mathfrak{p}_i^h \subseteq \mathfrak{q}_i$.
Hence, we have

$$(\bigcap_{j\neq i} \mathfrak{q}_j)\mathfrak{p}_i^h \subseteq \bigcap_{j\neq i} \mathfrak{q}_j \cap \mathfrak{q}_i = I.$$

Let $k$ be the smallest positive integer such that $(\bigcap_{j\neq i} \mathfrak{q}_j)\mathfrak{p}_i^k \subseteq I$, and take

$$a \in (\bigcap_{j\neq i} \mathfrak{q}_j)\mathfrak{p}_i^{k-1} \setminus I.$$

We have $a\mathfrak{p}_i \subseteq I$, which implies that $\mathfrak{p}_i \subseteq I : (a)$.

To prove the opposite inclusion, we can simply observe that $a \in \bigcap_{j\neq i} \mathfrak{q}_j \setminus \mathfrak{q}_i$, and then proceed as in the proof of **T.7.8** to deduce that

$$I : (a) \subseteq \sqrt{I : (a)} = \mathfrak{p}_i.$$

□

**Proof T. 7.10.** 1. Let $I = \bigcap_{i=1}^t \mathfrak{q}_i$ be a minimal primary decomposition of $I$. To simplify notation, let $\operatorname{MinAss} I$ denote the set of minimal primes of $I$. We want to prove that

$$\operatorname{MinAss} I = \operatorname{Min} I.$$

Let $\mathfrak{p} \in \operatorname{Min} I$. Then, $\mathfrak{p} \supseteq \sqrt{I} = \bigcap_{\mathfrak{p}_i \in \operatorname{Ass} I} \mathfrak{p}_i = \bigcap_{\mathfrak{p}_i \in \operatorname{MinAss} I} \mathfrak{p}_i.$
By applying **T.1.12**.2, we obtain $\mathfrak{p} \supseteq \mathfrak{p}_{i_0}$ for some $\mathfrak{p}_{i_0}$ in $\operatorname{MinAss} I$. Since $\mathfrak{p} \in \operatorname{Min} I$, we have $\mathfrak{p} = \mathfrak{p}_{i_0}$.
We have shown that $\operatorname{Min} I \subseteq \operatorname{MinAss} I$, and, in particular, $\operatorname{Min} I$ is finite.

To prove the opposite inclusion, let $\mathfrak{p} \in \operatorname{MinAss} I$.
Since $\mathfrak{p} \supseteq I$, there exists $\mathfrak{p}' \in \operatorname{Min} I$ such that

$$\mathfrak{p} \supseteq \mathfrak{p}' \supseteq \sqrt{I} = \bigcap_{\mathfrak{p}_i \in \operatorname{MinAss} I} \mathfrak{p}_i.$$

Again by **T.1.12**.2, we have $\mathfrak{p} \supseteq \mathfrak{p}' \supseteq \mathfrak{p}_{i_0}$ for some $\mathfrak{p}_{i_0} \in \operatorname{MinAss} I$. Hence, $\mathfrak{p} = \mathfrak{p}_{i_0} = \mathfrak{p}'$ and $\mathfrak{p} \in \operatorname{Min} I$.

2. It is a corollary of part 1.

3. The first equality follows directly from part 1, since $\mathcal{N}(A) = \sqrt{(0)}$.

To prove the second equality, we recall that, by **E.8.33**,

$$\mathcal{D}(A) = \bigcup_{0 \neq a \in A} \sqrt{0 : (a)}.$$

Let $\mathfrak{p}_i \in \operatorname{Ass}(0)$. Then, from the Uniqueness Theorem **T.7.8**, we obtain $\mathfrak{p}_i = \sqrt{0 : (a_i)}$ for some $a_i \in A$, which shows $\supseteq$.

To prove the opposite inclusion, we observe that if $0 \neq a \in \mathcal{D}(A)$, then $a \notin \bigcap_{i=1}^t \mathfrak{q}_i$. Hence, by **T.7.7**, we have

$$\sqrt{0 : (a)} = \bigcap_{i=1}^t \sqrt{\mathfrak{q}_i : (a)} = \bigcap_{i:\, a \notin \mathfrak{q}_i} \mathfrak{p}_i.$$

Thus, $\sqrt{0 : (a)}$ is contained in some $\mathfrak{p}_i \in \operatorname{Ass}(0)$. □

**Proof T. 7.11.** 1. We only need to prove that $S^{-1}\mathfrak{q}$ is primary in $S^{-1}A$ when $S \cap \mathfrak{p} = \emptyset$. The other statements follow directly from **T.6.7**.1b and the fact that $S^{-1}\sqrt{\mathfrak{q}} = \sqrt{S^{-1}\mathfrak{q}}$, see **T.6.9**.4.
Suppose $S \cap \mathfrak{p} = \emptyset$, and let $\frac{a}{s}\frac{b}{t} = \frac{q}{u} \in S^{-1}\mathfrak{q}$ with $\frac{a}{s} \notin S^{-1}\mathfrak{q}$. Then, $wa \notin \mathfrak{q}$ for all $w \in S$. Therefore, there exists $w \in S$ such that $wuab = wstq \in \mathfrak{q}$, and since $wua \notin \mathfrak{q}$, we have $b \in \sqrt{\mathfrak{q}}$ and $\frac{b}{t} \in S^{-1}\sqrt{\mathfrak{q}} = \sqrt{S^{-1}\mathfrak{q}}$.

2. It is a straightforward application of **T.6.9**.3 and part 1.

3. It is a special case of part 2. □

**Proof T. 7.12.** Let $S_i = A \setminus \mathfrak{p}_i$, with $\mathfrak{p}_i \in \operatorname{Min} I$. Then, $S_i \cap \mathfrak{p}_j \neq \emptyset$ for all $j \neq i$. Otherwise, we would have $\mathfrak{p}_j \subset \mathfrak{p}_i$, against the minimality of $\mathfrak{p}_i$.
From **T.7.11**.1, it follows that $IA_{\mathfrak{p}_i} = S_i^{-1}I = S_i^{-1}\mathfrak{q}_i$, and $(IA_{\mathfrak{p}_i})^c = \mathfrak{q}_i$. □

**Proof T. 7.14.** In a local ring $(A, \mathfrak{m})$, every element of $A \setminus \mathfrak{m}$ is invertible. Therefore, we only need to show that every element of $\mathfrak{m}$ is nilpotent.
Since $A$ is Artinian, every prime ideal is maximal by **T.7.13**.1.
Consequently, $\mathcal{N}(A) = \mathfrak{m}$, and the conclusion follows from **T.7.13**.3. □

**Proof T. 7.16.** By **T.7.13**, if $A$ is an Artinian ring, then every prime ideal is maximal, hence $\dim A = 0$.
Moreover, there exist only finitely many maximal ideals $\mathfrak{m}_1, \ldots, \mathfrak{m}_s$ in $A$, and there exists a positive integer $k$ such that

$$0 = \mathcal{N}(A)^k = \Big(\bigcap_{i=1}^{s} \mathfrak{m}_i\Big)^k \supseteq \prod_{i=1}^{s} \mathfrak{m}_i^k.$$

By **T.7.15**, we conclude that $A$ is Noetherian.

Conversely, assume that $A$ is Noetherian and 0-dimensional.
We can write $\sqrt{(0)} = \bigcap_{i=1}^{s} \mathfrak{m}_i$ as a finite intersection of maximal ideals, see **T.7.10**.3. Therefore, there exists positive integer $k$ such that

$$\prod_{i=1}^{s} \mathfrak{m}_i^k \subseteq \sqrt{(0)}^{\,k} \subseteq (0)$$

by **T.7.5**. Thus, again by **T.7.15**, $A$ is Artinian. □

# Chapter 17
# Solutions to the Exercises

## 17.1 Chapter 8

**Solution E. 8.1.** Let $I$ be an ideal containing $S$. Then, $as_1 + bs_2 \in I$ for any $s_1$, $s_2 \in S$ and any $a$, $b \in A$. Thus, $I \supseteq (S)$ and the intersection of all these ideals is an ideal containing $S$.

To prove the opposite inclusion, we can simply observe that $(S)$ is an ideal of $A$, and by construction, it contains $S$.

**Solution E. 8.2.** $1 \Rightarrow 2$. Obviously, if $b = c$, then $ab = ac$ for any $a \in A$.

Conversely, assume $ab = ac$ with $a \neq 0$.
Then, since $a \neq 0$ and $A$ is a domain, $a(b - c) = 0$ implies $b - c = 0$.

$2 \Rightarrow 3$. Let $b$, $c \in A \setminus \{0\}$, and assume, by contradiction, that $bc \notin A \setminus \{0\}$, that is, $bc = 0$. Then, either $b = 0$ or the hypothesis yields $c = 0$, because $bc = 0 = b0$ with $b \neq 0$.
In both cases we obtain a contradiction.

$3 \Rightarrow 1$. For any $b$, $c \in A$, the statement: if $b$, $c \neq 0$, then $bc \neq 0$ is equivalent to the statement: if $bc = 0$, then either $b = 0$ or $c = 0$, which is the definition of domain.

**Solution E. 8.3.** 1. The ring $A$ is finite, hence, by **T.1.1**, we have

$$A = \mathcal{D}(A) \sqcup A^*.$$

An element $\overline{h} \in A$ is invertible if and only if there exists $k \in \mathbb{Z}$ such that $hk \equiv 1 \mod 24$, that is, $(h, 24) = 1$.
Therefore,

$$A^* = \{\overline{h}\colon\ h \in \mathbb{Z},\ (h, 24) = 1\} \ \text{ and } \ \mathcal{D}(A) = \{\overline{h}\colon\ h \in \mathbb{Z},\ (h, 24) \neq 1\}.$$

The ideals of $\mathbb{Z}/(24)$ correspond to ideals $(a)$ of $\mathbb{Z}$ such that $(a) \supseteq (24)$, *i.e.*, such that $a \mid 24$.
The prime ideals are $(\overline{2})$ and $(\overline{3})$ and they are also maximal.

 
A. Bandini et al., *Commutative Algebra through Exercises*, UNITEXT 159,
https://doi.org/10.1007/978-3-031-56910-4_17

2. Reasoning as in part 1, we find

$$A^* = \{\overline{h}\colon\ h \in \mathbb{Z},\ (h, 17) = 1\} = A \setminus \{\overline{0}\} \ \text{ and } \ \mathcal{D}(A) = \{\overline{0}\}.$$

Therefore, $(\overline{0})$ is the unique prime ideal and it is also maximal.
3. The ring $A = \mathbb{Z}/(n)$ is a domain if and only if $\mathcal{D}(A) = \{\overline{0}\}$, *i.e.*, if and only if $A^* = A \setminus \{\overline{0}\}$, which happens if and only if $A$ is a field.
This holds if and only if $n$ is prime in $\mathbb{Z}$, because in that case $(m, n) = 1$ for all $m \in \mathbb{N}_+$ with $m \leq n - 1$.

**Solution E. 8.4.** 1. Since $a$ is nilpotent, there exists $n \in \mathbb{N}_+$ such that $a^n = 0$. Thus,

$$1 = 1 - a^n = (1 - a)\sum_{i=0}^{n-1} a^i,$$

which proves the first statement.

2. Let $b \in A^*$. We have to prove that $b + a \in A^*$.
Write

$$b + a = b(1 + ab^{-1}).$$

Since $-ab^{-1}$ is nilpotent, the claim follows from part 1, because the product of invertible elements is invertible.

**Solution E. 8.5.** 1. Assume $f = \sum_{i=0}^{n} a_i x^i$ is invertible, and denote its inverse by $g(x) = \sum_{i=0}^{m} b_i x^i$. Then, $fg = \sum_{i=0}^{n+m} c_i x^i = 1$.
We immediately obtain $c_0 = a_0 b_0 = 1$, thus $a_0$ and $b_0$ are invertible.

Moreover, we have $c_{n+m} = a_n b_m = 0$ and $c_{n+m-1} = a_n b_{m-1} + a_{n-1} b_m = 0$. Multiplying by $a_n$ we obtain

$$a_n^2 b_{m-1} = -a_n a_{n-1} b_m = 0.$$

By repeating this process, from $c_{n+m-r} = 0$ we obtain the relation

$$a_n^{r+1} b_{m-r} = 0.$$

When $r = m$, this becomes $a_n^{m+1} b_0 = 0$.
Since $b_0$ is invertible, we conclude that $a_n$ is nilpotent.
Now, consider the polynomial $f - a_n x^n$, which has degree $< n$ and is still invertible by **E.8.4**.
By repeating the same argument we complete the first part of the proof.

The opposite implication immediately follows from **E.8.4**, since we can write $f = a_0 + (a_1 x + \ldots + a_n x^n)$, where $a_0$ is invertible, and the sum of the nilpotents $a_i x^i$, for $i = 1, \ldots, n$, is also nilpotent.

2. We have observed above that if $a_0, \ldots, a_n$ are nilpotent, then $f$ is nilpotent.

Conversely, if $f$ is nilpotent, then $xf$ is nilpotent, and $1 + xf$ is invertible. The conclusion follows from part 1.

3. One implication is simply the definition of zero-divisor.

Suppose that $f = \sum_{i=0}^{n} a_i x^i$ is a zero-divisor, and let $0 \neq g = \sum_{i=0}^{m} b_i x^i$ be a polynomial of minimal degree $m$ such that $fg = 0$.
Since $a_n b_m = 0$ and $(a_n g) f = 0$, the minimality of $\deg g$ implies $a_n g = 0$. Therefore, we have

$$a_n b_i = 0 \text{ for all } i.$$

Next, consider the coefficient of the term of degree $n + m - 1$ of $fg$.
We find that

$$a_n b_{m-1} + a_{n-1} b_m = 0,$$

which implies that $a_{n-1} b_m = 0$. By iterating this process on the coefficients of the terms of $fg$ of degree less than $n + m - 1$, we get $a_i b_m = 0$ for all $i$. The conclusion follows taking $a = b_m$.

**Solution E. 8.6.** The Jacobson radical always contains the nilradical.

To prove the opposite inclusion, let $f \in \mathcal{J}(A[x])$. Then, $1 + xf$ is invertible by the characterization of the elements of the Jacobson radical **T.1.15**.
By **E.8.5**.1, the coefficients of $f$ are nilpotent, hence, by **E.8.5**.2, the polynomial $f$ is nilpotent.

**Solution E. 8.7.** Take $a \in I$, and assume that $b(1 + a) = 0$ for some $b \in A$. Then, $b = -ba$ and $b$ is an element of $I$. Multiplying by $-a$ on both sides, we get

$$b = -ba = ba^2 \in I^2.$$

Iterating, we obtain $b \in I^n$ for any $n$, hence $b = 0$.

**Solution E. 8.8.** 1. Clearly, $0 \in \operatorname{Ker} f$. If $a, b \in \operatorname{Ker} f$, then $f(a) = f(b) = 0$. Therefore, $f(a - b) = f(a) - f(b) = 0$ and $f(ca) = f(c)f(a) = 0$ for any $c \in A$, that is, $a - b$, $ca \in \operatorname{Ker} f$.

2. Assume $\operatorname{Ker} f = (0)$ and let $a, b \in A$ be such that $f(a) = f(b)$. Then, $f(a - b) = 0$, that is, $a - b \in (0)$, and hence, $a = b$.

Conversely, if $\operatorname{Ker} f \neq (0)$, then there exists $0 \neq a$ such that $f(a) = f(0)$, and $f$ is not injective.

3. We have to verify that the sum of two elements in the image of $f$ is still in $\operatorname{Im} f$, and this holds because of the linearity of $f$.
Since $f(a)f(b) = f(ab)$ for every $a, b \in A$, the image of $f$ is closed under multiplication.
Finally, $1_B = f(1_A) \in \operatorname{Im} f$.

**Solution E. 8.9.** By definition, we have $f(0) = 0$ and $f(1) = 1$.
If $n$ is a positive integer, then

$$f(n) = \underbrace{f(1) + \ldots + f(1)}_{n \text{ times}} = nf(1) = n.$$

If $n$ is a negative integer, then $f(n) = f(-(-n)) = -f(-n) = -(-n) = n$. Therefore, we have $f(n) = n$ for any $n \in \mathbb{Z}$, *i.e.*, the only ring homomorphism is the identity.

**Solution E. 8.10.** 1. Consider the evaluation homomorphism

$$\varphi_a\colon A[x] \longrightarrow A \;\text{ defined by }\; \varphi_a(f) = f(a).$$

Then, $J = \varphi_a^{-1}(I)$ is an ideal of $A[x]$.
2. Composing $\varphi_a$ with the canonical projection onto $A/I$ we find a surjective homomorphism $\pi \circ \varphi_a\colon A[x] \longrightarrow A/I$, whose kernel is $J$.
Hence,

$$A[x]/J \simeq A/I,$$

and it follows that $J$ is prime if and only if $I$ is prime.
3. We have $\varphi_a(x-1) = \varphi_a(y-2) = y-2 \in I$, and thus $(x-1,\, y-2) \subseteq J$. Since $J$ is a proper ideal, *e.g.*, because $x \notin J$, and since $(x-1,\, y-2)$ is a maximal ideal, we have $J = (x-1,\, y-2)$.

**Solution E. 8.11.** $1 \Rightarrow 2$. Assume $A$ is a field and let $I$ be an ideal of $A$. If $I \neq 0$, then there exists $0 \neq a \in I$ and such element is invertible.
Thus, $I = (a) = (1)$.

$2 \Rightarrow 1$. Assume $A$ has no non-trivial ideals. Then, for any $0 \neq a \in A$, the ideal generated by $a$ is equal to $(1)$, *i.e.*, every non-zero element of $A$ is invertible.

$2 \Rightarrow 3$. Let $f\colon A \longrightarrow B$ be a ring homomorphism. Then, by hypothesis, $\operatorname{Ker} f$ can only be $(0)$ or $(1)$.
If $\operatorname{Ker} f = (1)$, then $f$ is the zero homomorphism. Hence, the condition $f(1_A) = 1_B$ implies $B = 0$, which is not possible.
Therefore, $\operatorname{Ker} f = (0)$.

$3 \Rightarrow 2$. Since $A$ is non-zero, it has a maximal ideal $\mathfrak{m}$.
Consider the canonical projection $\pi\colon A \longrightarrow A/\mathfrak{m} \neq 0$. By hypothesis, $\pi$ is injective, hence $\mathfrak{m} = \operatorname{Ker} \pi = (0)$.
This shows that $(0)$ and $(1)$ are the only ideals of $A$.

**Solution E. 8.12.** Since $I$ is an ideal of $A$, it is easy to prove that $I[x]$ is an ideal of $A[x]$.
Consider the map

$$\varphi\colon A[x] \longrightarrow (A/I)\,[x], \quad \text{defined by } \varphi(\sum_i a_i x^i) = \sum_i \overline{a_i} x^i.$$

It is easy to verify that $\varphi$ is a surjective homomorphism.
Moreover, $f(x) \in \operatorname{Ker} \varphi$ if and only if $\sum_i \overline{a_i} x^i = 0$, *i.e.*, if and only if $\overline{a_i} = \overline{0}$ for all $i$. This is equivalent to $a_i \in I$ for all $i$, that is, to $f(x) \in I[x]$.
The final claim follows from the homomorphism theorem **T.1.9**.1.

**Solution E. 8.13.** For any polynomial $f = \sum_{i=0}^n f_i x^i \in A[x]$, we denote by $I(f) = (f_0, \dots, f_n)$ the ideal of $A$ generated by the coefficients of $f$.
If $fg$ is primitive, then $(1) = I(fg) \subseteq I(f)I(g)$. Thus, $I(f) = I(g) = (1)$.

Conversely, let $f = \sum_{i=0}^n f_i x^i$ and $g = \sum_{j=0}^m g_j x^j$ be primitive polynomials, and let $h = fg = \sum_{l=0}^t h_l x^l$.

If $I(h)$ is a proper ideal of $A$, then there exists a maximal ideal $\mathfrak{m} \supseteq I(h)$. By hypothesis, there exist coefficients $f_i$ and $g_j$ which do not belong to $\mathfrak{m}$.
Let $r$ and $s$ be the smallest indices such that $f_r$, $g_s \notin \mathfrak{m}$. Then,

$$h_{r+s} = \sum_{i=0}^{r+s} f_i g_{r+s-i} = \sum_{i=0}^{r-1} f_i g_{r+s-i} + f_r g_s + \sum_{i=r+1}^{r+s} f_i g_{r+s-i}$$

yields $f_r g_s \in \mathfrak{m}$, a contradiction.

Alternatively, if $\mathfrak{m} \supseteq I(h)$, then

$$h \in \mathfrak{m}[x] \text{ and } \overline{fg} = \overline{h} = \overline{0} \text{ in } A[x]/\mathfrak{m}[x],$$

which is a domain because it is isomorphic to $(A/\mathfrak{m})[x]$, see **E.8.12**.
Thus, either $\overline{f} = \overline{0}$ or $\overline{g} = \overline{0}$, *i.e.*, either $\mathfrak{m} \supseteq I(f)$ or $\mathfrak{m} \supseteq I(g)$, which is a contradiction.

**Solution E. 8.14.** By the characterization of local rings **T.1.16**, it suffices to prove that every element $a \notin \mathcal{J}(A)$ is invertible.
By ii), there exists $0 \neq b \in A$ such that $\mathcal{J}(A) = (b)$.
Moreover, if $a \notin \mathcal{J}(A)$, then $(\mathcal{J}(A), a) = (c)$ for some $c \notin \mathcal{J}(A)$. Hence, $b = cd$ for some $d \in A$. Since $\mathcal{J}(A)$ is a prime ideal and $c \notin \mathcal{J}(A)$, we have $d \in \mathcal{J}(A)$. Thus, $d = rb$ for some $r \in A$. Therefore, $b = cd = crb$, which yields $b(1 - cr) = 0$.
By iii), we obtain $1 - cr \in \mathcal{J}(A)$, and, by **T.1.15**, $cr = 1-(1-cr)$ is invertible. Accordingly, $c$ is also invertible. Therefore, $(\mathcal{J}(A), a) = (1)$ and there exists $s \in A$ such that $1 - sa \in \mathcal{J}(A)$.
Repeating the previous argument, we find that $a$ is invertible, as desired.

**Solution E. 8.15.** 1. If $(a) = (b)$, then there exist elements $r$, $s \in A$ such that $a = bs = ars$. Assume that $s$ is not invertible. Then, $s \in \mathfrak{m} = \mathcal{J}(A)$, and hence, $1 - rs$ is invertible.
From the equation $a(1 - rs) = 0$, we obtain $a = 0$, which contradicts the hypothesis.

The converse is obvious.

2. Let $m = cd$, and assume $c$ is not invertible. If $(m) = (d)$, then, by part 1, there exists $u \in A^*$ such that $d = um$, and this leads to $m(1 - uc) = 0$.
Since $A$ is local, $c \in \mathfrak{m}$. Thus, $1 - uc \in A^*$, and finally $m = 0$, which contradicts the hypothesis. Therefore, $(m) \subsetneq (d)$, and the maximality of $(m)$ yields $(d) = A$. Thus, $d$ is invertible and $m$ is irreducible.

**Solution E. 8.16.** Let $a \in I$ and $b \in J$ be such that $a + b = 1$.
Since $I \subseteq \mathcal{J}(A)$, the element $1 - a = b \in J$ is invertible, and the conclusion follows.

**Solution E. 8.17.** Let $a$ be idempotent, *i.e.*, $a(1 - a) = 0$.
If $a$ is not invertible, then it has to be contained in the maximal ideal of $A$.
Thus, $1 - a$ is invertible and this implies $a = 0$.
Otherwise, $a$ is invertible, and we obtain $a = 1$.

**Solution E. 8.18.** 1. Since $2a = (2a)^2 = 4a^2 = 4a$, we obtain $2a = 0$.

2. By hypothesis, $A/\mathfrak{p}$ is a domain. Take $\overline{a} \neq \overline{0}$ in $A/\mathfrak{p}$. Since $\overline{a}^2 = \overline{a^2} = \overline{a}$, we have $\overline{a} = \overline{1}$. Thus, $A/\mathfrak{p}$ is a field with two elements.
In particular, $\mathfrak{p}$ is also a maximal ideal.

3. Note that, for any two elements $a, b \in A$, we can write $a = a(a+b-ab)$ and $b = b(a+b-ab)$.
Hence,

$$(a, b) = (a + b - ab)$$

which proves the result.

**Solution E. 8.19.** The ring $A$ is a domain, because $(0)$ is prime.
Let $b \in A$ be not invertible. Since $(b^2)$ is prime, we have $b \in (b^2)$ and $b = ab^2$ for some $a \in A$. Since $b$ is not invertible, we have $ab \neq 1$. The equation $b(1 - ab) = 0$ then yields $b = 0$.
This shows that $A$ is a field.

**Solution E. 8.20.** 1. For each $m$, $n \in \mathbb{Z}$ there exist $\alpha$, $\beta \in \mathbb{Z}$ such that $\alpha m + \beta n = \gcd(m, n)$. This shows that $I + J = (m, n) \supseteq (\gcd(m, n))$.

To prove the opposite inclusion, note that $\gcd(m, n)$ divides $m$ and $n$. Thus, $(\gcd(m, n)) \supseteq I + J$.

2. If $a \in I \cap J$, then there exist $\alpha$, $\beta \in \mathbb{Z}$ such that $a = \alpha m = \beta n$. Then, by definition, $\operatorname{lcm}(m, n) \mid a$.

To prove the opposite inclusion, note that both $m$ and $n$ divide $\operatorname{lcm}(m, n)$. Thus, $\operatorname{lcm}(m, n) \in I \cap J$.

3. Since $IJ = (ij \colon\ i \in I,\ j \in J)$ is the ideal generated by products of elements of $I$ and $J$, it is clear that $mn \in IJ$.

To prove the opposite inclusion, let $a \in IJ$, and write

$$a = \sum_{k=1}^{r} a_k(\alpha_k m)(\beta_k n)$$

for some integers $a_k$, $\alpha_k$ and $\beta_k$.
Therefore, we have $a = (\sum_{k=1}^{r} a_k \alpha_k \beta_k) mn \in (mn)$.

4. Note that $I : J = (m) : (n)$ is the set of elements $a \in \mathbb{Z}$ such that $an \in (m)$. It is clear that $\frac{m}{\gcd(m,n)} \cdot n$ is a multiple of $m$, and therefore the inclusion $\supseteq$ holds.

Now, write $m = m' \gcd(m, n)$ and $n = n' \gcd(m, n)$. Then, $a \in (m) : (n)$ implies that $an'$ is a multiple of $m'$. This proves the opposite inclusion.

5. By the Fundamental Theorem of Arithmetic, we can write any $d \in \mathbb{Z}$ as a product

$$d = \pm \prod_{p \in \mathbb{N},\, p \text{ prime}} p^{d_p},$$

where $d_p = 0$ for almost all $p$.

Using parts 1 and 2, we see that the statement is equivalent to proving the equality

$$\operatorname{lcm}(m, \gcd(n, h)) = \gcd(\operatorname{lcm}(m, n), \operatorname{lcm}(m, h)),$$

that holds if and only if for any positive prime $p$, we have

$$\max\{m_p, \min\{n_p, h_p\}\} = \min\{\max\{m_p, n_p\}, \max\{m_p, h_p\}\}.$$

This final equality is immediate.

6. Using parts 1, 2 and 3, we have

$$(I + J)(I \cap J) = (\gcd(m, n) \operatorname{lcm}(m, n)) = (mn) = IJ.$$

**Solution E. 8.21.** Parts 1 and 2 immediately follow from the definition of quotient ideal.

3. By definition, $a \in (I : J) : H$ if and only if $aH \subseteq I : J$. Hence, if and only if $aHJ \subseteq I$, *i.e.*, if and only if $aJH \subseteq I$.
This proves both equalities.

4. Note that $aJ \subseteq I_\alpha$ for every $\alpha$ if and only if $a \in I_\alpha : J$ for every $\alpha$.

5. If $a \sum_\alpha I_\alpha \subseteq I$, then $aI_\alpha \subseteq a \sum_\alpha I_\alpha \subseteq I$ for all $\alpha$. This proves $\subseteq$.

To prove the opposite inclusion, we observe that if $aI_\alpha \subseteq I$ for every $\alpha$. Then, for any finite sum $\sum_\alpha j_\alpha$, where $j_\alpha \in I_\alpha$ for all $\alpha$, we immediately have $a \sum_\alpha j_\alpha = \sum_\alpha aj_\alpha \in I$.

**Solution E. 8.22.** If $g \in \frac{1}{f}(I \cap (f))$, then $gf \in I$, that is, $g \in I : (f)$.

Conversely, if $g \in I : (f)$, then $gf \in I$ and, in particular, $gf \in I \cap (f)$. Hence, $g \in \frac{1}{f}(I \cap (f))$.

**Solution E. 8.23.** By hypothesis, $(\overline{f}, \overline{g}) = (\overline{1})$ in $A/I$. Hence, by **T.1.4**,

$$(\overline{f}) \cap (\overline{g}) = (\overline{f})(\overline{g}) = (\overline{fg}).$$

This is equivalent to $(I, f) \cap (I, g) = (I, fg)$.

**Solution E. 8.24.** 1. By hypothesis, there exist $a \in I$, $b \in J$, and $h_1$, $h_2 \in H$ such that $a + h_1 = b + h_2 = 1$. We can write $1 = (a + h_1)(b + h_2) = ab + q$ where $q = ah_2 + bh_1 + h_1h_2 \in H$.
Thus, for each $n \in \mathbb{N}$,

$$1 = (ab + q)^n = abp + q^n \in I \cap J + H^n$$

where $p = \sum_{i=1}^n \binom{n}{i} (ab)^{i-1} q^{n-i} \in A$.

2. It is sufficient to prove that $H \subseteq I$.
Take $h \in H$. By hypothesis, there exist $j_1$, $j_2 \in H \cap J = I \cap J$ and $i \in I$ such that $h + j_1 = i + j_2$.
Thus, $h = i - j_1 + j_2 \in I$, as desired.

**Solution E. 8.25.** 1. We prove the statement by induction on $k$. For $k = 1$, there is nothing to prove.
Let $k = 2$. For $i = 1, 2$, let $a_i \in I$ and $b_i \in H_i$ be such that $1 = a_i + b_i$. Then,

$$1 = (a_1 + b_1)(a_2 + b_2) = a_1a_2 + a_1b_2 + a_2b_1 + b_1b_2 \in I + H_1H_2.$$

Now, let $k \geq 3$. Since the inductive assumption $I + H_1 \cdots H_{k-1} = A$ holds, and, by hypothesis, $I + H_k = A$ the conclusion immediately follows from the case $k = 2$.

2. Simply let $H_i = J$ for all $i$, then exchange the roles of $I$ and $J$.

**Solution E. 8.26.** 1. By **T.1.6**.5, we have $\sqrt{I} = (1)$ if and only if $I = (1)$, therefore, $\sqrt{IJ} = A$ if and only if $IJ = A$.
Since $IJ \subset I$ and $IJ \subset J$, the conclusion is immediate.

2. Clearly, $\mathfrak{p} \subseteq I$ and $\mathfrak{p} \subseteq J$.
If $\mathfrak{p}$ contains neither $I$ nor $J$, then there exist $i \in I \setminus \mathfrak{p}$ and $j \in J \setminus \mathfrak{p}$ such that $ij \in IJ = \mathfrak{p}$, but this contradicts the primality of $\mathfrak{p}$.
Thus, either $I = \mathfrak{p}$ or $J = \mathfrak{p}$.

**Solution E. 8.27.** 1. Since every ideal is contained in its radical, if $I + J = (1)$, then $\sqrt{I} + \sqrt{J} = (1)$ as well.

Conversely, if $a \in \sqrt{I}$ and $b \in \sqrt{J}$ are such that $a + b = 1$, with $a^m \in I$ and $b^n \in J$, then we have

$$\begin{aligned} 1 &= (a+b)^{m+n} \\ &= a^m \sum_{i=0}^{n-1} \binom{m+n-i}{i} a^{n-i} b^i + b^n \sum_{i=n}^{m+n} \binom{m+n-i}{i} a^{m+n-i} b^{i-n} \in I + J. \end{aligned}$$

2. Since

$$I + \sqrt{J} \supseteq I + J,$$

taking radicals we obtain the inclusion $\supseteq$.

To prove the opposite inclusion, let $a \in \sqrt{I + \sqrt{J}}$. Then, there exists $m \in \mathbb{N}$ such that $a^m = i + j$, with $i \in I$ and $j \in \sqrt{J}$.
If $n \in \mathbb{N}$ is such that $j^n \in J$, then

$$a^{nm} = (i + j)^n = c + j^n,$$

with $c \in I$ and $j^n \in J$, and the conclusion follows.

**Solution E. 8.28.** Consider the ideals $I = (x^2 + y)$ and $J = (x^2 - y)$ in $\mathbb{Q}[x, y]$. Since $x^2 + y$ and $x^2 - y$ are both irreducible polynomials in $\mathbb{Q}[x, y]$, we have that $\sqrt{I} = I$ and $\sqrt{J} = J$.
Therefore, $\sqrt{I} + \sqrt{J} = (x^2, y)$, while $\sqrt{I + J} = (x, y)$.

**Solution E. 8.29.** We claim that $\sqrt{I} = (x)$, which is a prime ideal of $A$.
Since $x^2 \in I$, the inclusion $(x) \subseteq \sqrt{I}$ holds.

To prove the opposite inclusion, take $a \in A$ such that $a^k \in I$. Then,

$$a^k = \alpha x^2 + \beta xy \in (x)$$

for some $\alpha, \beta \in A$.
Hence, the primality of $(x)$ implies that $a \in (x)$.

To prove that $I$ is not primary, note that $xy \in I$, but $x \notin I$ and $y^n \notin (x)$ for all $n \in \mathbb{N}$.

**Solution E. 8.30.** The inclusion $\mathcal{N}(A) \subseteq \mathcal{J}(A)$ holds in any ring.

To prove the opposite inclusion, suppose that $\mathcal{J}(A) \nsubseteq \mathcal{N}(A)$.
Then, by hypothesis, there exists an idempotent $0 \neq a \in \mathcal{J}(A)$, *i.e.*, such that $a(a-1) = 0$. Since $a \in \mathcal{J}(A)$, the element $a - 1$ is invertible, see **T.1.15**, and hence $a = 0$.
This is a contradiction.

**Solution E. 8.31.** $1 \Rightarrow 2$. By hypothesis, $A$ is a local ring and $\mathcal{N}(A)$ is its maximal ideal. Thus, every element not in $\mathcal{N}(A)$ is invertible.

$2 \Rightarrow 3$. If $a$ is nilpotent, then $\overline{a} = \overline{0}$ in $A/\mathcal{N}(A)$.
Otherwise, $a$ is invertible in $A$, and therefore, $\overline{a}$ is invertible in $A/\mathcal{N}(A)$.

$3 \Rightarrow 1$. If $\mathfrak{p} \subset A$ is a prime ideal, then $\mathcal{N}(A) \subseteq \mathfrak{p}$.
Since $\mathcal{N}(A)$ is maximal, we have $\mathcal{N}(A) = \mathfrak{p}$, *i.e.*, $A$ has only one prime ideal.

**Solution E. 8.32.** An element $a$ belongs to $\sqrt{\bigcup_\alpha E_\alpha}$ if and only if there exist $n \in \mathbb{N}$ and an index $\alpha_0 \subset A$ such that $a^n \in E_{\alpha_0}$, *i.e.*, if and only if $a \in \sqrt{E_{\alpha_0}} \subseteq \bigcup_\alpha \sqrt{E_\alpha}$.

The opposite inclusion immediately follows from $E_\alpha \subseteq \bigcup_\alpha E_\alpha$ for all $\alpha$.

**Solution E. 8.33.** The definitions of zero-divisor and of annihilator of an element immediately imply that

$$\mathcal{D}(A) = \bigcup_{0 \neq a \in A} \operatorname{Ann} a.$$

By **E.8.32**, the operations of taking radicals and unions commute.
Therefore, it is sufficient to prove that $\mathcal{D}(A) = \sqrt{\mathcal{D}(A)}$.
Take $0 \neq a \in \sqrt{\mathcal{D}(A)}$, and let $n$ be the smallest positive integer such that $a^n \in \mathcal{D}(A)$. Let $b \neq 0$ be such that $a^n b = 0$.
Since $a^{n-1} b \neq 0$, we conclude that $a$ is a zero-divisor.

**Solution E. 8.34.** 1. Let $\pi$ be the canonical projection $A \longrightarrow A/\mathcal{J}(A)$.
If $a \in A$ is invertible with inverse $b$, then

$$\overline{a}\overline{b} = \pi(a)\pi(b) = \pi(1) = \overline{1}.$$

Conversely, if $a \in A$ and there exists $b \in A$ such that $\overline{a}\overline{b} = \overline{1}$, then $1 - ab \in \mathcal{J}(A)$.
Thus, $ab = 1 - (1 - ab)$ is invertible in $A$, and $a$ is also invertible.

2. Let $a \in \mathcal{J}(A)$ be such that $a^2 = a$. Then, $1-a$ is invertible and the equality $a(1-a) = 0$ implies that $a = 0$.

**Solution E. 8.35.** By hypothesis, $\overline{a}^2 = \overline{a}$ in $A/I$, that is, $a(1-a) \in I$. Since $a \in \mathcal{J}(A)$, the element $1-a$ is invertible, therefore $a \in I$.

**Solution E. 8.36.** Note that, by hypothesis, $A$ is not a field.
Assume, by contradiction, that $A$ has only finitely many maximal ideals, namely $\mathfrak{m}_1, \ldots, \mathfrak{m}_k$.
For each $i = 1, \ldots, k$, let $a_i$ be a non-zero element of $\mathfrak{m}_i$ and let $a = \prod_i a_i$. Then,

$$a \in \bigcap_i \mathfrak{m}_i = \mathcal{J}(A),$$

and therefore, $1 - ab$ is invertible for all $b \in A$.
Since $A$ has infinitely many elements, but only finitely many of them are invertible, there exist $b_1 \neq b$ such that $1 - ab = 1 - ab_1$.
Since $A$ is a domain, this last equality leads to $b = b_1$, a contradiction.

**Solution E. 8.37.** We recall that operations on the ring $A \times B$ are defined componentwise.
Let $I$ and $J$ be ideals of $A$ and $B$, respectively. Then, $(0_A, 0_B) \in I \times J$.
For any $(a, b) \in A \times B$ and $(i, j)$, $(h, k) \in I \times J$, we have

$$(i, j) - (h, k) = (i - h, j - k) \in I \times J \quad \text{and} \quad (a, b)(i, j) = (ai, bj) \in I \times J.$$

This shows that $I \times J$ is an ideal of $A \times B$.

Conversely, let $H \subseteq A \times B$ be an ideal.
We claim that there exist ideals $I$ of $A$ and $J$ of $B$ such that $H = I \times J$.
Consider the projections $\pi_1 \colon A \times B \longrightarrow A$ and $\pi_2 \colon A \times B \longrightarrow B$, defined by $(a, b) \mapsto a$ and $(a, b) \mapsto b$, respectively. It is easy to verify that these are surjective ring homomorphisms. Thus, $\pi_1(H)$ and $\pi_2(H)$ are ideals of $A$ and $B$, respectively.
Moreover, if $(a, b) \in H$, then $a = \pi_1(a, b) \in \pi_1(H)$ and $b = \pi_2(a, b) \in \pi_2(H)$, which implies $H \subseteq \pi_1(H) \times \pi_2(H)$.

To prove the opposite inclusion, take $(a, b) \in \pi_1(H) \times \pi_2(H)$. Then, there exist $a_1 \in A$ and $b_1 \in B$ such that $(a_1, b)$, $(a, b_1) \in H$.
Therefore, $(a, 0) = (1, 0)(a, b_1)$ and $(0, b) = (0, 1)(a_1, b)$ belong to $H$.
It follows that $(a, b) = (a, 0) + (0, b) \in H$.

Note that if $A$, $B \neq 0$, then $A \times B$ is not a domain, because $(1, 0)(0, 1) = (0, 0)$.

Finally, let $H = I \times J$ be an ideal of $A \times B$ and consider the homomorphism $f \colon A \times B \longrightarrow A/I \times B/J$ given by $(a, b) \mapsto (\overline{a}, \overline{b})$.
Since both projections are surjective, $f$ is also surjective. Moreover, its kernel is $I \times J = H$.
Using homomorphism theorem **T.1.9**.1, we obtain that

$$(A \times B)/H = (A \times B)/(I \times J) \simeq A/I \times B/J.$$

From the previous observations, it follows that for $H$ to be prime, respectively maximal, it must be of the form $I \times (1)$ or $(1) \times J$, with $I$ prime, respectively maximal, in $A$, and $J$ prime, respectively maximal, in $B$.

**Solution E. 8.38.** The first statement is a simple generalization of the case $n = 2$, see **E.8.37**.
Moreover, an ideal $H = I_1 \times \cdots \times I_n$ is prime or maximal, if and only if

$$H = (1) \times \cdots \times (1) \times I_j \times (1) \times \cdots \times (1)$$

with $j \in \{1, \ldots, n\}$ and $I_j$ prime or maximal in $A_j$.

**Solution E. 8.39.** 1. Assume $A \simeq \prod_{i=1}^n K_i$ for some $n \in \mathbb{N}_+$ and for some fields $K_i$.
By **E.8.38**, the ideals of $A$ are all of the form $I_1 \times \cdots \times I_n$ with $I_j = (0)$ or $I_j = K_j$, hence there are only finitely many ideals.
Moreover, the maximal ideals are exactly those with a unique $I_j = (0)$, and therefore

$$\mathcal{J}(A) = \bigcap_{\mathfrak{m} \in \operatorname{Max} A} \mathfrak{m} = (0).$$

Conversely, assume that $A$ has only finitely many maximal ideals, which are then pairwise comaximal, and that $\mathcal{J}(A) = (0)$.
Then, by the Chinese Remainder Theorem, we have

$$A \simeq A/\mathcal{J}(A) \simeq \prod_{i=1}^{n} A/\mathfrak{m}_i,$$

*i.e.*, $A$ is isomorphic to a finite direct product of fields.
2. Assume $A \simeq \prod_{i=1}^n K_i$ for some $n \in \mathbb{N}_+$ and for some fields $K_i$.
Then, a nilpotent element in $A$ is of the form $(a_1, \ldots, a_n)$, where $a_i \in K_i$ is nilpotent for all $i$. Since $K_i$ is a field, this yields $a_i = 0$ for all $i$.

Conversely, in a finite ring there are only finitely many ideals.
Moreover, for any prime ideal $\mathfrak{p}$, the quotient $A/\mathfrak{p}$ is a finite domain, *i.e.*, a field. Thus, all primes are maximal and $\mathcal{J}(A) = \mathcal{N}(A) = (0)$.
The conclusion follows from part 1.

**Solution E. 8.40.** We recall that sum and product in a direct sum of rings are defined componentwise and that the ideals in a direct sum are direct sums of ideals of the components, see **E.8.38**.

1. Since 0 is the only nilpotent of $\mathbb{Z}$ and $\mathbb{Q}$, the nilradical of $A$ contains only elements of the form $(0, \overline{a}, 0)$, where $\overline{a}$ is nilpotent in $\mathbb{Z}/(36)$.
Thus,

$$\mathcal{N}(A) = (0) \times (\overline{6}) \times (0).$$

2. An element is idempotent if and only if $(a^2, \overline{b}^2, c^2) = (a, \overline{b}, c)$, which is equivalent to $a, c \in \{0, 1\}$ and $b^2 \equiv b \mod 36$.

Since $\mathbb{Z}/(36) \simeq \mathbb{Z}/(4) \times \mathbb{Z}/(9)$ and the only idempotents in $\mathbb{Z}/(4)$ and $\mathbb{Z}/(9)$ are $\overline{0}$ and $\overline{1}$, we obtain

$$b \equiv 0,\ 1,\ 9,\ 28 \mod 36,$$

and we are done.

3. As remarked above, the ideals of $A$ are of the form $I \times J \times H$ where $I$, $J$, and $H$ are principal ideals of $\mathbb{Z}$, $\mathbb{Z}/(36)$, and $\mathbb{Q}$, respectively.
Thus, all ideals of $A$ are principal, *i.e.*, $A$ is a PIR, but it is not a PID, because it is not a domain.

4. An ideal is prime if and only if $A/\mathfrak{p}$ is a domain, hence one of the following must hold:

i) $\mathfrak{p} = ((p, \overline{1}, 1))$, with $p$ prime of $\mathbb{Z}$ or $p = 0$;
ii) $\mathfrak{p} = ((1, \overline{q}, 1))$, with $\mathfrak{q} = (\overline{q})$ prime ideal of $\mathbb{Z}/(36)$, *i.e.*, $q = 2, 3$;
iii) $\mathfrak{p} = ((1, \overline{1}, 0))$.

It is easy to verify that these ideals are all maximal except for the one generated by $(0, \overline{1}, 1)$.

**Solution E. 8.41.** We assume that all rational numbers $\frac{a}{b}$ are reduced, *i.e.*, $\gcd(a, b) = 1$.

1. Let

$$I = \left\{ \frac{a}{b} \in A_{(p)} \colon a \equiv 0 \mod p \right\}.$$

Then, $0 \in I$, and

$$\frac{a}{b} - \frac{c}{d} = \frac{ad - bc}{bd} \in I \text{ and } \frac{\alpha}{\beta}\frac{a}{b} = \frac{\alpha a}{\beta b} \in I, \text{ for any } \frac{a}{b}, \frac{c}{d} \in I \text{ and } \frac{\alpha}{\beta} \in A_{(p)}.$$

Thus, $I$ is an ideal of $A_{(p)}$.

If $\frac{\alpha}{\beta} \notin I$, then $\alpha \not\equiv 0 \mod p$, and hence $\frac{\beta}{\alpha} \in A_{(p)}$, *i.e.*, every element not in $I$ is invertible.
Therefore, $A_{(p)}$ is local with maximal ideal $I$, see **T.1.16**.1.

Finally, consider the map $f \colon A_{(p)} \longrightarrow \mathbb{Z}/(p)$ defined by $f\left(\frac{a}{b}\right) = \overline{a}\overline{b}^{-1}$.
It is a well-defined homomorphism, because

$$f\left(\frac{a}{b} + \frac{c}{d}\right) = f\left(\frac{ad + bc}{bd}\right) = (\overline{ad + bc})\overline{bd}^{-1} = \overline{a}\overline{b}^{-1} + \overline{c}\overline{d}^{-1} = f\left(\frac{a}{b}\right) + f\left(\frac{c}{d}\right)$$

$$f\left(\frac{a}{b} \cdot \frac{c}{d}\right) = f\left(\frac{ac}{bd}\right) = \overline{ac}\overline{bd}^{-1} = \overline{a}\overline{b}^{-1}\overline{c}\overline{d}^{-1} = f\left(\frac{a}{b}\right) f\left(\frac{c}{d}\right),$$

for any $\frac{a}{b}$, $\frac{c}{d} \in A_{(p)}$.
For each $\overline{a} \in \mathbb{Z}/(p)$, we have that $f\left(\frac{a}{1}\right) = \overline{a}$, and hence, $f$ is surjective.
Obviously, $f\left(\frac{a}{b}\right) = \overline{0}$ if and only if $a \equiv 0 \mod p$, therefore $\operatorname{Ker} f = I$.
It follows that $f$ induces an isomorphism $A_{(p)}/I \simeq \mathbb{Z}/(p)$.

2. Arguing as above, we can prove that the sets

$$I_j = \left\{ \frac{a}{b} \in A_{(p_1,\dots,p_n)} \colon \ a \equiv 0 \mod p_j \right\}$$

are ideals of $A_{(p_1,\dots,p_n)}$, and that the homomorphisms

$$f_j \colon A_{(p_1,\dots,p_n)} \longrightarrow \mathbb{Z}/(p_j), \qquad \frac{a}{b} \mapsto \overline{a}\overline{b}^{-1}$$

induce isomorphisms

$$\widetilde{f}_j \colon A_{(p_1,\dots,p_n)}/I_j \longrightarrow \mathbb{Z}/(p_j),$$

as $j$ varies in $\{1,\dots,n\}$.
Thus, all ideals $I_j$ are maximal in $A_{(p_1,\dots,p_n)}$.

Let $0 \neq J$ be an ideal of $A_{(p_1,\dots,p_n)}$.
If $J \not\subseteq I_j$ for every $j$, then, by the Prime Avoidance Lemma **T.1.12**.1, we have $J \not\subseteq \bigcup_{j=1}^n I_j$. Hence, there exists $\frac{a}{b} \in J$ such that $a \not\equiv 0 \mod p_j$ for every $j$, that is, $\frac{b}{a} \in A_{(p_1,\dots,p_n)}$ and $J = A_{(p_1,\dots,p_n)}$.
Therefore,

$$\operatorname{Max} A_{(p_1,\dots,p_n)} = \{I_1,\dots,I_n\}.$$

**Solution E. 8.42.** 1. Assume $A = \prod_{i=1}^n A_i$ for some local rings $A_i$, and let $I = \prod_{i=1}^n I_i$ be an ideal of $A$. Since $A/I \simeq \prod_{i=1}^n A_i/I_i$, the maximal ideals of $A$ are all of the form $A_1 \times \cdots \times A_{i-1} \times \mathfrak{m}_i \times A_{i+1} \times \cdots \times A_n$, where $\mathfrak{m}_i$ is the maximal ideal of $A_i$.
Thus, there are only finitely many maximal ideals.

2. The proof is essentially the same as the one of part 1.

3. For $n \geq 2$, the ring $A_{(p_1,\dots,p_n)}$ defined in **E.8.41** is semilocal and non-local. Since it is a domain, $A_{(p_1,\dots,p_n)}$ cannot be a direct product of local rings.

**Solution E. 8.43.** We prove the statement by showing that if $I$ is a prime ideal, then every non-zero element of $A/I$ is invertible.
Take $\overline{a} \in A/I \setminus \{\overline{0}\}$. Then, by hypothesis, there exists $n > 1$ such that $\overline{a}(\overline{a}^{n-1} - 1) = \overline{0}$.
The conclusion follows from the fact that $A/I$ is a domain.

**Solution E. 8.44.** 1. Since $(0) \in \Sigma$, the set $\Sigma$ is non-empty.

For any chain $\mathcal{C}$ of elements of $\Sigma$, the union of the ideals of $\mathcal{C}$ is an ideal contained in $\mathcal{D}(A)$, *i.e.*, an element of $\Sigma$, and hence it is an upper bound for $\mathcal{C}$ in $\Sigma$. By Zorn's Lemma, $\Sigma$ has maximal elements.

Now, let $P$ be a maximal element of $\Sigma$. We show that $P$ is a prime ideal. Let $a, b \notin P$ and consider the ideals $(P, a)$ and $(P, b)$, which, by hypothesis, properly contain $P$.
Therefore, there exist $\alpha = p + ka \in (P, a)$ and $\beta = q + hb \in (P, b)$ which are not zero-divisors. Thus, the element $\alpha\beta \in (P, ab)$ is not a zero-divisor, and then $P \subsetneq (P, ab)$, which means that $ab \notin P$.

2. Note that if $a \neq 0$, then $\operatorname{Ann} a \in \Sigma$.
Since $\mathcal{D}(A) = \bigcup_{0 \neq a \in A} \operatorname{Ann} a$, we can write $\mathcal{D}(A) \subseteq \bigcup_\alpha \mathfrak{p}_\alpha$, where each $\mathfrak{p}_\alpha$ is a maximal element of $\Sigma$. By part 1, every $\mathfrak{p}_\alpha$ is a prime ideal.
Moreover, by definition of $\Sigma$, all such $\mathfrak{p}_\alpha$ are contained in $\mathcal{D}(A)$.
Thus,

$$\mathcal{D}(A) = \bigcup_{\substack{\mathfrak{p} \in \operatorname{Spec} A \\ \text{maximal in } \Sigma}} \mathfrak{p}.$$

**Solution E. 8.45.** Let

$$\Sigma = \{J \subset A\colon\ J \text{ is not principal}\}$$

and assume, by contradiction, that $\Sigma \neq \emptyset$.
Consider $\Sigma$ partially ordered by set inclusion, and note that for any chain $\mathcal{C}$ of elements of $\Sigma$, the union of the elements of $\mathcal{C}$ is still an ideal which is not principal (verify it!). By Zorn's Lemma, we get that $\Sigma$ has at least a maximal element $I$.

Since $I$ is not principal, by hypothesis, $I$ is not prime and there exist elements $a$, $b \notin I$ such that $ab \in I$. Moreover, by the maximality of $I$ in $\Sigma$, the ideal $(I, a)$ is principal, and we set $(I, a) = (c)$.
Note that $b \in I : (c)$, because $bI \subseteq I$ and $ab \in I$. Hence $I \subsetneq I : (c)$, and, again by the maximality of $I$, we get that $I : (c) = (d)$ is also principal.
We prove that $I = (cd)$, thus providing the required contradiction.
Obviously, $(cd) \subseteq I$.

To prove the opposite inclusion, take $j \in I \subset (I, a) = (c)$. Then, $j = ck$ for some $k \in I : (c) = (d)$, that is, $k = hd$ for some $h \in A$ and $j = hcd \in (cd)$, as desired.

**Solution E. 8.46.** 1. Since the sum of ideals contains its addends, the inclusion $\supseteq$ follows from **T.1.17**.1.

To prove the opposite inclusion, take

$$a = \sum_{i=1}^{n} b_i f(a_i) \in (I_1 + I_2)^e$$

with $b_i \in B$ and $a_i \in I_1 + I_2$ for any $i$. Then, for any $i$, there exist $c_i \in I_1$ and $d_i \in I_2$ such that $a_i = c_i + d_i$.
Thus,

$$\begin{aligned} a &= \sum_{i=1}^{n} b_i f(c_i + d_i) = \sum_{i=1}^{n} b_i (f(c_i) + f(d_i)) \\ &= \sum_{i=1}^{n} b_i f(c_i) + \sum_{i=1}^{n} b_i f(d_i) \in I_1^e + I_2^e. \end{aligned}$$

2. Let $G$ and $H$ be sets of generators of $I_1$ and $I_2$, respectively. Then,

$$\begin{aligned}(I_1I_2)^e &= (f(gh)\colon\ g \in G,\ h \in H)\\ &= (f(g)f(h)\colon\ g \in G,\ h \in H)\\ &= (f(g)\colon\ g \in G)(f(h)\colon\ h \in H) = I_1^eI_2^e.\end{aligned}$$

3. It follows immediately from $I_1 \cap I_2 \subseteq I_h$ for $h = 1, 2$ and **T.1.17**.1.

4. Take $a_1 \in J_1^c$ and $a_2 \in J_2^c$. Then, $f(a_i) \in J_i$ for $i = 1, 2$.
Thus, $f(a_1 + a_2) = f(a_1) + f(a_2) \in J_1 + J_2$, that is, $a_1 + a_2 \in (J_1 + J_2)^c$.

5. Take $a_1 \in J_1^c$ and $a_2 \in J_2^c$. Then, $f(a_i) \in J_i$ for $i = 1, 2$.
Thus, $f(a_1a_2) = f(a_1)f(a_2) \in J_1J_2$, that is, $a_1a_2 \in (J_1J_2)^c$. Since $J_1^cJ_2^c$ is generated by such elements, the conclusion follows.

6. The inclusion $\subseteq$ follows from $J_1 \cap J_2 \subseteq J_h$ for $h = 1, 2$ and **T.1.17**.2.

To prove the opposite inclusion, take $a \in J_1^c \cap J_2^c$. Then, $f(a) \in J_1$ and $f(a) \in J_2$, that is, $f(a) \in J_1 \cap J_2$.
Thus, $a \in (J_1 \cap J_2)^c$.

Now let us show some examples.
Consider the ring homomorphism $f\colon \mathbb{Z}[x, y] \longrightarrow \mathbb{Z}$ defined by $f(x) = 10$ and $f(y) = 15$. If $I_1 = (x)$ and $I_2 = (y)$, then we have

$$(I_1 \cap I_2)^e = (xy)^e = (150) \subsetneq (30) = (10) \cap (15) = I_1^e \cap I_2^e.$$

Next, consider the inclusion homomorphism $\mathbb{Z} \longrightarrow \mathbb{Z}[x]$. If $J_1 = (x)$ and $J_2 = (x + 1)$, then we have

$$J_1^c + J_2^c = (0) + (0) = (0) \subsetneq \mathbb{Z} = (1)^c = (J_1 + J_2)^c.$$

Finally, consider the inclusion homomorphism $\mathbb{Z} \longrightarrow \mathbb{Z}[i]$, and take $J = (1+i)$.
Then, $2i = (1 + i)^2$ and $i \in \mathbb{Z}[i]^*$, hence $2 \in J^c$ and $J^2 = (2)$.
Since $(2)$ is maximal in $\mathbb{Z}$ and $J \subsetneq \mathbb{Z}[i]$, we have $J^c = (2)$.
Thus,

$$J^cJ^c = (2)(2) = (4) \subsetneq (2) = (2)^c = (JJ)^c.$$

**Solution E. 8.47.** 1. Since $I[x] \subseteq IA[x] = i(I)A[x]$, we have $I[x] \subseteq I^e$.

To prove the opposite inclusion, we simply observe that $I[x]$ is an ideal containing $I$. Then, $I^e$, the smallest ideal of $A[x]$ containing $I$, is contained in $I[x]$.

2. By **E.8.12**, $A[x]/I[x] \simeq (A/I)[x]$, therefore, $A[x]/I[x]$ is a domain if and only if $A/I$ is a domain.

3. It is not true in general. For example, consider a prime $p$ of $\mathbb{Z}$.
Then, $(p)$ is maximal in $\mathbb{Z}$, but its extension is not maximal in $\mathbb{Z}[x]$, because

$$(p)^e = (p)[x] \subsetneq (p,\ x) \subsetneq \mathbb{Z}[x].$$

**Solution E. 8.48.** 1. Take $b \in f(\sqrt{I})$. Then, $b = f(c)$ with $c^m \in I$ for some $m$. Hence, $b^m = f(c)^m = f(c^m) \in f(I)$, that is, $b \in \sqrt{f(I)}$.

Thus,

$$(f(\sqrt{I})) \subseteq \sqrt{f(I)} \subseteq \sqrt{(f(I))},$$

as desired.

2. The surjectivity of $f$ yields $f(I) = (f(I)) = I^e$ for every ideal $I$.

Since we proved one inclusion in part 1, we will now show the reverse. Take $b \in \sqrt{f(I)}$ and let $m \in \mathbb{N}$ be such that $b^m = f(c)$ for some $c \in I$. The surjectivity of $f$ yields $b = f(a)$ for some $a \in A$.
Then,

$$f(a^m - c) = f(a^m) - f(c) = f(a)^m - f(c) = 0,$$

that is, $a^m - c \in \operatorname{Ker} f \subseteq I$.
Therefore, $a^m \in I$, which yields $a \in \sqrt{I}$, and finally, $b \in f(\sqrt{I}\,)$.

3. It holds that $a \in \sqrt{f^{-1}(J)}$ if and only if $a^m \in f^{-1}(J)$ for some $m \in \mathbb{N}$, *i.e.*, if and only if $f(a)^m = f(a^m) \in J$ for some $m$.
The last statement is equivalent to $f(a) \in \sqrt{J}$, that is, $a \in f^{-1}(\sqrt{J}\,)$.

**Solution E. 8.49.** We recall that the ring of Gaussian integers $\mathbb{Z}[i]$ is an Euclidean ring, hence a PID and a UFD.

For $p = 2$, simply note that $2 = (1+i)(1-i)$ and that the ideals $(1+i)$ and $(1-i)$ are equal, because $1 + i = (1-i)i$, that is, their generators are associate.
Thus,

$$(2)^e = (1+i)^2.$$

Now, assume $p$ is odd. We will show that $p$ is reducible in $\mathbb{Z}[i]$ if and only if $p \equiv 1 \mod 4$.
If $p$ is reducible, then there exist non-invertible elements $a + ib,\ c + id \in \mathbb{Z}[i]$ such that $p = (a+ib)(c+id)$. Since $p \in \mathbb{Z}$, we also have $p = (a-ib)(c-id)$, and therefore, $p^2 = (a^2+b^2)(c^2+d^2)$.
Note that if $\alpha^2+\beta^2 = 1$, then either $\alpha = \pm 1$ or $\beta = \pm 1$, that is, $\alpha + i\beta \in \mathbb{Z}[i]^*$. Hence, $a^2 + b^2 = p$. Since the only squares modulo 4 are 0 and 1, we have $p \equiv 1 \mod 4$.

Conversely, if $p \equiv 1 \mod 4$, then $|(\mathbb{Z}/(p))^*| = p - 1$ is divisible by 4, and thus, there exists an element $a \in \{1, \ldots, p-1\}$ such that $a^2 \equiv -1 \mod p$. Therefore,

$$(a+i)(a-i) = a^2 + 1 = kp \ \text{ for some integer } \ 0 < k < p.$$

Since $p^2 \nmid a^2 + 1$, we also have that $p$ does not divide $a \pm i$ in $\mathbb{Z}[i]$. Therefore, the element $p$ is not prime, and thus, not irreducible in $\mathbb{Z}[i]$.

In conclusion,

$$(p)^e \ \text{ is prime if and only if } \ p \equiv 3 \mod 4.$$

**Solution E. 8.50.** Obviously, $\frac{1-\zeta_p^a}{1-\zeta_p} = 1 + \zeta_p + \ldots + \zeta_p^{a-1} \in \mathbb{Z}[\zeta_p]$.
We look for its inverse. Let $b$ be such that $ab \equiv 1 \mod p$. Then,

$$\frac{1-\zeta_p}{1-\zeta_p^a} = \frac{1-\zeta_p^{ab}}{1-\zeta_p^a} = 1+\zeta_p^a+\ldots+\zeta_p^{a(b-1)} \in \mathbb{Z}[\zeta_p].$$

We recall that the elements $\zeta_p^a$ with $a = 1, \ldots, p-1$ are precisely all the roots of the polynomial $1+x+\ldots+x^{p-1}$.
Therefore, $1+x+\ldots+x^{p-1} = \prod_{a=1}^{p-1}(x-\zeta_p^a)$, and, evaluating at $x=1$, we obtain $p = \prod_{a=1}^{p-1}(1-\zeta_p^a)$. By what we proved above, $(1-\zeta_p) = (1-\zeta_p^a)$ for all $a = 1, \ldots, p-1$.
Hence,

$$(p)^e = p\mathbb{Z}[\zeta_p] = \left(\prod_{a=1}^{p-1}(1-\zeta_p^a)\right) = \prod_{a=1}^{p-1}(1-\zeta_p^a) = \prod_{a=1}^{p-1}(1-\zeta_p) = (1-\zeta_p)^{p-1}.$$

**Solution E. 8.51.** 1. Let $a \in \mathcal{N}(A)$. Then, there exists $n \in \mathbb{N}$ such that $a^n = 0$. Thus, $f(a)^n = f(a^n) = 0$, which yields $f(a) \in \mathcal{N}(B)$, and hence, $f(\mathcal{N}(A)) \subseteq \mathcal{N}(B)$.

2. Take $a \in \mathcal{J}(A)$. Then, $1-ba$ is invertible for each $b \in A$. Therefore, $f(1-ba) = 1 - f(b)f(a)$ is invertible in $B$.
The surjectivity of $f$ yields $f(a) \in \mathcal{J}(B)$.

3. We already saw that $A_{(2)} = \left\{\frac{a}{b} \in \mathbb{Q} \colon b \not\equiv 0 \mod 2\right\}$ is a local subring of $\mathbb{Q}$ and its maximal ideal is $(2) = \left\{\frac{a}{b} \in A_{(2)} \colon a \equiv 0 \mod 2\right\}$, see **E.8.41**.1. Consider the inclusion homomorphism of $A_{(2)}$ in $\mathbb{Q}$, which is clearly injective and not surjective.
We have $\mathcal{J}(\mathbb{Q}) = 0$, while $\mathcal{J}(A_{(2)}) = (2)$.

4. Consider the canonical projection $f \colon \mathbb{Z} \longrightarrow \mathbb{Z}/(4)$.
Then,

$$f(\mathcal{J}(\mathbb{Z})) = (0) \subsetneq \mathcal{J}(\mathbb{Z}/(4)) = (2).$$

5. Assume $A$ is semilocal and let $\mathfrak{m}_1, \ldots, \mathfrak{m}_k$ be its maximal ideals.
By **T.1.18**.3, the ideal $f(\mathfrak{m}_i)$ is maximal in $B$ whenever $\mathfrak{m}_i \supseteq \operatorname{Ker} f$, otherwise $f(\mathfrak{m}_i) = B$ (prove it!).
Since $\mathcal{J}(A) = \bigcap_{i=1}^k \mathfrak{m}_i = \prod_{i=1}^k \mathfrak{m}_i$, it follows that

$$f(\mathcal{J}(A)) = f\Big(\prod_{i=1}^k \mathfrak{m}_i\Big) = \prod_{i=1}^k f(\mathfrak{m}_i) = \bigcap_{i=1}^k f(\mathfrak{m}_i) \supseteq \mathcal{J}(B).$$

The conclusion follows from part 2.

**Solution E. 8.52.** The one-to-one correspondence between ideals of $A/I$ and ideals of $A$ containing $I$ immediately shows that, when $A$ is local, also $A/I$ is local for any ideal $I$.

Conversely, assume that $A/I$ is local with maximal ideal $\overline{\mathfrak{m}}$, and let $\mathfrak{m}$ be its preimage in $A$. Take $a \notin \mathfrak{m}$. To prove that $A$ is local with maximal ideal $\mathfrak{m}$, we show that $a$ is invertible.

Since $\overline{a} \notin \overline{\mathfrak{m}}$, the element $\overline{a}$ is invertible because $A/I$ is local. Then, there exist $b \in A$ and $i \in I$ such that $ab = 1 + i$. Since $I \subseteq \mathcal{N}(A)$, from **E.8.4** we obtain that $1 + i$ is invertible, and hence $a$ is invertible as well.

**Solution E. 8.53.** We can assume $a, b \neq 0$. If $(a) = (b)$, then there exist $c, d \in A$ such that $a = bd = acd$. This yields $a(1 - cd) = 0$, and therefore, $1 - cd \in \mathcal{D}(A) \subseteq \mathcal{J}(A)$. It follows that $cd$ is invertible, and hence, $d$ is also invertible.

**Solution E. 8.54.** 1. Let $(a, b) = (\delta)$. Then, $\delta \mid a$ and $\delta \mid b$, and there exist elements $u, v \in A$ such that $\delta = ua + vb$. Moreover, any $c \in A$ dividing both $a$ and $b$, divides also $ua + vb = \delta$. Thus, by definition of greatest common divisor, $\delta = d$.

2. Consider the ring $A = \mathbb{Z}[x]$, which is a UFD but not a PID.
We have $\gcd(3, x) = 1$, however, $3f + xg \neq 1$, for all $f, g \in A$, because its constant term belongs to $(3) \subsetneq (1)$.

**Solution E. 8.55.** The inclusion $I^2 + J^2 \subseteq (I + J)^2$ always holds.

To prove the opposite inclusion, let $I = (a)$, $J = (b)$, and $I + J = (d)$. Then, by **E.8.54**.1, we have $\gcd(a, b) = d$, and we can write $a = da_1$ and $b = db_1$ for some $a_1, b_1 \in A$ such that $\gcd(a_1, b_1) = 1 = \gcd(a_1^2, b_1^2)$. Hence, there exist $\alpha, \beta \in A$ such that $1 = \alpha a_1^2 + \beta b_1^2$.
Thus,

$$d^2 = d^2 \cdot 1 = \alpha a^2 + \beta b^2 \in I^2 + J^2.$$

**Solution E. 8.56.** We begin by proving that the ideal $\mathcal{J}(A)$ is prime.
If $ab \in \mathcal{J}(A)$, then there exists $c \neq 0$ such that $cab = 0$.
Therefore, at least one of $a$ and $b$ is an element of $\mathcal{D}(A) = \mathcal{J}(A)$.

Now, let $\mathcal{J}(A) = (j)$ for some $j \neq 0$. In order to prove that $\mathcal{J}(A)$ is maximal, it suffices to show that if $a \notin \mathcal{J}(A)$, then $(a) + \mathcal{J}(A) = A$.
Let $(a) + \mathcal{J}(A) = (a, j) = (b)$, where $b \notin \mathcal{J}(A)$. Then, $j = bc$ for some $c \in \mathcal{J}(A)$, because $\mathcal{J}(A)$ is prime. We can write $c = jd$ for some $d \in A$ and obtain $j(1 - bd) = bc - bc = 0$. Thus, $1 - bd \in \mathcal{D}(A) = \mathcal{J}(A)$.
Therefore, $bd$ is invertible, which shows that $b$ is also invertible, as desired.

**Solution E. 8.57.** Since an ideal $\mathfrak{m}$ is maximal if and only if $A/\mathfrak{m}$ is a field, parts 1 and 2 are equivalent. Thus, it is enough to prove 2.
2. If $a$ is irreducible, then its only divisors are invertible or associate to $a$. Therefore, the only ideals containing $(a)$ are $(1)$ and $(a)$ itself.
Since all ideals of $A$ are principal, this proves that irreducible elements generate maximal ideals.

Conversely, let $b \in A$ be reducible and write $b = ac$ with $a, c \notin A^*$. Then, $(b) \subsetneq (a) \subsetneq (1)$, and the ideal $(b)$ is not maximal.

3. By **T.1.27**, the ring $K[x]$ is a PID.
Since a non-trivial element in a PID is prime if and only if it is irreducible, the conclusion follows from part 2, see also **T.1.22** and **T.1.23**.

**Solution E. 8.58.** Assume $f = \prod_i f_i$ for some distinct and irreducible $f_i$, and, for each $i$, let $F_i$ be the field $K[x]/(f_i)$, see **E.8.57**.

By the Chinese Remainder Theorem,

$$A = K[x]/\prod_i (f_i) \simeq \prod_i F_i,$$

and therefore, $\mathcal{N}(\prod_i (F_i)) = \prod_i \mathcal{N}(F_i) = (0)$ yields $\mathcal{N}(A) = (0)$, see **E.8.48**.2.

Conversely, assume $f = \prod_i f_i^{s_i}$ with $s_i > 1$ for some $i$.
The coset of the element $\prod_i f_i$ is a non-zero element of $\mathcal{N}(A)$, therefore the ring $A$ is not reduced.

**Solution E. 8.59.** 1. If $f$ is not squarefree, then there exists an irreducible $f_1$ such that $f = f_1^e h$ with $e > 1$ and $h \in K[x]$. Since $f' = f_1^{e-1}(ef_1'h + f_1h')$, we have $\gcd(f, f') \neq 1$.

Conversely, let $g$ be an irreducible factor of $\gcd(f, f')$. Then, $\deg g > 0$, $f = gh$, and $f' = g'h + gh' = gq$ for some $h, q \in K[x]$. Thus, $g \mid g'h$. If $g' \neq 0$, then $\deg g' < \deg g$, and therefore $g \mid h$.
Hence, $g^2 \mid f$ and $f$ is not squarefree.

To conclude we note that, since $g$ is irreducible, its derivative $g'$ can never be 0. This is obvious in characteristic 0.
Now, let $\operatorname{char} K = p$, and assume, by contradiction, $g' = 0$. It is easy to verify that there exists $r \in K[x]$ such that $g(x) = r(x^p)$. Since $K = K^{(p)}$, there exists $s \in K[x]$ such that

$$g(x) = r(x^p) = (s(x))^p,$$

that is not possible.
2. We have $\gcd(f, f') = 1$ in $K[x]$ if and only if there exist $h, g \in K[x] \subseteq L[x]$ such that $hf + gf' = 1$. Thus, $\gcd(f, f') = 1$ in $L[x]$ as well.

The opposite implication is trivial.

**Solution E. 8.60.** Reordering the variables, we can always assume that $i_j = j$ for any $j$.
Let $r = 1$. By **E.8.58**, the ideal $(h_1)$ is radical in $K[x_1]$. From **E.8.12** and the Chinese Remainder Theorem, we obtain

$$K[x_1, \dots, x_n]/(h_1) \simeq (K[x_1]/(h_1))\,[x_2, \dots, x_n] \simeq \left(\prod\nolimits_i F_i\right)[x_2, \dots, x_n],$$

where $K \subseteq F_i$ are field extensions, and the direct product is finite, see also the proof of **E.8.58**.
Since $\prod_i F_i$ has no non-trivial nilpotents, by repeatedly using **E.8.5**.2, we obtain $\mathcal{N}(K[x_1, \dots, x_n]/(h_1)) = 0$.

Now, fix $r = 2$ and consider $K[x_1, \dots, x_n]/(h_1, h_2)$.
We have

$$\begin{aligned} K[x_1, \dots, x_n]/(h_1, h_2) &\simeq (K[x_1]/(h_1))\,[x_2, \dots, x_n]/(h_2) \\ &\simeq \left(\prod\nolimits_i (F_i[x_2]/(h_2))\right)[x_3, \dots, x_n], \end{aligned}$$

since $(\prod_i F_i)[x] \simeq \prod_i (F_i[x])$.

By **E.8.59**.2, the polynomial $h_2$ is still squarefree in $F_i[x_2]$ for all $i$, and hence, $F_i[x_2]/(h_2) \simeq \prod_j F_{ij}$ is a product of field extensions of $F_i$.
Thus,

$$K[x_1,\ldots,x_n]/(h_1,h_2) \simeq \Big(\prod\nolimits_{i,j} F_{ij}\Big)[x_3,\ldots,x_n]$$

is reduced.
We complete the proof by repeating this process.

**Solution E. 8.61.** For any $i = 1,\ldots,n$, let $I_i = (x-\alpha_i)$.
Constructing a polynomial $f(x)$ such that $f(\alpha_i) = \beta_i$ for all $i$, is equivalent to finding an element $f(x) \in K[x]$ such that $f(x) \equiv \beta_i \mod I_i$ for all $i$.

In the Euclidean ring $K[x]$ the ideals $I_i$ and $I_j$ are maximal, hence comaximal whenever $i \neq j$.
Clearly,

$$\frac{x-\alpha_i}{\alpha_j-\alpha_i} + \frac{x-\alpha_j}{\alpha_i-\alpha_j} = 1.$$

If we define

$$L_i = \prod_{j\neq i} \frac{x-\alpha_j}{\alpha_i-\alpha_j},$$

we have $L_i \equiv 0 \mod I_j$, for any $j \neq i$, and $L_i \equiv 1 \mod I_i$.
Therefore, the polynomial

$$f = \sum_{i=1}^{n} \beta_i L_i = \sum_{i=1}^{n} \beta_i \prod_{j\neq i} \frac{x-\alpha_j}{\alpha_i-\alpha_j}$$

has the required properties.

**Solution E. 8.62.** Write $f = \prod\limits_{i=1}^{n} f_i$ for some irreducible and distinct $f_i \in A$.
By the Chinese Remainder Theorem, there exists an isomorphism

$$\psi\colon B \longrightarrow \prod_{i=1}^{n} A/(f_i),$$

where the $A/(f_i)$ are finite fields containing $\mathbb{Z}/(p)$ for all $i$.
Note that $B$ and the fields $A/(f_i)$ are finite dimensional $\mathbb{Z}/(p)$-vector spaces.
Moreover, both $\varphi_p$ and $\psi$ are $\mathbb{Z}/(p)$-linear maps. Thus, $\varphi_p - \mathrm{id}_B$ is also a linear map between finite dimensional vector spaces, which is not injective.

1. Take $\overline{g} \in \mathrm{Ker}(\varphi_p - \mathrm{id}_B)$ and write

$$\psi(\overline{g}) = (\overline{g}_1,\ldots,\overline{g}_n), \text{ with } \overline{g}_i \equiv \overline{g} \mod (f_i) \text{ for all } i.$$

Then, $\overline{g}_i^p = \overline{g}_i$ in $A/(f_i)$ for all $i$.

The unique roots of $y^p - y$ in $A/(f_i)$ are the elements of $\mathbb{Z}/(p)$. Therefore, $\overline{g} \in \operatorname{Ker}(\varphi_p - \mathrm{id}_B)$ if and only if $\overline{g}_i \in \mathbb{Z}/(p)$ for all $i$, that is, if and only if $\psi(\overline{g}) \in (\mathbb{Z}/(p))^n$.
Since $\psi$ is also a $\mathbb{Z}/(p)$-linear isomorphism, we have that

$$\operatorname{Ker}(\varphi_p - \mathrm{id}_B) \simeq (\mathbb{Z}/(p))^n.$$

2. Take $\overline{g} \in \operatorname{Ker}(\varphi_p - \mathrm{id}_B)$, and let $g$ be a representative of $\overline{g}$ in $A$.
Let $\psi(\overline{g}) = (a_1, \dots, a_n) \in (\mathbb{Z}/(p))^n$. Then, for any $a \in \mathbb{Z}/(p)$, we have $f_i \mid g-a$ if and only if the $i$-th coordinate of

$$\psi(\overline{g-a}) = \psi(\overline{g}) - \psi(a) = (a_1 - a, \dots, a_n - a)$$

is trivial, *i.e.*, if and only if $a_i = a$.
Thus,

$$\gcd(f, g-a) = \prod_{\substack{i=1,\dots,n \\ a_i = a}} f_i,$$

and the last identity immediately follows.

**Solution E. 8.63.** Let $ab \in \mathfrak{q} \subseteq \mathfrak{q}_1$, with $a \notin \mathfrak{q}_1$. Since $\mathfrak{q}_1$ and $\mathfrak{q}_2$ are primary, $b \in \sqrt{\mathfrak{q}_1} = \sqrt{\mathfrak{q}_2}$. Thus, $b^n \in \mathfrak{q}_1$ and $b^m \in \mathfrak{q}_2$ for some $n, m \in \mathbb{N}_+$. Therefore $b^{n+m} \in \mathfrak{q}_1\mathfrak{q}_2 \subseteq \mathfrak{q}$. (What happens if $a \in \mathfrak{q}_1 \setminus \mathfrak{q}$?)

Furthermore, we have

$$\sqrt{\mathfrak{q}} = \sqrt{\mathfrak{q}_1 \cap \mathfrak{q}_2} = \sqrt{\mathfrak{q}_1} \cap \sqrt{\mathfrak{q}_2} = \sqrt{\mathfrak{q}_1}.$$

**Solution E. 8.64.** Let $I = (a)$ be a primary ideal with $\sqrt{I} = \mathfrak{p} = (p)$. Then, there exists $k \in \mathbb{N}_+$ such that $p^k \in I$. Thus, $a \mid p^k$ and, since $A$ is a UFD, $p$ is the only irreducible element dividing $a$.
Therefore, $a = up^t$ for some $u \in A^*$ and $t \in \mathbb{N}$, that is, $(a) = (p^t)$.

**Solution E. 8.65.** 1. By **T.1.23**.1, the element $p$ is irreducible.
If $ab \in (p^i)$ and $a \notin (p^i)$, then $p \mid b$ by the uniqueness of the factorization. Hence, $b \in (p) = \sqrt{(p^i)}$.

2. Let $A = K[x, y]$ and $\mathfrak{q} = (x, y^2)$. Then, $\sqrt{\mathfrak{q}} = (x, y)$ is maximal, thus $\mathfrak{q}$ is primary. Clearly,

$$(x, y)^2 \subsetneq \mathfrak{q} \subsetneq (x, y),$$

and therefore, $\mathfrak{q}$ is not a power of $(x, y)$ or of any other prime.

3. Since $B/\mathfrak{p} \simeq A/(x, z) \simeq K[y]$, the ideal $\mathfrak{p}$ is prime.

To prove that $\mathfrak{p}^2$ is not primary note that

$$\overline{xy} = \overline{z}^2 \in (\overline{x}^2, \overline{xz}, \overline{z}^2) = \mathfrak{p}^2,$$

but $\overline{x} \notin \mathfrak{p}^2$ and $\overline{y}^k \notin \mathfrak{p}^2$ for all integers $k$.

**Solution E. 8.66.** If $A$ is a field, then $A[x]$ is an Euclidean ring, hence a PID by **T.1.27**.

Conversely, consider the evaluation homomorphism $\varphi_0 \colon A[x] \longrightarrow A$ defined by $\varphi_0(f) = f(0)$.
It is easy to verify that $\varphi_0$ is surjective and that $\operatorname{Ker}\varphi_0 = (x)$, hence $A[x]/(x) \simeq A$. By hypothesis, $A[x]$ is a domain, hence $A$ is also a domain. Therefore, the ideal $(x)$ is prime, and $x$ is a prime element by **T.1.22**.5. Moreover, by **T.1.23**.1, we have that $x$ is irreducible. Then, since $A[x]$ is a PID, $(x)$ is maximal by **E.8.57**.
Hence, $A$ is a field.

**Solution E. 8.67.** We recall that the norm of $\alpha = a + b\sqrt{-5}$ is given by $N(\alpha) = a^2 + 5b^2$, and that $N(\alpha\beta) = N(\alpha)N(\beta)$, for any $\alpha$, $\beta \in \mathbb{Z}[\sqrt{-5}]$.
It is not hard to prove that $\alpha$ is invertible if and only if $N(\alpha) = 1$ and that no element of $\mathbb{Z}[\sqrt{-5}]$ has norm 3.

Let $a+b\sqrt{-5}$, $c+d\sqrt{-5} \in \mathbb{Z}[\sqrt{-5}]$ be such that $(a+b\sqrt{-5})(c+d\sqrt{-5}) = 3$. Taking norms, we obtain $(a^2 + 5b^2)(c^2 + 5d^2) = 9$. This is possible if and only if one of the two factors is 1, *i.e.*, if and only if either $a + b\sqrt{-5}$ or $c + d\sqrt{-5}$ is invertible.
It follows that 3 is an irreducible element.

The ideals $(3, 1 + \sqrt{-5})$ and $(3, 1 - \sqrt{-5})$ are comaximal, hence

$$(3, 1 + \sqrt{-5}) \cap (3, 1 - \sqrt{-5}) = (3, 1 + \sqrt{-5})(3, 1 - \sqrt{-5}) = (3).$$

Since the norm of 3 does not divide $N(1 \pm \sqrt{-5})$, we have $(3) \subsetneq (3, 1 + \sqrt{-5})$ and $(3) \subsetneq (3, 1 - \sqrt{-5})$.
Therefore, $(3)$ is not irreducible.

**Solution E. 8.68.** Assume, by contradiction, that there exists a non-trivial element $a_1 \in A$ which is not invertible and does not have a factorization as a finite product of irreducible elements.
Then, $a_1$ is not irreducible and can be written as a product $a_2 b_1$, where both $a_2$ and $b_1$ are not invertible and at least one of them does not have a factorization as a finite product of irreducible elements. Without loss of generality, assume it is $a_2$.
We can repeat the previous argument on $a_2$ and, iterating this process, we find an infinite ascending chain of ideals $(a_1) \subsetneq (a_2) \subsetneq \dots$ which contradicts the hypothesis.

**Solution E. 8.69.** We look for a UFD ring $A$ and an ideal $I$ of $A$ such that $A/I$ is a domain but without unique factorization.
Let

$$A = \mathbb{Q}[x, y, z, t]/(xy - zt).$$

Then, $A$ is a domain (prove it!) and it is not a UFD.
Clearly, $A$ inherits (UFD1) from $\mathbb{Q}[x, y, z, t]$ and, intuitively, we have "spoiled" the uniqueness of the factorization of the element $\overline{xy}$, *i.e.*, we have built a ring in which (UFD2) does not hold.

With direct computations it can be verified that $\overline{x}$, $\overline{y}$, $\overline{z}$, $\overline{t}$ are irreducible elements that are not associate. This shows that the factorizations $\overline{xy} = \overline{zt}$ are distinct.

Alternatively, we can observe that $\overline{x}$ is irreducible, but $(\overline{x})$ is not a prime ideal, see **T.1.25**, and the same holds for $\overline{y}$, $\overline{z}$ and $\overline{t}$.

**Solution E. 8.70.** 1. Let $c_1 = ab_1 - b_1$, $c_2 = ab_2 - b_2 \in I_a$, and $d \in A$. Then, $0 \in I_a$, $c_1 + c_2 = a(b_1 + b_2) - (b_1 + b_2)$, and $dc_1 = a(db_1) - (db_1)$ are elements of $I_a$, and then, $I_a$ is an ideal.

Alternatively, it suffices to note that $I_a$ is the ideal of $A$ generated by $a - 1$. In particular, $a$ is quasi-regular if and only if $a - 1$ is invertible.

2. If $a$ is quasi-regular, then $a \in I_a$. Hence, there exists $c \in A$ such that $a = ac - c$.

Conversely, assume $a = ac - c \in I_a$.
We have to prove that any $d \in A$ belongs to $I_a$. We have $ad \in I_a$ and, by definition of $I_a$, also $ad - d \in I_a$. Thus, $d = ad - (ad - d) \in I_a$.

3. If $a$ is nilpotent, then $1 - a$ is invertible. Therefore, $a$ is quasi-regular.

4. Let $a \in A \setminus \{0, 1\}$. By hypothesis, $a$ is quasi-regular, *i.e.*, $1 - a$ is invertible. Hence, there exists $b \in A \setminus \{0\}$ such that $b - ab = b(1 - a) = 1$, which yields $ab = b - 1$. Since $b \neq 1$ is quasi-regular, $b - 1$ is invertible.
It follows that $a$ is invertible, as desired.

**Solution E. 8.71.** 1. Since $A$ is not a field, every maximal ideal $\mathfrak{m}$ is non-zero. Let $a \in \mathfrak{m} \setminus \{0\}$.
By hypothesis,

$$(a) = \prod_{i=1}^{k} \mathfrak{m}_i^{s_i} = \bigcap_{i=1}^{k} \mathfrak{m}_i^{s_i}$$

for some distinct maximal ideals $\mathfrak{m}_i$ and some positive integers $s_i$, where the equality follows from the fact that powers of distinct maximal ideals are comaximal, see **E.8.25**.2.
Therefore, $\bigcap_{i=1}^{k} \mathfrak{m}_i^{s_i} \subseteq \mathfrak{m}$, and there exists $i$ such that $\mathfrak{m}_i^{s_i} \subseteq \mathfrak{m}$ by **T.1.12**.2.
By the maximality of $\mathfrak{m}$ and of $\mathfrak{m}_i$, we have $\mathfrak{m}_i = \mathfrak{m}$.
To conclude, simply take

$$I = \mathfrak{m}_i^{s_i - 1} \cdot \prod_{j=1,\ j \neq i}^{k} \mathfrak{m}_j^{s_j}.$$

2. By part 1, let $I\mathfrak{m} = (a)$ with $a \neq 0$.
Multiplying this equality by $J$, we obtain $J(a) = J\mathfrak{m}I = H\mathfrak{m}I = H(a)$. Thus, for any $j \in J$, there exists $h \in H$ such that $ja = ha$, that is, $a(j - h) = 0$. Since $a \neq 0$ and $A$ is a domain, we have $j = h$, which implies $J \subseteq H$.
By reversing the roles of $J$ and $H$, we can conclude the proof.

**Solution E. 8.72.** 1. Since $\sqrt{I} \subseteq \sqrt{I : J^m}$, it is sufficient to prove the opposite inclusion.

By hypothesis, $J \not\subseteq \sqrt{I}$, hence there exists $j \in J$ such that $j^k \notin I$ for all $k \in \mathbb{N}$. Let $a \in \sqrt{I : J^m}$. Then, $a^n j^m \in I$ for some $n \geq 1$. Since $I$ is primary and $j^m \notin I$, we have $a \in \sqrt{I}$.

2. It is sufficient to show that $\sqrt{I : (h)} \subseteq I : (h)$.
Let $a \in \sqrt{I : (h)}$. Then, $a^n h \in I$ for some positive integer $n$, and this yields $(ah)^n \in I$.
Therefore, $ah \in \sqrt{I} = I$, and then $a \in I : (h)$.

**Solution E. 8.73.** 1. If $p$ is invertible, then there exists $q$ such that $pq = 1$, and, in particular, $a_0$ is invertible.

Conversely, assuming $a_0$ is invertible, we explicitly construct the inverse $q = \sum_{i \in \mathbb{N}} b_i x^i \in A[[x]]$ of $p$.
From $pq = 1$ we get

$$a_0 b_0 = 1,\ a_0 b_1 + a_1 b_0 = 0,\ \ldots,\ \sum_{i=0}^{k} a_i b_{k-i} = 0,\ \ldots,$$

which imply

$$b_0 = {a_0}^{-1},\ b_1 = -{a_0}^{-1} a_1 b_0,\ \ldots,\ b_k = -{a_0}^{-1} \sum_{i=1}^{k} a_i b_{k-i}.$$

Therefore, we have a recursive formula to compute $b_i$ in terms of the coefficients $a_j$, which only requires $a_0$ to be invertible.

2. If $p$ is nilpotent, then $p^n = 0$ for some integer $n$. This immediately implies that $a_0$ is nilpotent. Moreover, $p - a_0 = x \sum_{i \in \mathbb{N}_+} a_i x^{i-1}$ is also nilpotent, because it is a sum of nilpotents. Therefore, $a_1$ is nilpotent.
By iterating this process, we prove that all $a_i$ are nilpotent.

Consider the ring

$$A = \prod_{n \in \mathbb{N}_+} \mathbb{Z}/(2^n),$$

with operations defined componentwise.
We recall that $(2)$ is the unique maximal ideal of $\mathbb{Z}/(2^n)$.
Let

$$p = (0, 0, \ldots) + (0, 2, 0, \ldots)x + (0, 0, 2, \ldots)x^2 + \ldots \in A[[x]].$$

Then, all the coefficients of $p$ are nilpotent, but it is not hard to verify that $p$ is not nilpotent.

3. An element $p$ belongs to $\mathcal{J}(A[[x]])$ if and only if $1 - pq$ is invertible for all $q \in A[[x]]$. By part 1, this is equivalent to saying that $1 - a_0 b_0$ is invertible for all $b_0 \in A$. This holds if and only if $a_0 \in \mathcal{J}(A)$.

4. Let $\mathfrak{m} \subset A[[x]]$ be a maximal ideal.

We begin by proving that $x \in \mathfrak{m}$. Indeed, if $x \notin \mathfrak{m}$, then, by the maximality of $\mathfrak{m}$, we have $(\mathfrak{m}, x) = (1)$. Thus, there exist $f \in \mathfrak{m}$ and $h \in A[[x]]$ such that $1 = f + xh = f_0$, where $f_0$ denotes the constant term of $f$.
Hence, $f_0$ is invertible and, by part 1, $f$ is invertible in $A[[x]]$, a contradiction.
Now, set

$$\mathfrak{n} = \{a \in A\colon\ a \text{ constant term of an element of } \mathfrak{m}\}.$$

It is easy to verify that $\mathfrak{n}$ is an ideal of $A$, and that $(\mathfrak{n}, x) \subseteq \mathfrak{m}$. The opposite inclusion follows immediately from the definition of $\mathfrak{n}$.
Therefore, $\mathfrak{m} = (\mathfrak{n}, x)$ and this yields $\mathfrak{m}^c = \mathfrak{n}$.

Finally, we note that $\mathfrak{n} = \mathfrak{m}^c$ is maximal, because the evaluation homomorphism $\varphi_0\colon A[[x]] \longrightarrow A$ defined by $x \mapsto 0$, composed with the canonical projection $\pi\colon A \longrightarrow A/\mathfrak{n}$, is surjective and induces an isomorphism

$$A[[x]]/\mathfrak{m} \simeq A/\mathfrak{n}.$$

Thus, $A/\mathfrak{n}$ is a field.

**Solution E. 8.74.** 1. By hypothesis the equality holds for $s = 1$.
To apply induction, we assume it holds for $s \geq 1$ and prove it for $s + 1$.
One inclusion is always true, thus we focus on the non-trivial one.

Let $b \in I : (g^{m+s+1})$. Then $bg \in I : (g^{m+s}) = I : (g^m)$, and therefore, $bg^{m+1} \in I$. This leads to $b \in I : (g^{m+1}) = I : (g^m)$, as desired.

2. It is sufficient to prove that if $a \in (I : (g^m)) \cap (I, g^m)$, then $a \in I$.
Write $a = i + hg^m$ for some $i \in I$ and $h \in A$. Since $ig^m + hg^{2m} = ag^m \in I$, we have $h \in I : (g^{2m})$.
By the previous part, $h \in I : (g^m)$, and this yields $a \in I$.

**Solution E. 8.75.** Let $\Sigma$ be the set of prime ideals of $A$ partially ordered by set inclusion $\supseteq$. The set $\Sigma$ is not empty because $A \neq 0$ has at least a maximal ideal.
Let $\{\mathfrak{p}_\lambda\}_{\lambda \in \Lambda}$ be a descending chain of prime ideals. If we prove that $\mathfrak{p} = \bigcap_\lambda \mathfrak{p}_\lambda$ is a prime ideal, then the first statement immediately follows from Zorn's Lemma.

The intersection of any family of ideals is an ideal. Let $a, b \in A$ be such that $ab \in \mathfrak{p}$. Then, $ab \in \mathfrak{p}_\lambda$ for all $\lambda$, and we assume, by contradiction, that $a, b \notin \mathfrak{p}$. Then, there exist $\alpha, \beta \in \Lambda$ such that $a \notin \mathfrak{p}_\alpha$ and $b \notin \mathfrak{p}_\beta$. Without loss of generality, we can assume $\alpha \leq \beta$. Thus, $\mathfrak{p}_\alpha \supseteq \mathfrak{p}_\beta$ and $ab \notin \mathfrak{p}_\beta$, a contradiction.

To prove the statement regarding primes containing $I$, we can simply repeat the proof with the set $\Sigma_I$ consisting of prime ideals of $A$ that contain $I$, partially ordered by set inclusion $\supseteq$.

Additionally, we observe that the previous proof implies that the intersection of all minimal prime ideals of $A$ is equal to $\mathcal{N}(A)$, by **T.1.14**.1.

**Solution E. 8.76.** Let $b \in A \setminus \{0\}$ be such that $ab = 0$.
Since $\mathcal{N}(A) = (0)$, there exists a minimal prime $\mathfrak{p}$ which does not contain $b$. Therefore, $ab = 0 \in \mathfrak{p}$ yields $a \in \mathfrak{p}$, as required.

**Solution E. 8.77.** Since $63 = 3^2 7$, by **T.1.6**.7, we can write

$$\sqrt{I} = \sqrt{(I,9)} \cap \sqrt{(I,7)}.$$

The ideal

$$(I,9) = (9x^2 - y,\, 7y^2 + 2x + y,\, 9) = (y,\, 2x,\, 9) = (x,\, y,\, 9),$$

is primary by **T.1.7**.2, because $\sqrt{(I,\,9)} = (x,\, y,\, 3)$ is maximal.
Since

$$(I,7) = (2x^2 - y,\, 2x + y,\, 7) = (2x^2 + 2x,\, 2x + y,\, 7) = (x^2 + x,\, 2x + y,\, 7),$$

by **T.1.6**.3 and 7, we also have that

$$\begin{aligned}\sqrt{(I,7)} &= \sqrt{(x,\, 2x + y,\, 7) \cap (x + 1,\, 2x + y,\, 7)} \\ &= \sqrt{(x,\, y,\, 7)} \cap \sqrt{(x + 1,\, y - 2,\, 7)} \\ &= (x,\, y,\, 7) \cap (x + 1,\, y - 2,\, 7),\end{aligned}$$

where the last equality holds because both ideals under the radical are maximal.

1. The previous remarks, **T.1.14**.2, **E.8.75**, and **T.1.12**.2 immediately show that the minimal primes of $I$ are also maximal.
The conclusion follows from the one-to-one correspondence between ideals of $A/I$ and ideals of $A$ containing $I$.

2. Since (9) and (7) are comaximal, $(I,9)$ and $(I,7)$ are also comaximal, and **T.1.4** yields

$$I \subseteq (I,\, 9) \cap (I,\, 7) = (I,\, 9)(I,\, 7) \subseteq I.$$

Then, by the Chinese Remainder Theorem,

$$A/I \simeq A/(I,\, 9) \times A/(I,\, 7) = A/(x,\, y,\, 9) \times A/(x^2 + x,\, 2x + y,\, 7).$$

Note that $(x + 1)$ and $(x)$ are also comaximal. Thus,

$$\begin{aligned}(I,\, 7) &= (x + 1,\, 2x + y,\, 7)(x,\, 2x + y,\, 7) = (x + 1,\, y - 2,\, 7)(x,\, y,\, 7) \\ &= (x + 1,\, y - 2,\, 7) \cap (x,\, y,\, 7),\end{aligned}$$

and we can conclude that

$$A/I \simeq A/(x,\, y,\, 9) \times A/(x + 1,\, y - 2,\, 7) \times A/(x,\, y,\, 7) \simeq \mathbb{Z}/(9) \times (\mathbb{Z}/(7))^2.$$

3. By part 2, we have

$$I = (I, 9) \cap (I, 7) = (x,\, y,\, 9) \cap (x+1,\, y-2,\, 7) \cap (x,\, y,\, 7),$$

which is the desired decomposition.

4. Since $A/I$ is finite, if such an $f$ exists, then $B$ is a finite domain, hence a field.

**Solution E. 8.78.** Obviously, $(0)$ is a prime ideal of $\mathbb{Z}[x]$.
Let $\mathfrak{p} \neq (0)$ be a prime ideal of $\mathbb{Z}[x]$.
Consider the inclusion homomorphism $\mathbb{Z} \longrightarrow \mathbb{Z}[x]$. The ideal $\mathfrak{p}^c = \mathfrak{p} \cap \mathbb{Z}$ is a prime ideal, hence either $\mathfrak{p}^c = (0)$ or $\mathfrak{p}^c = (p)$ for some prime $p$ of $\mathbb{Z}$.

If $\mathfrak{p}^c = (0)$, then let $f \in \mathfrak{p} \setminus \{0\}$ be an element with the smallest degree. Note that, since $\mathfrak{p}$ is a prime ideal, $f$ must be irreducible. If there exists a polynomial $g \in \mathfrak{p} \setminus (f)$, then, viewing $f$ and $g$ as polynomials in $\mathbb{Q}[x]$, we have $\gcd(f, g) = 1$. Thus, there exist $r, s \in \mathbb{Q}[x]$ such that $rf + sg = 1$.
Let $m$ be the least common multiple of the denominators of the coefficients of $r$ and $s$.
Then, multiplying the equation $rf + sg = 1$ by $m$, we obtain

$$\tilde{r}f + \tilde{s}g = m \quad \text{for some} \quad \tilde{r},\, \tilde{s} \in \mathbb{Z}[x].$$

This implies $m \in \mathfrak{p} \cap \mathbb{Z} = (0)$, which is a contradiction.
Thus, $\mathfrak{p} = (f)$ with $f$ irreducible in $\mathbb{Z}[x]$.

Now, assume $\mathfrak{p}^c = (p)$. Then, either $\mathfrak{p} = (p)$, or there exists a polynomial $g \in \mathfrak{p} \setminus (p)$ of smallest degree.
Suppose, by contradiction, that $\overline{g}$ is reducible in $(\mathbb{Z}/(p))[x]$. Then, there exist $f,\, h \in \mathbb{Z}[x]$ such that $\overline{g} = \overline{f}\,\overline{h}$.
Without loss of generality, we can assume that $\deg f = \deg \overline{f}$, $\deg h = \deg \overline{h}$, and that both degrees are smaller than $\deg \overline{g} = \deg g$. Hence, there exists $r \in \mathbb{Z}[x]$ such that $fh = g + pr \in \mathfrak{p}$. This yields either $f \in \mathfrak{p} \setminus (p)$ or $h \in \mathfrak{p} \setminus (p)$, which contradicts the minimality of $\deg g$.
Therefore, $\overline{g}$ is irreducible in $\mathbb{Z}[x]/(p)$, and

$$\mathbb{Z}[x]/(p, g) \simeq (\mathbb{Z}/(p))[x]/(\overline{g})$$

is a field.
Hence, $(p, g)$ is a maximal ideal contained in $\mathfrak{p}$, thus it is equal to $\mathfrak{p}$.

Note that the latter is the only case where we obtain a maximal ideal.

## 17.2 Chapter 9

**Solution E. 9.1.** Clearly, a non-stationary descending chain in $\mathbb{N}^n$ is a non-empty subset of $\mathbb{N}^n$ without minimum.

Conversely, if $>$ is not a well-order, then there exists a non-empty subset $V \subset \mathbb{N}^n$ that does not have a minimum. Thus, for any given element $\mathbf{a}_1 \in V$, there exists $\mathbf{a}_2 \in V$ such that $\mathbf{a}_1 > \mathbf{a}_2$.
Continuing this reasoning, we can construct a descending chain of elements of $\mathbb{N}^n$ that is non-stationary.

**Solution E. 9.2.** All three relations are total on $\mathbb{N}^n$.
Let $\mathbf{a} \neq \mathbf{b}$. Then, either $|\mathbf{a}| \neq |\mathbf{b}|$, and the smallest between $\mathbf{a}$ and $\mathbf{b}$ is the one with smaller degree for the deglex and degrevlex orders, or $|\mathbf{a}| = |\mathbf{b}|$, and in this case, $a_i \neq b_i$ for some $i$.
In all cases there exists a first non-zero component in $\mathbf{a} - \mathbf{b}$, from the left side for the lex order and from the right side for the degrevlex order, which determines whether $\mathbf{a} > \mathbf{b}$ or $\mathbf{a} < \mathbf{b}$.

Now, we show that they are well-orders. Let $S$ be a non-empty subset of $\mathbb{N}^n$. We will prove that it has a minimum.
Consider first the lex order. Since the Well-order Principle holds for $\mathbb{N}$, there exists $\alpha_1 = \min\{a_1 : \mathbf{a} \in S\}$ which is the minimum of all the first coordinates of elements of $S$.
Define

$$S_1 = \{\mathbf{a} \in S : a_1 = \alpha_1\},$$

and consider $\alpha_2 = \min\{a_2 : \mathbf{a} \in S_1\}$.
Continuing this process, we find $\boldsymbol{\alpha} = (\alpha_1, \dots, \alpha_n) \in S$ that is the minimum of $S$ with respect to lex.

For the deglex order, we can follow a similar approach, but first we need to reduce the set $S$ to the subset

$$S_0 = \{\mathbf{a} \in S : |\mathbf{a}| = \min\{|\mathbf{b}| : \mathbf{b} \in S\}\}.$$

For the degrevlex order, we proceed with the reduction from $S$ to $S_0$ as we did above. Then, we define $\beta_n = \max\{c_n : \mathbf{c} \in S_0\}$ to be the maximum of all the last coordinates of the elements of $S_0$, and observe that this maximum exists because $S_0$ is finite.
Next, we consider $S_1' = \{\mathbf{a} \in S_0 : a_n = \beta_n\}$ and continue this process. Eventually, we find $\boldsymbol{\beta} = (\beta_1, \dots, \beta_n)$, which is the minimum of $S$ with respect to degrevlex.

Finally, let $\mathbf{a}, \mathbf{b}, \mathbf{c} \in \mathbb{N}^n$ be such that $\mathbf{a} > \mathbf{b}$ with respect to any of the given orders lex, deglex, degrevlex.
Since $|\mathbf{a}+\mathbf{c}| = |\mathbf{a}| + |\mathbf{c}|$, $|\mathbf{b}+\mathbf{c}| = |\mathbf{b}| + |\mathbf{c}|$ and $(\mathbf{a}+\mathbf{c}) - (\mathbf{b}+\mathbf{c}) = \mathbf{a} - \mathbf{b}$, addition preserves the ordering of $\mathbf{a}$ and $\mathbf{b}$.

**Solution E. 9.3.** By definition of monomial ordering, we only need to prove that $>$ is a well-order if and only if $\mathbf{a} \geq \mathbf{0}$ for all $\mathbf{a} \in \mathbb{N}^n$.

If $>$ is a well-order, then there exists a minimum $\overline{\mathbf{a}} \in \mathbb{N}^n$. If $\mathbf{0} > \overline{\mathbf{a}}$, then we have a descending chain $\mathbf{0} > \overline{\mathbf{a}} > 2\overline{\mathbf{a}} > 3\overline{\mathbf{a}} > \ldots$ which is non-stationary. This contradicts the well-order property of $>$, see **E.9.1**.

Conversely, suppose that $\mathbf{a} \geq \mathbf{0}$ for all $\mathbf{a} \in \mathbb{N}^n$.
Let $\Sigma \subset \mathbb{N}^n$ be a non-empty subset, and let

$$E = \Sigma + \mathbb{N}^n = \{\mathbf{a}+\mathbf{b}\colon\ \mathbf{a} \in \Sigma,\, \mathbf{b} \in \mathbb{N}^n\}$$

be the $\mathcal{E}$-subset generated by the elements of $\Sigma$.
By Dickson's Lemma **T.2.3**, the $\mathcal{E}$-subset $E$ has a minimal finite border $F = \{\mathbf{a}_1, \ldots, \mathbf{a}_m\}$. Clearly, $F \subseteq \Sigma$. By reordering the elements of $F$, if necessary, we can assume $\mathbf{a}_1 > \ldots > \mathbf{a}_m$.
Now, we will prove that $\mathbf{a}_m$ is the minimum of $\Sigma$. For any $\mathbf{b} \in E$ there exist $i \in \{1, \ldots, m\}$ and $\mathbf{c} \in \mathbb{N}^n$ such that $\mathbf{b} = \mathbf{a}_i + \mathbf{c}$.
Since, by hypothesis, $\mathbf{c} \geq \mathbf{0}$, we have $\mathbf{b} = \mathbf{a}_i + \mathbf{c} \geq \mathbf{a}_i \geq \mathbf{a}_m$.

Note that this result can be used to simplify the proof of **E.9.2**.

**Solution E. 9.4.** Consider a fixed monomial ordering on the set of monomials in the variables $x_1, \ldots, x_n$, and let $G = \{g_1, \ldots, g_s\}$ be a Gröbner basis of $I \subseteq K[x_1, \ldots, x_n]$ with respect to this order. Observe that $\{g_1, \ldots, g_s\}$ is a generating set of $I$ by **T.2.10**, and therefore, it is also a generating set of $I^e$.

We will show that $\mathrm{Lt}(I)^e = \mathrm{Lt}(I^e)$.
We already know that $\mathrm{Lt}(I)^e = \mathrm{Lt}(G)^e$. Since all monomials of $\mathrm{Lt}(G)$ are clearly elements of $\mathrm{Lt}(I^e)$, we immediately have $\mathrm{Lt}(I)^e \subseteq \mathrm{Lt}(I^e)$.

To prove the opposite inclusion, consider a basis $\{e_\lambda\}_{\lambda \in \Lambda}$ of $K'$ over $K$. For any $f \in K'[x_1, \ldots, x_n]$, we can write

$$f = \sum_{\mathbf{a} \in \mathbb{N}^n} c_{\mathbf{a}} X^{\mathbf{a}},$$

where $X^{\mathbf{a}} = x_1^{a_1} \cdots x_n^{a_n}$ and $c_{\mathbf{a}} \neq 0$ for finitely many $\mathbf{a}$.
Since $c_{\mathbf{a}} \in K'$, we can write $c_{\mathbf{a}} = \sum_{\lambda \in \Lambda} c_{\mathbf{a},\lambda} e_\lambda$, with $c_{\mathbf{a},\lambda} \in K$ and $c_{\mathbf{a},\lambda} \neq 0$ for finitely many $\lambda$.
Thus,

$$f = \sum_{\mathbf{a} \in \mathbb{N}^n} c_{\mathbf{a}} X^{\mathbf{a}} = \sum_{\mathbf{a} \in \mathbb{N}^n} \sum_{\lambda \in \Lambda} c_{\mathbf{a},\lambda} e_\lambda X^{\mathbf{a}} = \sum_{\lambda \in \Lambda} e_\lambda \sum_{\mathbf{a} \in \mathbb{N}^n} c_{\mathbf{a},\lambda} X^{\mathbf{a}} = \sum_{\lambda \in \Lambda} e_\lambda f_\lambda$$

for some $f_\lambda \in K[x_1, \ldots, x_n]$.
Therefore, every element of $I^e$ can be expressed as

$$f = \sum_{i=1}^{s} f_i g_i = \sum_i \left( \sum_\lambda e_\lambda f_{i,\lambda} \right) g_i = \sum_\lambda e_\lambda \sum_i f_{i,\lambda} g_i$$

with $f_i \in K'[x_1, \ldots, x_n]$ and $f_{i,\lambda} \in K[x_1, \ldots, x_n]$.

We note that there exists $\lambda_0$ such that $\mathrm{lt}(f) = \alpha h$, with $\alpha \in K'$ and $h = \mathrm{lm}(\sum_i f_{i,\lambda_0} g_i) \in \mathrm{Lt}(I)$. In fact, $\mathrm{lm}(f)$ is given by the greatest leading monomial of the polynomials $\sum_i f_{i,\lambda} g_i$, because, if some of these polynomials share the maximal leading monomial, then no cancellation can occur due to the linear independence of the $e_\lambda$ over $K$.
Therefore, we have shown the opposite inclusion.

Alternatively, by Buchberger's Criterion, all of the $S$-polynomials constructed using $g_1, \ldots, g_s$ reduce to 0. Thus, they also reduce to 0 as elements of $K'[x_1, \ldots, x_n]$. Again by Buchberger's Criterion, we can conclude that $\{g_1, \ldots, g_s\}$ is a Gröbner basis of $I^e$.

The final assertion follows immediately from the previous statement and the definition of staircase of an ideal.

**Solution E. 9.5.** We divide $f$ by $F = \{f_1, f_2\}$. If we divide $f$ by $f_1$ we obtain $f = x_1 x_2 f_1 + 0$. If we start dividing by $f_2$, then the division yields $f = (x_1^2 + x_2) f_2 + x_2^3$.
Thus, the remainder of the division of $f$ by $F$ is not unique.

**Solution E. 9.6.** We provide a proof that does not use Buchberger's Criterion. First, let us show that $G$ is a Gröbner basis of $I$ with respect to $>_1$.
By contradiction, assume there exists $f \in I$ such that

$$\mathrm{lt}_{>_1}(f) \notin \mathrm{Lt}_{>_1}(G) = (\mathrm{lt}_{>_1}(g_1), \mathrm{lt}_{>_1}(g_2)) = (z,\, y).$$

Then, $z, y \nmid \mathrm{lt}_{>_1}(f)$. Since $>_1$ is a lex order, $z$ and $y$ do not appear as divisors of the remaining terms of $f$, that is, $f = f(x)$. We can also write $f = u_1(z + x) + u_2(y - x)$ for some $u_1, u_2 \in \mathbb{Q}[x, y, z]$.
By substituting $y = x$, we obtain

$$f = f(x) = u_1(x, x, z)(z + x).$$

Thus, $z + x$ divides $f$, which is not possible since $f$ does not have terms containing $z$.

As for the second order, let us rewrite $g_1 = x + z$ and $g_2 = -x + y$. Then, $y + z = (x + z) + (-x + y) = g_1 + g_2 \in I$ and $y \in \mathrm{Lt}_{>_2}(I)$.
On the other hand, $\mathrm{lm}_{>_2}(g_1) = \mathrm{lm}_{>_2}(g_2) = x$, and accordingly,

$$\mathrm{Lt}_{>_2}(I) \supseteq (x,\, y) \supsetneq (x) = (\mathrm{lt}_{>_2}(g_1), \mathrm{lt}_{>_2}(g_2)) = \mathrm{Lt}_{>_2}(G).$$

**Solution E. 9.7.** Assume that $I$ is monomial.
Then, $f = \sum_{\mathbf{a}} c_{\mathbf{a}} X^{\mathbf{a}} \in I$ if and only if $X^{\mathbf{a}} \in I$ for all $\mathbf{a}$ such that $c_{\mathbf{a}} \neq 0$, see **T.2.1**. Thus, given any monomial ordering $>$, we have $\mathrm{Lt}_>(I) = I$.
The minimal generating set $G(I)$ of $\mathrm{Lt}_>(I)$ is a minimal Gröbner basis of $I$, and since $G(I)$ consists of monomials, it is obviously reduced.

Conversely, if there exists a Gröbner basis of $I$ consisting of monomials, then $I$ surely has a monomial generating set.

**Solution E. 9.8.** We will show that the reduced Gröbner basis with respect to the given order is

$$G = \{x^2 - xy,\ xz - y^2,\ yz^2 - z^4,\ xy^2 - y^3,\ y^4 - z^7,\ y^3z - z^7,\ z^9 - z^8\}.$$

Let $f_1 = x^2 - xy,\ f_2 = xz - y^2,\ f_3 = yz^2 - z^4$, and let $G_0 = \{f_1, f_2, f_3\}$. In the following, let $S_{ij}$ denote the $S$-polynomial of $f_i$ and $f_j$.

$S_{12} = zf_1 - xf_2 = xy^2 - xyz \xrightarrow{f_2} xy^2 - y^3 = f_4$,
which is reduced with respect to $G_0$. Set $G_1 = G_0 \cup \{f_4\}$.
$S_{14} = y^2 f_1 - xf_4 = xy^3 - xy^3 = 0$.
$S_{24} = y^2 f_2 - zf_4 = -y^4 + y^3z$,
which is reduced with respect to $G_1$. Set $f_5 = y^4 - y^3z$ and $G_2 = G_1 \cup \{f_5\}$.
$S_{35} = y^3 f_3 - z^2 f_5 = y^3z^3 - y^3z^4 \xrightarrow{f_3} -z^{10} + z^9$,
which is reduced with respect to $G_2$. Set $f_6 = z^{10} - z^9$ and $G_3 = G_2 \cup \{f_6\}$.
$S_{23} = yzf_2 - xf_3 = xz^4 - y^3z \xrightarrow{f_2, f_3} yz^5 - y^3z \xrightarrow{f_3} -y^3z + z^7$,
which is reduced with respect to $G_3$. Set $f_7 = y^3z - z^7$ and $G_4 = G_3 \cup \{f_7\}$.

Now, we observe that we can reduce $f_5$ with respect to $G_4$ by using $f_7$. We obtain $f_{5'} = y^4 - z^7$ and set

$$G_5 = (G_4 \cup \{f_{5'}\}) \setminus \{f_5\}.$$

We continue constructing the basis using the criterion of $S$-polynomials and observe that all $S$-polynomials computed so far reduce to 0 with respect to $G_5$. We also have
$S_{26} = z^9 f_2 - xf_6 = xz^9 - y^2z^9 \xrightarrow{f_2, f_3, f_6}_* 0$
$S_{27} = y^3 f_2 - xf_7 \xrightarrow{f_2, f_3, f_{5'}, f_3, f_6}_* 0$.
$S_{34} = xyf_3 - z^2 f_4 = -xyz^4 + y^3z^2 \xrightarrow{f_2, f_3} z^9 - z^8 = f_{6'}$,
which is reduced with respect to $G_5$ and divides $f_6$. Set

$$G_6 = (G_5 \cup \{f_{6'}\}) \setminus \{f_6\}.$$

We need to continue computing $S$-polynomials.
We know that $S_{12},\ S_{14},\ S_{23},\ S_{24},\ S_{27},\ S_{34} \xrightarrow{G_6}_* 0$.
Also,

$S_{26'} \xrightarrow{f_2, f_3, f_{6'}}_* 0$,

$S_{35'} \xrightarrow{f_3, f_{6'}}_* 0$,

$S_{36'} \xrightarrow{f_3, f_{6'}}_* 0$,

$S_{37} \xrightarrow{f_3}_* 0$,

$S_{45'} \xrightarrow{f_2, f_{5'}, f_3, f_{6'}}_* 0$,

$S_{47} \xrightarrow{f_2, f_{5'}, f_3, f_{6'}}_* 0,$

$S_{5'7} \xrightarrow{f_3, f_{6'}}_* 0,$

$S_{6'7} \xrightarrow{f_3, f_{6'}}_* 0.$

Moreover, since the $S$-polynomial of two polynomials with coprime leading monomials reduces to 0, see **T.2.16**, we have

$$S_{13},\ S_{15'},\ S_{16'},\ S_{17},\ S_{25'},\ S_{46'},\ S_{5'6'} \xrightarrow{G_6}_* 0.$$

Thus, $G_6 = G$ is the desired basis, which is already reduced by construction.

**Solution E. 9.9.** The reduced Gröbner basis of $I$ with respect to the lex order with $x > y > z$ is

$$G = \{xz + 2z,\ y - z,\ z^2 - z\}.$$

Since $f \xrightarrow{G}_* -9z \neq 0$, the polynomial $f$ does not belong to $I$.

Alternatively, we can observe that $g(-2, 1, 1) = 0$ for all $g \in G$. Therefore, $(-2, 1, 1) \in \mathbb{V}(I)$. On the other hand, since $f(-2, 1, 1) = -9 \neq 0$, we have $f \notin \mathbb{I}(\mathbb{V}(I)$, which implies $f \notin I$ by **T.3.1**.2.

**Solution E. 9.10.** The reduced Gröbner basis of $I$ with respect to the lex order with $x > y$ is

$$G = \{x + y^4 + y,\ y^6 + y^3 + 1\}.$$

The reduction of $f_1 - f_2 = x^3 - x^2y + xy^2 - xy + y^5 - y^2$ modulo $G$ is 0, and therefore, $\overline{f_1} = \overline{f_2}$.

**Solution E. 9.11.** The generating set

$$G = \{f_1 = yx^2 - y + x,\ f_2 = y^2 - yx - x^2,\ f_3 = x^3 + y - 2x\}$$

of $I$ is already a Gröbner basis with respect to the deglex order with $y > x$, and $f$ is reduced modulo $G$.

To verify whether $\overline{f}$ is invertible in $\mathbb{Q}[x, y]/I$ it is sufficient to apply **T.2.24**. Let then $H = G \cup \{f\}$.

Since

$$\begin{aligned}
S(f_1, f) &\xrightarrow{G} x^2 + x = -(f_1 - x^2 f + f_3) = g_1,\\
S(f_2, f) &\xrightarrow{G\cup\{g_1\}} 2x + 1 = f_2 - yf + 2xf - g_1 + f\\
&\qquad\qquad = f_2 - (y - 2x - 1)f - g_1 = g_2,\\
S(g_1, g_2) &\xrightarrow{G\cup\{g_1, g_2\}} -\tfrac{1}{2} = 2g_1 - xg_2 - \tfrac{1}{2}g_2\\
&\qquad\qquad = 2g_1 - (x + \tfrac{1}{2})g_2,
\end{aligned}$$

we can conclude that $1 \in (I, f)$ and $\overline{f}$ is invertible in $\mathbb{Q}[x, y]/I$.

Now, we compute its inverse working backwards:

$$\begin{aligned}1 &= (-2)\cdot\left(-\tfrac{1}{2}\right) = -2\left(2g_1 - \left(x+\tfrac{1}{2}\right)g_2\right)\\ &= -2\left[2\left(x^2f - f_1 - f_3\right) - \left(x+\tfrac{1}{2}\right)\left((2x-y+1)f + f_2 - g_1\right)\right] =\\ &= -2\left[f\left(2x^2 - \left(x+\tfrac{1}{2}\right)(2x-y+1)\right) + \left(x+\tfrac{1}{2}\right)\left(x^2f - f_1 - f_3\right) + h_1\right]\\ &= f(-2x^3 - 2xy - x^2 - y + 4x + 1) + h_2\\ &= f(-2xy - x^2 + y + 1) + h_3,\end{aligned}$$

where $h_1$, $h_2$, $h_3 \in I$ and, in the last equality, we have reduced the polynomial that multiplies $f$ modulo $I$ by adding $2f_3$.
Thus,

$$\overline{f}^{-1} = \overline{-2xy - x^2 + y + 1}.$$

**Solution E. 9.12.** 1. The desired Gröbner basis is

$$G = \{f_1 = x^2y + z,\ f_2 = xz + y,\ f_3 = xy^2 - z^2,\ f_4 = y^3 + z^3\}.$$

2. Given that

$$S(f_1, f_2) = zf_1 - xyf_2 = -f_3,$$

$$S(f_2, f_3) = y^2f_2 - zf_3 = f_4 = y^2f_2 + z^2f_1 - xyzf_2 = z^2f_1 + (-xyz + y^2)f_2,$$

the matrix $M$ such that $M(f_1, f_2)^{\mathrm{t}} = (f_1, f_2, f_3, f_4)^{\mathrm{t}}$ is

$$M = \begin{pmatrix} 1 & 0\\ 0 & 1\\ -z & xy\\ z^2 & -xyz + y^2\end{pmatrix}.$$

3. It is well-known that the quotients of the division of $f$ by $G$ are not unique. Clearly, $f = y^2f_2 = 0f_1 + y^2f_2 + 0f_3 + 0f_4 = 0f_1 + y^2f_2$.

On the other hand, $f = 0f_1 + 0f_2 + zf_3 + f_4$, and in order to compute the coefficients of $f$ with respect to $f_1$, $f_2$, we can use the transition matrix obtained earlier

$$M^{\mathrm{t}}\begin{pmatrix}0\\0\\z\\1\end{pmatrix} = \begin{pmatrix}0\\y^2\end{pmatrix}.$$

A third possible way to express $f$ is $f = yf_1 - zf_2 - (x - z)f_3 + f_4$ (are there infinite ways of writing $f$ as such a combination? Why?).
From this representation we can recover the coefficients of $f$ with respect to the original generators using $M$

$$M^{\mathrm{t}}\begin{pmatrix}y\\-z\\-x+z\\1\end{pmatrix} = \begin{pmatrix}xz + y\\-x^2y + y^2 - z\end{pmatrix}.$$

**Solution E. 9.13.** The reduced Gröbner basis of $I$ with respect to the lex order with $x > y$ is

$$\{x^2 + 2y^2 - 3,\ xy - y^2,\ y^3 - y\}.$$

From the Elimination Theorem **T.2.25** it follows that $I \cap \mathbb{C}[y] = (y^3 - y)$.

**Solution E. 9.14.** We outlined a strategy to perform such a computation right after **T.2.26**.
Let $g_1 = x(x+y)^2$, $g_2 = y$, $f_1 = x^2$, and $f_2 = x + y$.
By **E.8.21**.5, we have

$$I : J = (I : (f_1)) \cap (I : (f_2)).$$

The ideal $I = (x^3,\ y)$ is monomial, then by **T.2.6**.3, $I : (x^2) = (x,\ y)$.

By **E.8.22**, we can obtain $I : (f_2)$ by computing $\frac{1}{f_2}(I \cap (f_2))$. We do this using **T.2.26**.1. A Gröbner basis of $(tI, (1-t)f_2)$ with respect to the lex order with $t > x > y$ is

$$\{tx - x - y,\ ty,\ x^3 + y^3,\ xy + y^2\}.$$

Therefore,

$$\frac{1}{x+y}(I \cap (f_2)) = \frac{1}{x+y}(x^3 + y^3,\ xy + y^2) = (x^2 - xy + y^2,\ y) = (x^2,\ y).$$

Thus,

$$I : J = (x,\ y) \cap (x^2,\ y) = (x^2,\ y).$$

**Solution E. 9.15.** By applying **T.1.6**.7 twice, we have

$$\begin{aligned}\sqrt{I} &= \sqrt{(x^2 + y^2,\ y^3)} \cap \sqrt{(x^2 + y^2,\ x^3 + y)}\\ &= (x,\ y) \cap \sqrt{(x^6 + x^2,\ x^3 + y)}\\ &= (x,\ y) \cap \sqrt{(x^4 + 1,\ x^3 + y)}.\end{aligned}$$

The ideal $(x^4 + 1,\ x^3 + y)$ is radical.
In fact, $K[x,y]/(x^4 + 1,\ x^3 + y) \simeq K[x]/(x^4 + 1)$ has no non-trivial nilpotents, since $x^4 + 1$ is squarefree by the hypothesis on the characteristic.
Thus,

$$\sqrt{I} = (x,\ y) \cap (x^4 + 1,\ x^3 + y)$$

and $f \notin \sqrt{I}$ because it does not belong to $(x^4 + 1,\ x^3 + y)$.
Indeed, the polynomials $x^4 + 1$, $x^3 + y$ form a Gröbner basis for the ideal they generate with respect to the lex order with $y > x$, and we conclude because $f \neq 0$ is reduced with respect to such a basis.

**Solution E. 9.16.** 1. Before proceeding with the computation of a Gröbner basis of $I$, observe that we may reduce $x^2y^2z^4$ by using $xy - 2$ and obtain

$z^4 \in I$. In this way we have found a simpler set of generators of $I$, namely

$$I = (f_1,\, f_2,\, f_3), \text{ with } f_1 = z^4,\ f_2 = x^2 + y^2 + z^2 - 1, \text{ and } f_3 = xy - 2.$$

It is sufficient to compute the following $S$-polynomials, with respect to the lex order with $x > y > z$:
$S(f_2, f_3) = yf_2 - xf_3 = 2x + y^3 + yz^2 - y = f_4;$
$S(f_2, f_4) = 2f_2 - xf_4 \xrightarrow{f_3} 0;$
$S(f_3, f_4) = 2yf_4 - 2f_3 = y^4 + y^2z^2 - y^2 + 4 = f_5.$
Since the leading monomials of $f_1$, $f_4$, $f_5$ are pairwise coprime,

$$G = \left\{f_1,\, \tfrac{1}{2}f_4,\, f_5\right\}$$

is a Gröbner basis of $I$, see **T.2.16**.
Moreover, it can be easily verified that $G$ is reduced.

We have $\dim_{\mathbb{C}} \mathbb{C}[x, y, z]/I = 16$, and a basis of this vector space is given by the images of the elements

$$1,\, y,\, z,\, y^2,\, yz,\, z^2,\, y^3,\, y^2z,\, yz^2,\, z^3,\, y^3z,\, y^2z^2,\, yz^3,\, y^3z^2,\, y^2z^3,\, y^3z^3.$$

2. Since $\mathbb{C}$ is algebraically closed, to prove $I + J = (1)$, we can show that $\mathbb{V}(I + J) = \emptyset$ by the Weak Nullstellensatz **T.3.13**. Thus, we can solve the system of polynomial equations found in part 1, *i.e.*, $f_1 = f_4 = f_5 = 0$, and verify that no solution satisfies the equations of $J$.

Alternatively, we can observe that since $z \in \sqrt{I + J}$, the polynomial $g_1 = 3x^3 - 2$ is in $\sqrt{I + J}$.
We also have

$$y^3 g_1 = 3x^3y^3 - 2y^3 \in \sqrt{I + J}.$$

Using the polynomial $xy - 2$, we obtain $2y^3 - 24 \in \sqrt{I + J}$.
Since $z \in \sqrt{I + J}$, we obtain

$$1 = \gcd(y^3 - 12,\, f_5 - y^2z^2) \in (y^3 - 12,\, f_5 - y^2z^2) \subseteq \sqrt{I + J},$$

and the conclusion follows from **T.1.6**.5.

**Solution E. 9.17.** Consider the lex order with $x > y > a$ and write

$$I = (f_1 = x + y - a,\ f_2 = x^2 + y^2 - a^2,\ f_3 = x^3 + y^3 - a^5).$$

Reducing $f_2$ by $f_1$, we obtain $f_2 \xrightarrow{f_1}_* 2y^2 - 2ya$.
Setting $f_4 = y^2 - ya$, we also have $f_2 = (x - y + a)(x + y - a) + 2f_4$ and $I = (f_1,\, f_4,\, f_3)$.
Now, reduce $f_3 \xrightarrow{f_1, f_4}_* -a^5 + a^3$ and let $f_5 = a^5 - a^3$. Then,

$$I = (f_1,\, f_4,\, f_5).$$

Observe that these generators form a Gröbner basis since their leading monomials $x$, $y^2$ and $a^5$ are pairwise coprime, see **T.2.16**.

Note that, by the Weak Nullstellensatz, the system certainly has solutions since $I \neq (1)$, and there are a finite number of them since $\mathrm{Lt}(I)$ contains pure powers of each variable, see **T.3.17**.
Now, it is easy to find the solutions of the equivalent upper-triangular system of polynomial equations $f_1 = f_4 = f_5 = 0$. In conclusion:

i) if $a = 0$, the only solution is $(0, 0)$;

ii) if $a = -1$ there are two solutions, namely $(-1, 0)$ and $(0, -1)$;

iii) if $a = 1$ there are two solutions, namely $(1, 0)$ and $(0, 1)$.

**Solution E. 9.18.** 1. The reduced Gröbner basis of $I$ is

$$G = \{x^2y + xz + yz,\, xyz^2,\, xz^3 + yz^3,\, y^2z\}.$$

2. The nilpotents of $A$ are the images in $A$ of the elements in

$$\begin{aligned}\sqrt{I} &= \sqrt{(x^2y,\, z)} \cap \sqrt{(x^2y + xz + yz,\, xyz^2,\, xz^3 + yz^3,\, y^2)}\\ &= (xy,\, z) \cap (xz,\, y) = (x,\, z) \cap (y,\, z) \cap (x,\, y) = (xy,\, xz,\, yz).\end{aligned}$$

3. Since $x^2y^3,\, y^3z \xrightarrow{G}_* 0$, we have $J = (x^2y^3,\, y^3z) \subseteq I$.
Also, for all $g \in G$, we have $g \xrightarrow{x^2,\, z}_* 0$, and thus, $I \subseteq (x^2,\, z)$.
Since

$$\begin{aligned}\mathrm{Lt}(J) = (x^2y^3,\, y^3z) &\subsetneq \mathrm{Lt}(I) = \mathrm{Lt}(G)\\ &= (x^2y,\, xyz^2,\, xz^3,\, y^2z) \subsetneq (x^2,\, z)\\ &= \mathrm{Lt}(x^2,\, z),\end{aligned}$$

the above inclusions are strict, *i.e.*, $J \subsetneq I \subsetneq (x^2,\, z)$.

**Solution E. 9.19.** 1. Let $\alpha_1, \ldots, \alpha_m \in \overline{K}$ be the roots of $f$.
From **T.3.9**.1 we easily obtain

$$\begin{aligned}\mathrm{Res}(f, g_1\, g_2) &= a_m^{\deg g_1 + \deg g_2} \prod_{i=1}^{m} g_1(\alpha_i) g_2(\alpha_i)\\ &= \left(a_m^{\deg g_1} \prod_{i=1}^{m} g_1(\alpha_i)\right) \left(a_m^{\deg g_2} \prod_{i=1}^{m} g_2(\alpha_i)\right)\\ &= \mathrm{Res}(f, g_1)\, \mathrm{Res}(f, g_2).\end{aligned}$$

2. By applying **T.3.9**.1 once again, we compute

$$\begin{aligned}\operatorname{Res}(f, g_1 f + g_2) &= a_m^N \prod_{i=1}^{m} (g_1 f + g_2)(\alpha_i) \\ &= a_m^N \prod_{i=1}^{m} g_2(\alpha_i) = a_m^{N-\deg(g_2)} \operatorname{Res}(f, g_2).\end{aligned}$$

**Solution E. 9.20.** All the statements are consequences of **T.3.9**.
Let $f = a_m \hat{f}$ and $g = b_n \hat{g}$, where $\hat{f} = \prod_{i=1}^{m}(x - \alpha_i)$ and $\hat{g} = \prod_{j=1}^{n}(x - \beta_j)$.

1. $$\begin{aligned}\operatorname{Res}_y(f(x-y),\, g(y)) &= (-1)^{mn} b_n^m \prod_{j=1}^{n} f(x - \beta_j) \\ &= (-1)^{mn} a_m^n b_n^m \prod_{j=1}^{n} \hat{f}(x - \beta_j) \\ &= (-1)^{mn} a_m^n b_n^m \prod_{i=1}^{m} \prod_{j=1}^{n} (x - (\alpha_i + \beta_j)).\end{aligned}$$

2. $$\begin{aligned}\operatorname{Res}_y(f(x+y),\, g(y)) &= (-1)^{mn} a_m^n b_n^m \prod_{j=1}^{n} \hat{f}(x + \beta_j) \\ &= (-1)^{mn} a_m^n b_n^m \prod_{i=1}^{m} \prod_{j=1}^{n} (x - (\alpha_i - \beta_j)).\end{aligned}$$

3. $$\begin{aligned}\operatorname{Res}_y\left(y^m f\left(\frac{x}{y}\right),\, g(y)\right) &= (-1)^{mn} b_n^m \prod_{j=1}^{n} \beta_j^m f\left(\frac{x}{\beta_j}\right) \\ &= (-1)^{mn} a_m^n b_n^m \prod_{i=1}^{m} \prod_{j=1}^{n} \beta_j \left(\frac{x}{\beta_j} - \alpha_i\right) \\ &= (-1)^{mn} a_m^n b_n^m \prod_{i=1}^{m} \prod_{j=1}^{n} (x - \alpha_i \beta_j).\end{aligned}$$

4. $$\begin{aligned}\operatorname{Res}_y(f(xy),\, g(y)) &= (-1)^{mn} b_n^m \prod_{j=1}^{n} f(x\beta_j) \\ &= (-1)^{mn} a_m^n b_n^m \prod_{i=1}^{m} \prod_{j=1}^{n} (x\beta_j - \alpha_i) \\ &= (-1)^{mn} a_m^n b_n^m \prod_{i=1}^{m} \prod_{j=1}^{n} \beta_j (x - \frac{\alpha_i}{\beta_j}),\end{aligned}$$

since $g(0) \neq 0$ implies $\beta_j \neq 0$ for all $j$.

**Solution E. 9.21.** By **E.9.19**.1, we have

$$\operatorname{Res}(f, x^k g) = \operatorname{Res}(f, x^k) \operatorname{Res}(f, g) = (\operatorname{Res}(f, x))^k \operatorname{Res}(f, g).$$

On the other hand, $\operatorname{Res}(f, x) = (-1)^{\deg f} f(0)$ by **T.3.9**.1.

Therefore, we can conclude that

$$\operatorname{Res}(f, x^k g) = (-1)^{k \deg f} f(0)^k \operatorname{Res}(f, g) = \operatorname{Res}(f, g),$$

since $k$ is even.

**Solution E. 9.22.** 1. From **T.3.7**, we obtain $(p) \subseteq (f, g) \cap \mathbb{Z}$. Then, since $(p)$ is maximal, it is enough to prove that $(f, g) \neq (1)$.

Assume, by contradiction, that there exist $a$, $b \in A$ such that $af + bg = 1$. Then, $\overline{a}\overline{f} + \overline{b}\overline{g} = \overline{1}$ in $(\mathbb{Z}/(p))[x]$. Since $f$ and $g$ are monic, the Sylvester matrix of $\overline{f}$ and $\overline{g}$ is exactly the reduction modulo $p$ of $\operatorname{Syl}(f, g)$.
Therefore, $\operatorname{Res}(\overline{f}, \overline{g}) = \overline{\operatorname{Res}(f, g)} = \overline{0}$, and this contradicts $\gcd(\overline{f}, \overline{g}) = \overline{1}$ by **T.3.9**.4.

2. If $f = x^2 - 4x + 1$ and $g = x^2 - x$, then

$$\operatorname{Res}(f, g) = \det \begin{pmatrix} 1 & -4 & 1 & 0 \\ 0 & 1 & -4 & 1 \\ 1 & -1 & 0 & 0 \\ 0 & 1 & -1 & 0 \end{pmatrix} = -2,$$

then, from the previous part, $I \cap \mathbb{Z} = (2)$.
Moreover, $I = I + (2) = (x - 1,\, 2)$, and $A/I \simeq \mathbb{Z}/(2)$.

**Solution E. 9.23.** Let $r = \operatorname{Res}(f, f') \in \mathbb{Z}$. Since $\gcd(f, f') = 1$, from **T.3.9**.4 we obtain $r \neq 0$. Furthermore, there exist polynomials $a$, $b \in \mathbb{Z}[x]$ such that $af + bf' = r$, see **T.3.7**. By considering the last equality modulo $p$, where $p$ is a prime of $\mathbb{Z}$ that does not divide $r$, we have

$$\overline{a}\overline{f} + \overline{b}\,\overline{f'} = \overline{r} \;\text{ in }\; (\mathbb{Z}/(p))[x],$$

that is, $\gcd(\overline{f}, \overline{f}') = 1$.
Thus, $(\mathbb{Z}/(p))[x]/(\overline{f})$ is reduced by **E.8.58** and **E.8.59**.1.

**Solution E. 9.24.** The reduced Gröbner basis with respect to the lex order with $x > y > z$ is

$$\{x - yz,\, yz^2 - y\}.$$

Therefore, $\sqrt{I} = (x,\, y) \cap (x + y,\, z + 1) \cap (x - y,\, z - 1)$ and

$$\mathbb{V}(I) = \mathbb{V}(\sqrt{I}) = \mathbb{V}(x,\, y) \cup \mathbb{V}(x + y,\, z + 1) \cup \mathbb{V}(x - y,\, z - 1).$$

Now, assume that $K$ is infinite, and let $f \in \mathbb{I}(\mathbb{V}(x, y))$. Then, we can write

$$f = xg(x, y, z) + yh(y, z) + r(z)$$

and $f(0, 0, a) = r(a) = 0$ for all $a \in K$ implies $r(z) = 0$, that is, $f \in (x,\, y)$. Since the opposite containment is obvious, $\mathbb{I}(\mathbb{V}(x,\, y)) = (x,\, y)$ is prime and $\mathbb{V}(x,\, y)$ is irreducible by **T.3.3**.
As for the other two components,

$$\mathbb{V}(x \pm y, z \pm 1) = \{(a, \mp a, \mp 1) : a \in K\} \subset K^3,$$

the proof is similar.
For all $f \in \mathbb{I}(\mathbb{V}(x \pm y, z \pm 1))$ we write

$$f = (z \pm 1)g(x, y, z) + (x \pm y)h(x, y) + r(y).$$

Since $0 = f(a, \mp a, \mp 1) = r(\mp a)$ for infinitely many $a$, we have $r = 0$.

Now, let $K$ be finite.
Obviously, $\mathbb{V}(x, y) = \{(0, 0, a) : a \in K\} \subset K^3$ decomposes as the finite union of $|K|$ points, which coincide with its irreducible components.
The same is true for $\mathbb{V}(x \pm y, z \pm 1)$.

**Solution E. 9.25.** The reduced Gröbner basis of $I$ with respect to the lex order with $x > y > z > t$ is

$$G = \{x^2 - zt,\ y - zt,\ zt^2 - t\}.$$

1. We have

$$\begin{aligned}\sqrt{I} &= \sqrt{(x^2 - zt,\ y - zt,\ t)} \cap \sqrt{(x^2 - zt,\ y - zt,\ zt - 1)} \\ &= (x,\ y,\ t) \cap (x + 1,\ y - 1,\ zt - 1) \cap (x - 1,\ y - 1,\ zt - 1).\end{aligned}$$

The ideals in this decomposition are prime and thus determine the irreducible components of $\mathbb{V}(I)$, since $\mathbb{I}(\mathbb{V}(I)) = \sqrt{I}$ by the Strong Nullstellensatz **T.3.14**.

2. With respect to the given monomial ordering

$$\mathrm{lt}(f) = xt \notin \mathrm{Lt}(I) = (x^2,\ y,\ zt^2).$$

Thus, $f \notin I$.

**Solution E. 9.26.** The reduced Gröbner basis of $I$ with respect to the lex order with $x > y > z$ is

$$\{x + y + z - 1,\ y^2 + yz - y + z^2 - z\}.$$

1. Since the initial ideal of $I$ does not contain any pure power of $z$, by **T.3.17** $\dim_{\mathbb{C}} \mathbb{C}[x, y, z]/I$ is infinite.

2. We know that $\mathbb{V}(I) \cap \mathbb{V}(z - 1) = \mathbb{V}(I, z - 1)$.
Since

$$(x + y + z - 1,\ y^2 + yz - y + z^2 - z,\ z - 1) = (x + y,\ y^2,\ z - 1),$$

the desired variety is $\{(0, 0, 1)\}$.

**Solution E. 9.27.** The reduced Gröbner basis of $I$ with respect to the lex order with $x > y > z$ is

$$\{xy + z^2,\ xz^2 + yz^2,\ y^2 - z^2, z^4\}.$$

1. From the Elimination Theorem **T.2.25**, we immediately have

$$I_1 = (y^2 - z^2,\, z^4) \text{ and } I_2 = (z^4).$$

2. Since $\mathbb{V}(I) = \{(a,0,0) : a \in \mathbb{C}\}$ and $\mathbb{V}(I_1) = \{(0,0)\}$, we have

$$\pi_1(\mathbb{V}(I)) = \mathbb{V}(I_1).$$

**Solution E. 9.28.** The reduced Gröbner basis of $I$ with respect to the lex order with $z > t > x > y$ is

$$G = \{z - x^2,\, t^2 - x,\, tx - y,\, ty - x^2,\, x^3 - y^2\}.$$

1. By **T.2.25**, we have $J = (x^3 - y^2)$.
2. Since $G$ contains monic polynomials in $t$ and $z$, the Extension Theorem **T.3.11** shows that every element of $\mathbb{V}(J)$ can be extended to an element of the variety $\mathbb{V}(I)$.

**Solution E. 9.29.** Since $I \subseteq J$, it is sufficient to verify whether $y^3z^2 - y^3 - y^2z$ belongs to $I$.
A direct computation of a Gröbner basis of $I$ with respect to the lex order with $x > y > z$ shows

$$I = (x^2 - y^2 - yz,\, xy - y^2z,\, y^3z^2 - y^3 - y^2z) = J.$$

By the Strong Nullstellensatz, $\mathbb{I}(\mathbb{V}(I)) = \sqrt{I}$. Hence, it is sufficient to determine whether $I$ is radical.
Consider the polynomial $y^2(yz^2 - y - z) \in I$.
Since $y^2z^2 \notin \mathrm{Lt}(I)$, we have

$$y(yz^2 - y - z) \in \sqrt{I} \setminus I.$$

Hence, $I \subsetneq \mathbb{I}(\mathbb{V}(I))$.

**Solution E. 9.30.** The reduced Gröbner basis of $I$ with respect to the lex order with $x > y > z$ is

$$G = \{x + y + z,\, y^2 + yz + z^2,\, z^3 - 1\}.$$

1. Note that $\alpha$ is a primitive third root of unity. Therefore,

$$\alpha \cdot \overline{\alpha} = \alpha \cdot \alpha^2 = \alpha^3 = 1 \text{ and } 1 + \alpha + \alpha^2 = 0.$$

Thus, it can be easily verified that the set consisting of the six elements obtained by permuting the coordinates of $(1, \alpha, \alpha^2)$ is contained in $\mathbb{V}(I)$.

Moreover, since $\mathrm{Lt}(I) = (x,\, y^2,\, z^3)$, the ideal $I$ is 0-dimensional and

$$|\,\mathbb{V}(I)| \leq \dim_{\mathbb{C}} \mathbb{C}[x,y,z]/I = 6,$$

by **T.3.17** and **T.3.19**.
Hence, $|\mathbb{V}(I)| = 6$ and $\mathbb{V}(I)$ is indeed the set described above.

2. By part 1 and **T.3.19**, we have

$$\dim_K(\mathbb{C}[x, y, z]/\sqrt{I}) = \dim_K(\mathbb{C}[x, y, z]/I) = 6.$$

Thus, $\mathrm{Lt}(\sqrt{I}) = \mathrm{Lt}(I)$, and $I$ is radical by **T.2.21**.

**Solution E. 9.31.** The reduced Gröbner basis of $I$ with respect to the lex order with $x > y > z$ is

$$\{x + y + z + 1,\ (z + 1)(y + z)(y + 1)\}.$$

1. By **T.3.17**, $\mathbb{V}(I)$ is not finite.
2. We have

$$\sqrt{I} = (x + y,\ z + 1) \cap (x + 1,\ y + z) \cap (x + z,\ y + 1),$$

and these ideals are prime.
Since the field is algebraically closed,

$$\mathbb{V}(I) = \mathbb{V}(x + y,\ z + 1) \cup \mathbb{V}(x + 1,\ y + z) \cup \mathbb{V}(x + z,\ y + 1)$$

is a decomposition as a finite union of irreducible varieties.

**Solution E. 9.32.** 1. We choose the lex order with $x > y > z$.
The reduced Gröbner basis of $I$ with respect to this monomial ordering is

$$G = \{x^2 + y^2 - yz,\ xyz - x,\ y(y - z)(yz - 1)\}.$$

2. Since $I^c$ is the first elimination ideal of $I$,

$$I^c = (y(y - z)(yz - 1)).$$

3. We have

$$\begin{aligned}\sqrt{I} &= \sqrt{(I, y)} \cap \sqrt{(I, y - z)} \cap \sqrt{(I, yz - 1)} \\ &= (x,\ y) \cap \sqrt{(x^2,\ xz^2 - x,\ y - z)} \cap \sqrt{(x^2 + y^2 - 1,\ yz - 1)}.\end{aligned}$$

Therefore, $\mathbb{V}_{\mathbb{Q}}(I) = \mathbb{V}_{\mathbb{Q}}(\sqrt{I}) \supset \mathbb{V}_{\mathbb{Q}}((x,\ y))$, and $\mathbb{V}_{\mathbb{Q}}(I)$ is infinite.

4. The ideals $(x,\ y)$ and $\sqrt{(x^2,\ xz^2 - x,\ y - z)} = (x,\ y - z)$ are clearly prime.
Moreover, $\sqrt{(x^2 + y^2 - 1,\ yz - 1)} = (x^2 + y^2 - 1,\ yz - 1)$ because the ring

$$\mathbb{Q}[x, y, z]/(x^2 + y^2 - 1,\ yz - 1) \simeq \mathbb{Q}[y, y^{-1}][x]/(x^2 + y^2 - 1)$$

is a domain, and therefore, $(x^2 + y^2 - 1,\ yz - 1)$ is prime.

Since there are no containment relations among the ideals we found above,

$$\operatorname{Min} I = \{(x,\, y),\, (x,\, y-z),\, (x^2+y^2-1,\, yz-1)\}.$$

**Solution E. 9.33.** 1. The reduced Gröbner basis of $I$ with respect to the lex order with $x > y > z > t$ is

$$G = \{x^2t^2,\, yt^2,\, z^2\}.$$

Therefore, $I$ is monomial.

2. We have

$$I = (x^2,\, yt^2,\, z^2) \cap (t^2,\, z^2) = (x^2,\, y,\, z^2) \cap (x^2,\, t^2,\, z^2) \cap (t^2,\, z^2),$$

and thus, the desired decomposition is $I = (x^2,\, y,\, z^2) \cap (z^2,\, t^2)$.

3. Since $\mathcal{N}(A) = \sqrt{I}/I$, from the previous discussion it immediately follows

$$\mathcal{N}(A) = (\overline{x},\, \overline{y},\, \overline{z}) \cap (\overline{z},\, \overline{t}).$$

4. Using $G$, it is easy to compute

$$\mathbb{V}(I) = \{(a,b,0,0)\colon\ a,b \in K\} \cup \{(0,0,0,c)\colon\ c \in K\}.$$

Therefore, $|\,\mathbb{V}(I)|$ is finite if and only if $K$ is finite.

**Solution E. 9.34.** The reduced Gröbner basis of $I$ with respect to the lex order with $x > y > z$ is

$$\{x^2-yz,\, xz-yz,\, y^2-yz\}.$$

1. We have $\mathbb{V}(I) = \mathbb{V}(\sqrt{I})$ and

$$\begin{aligned}\sqrt{I} &= \sqrt{(x^2-yz,\, xz-yz,\, y)} \cap \sqrt{(x^2-yz,\, xz-yz,\, y-z)}\\ &= (x,\, y) \cap (x-z,\, y-z).\end{aligned}$$

Finally, $\mathbb{V}(I) = \mathbb{V}(x,y) \cup \mathbb{V}(x-z, y-z)$ is a decomposition into irreducible components, see also the solution of **E.9.24**, and none of these components is finite.

2. If $f \in \sqrt{I}$, then $\mathbb{V}(f) \supseteq \mathbb{V}(I)$.
However, we have $(1,1,1) \in \mathbb{V}(I) \setminus \mathbb{V}(f)$, which implies that $f \notin \sqrt{I}$.

**Solution E. 9.35.** Since $(a,0,0) \in \mathbb{V}(I)$ for all $a \in \mathbb{C}$, clearly $\mathbb{V}(I)$ is not finite. If $I \subseteq (x^2,\, y+1,\, z-1)$, then

$$P = (0,-1,1) \in \mathbb{V}((x^2,\, y+1,\, z-1)) \subseteq \mathbb{V}(I).$$

Since $P$ is not a solution of $y^2z^2 - yz$, this is not possible.

**Solution E. 9.36.** Before computing a Gröbner basis $G$ of $I$ with respect to the lex order with $x > y > z$, observe that $I$ contains a monic polynomial in $x$, which we can use to reduce the other generators of $I$.
Thus, we can write

$$I = (f_1 = x - y^2 z,\ f_2 = y^3 z^2 - 2,\ f_3 = 3y^4 z^4 - y).$$

If $\operatorname{char} K = 3$, then $y \in I$. Therefore, we have $2 \in I$ and $I = (1)$.
Otherwise $\operatorname{char} K \neq 3$, and we have $S(f_2, f_3) = -6yz^2 + y$.
If $\operatorname{char} K = 2$, then $y \in I$ and $I = (x, y)$.
Finally, if $\operatorname{char} K \neq 2, 3$, we obtain

$$G = \{x - y^2 z,\ y^3 - 12,\ 6z^2 - 1\}.$$

1. Since $\mathbb{C}$ is algebraically closed, the conclusion follows directly from **T.3.17**.
2. When $p = 3$, we have $\mathbb{V}(I) = \emptyset$, while $\mathbb{V}(I)$ is clearly infinite for $p = 2$. In the remaining cases, $\mathbb{V}(I)$ is finite, again by **T.3.17**.

**Solution E. 9.37.** The reduced Gröbner basis $G$ of $I$ with respect to the lex order with $x > y > z$ is

$$G = \left\{x + 2z^3 - 3z,\ y^2 - z^2 + 1,\ z^4 - \tfrac{3}{2}z^2 + \tfrac{1}{2}\right\}.$$

1. We have $\operatorname{Lt}(I) = \operatorname{Lt}(G) = (x,\ y^2,\ z^4)$.
The images of the elements of

$$\mathcal{B} = \{1,\ y,\ yz,\ yz^2,\ yz^3,\ z,\ z^2,\ z^3\}$$

form a $\mathbb{Q}$-basis of $A$, see **T.2.23**.1.
2. Reducing $f$ modulo $G$, we obtain $2y + z^3 - 2z^2 - z + 4$.
Thus, its coordinate vector is

$$(4, 2, 0, 0, 0, -1, -2, 1).$$

3. The equality does not hold. In fact, we can factorize

$$z^4 - \tfrac{3}{2}z^2 + \tfrac{1}{2} = (z^2 - 1)(z^2 - \tfrac{1}{2}).$$

When $z = \pm 1$, there are only 2 points in $\mathbb{V}(I)$, namely $(1, 0, 1)$ and $(-1, 0, -1)$.
On the other hand, when $z = \pm\frac{1}{\sqrt{2}}$, we find the points

$$\left(\sqrt{2}, \pm\tfrac{i}{\sqrt{2}}, \tfrac{1}{\sqrt{2}}\right) \text{ and } \left(-\sqrt{2}, \pm\tfrac{i}{\sqrt{2}}, -\tfrac{1}{\sqrt{2}}\right).$$

Thus,

$$|\mathbb{V}_{\mathbb{C}}(I)| = 6 < \dim_{\mathbb{Q}} A = 8.$$

4. We have

$$\begin{aligned}\sqrt{I} &= \sqrt{\left(I, (z+1)(z-1)\left(z^2-\tfrac{1}{2}\right)\right)}\\ &= \sqrt{(I, z+1)} \cap \sqrt{(I, z-1)} \cap \sqrt{\left(I, z^2-\tfrac{1}{2}\right)}\\ &= \sqrt{(x-1,\, y,\, z-1)} \cap \sqrt{(x+1,\, y,\, z+1)} \cap \sqrt{\left(x-2z,\, y^2+\tfrac{1}{2},\, z^2-\tfrac{1}{2}\right)},\end{aligned}$$

where the ideals under the radical symbol are maximal in $\mathbb{Q}[x, y, z]$.
This is clear for the first two ideals. To see that the third ideal is maximal, it is enough to observe that

$$\mathbb{Q}[x, y, z]/\left(x-2z,\, y^2+\tfrac{1}{2},\, z^2-\tfrac{1}{2}\right) \simeq \mathbb{Q}[y, z]/\left(y^2+\tfrac{1}{2},\, z^2-\tfrac{1}{2}\right)$$

is isomorphic to the field $\mathbb{Q}(\sqrt{2}, i)$.

**Solution E. 9.38.** 1. Let $I = (f_1, f_2, f_3)$. With respect to the deglex order with $y > x$ we have $x^3,\ y^2 \in \mathrm{Lt}(I)$.
Therefore, $\Sigma$ has a finite number of solutions.

2. We use the lex order with $x > y$, and obtain

$$S(f_1, f_2) + 3f_3 = xy^2 - x = f \in I,$$

and $I = (I, f)$.
Thus,

$$\begin{aligned}\mathbb{V}(I) &= \mathbb{V}(I, x) \cup \mathbb{V}(I, y^2-1)\\ &= \mathbb{V}(x,\, y) \cup \mathbb{V}(x^2-3x+1,\, y-1) \cup \mathbb{V}(x^2+3x+1,\, y+1).\end{aligned}$$

Hence, the only rational solution is $(0, 0)$.

3. By further refining the decomposition of $\mathbb{V}(I)$ found earlier, we obtain

$$\begin{aligned}\mathbb{V}(I) =& \mathbb{V}(x, y) \cup \mathbb{V}\left(x-\tfrac{3-\sqrt{5}}{2},\, y-1\right) \cup \mathbb{V}\left(x-\tfrac{3+\sqrt{5}}{2},\, y-1\right)\\ &\cup \mathbb{V}\left(x-\tfrac{-3-\sqrt{5}}{2},\, y+1\right) \cup \mathbb{V}\left(x-\tfrac{-3+\sqrt{5}}{2},\, y+1\right).\end{aligned}$$

Since each of these 5 varieties consists of exactly one point of $\mathbb{R}^2$, we have obtained the desired decomposition.

**Solution E. 9.39.** The reduced Gröbner basis of $I$ with respect to the lex order with $x > z > y$ is

$$\{x-y,\, z-y^2,\, y^4-1\}.$$

1. We obtain $I \cap \mathbb{Q}[y] = (y^4-1)$.
Hence, $p(y) = y^4-1$ has the required properties.

2. By the Weak Nullstellensatz, $\mathbb{V}_{\mathbb{C}}((q, I)) = \emptyset$ if and only if $(q, I) = 1$. Therefore, in order for $q$ to be an element of $S$, it is necessary that

$$\gcd(q(y), y^4 - 1) \neq 1.$$

Clearly, $y + 1$ and $y - 1$ belong to $S$, while $y^2 + 1 \notin S$.

Finally, $S = (y - 1) \cup (y + 1)$ is not an ideal.

**Solution E. 9.40.** 1. By **T.3.15**, we have that

$$\mathbb{I}(V) = \bigcap_{i=1}^{m} \mathfrak{m}_{\alpha_i}.$$

Since $\alpha_i \neq \alpha_j$ when $i \neq j$, these ideals are pairwise comaximal. By the Chinese Remainder Theorem, we obtain that

$$A \simeq \prod_{i=1}^{m} A/\mathfrak{m}_{\alpha_i} \simeq \mathbb{C}^m.$$

It is straightforward to verify that the counterimages $a_i \in A$ of the elements $e_i$ of the canonical basis of $\mathbb{C}^m$ are idempotents satisfying the desired properties, see **T.1.19**.

2. The idempotents of $\mathbb{C}^m$ are precisely the $2^m$ vectors whose coordinates are either 0 or 1. By the above isomorphism, these idempotents correspond to the idempotents of $A$, which can be expressed as $a = \sum_{i=1}^{m} b_i a_i$, where $b_i \in \{0, 1\}$ for all $i$.

## 17.3 Chapter 10

**Solution E. 10.1.** We only need to prove that the product is well-defined. Once this is shown, the properties of the $A/I$-module structure on $M/IM$ immediately follow from the analogous ones of the $A$-module structure on $M$. If $\overline{a} = \overline{b}$ in $A/I$, then $a - b \in I$ and

$$\overline{am} - \overline{bm} = \overline{a}\,\overline{m} - \overline{b}\,\overline{m} = (\overline{a} - \overline{b})\overline{m} = \overline{(a-b)m} = \overline{0} \text{ in } M/IM.$$

**Solution E. 10.2.** It immediately follows from the definition of the $B$-module structure and from the fact that $f$ is a ring homomorphism, which, in particular, requires $f(1_A) = 1_B$.

**Solution E. 10.3.** Consider an ideal $I \subset B$. For the sum operation there is nothing to prove.
For the scalar product, for any $a \in A$, we have

$$a \cdot I = f(a)I \subseteq I.$$

Hence, $I$ is an $A$-submodule of $B$.
Now, let $M$ be an $A$-submodule of $B$.
Since, for any $b \in B$, there exists an $a \in A$ such that $f(a) = b$, we obtain

$$bM = f(a)M = a \cdot M \subseteq M.$$

Hence, $M$ is an ideal of $B$.

**Solution E. 10.4.** For parts 1, 2 and 3, consider the $\mathbb{Z}$-module $M = \mathbb{Z}$.
The sets

$$S = \{1\}, \ S_1 = \{2,3\}, \ \text{and} \ S_2 = \{6,10,15\}$$

are three minimal sets of generators with different cardinalities.
In particular, only $S$ is a basis of $M$, because $S_1$ and $S_2$ are not free.
Moreover, the set $S_3 = \{2\}$ is maximal and free, but it is not a basis of $M$.

4. Let $A = K[x_i \colon i \in \mathbb{N}_+]$ be the ring of polynomials in infinitely many variables with coefficients in a field $K$.
The $A$-module $A$ is finitely generated, *e.g.*, by 1, but the ideal $I = (x_i \colon i \in \mathbb{N}_+)$ is an $A$-submodule of $A$, which is not finitely generated.

5. Consider the $\mathbb{Z}$-module $N = \mathbb{Z}/(n)$ with $n \neq 0, \pm 1$.
Since $n \cdot m = 0$ for each $m \in N$, there are no linearly independent elements, hence there is no basis of $N$ over $\mathbb{Z}$.

6. Note that $\mathbb{Z}/(6)$ is free as a module over itself. However, the submodule $P = (2)\mathbb{Z}/(6) = (\overline{2})$ is not a free $\mathbb{Z}/(6)$-module, because $\overline{3}P = 0$.

**Solution E. 10.5.** Obviously, $0 \in N : P$.
For any $a, b \in N : P$, $c \in A$, and $p \in P$, the definition of $N : P$ and the fact that $N$ is a submodule of $M$ immediately imply that $(a-b)p = ap - bp \in N$ and $(ca)p = c(ap) \in N$.

**Solution E. 10.6.** Since every $n \in \langle m \rangle$ can be written as $bm$ for some $b \in A$, both inclusions follow immediately.

**Solution E. 10.7.** The statement is a consequence of the Homomorphism Theorems. Note that

$$IM \simeq I(A/J_1 \oplus A/J_2) \simeq (I + J_1)/J_1 \oplus (I + J_2)/J_2.$$

Moreover, the projection homomorphism

$$M \longrightarrow A/(J_1 + I) \oplus A/(J_2 + I)$$

is surjective, and it can be easily verified that its kernel is $IM$.

**Solution E. 10.8.** Consider the $A$-module homomorphism $f\colon A \longrightarrow aM$ defined by $1 \mapsto a\overline{1}$, which is obviously surjective.
An element $b \in A$ is in the kernel of $f$ if and only if $\overline{0} = f(b) = bf(1) = ba\overline{1}$, *i.e.*, if and only if $ab \in J$, which is equivalent to $b \in J : (a)$.
The conclusion follows from the homomorphism theorem **T.4.3**.1.

**Solution E. 10.9.** If $n > m$, then we can write $A^n = A^m \oplus A^{n-m}$.
Consider the surjective homomorphism

$$f \circ \pi\colon A^n = A^m \oplus A^{n-m} \longrightarrow A^m \longrightarrow A^n,$$

where $\pi$ is the natural projection.
By **T.4.14**, $f \circ \pi$ is an isomorphism.
Since $0 = \operatorname{Ker}(f \circ \pi) \supseteq A^{n-m}$, we obtain $A^{n-m} = 0$, *i.e.*, $A = 0$.

**Solution E. 10.10.** Let $\{m_1, \dots, m_r\}$ be a basis, and let $\{n_1, \dots, n_s\}$ be a set of generators of $M$. We assume, by contradiction, that $s < r$.
Consider the map that sends $m_i$ to $n_i$ for $i = 1, \dots, s$, and $m_i$ to $0$ for $i = s + 1, \dots, r$.
This map induces a surjective endomorphism $\tilde{f}$ on $M$, by **T.4.7**. According to **T.4.14**, $\tilde{f}$ is an isomorphism. This is a contradiction since $m_{s+1} \in \operatorname{Ker} \tilde{f}$.

**Solution E. 10.11.** The hypothesis implies that

$$N = \varphi(M) + IN.$$

Indeed, for any $n \in N$, there exists $m \in M$ such that $\overline{\varphi}(\overline{m}) = \overline{n}$, hence $n = \varphi(m) + h$ for some $h \in IN$. Thus, $N \subseteq \varphi(M) + IN$.
The opposite inclusion is trivial.

Since $I\varphi(M) \subset \varphi(M)$, we have

$$N = \varphi(M) + I(\varphi(M) + IN) = \varphi(M) + I^2N.$$

By iterating $k$ times, where $k$ is such that $I^k = 0$, we can conclude that $N = \varphi(M) + I^kN = \varphi(M)$, *i.e.*, $\varphi$ is surjective.

**Solution E. 10.12.** Let $\mathcal{B} = \{e_1, \dots, e_m\}$ be the canonical basis of $A^m$.

1. See **E.10.9**.

Alternatively, note that the surjectivity of $f$ implies that $f(\mathcal{B})$ is a set of generators of $A^n$. By **E.10.10**, we obtain $m \geq \operatorname{rank} A^n = n$.

2. Assume, by contradiction, that $m > n$ and consider the inclusion homomorphism

$$i\colon A^n \longrightarrow A^m \text{ defined by } (a_1, \dots, a_n) \mapsto (a_1, \dots, a_n, 0, \dots, 0).$$

Then, $\varphi = i \circ f \in \operatorname{End}_A(A^m)$ and, by the Cayley-Hamilton Theorem **T.4.10**, there exist elements $\alpha_i \in A$ such that

$$\varphi^k + \alpha_{k-1}\varphi^{k-1} + \ldots + \alpha_1\varphi + \alpha_0 = 0.$$

We can assume that $k$ is minimal with this property. In this case, we claim that $\alpha_0 \neq 0$. Indeed, if $\alpha_0 = 0$, then $\varphi(\varphi^{k-1} + \alpha_{k-1}\varphi^{k-2} + \ldots + \alpha_1) = 0$, and the injectivity of $\varphi$ implies $\varphi^{k-1} + \alpha_{k-1}\varphi^{k-2} + \ldots + \alpha_1 = 0$, which contradicts the minimality of $k$.

Evaluating at $e_m$ we find

$$(\varphi^k + \alpha_{k-1}\varphi^{k-1} + \ldots + \alpha_1\varphi)(e_m) = -\alpha_0 e_m,$$

but, by the definition of $\varphi$, the element on the left hand side has last coordinate 0, leading to a contradiction.

3. It immediately follows from parts 1 and 2.

**Solution E. 10.13.** 1. If, by contradiction, $\varphi\colon M/N \longrightarrow M$ is an isomorphism, then, composing with the canonical projection, we obtain a surjective homomorphism

$$\varphi \circ \pi\colon M \longrightarrow M/N \longrightarrow M.$$

By **T.4.14**, $\varphi \circ \pi$ is also injective, hence $N \subseteq \operatorname{Ker}(\varphi \circ \pi) = 0$, which contradicts the hypothesis.

2. Consider the $K$-module $M = K[x_i\colon i \in \mathbb{N}]$ and let $0 \neq N = \langle x_0 \rangle_K \subset M$. Then, the map

$$\varphi\colon M \longrightarrow M/N \simeq K[x_i\colon i \in \mathbb{N}_+] \text{ defined by } \varphi(x_i) = x_{i+1}$$

is a $K$-module isomorphism.

**Solution E. 10.14.** 1. If $M \simeq A/\mathfrak{m}$ for a maximal ideal $\mathfrak{m}$, then $M$ is a field and it is obviously simple.

Conversely, consider $0 \neq m \in M$ and let $f\colon A \longrightarrow M$ be the homomorphism defined by $1 \mapsto m$. Then, $0 \neq f(A) \subseteq M$ is a non-zero submodule of $M$. Since $M$ is simple, it follows that $f(A) = M$, *i.e.*, $f$ is a surjective homomorphism. The homomorphism theorem **T.4.3**.1 yields $M \simeq A/\operatorname{Ker} f$.

Finally, since every proper ideal strictly containing $\operatorname{Ker} f$ corresponds to a non-trivial submodule of $M$, the ideal $\operatorname{Ker} f$ is maximal.

2. We have $\operatorname{Ker}\varphi \subseteq M$ and $\operatorname{Im}\varphi \subseteq N$.
Since $M$ is simple, we have either $\operatorname{Ker}\varphi = M$, in which case $\varphi$ is the zero homomorphism, or $\operatorname{Ker}\varphi = 0$, implying that $\varphi$ is injective.
In the latter case we also have $0 \neq \operatorname{Im}\varphi$, and hence $\operatorname{Im}\varphi = N$.

3. By part 1, we have either $M = 0$, and the statement is trivial, or $M \simeq A/\mathfrak{m}$ for some maximal ideal $\mathfrak{m}$. In the latter case $\mathcal{J}(A) \subset \mathfrak{m} = \operatorname{Ann} M$.

**Solution E. 10.15.** 1. Assume $M$ is a simple module. Then, for any $m \in M$ the cyclic submodule $\langle m \rangle \subseteq M$ is either trivial or the whole $M$.

Conversely, assume $M \neq 0$ is cyclic, and let $0 \neq N \subseteq M$ be a submodule of $M$. By hypothesis, for any $0 \neq n \in N$ we have $\langle n \rangle = M$, hence $N = M$, as desired.

2. By part 1, a simple $\mathbb{Z}$-module is cyclic and every non-zero element is a generator. Such modules are of the form $\mathbb{Z}/(p)$ with $p$ prime, see also **E.10.14**.1.

**Solution E. 10.16.** We begin by proving that $M = pM \oplus qM$, and then we show $N = qM$ and $P = pM$.
Let $x, y \in \mathbb{Z}$ be such that $xp + yq = 1$.
For any $m \in M$, we have

$$m = (xp + yq)m = p(xm) + q(ym) \in pM + qM.$$

Since $\operatorname{Ann} M = (pq)$, for every $m \in pM \cap qM$, we have $\operatorname{Ann} m \supseteq (q)+(p) = (1)$. Thus, $m = 0$, *i.e.*, $pM \cap qM = 0$ and the sum is direct.

Now, we claim that $N = qM$.
Since $M = \langle m \rangle$ is a cyclic $\mathbb{Z}$-module and $N \subseteq M$, the module $N$ is also cyclic, and we can write $N = \langle n \rangle = \langle am \rangle$ for some $a \in \mathbb{Z}$. From the equations $apm = pn = 0$, we obtain $ap \in \operatorname{Ann} M = (pq)$.
Thus, $a = bq \in (q)$ for some $b \in \mathbb{Z}$, and we have proved that $N \subseteq qM$.

To prove the opposite inclusion, since $\operatorname{Ann} N = (p)$, we have $(b, p) = 1$ (why?). Therefore, there exist $c$, $d \in \mathbb{Z}$ such that $cb + dp = 1$.
Multiplying by $qm$, we obtain $qm = cbqm = cam \in N$.

The proof that $P = pM$ is essentially the same.

**Solution E. 10.17.** We prove the first isomorphism, the second one can be obtained similarly.
Define

$$\Phi\colon \operatorname{Hom}_A(M, P) \oplus \operatorname{Hom}_A(N, P) \longrightarrow \operatorname{Hom}_A(M \oplus N, P)$$

by $\Phi(\varphi_1, \varphi_2) = \lambda_{\varphi_1,\varphi_2}$, where

$$\lambda_{\varphi_1,\varphi_2}(m, n) = \varphi_1(m) + \varphi_2(n).$$

Let $i_M : M \longrightarrow M \oplus N$ and $i_N : N \longrightarrow M \oplus N$ be the natural inclusions, and define

$$\Psi\colon \mathrm{Hom}_A(M\oplus N, P) \longrightarrow \mathrm{Hom}_A(M,P)\oplus \mathrm{Hom}_A(N,P)$$

by

$$\Psi(\psi) = (\psi\circ i_M, \psi\circ i_N) = (i_M^*(\psi), i_N^*(\psi)).$$

It can be easily verified to that $\Phi$ and $\Psi$ are mutually inverse homomorphisms.

Alternatively, we can consider the exact sequence

$$0 \longrightarrow M \xrightarrow{i_M} M\oplus N \xrightarrow{\pi_N} N \longrightarrow 0,$$

and apply the functor $\mathrm{Hom}_A(\bullet, P)$ to obtain the exact sequence

$$0 \longrightarrow \mathrm{Hom}_A(N,P) \xrightarrow{\pi_N^*} \mathrm{Hom}_A(M\oplus N,P) \xrightarrow{i_M^*} \mathrm{Hom}_A(M,P).$$

For any $\varphi \in \mathrm{Hom}_A(M,P)$, consider the zero homomorphism $N \longrightarrow P$ and apply the universal property of direct sum **T.4.6**.1 to find an element $\psi \in \mathrm{Hom}(M\oplus N, P)$ such that $\psi\circ i_M = \varphi$, that is $i_M^*(\psi)=\varphi$.
Thus, $i_M^*$ is surjective.
Now, consider the projection $\pi_M\colon M\oplus N \longrightarrow M$.
It is easy to verify that

$$i_M^*\circ\pi_M^* = \mathrm{id}_{\mathrm{Hom}_A(M,P)}\,.$$

Hence $i_M^*$ has a section, the exact sequence of Hom splits, and the conclusion follows from **T.4.18**.

**Solution E. 10.18.** 1. For any $f\in\mathrm{Hom}_{\mathbb{Z}}(\mathbb{Q},\mathbb{Z})$ and any $\frac{a}{b}\in\mathbb{Q}$, we have $f\left(\frac{a}{b}\right) = af\left(\frac{1}{b}\right)$.
Therefore, it suffices to examine the admissible values for $f\left(\frac{1}{b}\right)$.
Since, for every $b\in\mathbb{Z}\setminus\{0\}$, we have $bf\left(\frac{1}{b}\right) = f(1)$, the value $f(1)\in\mathbb{Z}$ has to be divisible by every $b\neq 0$. It follows that $f(1)=0$.
Therefore, $f\left(\frac{1}{b}\right)=0$ for every $b\in\mathbb{Z}\setminus\{0\}$ and, as a result, $f=0$.

2. Let $f\in\mathrm{Hom}_{\mathbb{Z}}(\mathbb{Z}/(n),\mathbb{Z})$. Then, $\mathrm{Im}\, f$ is generated by $f(\overline{1})$.
Moreover, $nf(\overline{1}) = f(\overline{0}) = 0$ implies that $f(\overline{1})=0$, that is, $f=0$.

3. To obtain a non-trivial element of $\mathrm{Hom}_{\mathbb{Z}}(\mathbb{Z}/(n),\mathbb{Q}/\mathbb{Z})$, simply define $f(\overline{1})$ to be the coset of $\frac{1}{n}$ in $\mathbb{Q}/\mathbb{Z}$.

**Solution E. 10.19.** Note that by considering $\pi\colon A\longrightarrow A/I$ and defining $a\overline{b} = \pi(a)\overline{b} = \overline{a}\overline{b} = \overline{ab}$, we obtain an $A$-module structure on $A/I$ by restriction of scalars, see **E.10.2**.
Moreover, with a proof similar to the one of **E.10.5**, it can be easily verified that $0:_M I$ is a submodule of $M$.

1. Consider the map $\Phi\colon 0:_M I \longrightarrow \mathrm{Hom}_A(A/I, M)$ defined by

$$\Phi(m) = \varphi_m, \text{ where } \varphi_m(\overline{b}) = bm.$$

Note that $\Phi$ is well-defined, *i.e.*, $\varphi_m\in\mathrm{Hom}_A(A/I,M)$ for any $m\in 0:_M I$.

Indeed, if $\overline{b_1} = \overline{b_2} \in A/I$, then $b_1 - b_2 \in I$, and hence, $(b_1 - b_2)m = 0$ yields $\varphi_m(\overline{b_1}) = \varphi_m(\overline{b_2})$, showing that $\varphi_m$ is well-defined.

Moreover,

$$\varphi_m(\overline{b_1}) + \varphi_m(\overline{b_2}) = b_1 m + b_2 m = (b_1 + b_2)m = \varphi_m(\overline{b_1 + b_2}) = \varphi_m(\overline{b_1} + \overline{b_2})$$

and

$$a\varphi_m(\overline{b_1}) = a(b_1 m) = (ab_1)m = \varphi_m(\overline{ab_1}) = \varphi_m(a\overline{b_1})$$

for any $a, b_1, b_2 \in A$.
Therefore, $\varphi_m$ is an $A$-module homomorphism for any $m \in 0 :_M I$.

It is easy to verify that $\Phi$ is an $A$-module homomorphism. We claim that $\Phi$ is an isomorphism.
Indeed, $\Phi$ is injective because if $\Phi(m) = \varphi_m = 0$, then $0 = \varphi_m(\overline{1}) = m$.
It is surjective because each $f \in \operatorname{Hom}_A(A/I, M)$ is completely determined by $f(\overline{1})$, due to $A$-linearity. As above, it can be shown that $f(\overline{1})$ belongs to $0 :_M I$, and therefore $\Phi(f(\overline{1})) = \varphi_{f(\overline{1})} = f$.
2. By part 1, it is sufficient to prove that the $A$-module $0 :_M I$ is also an $A/I$-module.
Note that

$$I \subseteq 0 :_A (0 :_M I) = \operatorname{Ann}_A(0 :_M I),$$

and conclude using **T.4.1**.
3. By part 1, it is sufficient to prove that

$$0 :_{A/J} I \simeq (J : I)/J.$$

Consider the map

$$J : I \xrightarrow{\varphi} 0 :_{A/J} I \subseteq A/J$$

defined by $a \mapsto \overline{a}$.
This map is well-defined, because if $aI \subseteq J$, then $\overline{a}I = 0_{A/J}$.
Clearly, $\varphi$ is an $A$-module homomorphism and it is surjective, because if $\overline{a} \in 0 :_{A/J} I$, then $aI \subseteq J$, that is, $a \in J : I$.
The conclusion follows from the fact that $\operatorname{Ker}\varphi = J$.

**Solution E. 10.20.** Using the previous exercise, and the fact that $I \subset J$, we obtain

$$\operatorname{Hom}_A(A/I, A/J) \simeq (J : I)/J = A/J.$$

Since $\{x^2, yz\}$ is a Gröbner basis of $J$, the quotient $A/J$ is a $K$-vector space of infinite dimension. For example, the elements of the set $\{\overline{y}^n : n \in \mathbb{N}\}$ are all linearly independent, see **T.2.23**.1.

**Solution E. 10.21.** Since $A$ is local, and $M$ is finitely generated and non-zero, Nakayama's Lemma yields $\mathfrak{m}M \neq M$. As a consequence, $M/\mathfrak{m}M$ is a non-trivial finitely generated $A/\mathfrak{m}$-module, *i.e.*, a vector space of finite dimension.
Hence, there exists a non-zero linear map

$$f\colon M/\mathfrak{m}M \longrightarrow A/\mathfrak{m}.$$

Consider the canonical projection $\pi\colon M \longrightarrow M/\mathfrak{m}M$. Then, $f \circ \pi$ is a non-trivial element of $\operatorname{Hom}_A(M, A/\mathfrak{m})$.

**Solution E. 10.22.** Every non-zero element of $B$ is invertible in $K$. Let $\mathfrak{m}$ be the maximal ideal of $B$. Then, we have $\mathfrak{m}K = K$.
If $K$ is a finitely generated $B$-module, then $K = 0$ by Nakayama's Lemma, but this is not possible because $B \neq 0$.

**Solution E. 10.23.** Let $a \in \sqrt{\operatorname{Ann} M + I}$. Then, there exists $n \in \mathbb{N}$ such that $a^n = b + i$ for some $b \in \operatorname{Ann} M$ and $i \in I$. Hence, we have

$$a^n M = (b + i)M = 0 + iM \subset IM,$$

which implies $a^n \in \operatorname{Ann}(M/IM)$. Thus, $a \in \sqrt{\operatorname{Ann}(M/IM)}$.

To prove the opposite inclusion, take $a \in \sqrt{\operatorname{Ann}(M/IM)}$, and let $k \in \mathbb{N}$ be such that $a^k \in \operatorname{Ann}(M/IM)$, that is, $a^k M \subseteq IM$.
Consider the endomorphism $\varphi\colon M \longrightarrow M$ defined by $\varphi(m) = a^k m$.
We have $\varphi(M) = a^k M \subseteq IM$. Since $M$ is finitely generated, we can apply the Cayley-Hamilton Theorem **T.4.10** to find $a_0, \ldots, a_{n-1} \in I$ such that

$$\varphi^n + a_{n-1}\varphi^{n-1} + \ldots + a_0 = 0.$$

By setting

$$b = a^{kn} + a_{n-1}a^{k(n-1)} + \ldots + a_1 a^k + a_0,$$

we obtain $bM = 0$, that is, $b \in \operatorname{Ann} M$.
Finally, $a^{kn} = b - \sum_{i=0}^{n-1} a_i a^{ki}$ yields $a \in \sqrt{\operatorname{Ann} M + I}$, because $\sum_{i=0}^{n-1} a_i a^{ki}$ belongs to $I$.

**Solution E. 10.24.** Let $\{a_1, \ldots, a_n\}$ be a set of generators of $\mathcal{N}(A)$. For any $i = 1, \ldots, n$, there exists $r_i \in \mathbb{N}$ such that $a_i^{r_i} = 0$.
It immediately follows that there exists $s \in \mathbb{N}$ such that $\mathcal{N}(A)^s = 0$.
Since $\mathcal{N}(A)M = M$, we find

$$0 = \mathcal{N}(A)^s M = M.$$

**Solution E. 10.25.** Consider the ring $A = K[x_i\colon i \in \mathbb{N}]$ of polynomials in infinitely many variables and with coefficients in a field $K$.
Let $I \subset A$ be the ideal

$$I = (x_0^2, x_1^2 - x_0,\ x_2^2 - x_1, \ldots, x_n^2 - x_{n-1}, \ldots).$$

Since $\overline{x_0} \in \mathcal{N}(A/I)$ and $\overline{x_i}^{2^i} = \overline{x_0}$ for every $i$, we have $\mathcal{N}(A/I) \supseteq (\overline{x_i} : i \in \mathbb{N})$. On the other hand, since $(\overline{x_i} : i \in \mathbb{N})$ is maximal, the opposite inclusion holds as well.

Note that $\mathcal{N}(A/I)^2 = \mathcal{N}(A/I)$, and recall that $\mathcal{N}(A/I) \subseteq \mathcal{J}(A/I)$.
The $A/I$-module $M = \mathcal{N}(A/I) \neq 0$ provides the required counterexample.

**Solution E. 10.26.** We denote the homomorphisms in the first row by $f_i\colon M_i \longrightarrow M_{i+1}$, and the ones in the second row by $g_i\colon N_i \to N_{i+1}$.
1. Let $m_3 \in M_3$ be such that $\alpha_3(m_3) = 0$. We have to show that $m_3 = 0$.
We have
$$0 = \alpha_3(m_3) = g_3(\alpha_3(m_3)) = \alpha_4(f_3(m_3)).$$

Since $\alpha_4$ is injective, $m_3 \in \operatorname{Ker} f_3 = \operatorname{Im} f_2$. Thus, there exists $m_2 \in M_2$ such that $f_2(m_2) = m_3$ and $g_2(\alpha_2(m_2)) = \alpha_3(f_2(m_2)) = 0$. Hence, $\alpha_2(m_2)$ is in $\operatorname{Ker} g_2 = \operatorname{Im} g_1$, and there exists $n_1 \in N_1$ such that $g_1(n_1) = \alpha_2(m_2)$.
Since $\alpha_1$ is surjective, there exists $m_1 \in M_1$ such that $\alpha_1(m_1) = n_1$.
Moreover,
$$\alpha_2(f_1(m_1)) = g_1(\alpha_1(m_1)) = \alpha_2(m_2),$$

and the injectivity of $\alpha_2$ leads to $m_2 = f_1(m_1)$.
Finally,
$$m_3 = f_2(m_2) = f_2(f_1(m_1)) = 0.$$

2. Consider $n_3 \in N_3$. Then, $g_3(n_3) \in N_4$, and since $\alpha_4$ is surjective, there exists $m_4 \in M_4$ such that $\alpha_4(m_4) = g_3(n_3)$.
We have
$$\alpha_5(f_4(m_4)) = g_4(\alpha_4(m_4)) = g_4(g_3(n_3)) = 0.$$

The injectivity of $\alpha_5$ yields $f_4(m_4) = 0$. Hence, $m_4 \in \operatorname{Ker} f_4 = \operatorname{Im} f_3$ and we can write $m_4 = f_3(m_3)$ for some $m_3 \in M_3$.
Using $\alpha_4(f_3(m_3)) = g_3(\alpha_3(m_3))$, we obtain

$$g_3(\alpha_3(m_3) - n_3) = \alpha_4(m_4) - g_3(n_3) = 0,$$

that is, $\alpha_3(m_3) - n_3 \in \operatorname{Ker} g_3 = \operatorname{Im} g_2$.
Thus, $\alpha_3(m_3) - n_3 = g_2(n_2)$ for some $n_2 \in N_2$.
Finally, the surjectivity of $\alpha_2$ yields $n_2 = \alpha_2(m_2)$ for some $m_2 \in M_2$, and therefore, $\alpha_3(f_2(m_2)) = g_2(\alpha_2(m_2)) = \alpha_3(m_3) - n_3$.
It follows that $\alpha_3(m_3 - f_2(m_2)) = n_3$, and $\alpha_3$ is surjective.

**Solution E. 10.27.** We prove that there always exists a retraction of $f$, *i.e.*, a homomorphism $\alpha\colon M \longrightarrow N$ such that $\alpha \circ f = \mathrm{id}_N$.
Since $(p)$ and $(q)$ are comaximal, there exists $p' \in (p)$ and $q' \in (q)$ such that $p' + q' = 1$. For any $m \in M$, we can write $m = p'm + q'm$.
Thus, we have
$$g(m - p'm) = q'g(m) \in qP = 0.$$

By exactness, $m - p'm \in \operatorname{Ker} g = \operatorname{Im} f$, and there exists a unique $n \in N$ such that $f(n) = m - p'm = q'm$.
We define $\alpha(m) = q'n$. It is easy to verify that $\alpha$ is a homomorphism.
To verify that $\alpha \circ f = \mathrm{id}_N$, we observe that

$$\alpha(f(n)) = \alpha(q'm) = q'\alpha(m) = (1 - p')^2 n = n,$$

where the last equality follows from the fact that $p'n \in pN = 0$.

**Solution E. 10.28.** 1. Consider the sequence

$$0 \longrightarrow \mathbb{Z}/(2) \xrightarrow{f} \mathbb{Z}/(4) \xrightarrow{g} \mathbb{Z}/(2) \longrightarrow 0,$$

where $f$ is defined by $f(\overline{1}) = \overline{2}$, and $g = \pi$ is the canonical projection of $\mathbb{Z}/(4)$ onto $\mathbb{Z}/(2)$.
Clearly, $f$ is well-defined and injective, while $g$ is well-defined and surjective. The kernel of $g$ is $(2)\mathbb{Z}/(4)$, which is also the image of $f$.
2. Consider the sequence

$$0 \longrightarrow \mathbb{Z}/(2) \xrightarrow{f'} \mathbb{Z}/(2) \oplus \mathbb{Z}/(4) \xrightarrow{g'} \mathbb{Z}/(2) \oplus \mathbb{Z}/(2) \longrightarrow 0,$$

where $f' = (0, f)$ and $g' = \mathrm{id}_{\mathbb{Z}/(2)} \oplus g$, with $f$ and $g$ as in part 1.
Then, $g'$ is surjective because both its components are surjective, while $f'$ is injective because $f$ is injective.
Clearly, $\operatorname{Im} f' = 0 \oplus (2)\mathbb{Z}/(4) = \operatorname{Ker} g'$.
3. Consider a homomorphism $g_1 \colon \mathbb{Z}/(8) \longrightarrow \mathbb{Z}/(2) \oplus \mathbb{Z}/(2)$.
Since $\operatorname{Im} g_1$ is generated by $g_1(\overline{1})$, it is a cyclic $\mathbb{Z}$-module, hence $g_1$ cannot be surjective.

Alternatively, observe that a homomorphism $f_1 \colon \mathbb{Z}/(2) \longrightarrow \mathbb{Z}/(8)$ is uniquely determined by $f_1(\overline{1})$, and it verifies $2f_1(\overline{1}) = \overline{0}$.
Thus, the only possible injective homomorphism sends $\overline{1}$ to $\overline{4}$. Its image is $(4)\mathbb{Z}/(8)$, and its cokernel is isomorphic to $\mathbb{Z}/(4)$.
Hence, in this case, the only admissible exact sequence is

$$0 \longrightarrow M \longrightarrow P \longrightarrow N \longrightarrow 0.$$

**Solution E. 10.29.** Let $n_1, \dots, n_r$ and $p_1, \dots, p_s$ be sets of generators of $N$ and $P$, respectively. Moreover, let $m_1, \dots, m_s \in M$ be such that $g(m_i) = p_i$ for any $i$. These elements exist because $g$ is surjective.
Consider $m \in M$. Then,

$$g(m) = \sum_{i=1}^{s} a_i p_i = \sum_{i=1}^{s} a_i g(m_i).$$

Therefore, $g\left(m - \sum_{i=1}^{s} a_i m_i\right) = 0$, *i.e.*,

$$m - \sum_{i=1}^{s} a_i m_i \in \operatorname{Ker} g = \operatorname{Im} f = \langle f(n_1), \dots, f(n_r) \rangle.$$

Thus, we can write any element of $M$ as a linear combination of the elements $f(n_1), \dots, f(n_r), m_1, \dots, m_s$.

**Solution E. 10.30.** Consider $n \in N$, and write

$$n = (n - f(g(n))) + f(g(n)).$$

Since, by hypothesis, $g \circ f = \mathrm{id}_M$, we have $n - f(g(n)) \in \operatorname{Ker} g$, and obviously, $f(g(n)) \in \operatorname{Im} f$.
Moreover, if $n = f(m) \in \operatorname{Ker} g$, then $0 = g(n) = g(f(m)) = m$.
This proves that the sum is direct.

Alternatively, note that since $g \circ f = \mathrm{id}_M$, we have that $f$ is injective. Consider the exact sequence

$$0 \longrightarrow M \xrightarrow{f} N \xrightarrow{\pi} \operatorname{Coker} f \longrightarrow 0.$$

Since the sequence splits by hypothesis, we have

$$N \simeq M \oplus \operatorname{Coker} f \simeq \operatorname{Im} f \oplus \operatorname{Ker} g.$$

Indeed, it is easy to verify that $\pi_{|\operatorname{Ker} g}\colon \operatorname{Ker} g \longrightarrow \operatorname{Coker} f$ is an isomorphism, see also the proof of **T.4.18**.

**Solution E. 10.31.** The sequence is obviously exact at $M$ and $W$. We need to show exactness at $N$ and $T$, *i.e.*, prove that $\operatorname{Ker}(g \circ f) = \operatorname{Im} \varphi$ and $\operatorname{Im}(g \circ f) = \operatorname{Ker} \psi$.
Since $g$ is injective, we have $\operatorname{Ker}(g \circ f) = \operatorname{Ker} f = \operatorname{Im} \varphi$.
The surjectivity of $f$ yields $\operatorname{Im}(g \circ f) = \operatorname{Im} g = \operatorname{Ker} \psi$.

**Solution E. 10.32.** Clearly, $\mathbb{Z}/(n)$ is a free $\mathbb{Z}/(n)$-module, hence it is projective by **T.4.21**.

Now, assume, by contradiction, that $\mathbb{Z}/(n)$ is a projective $\mathbb{Z}$-module. Then, by **T.4.22**, the sequence

$$0 \longrightarrow \mathbb{Z} \xrightarrow{n} \mathbb{Z} \xrightarrow{\pi} \mathbb{Z}/(n) \longrightarrow 0$$

splits.
By **T.4.18**, there exists a section $s$ of $\pi$, *i.e.*, a non-trivial element of $\operatorname{Hom}_{\mathbb{Z}}(\mathbb{Z}/(n), \mathbb{Z})$.
This is not possible by **E.10.18**.2.

**Solution E. 10.33.** By the Chinese Remainder Theorem, we can express $A \simeq \mathbb{Z}/(4) \oplus \mathbb{Z}/(3)$, and, clearly, $A$ is a free $A$-module.
Thus, since $\mathbb{Z}/(4)$ is a direct summand of a free $A$-module, it is also a projective $A$-module, see **T.4.22**.

However, by considering cardinalities, it is clear that $\mathbb{Z}/(4)$ cannot be isomorphic to $A^n$ for any $n$, and hence, it cannot be free.

**Solution E. 10.34.** In $A$, the unique non-trivial submodule is $(\overline{2}) \simeq \mathbb{Z}/(2)$ and it is not projective because it cannot be a direct summand of $(\mathbb{Z}/(4))^n$ for any $n$. Indeed, that cannot happen because $\mathbb{Z}/(4) \not\simeq \mathbb{Z}/(2) \oplus \mathbb{Z}/(2)$.

The only non-trivial submodules of $B$ are $(\overline{3}) \simeq \mathbb{Z}/(2)$ and $(\overline{2}) \simeq \mathbb{Z}/(3)$. Since $\mathbb{Z}/(6) \simeq \mathbb{Z}/(2) \oplus \mathbb{Z}/(3)$, both submodules are projective.

**Solution E. 10.35.** 1. Consider the sequence

$$0 \longrightarrow I \cap J \xrightarrow{f} I \oplus J \xrightarrow{g} I + J \longrightarrow 0,$$

where $f(a) = (a, -a)$ and $g(a, b) = a + b$.
It is easy to verify that it is a short exact sequence.
Since, by hypothesis, $I + J = A$, we have $I \cap J = IJ$, and we can rewrite the sequence as

$$0 \longrightarrow IJ \xrightarrow{f} I \oplus J \xrightarrow{g} A \longrightarrow 0.$$

Since $A$ is projective, the sequence splits, and we find $I \oplus J \simeq IJ \oplus A$.
2. Let $IJ = (d)$. If $d = 0$, then $I \oplus J = A$. Otherwise, $d \neq 0$, and since $A$ is a domain, $IJ = (d) \simeq A$.
Therefore, part 1 immediately yields

$$I \oplus J \simeq A^2.$$

In both cases $I$ and $J$ are direct summands of a free module, hence they are projective.

**Solution E. 10.36.** 1. If $M \simeq N \simeq \mathbb{Z}$ and $f(1) = n$ with $n \neq 0$, then $P \simeq \mathbb{Z}/(n)$. In particular, when $n = \pm 1$, we have $P = 0$.
Otherwise, $P \simeq \mathbb{Z}$ is projective and the sequence

$$0 \longrightarrow M \longrightarrow M \oplus \mathbb{Z} \longrightarrow \mathbb{Z} \longrightarrow 0$$

splits.
Thus, $M \simeq \mathbb{Z}$ yields $N \simeq \mathbb{Z}^2$, while $N \simeq \mathbb{Z}$ yields $M = 0$.
2. If $N \simeq \mathbb{Z}$, then either $M = 0$, or $M \simeq \mathbb{Z}$ and the answer follows from the previous part.
If $P \simeq \mathbb{Z}$, then the sequence splits and, for any $\mathbb{Z}$-module $M$, we obtain

$$0 \longrightarrow M \longrightarrow M \oplus \mathbb{Z} \longrightarrow \mathbb{Z} \longrightarrow 0.$$

Finally, if $M \simeq \mathbb{Z}$, then we cannot draw any conclusion about the sequence without additional hypotheses.
3. In the setting of part 1, where $M \simeq N \simeq A$ and $f(1_A) = a \neq 0$, we find that $P \simeq A/(a)$. In particular, when $a \in A^*$, we have $P = 0$.
Throughout all the remaining proofs we used the following facts:

i) $\mathbb{Z}$ is a projective $\mathbb{Z}$-module;
ii) a submodule of $\mathbb{Z}$ is trivial or isomorphic to $\mathbb{Z}$.

Both of these facts are still true in any PID, see **T.4.23**.
Therefore, all the proofs and statements remain valid when considering a general PID.

**Solution E. 10.37.** 1. By hypothesis, $M/N$ and $M'/N'$ are free.
Hence, they are also projective, and the sequences split.

2. By part 1, we have that $M \simeq N \oplus M/N$ and $M' \simeq N' \oplus M'/N'$.
Since $N \simeq N'$ and $M/N \simeq M'/N'$, it immediately follows that $M \simeq M'$.

**Solution E. 10.38.** 1. Every element $m \in M$ can be expressed as

$$m = (m - \varphi(m)) + \varphi(m),$$

hence $M = (\mathrm{id}_M - \varphi)(M) + \varphi(M)$.

To verify that the sum is direct, note that if

$$m = \varphi(m_1) = m_2 - \varphi(m_2) \in \varphi(M) \cap (\mathrm{id}_M - \varphi)(M),$$

then $m = \varphi^2(m_1) = \varphi(m_2) - \varphi(m_2) = 0$.

2. Since $M$ is finitely generated, there exist $n \in \mathbb{N}_+$ and a surjective homomorphism $g\colon A^n \longrightarrow M$. If $M$ is projective, then there exists a section $\sigma\colon M \longrightarrow A^n$ such that $g \circ \sigma = \mathrm{id}_M$.
Define $f\colon A^n \longrightarrow A^n$ as $f = \sigma \circ g$.
It is a homomorphism such that $f^2 = \sigma \circ g \circ \sigma \circ g = f$.
Moreover,

$$f(A^n) = \sigma(g(A^n)) = \sigma(M) \simeq M,$$

where the isomorphism is due to the injectivity of $\sigma$.

Conversely, assume there exists $f \in \mathrm{End}_A(A^n)$ such that $f^2 = f$ and $f(A^n) \simeq M$.
By part 1,

$$A^n \simeq f(A^n) \oplus (\mathrm{id}_{A^n} - f)(A^n) \simeq M \oplus (\mathrm{id}_{A^n} - f)(A^n).$$

Therefore, $M$ is a direct summand of a free module, *i.e.*, it is projective.

**Solution E. 10.39.** If $A$ is a field, then every $A$-module is a vector space, and every short exact sequence of vector spaces splits.
Thus, every $A$-module is projective.

Conversely, let $0 \neq a \in A$, and consider the exact sequence

$$0 \longrightarrow A \xrightarrow{f} A \longrightarrow A/(a) \longrightarrow 0,$$

where $f$ is the multiplication by $a$, which is injective because $A$ is a domain. By hypothesis, the sequence splits. Hence, there exists $g\colon A \longrightarrow A$ such that $g \circ f = \mathrm{id}_A$. Thus, $1 = g(f(1)) = g(a)$. Since $g$ is an $A$-module homomorphism, we have $1 = g(a) = ag(1)$.
Therefore, $a$ is invertible in $A$, *i.e.*, $A$ is a field.

**Solution E. 10.40.** 1. It is easy to verify that

$$A/I \simeq A/J \simeq \mathbb{Z}/(3)$$

is a field.
Thus, $I$ and $J$ are maximal ideals.
Moreover, if $I = J$, then $1 \in I$, which is not possible.

If we assume, by contradiction, that $\alpha = a + b\sqrt{-5} \in A$ is a generator of $I = (\alpha)$, then $\alpha \mid 3$ and $\alpha \mid 1 - \sqrt{-5}$.
Denote by $\overline{\alpha}$ the complex conjugate of $\alpha$. We have that $\alpha\overline{\alpha} \in \mathbb{Z}$ is such that $\alpha\overline{\alpha} \mid 9$ and $\alpha\overline{\alpha} \mid 6$, that is, either $\alpha\overline{\alpha} = 1$ or $\alpha\overline{\alpha} = 3$.
Since $\alpha\overline{\alpha} = a^2 + 5b^2$, the case $\alpha\overline{\alpha} = 3$ is not possible, while $\alpha\overline{\alpha} = 1$ leads to $I = A$, that is a contradiction.

With a similar proof we show that the same conclusion holds for $J$.

2. The ideals $I$ and $J$ are comaximal, hence $I \cap J = IJ$.
Moreover, $3 \in I \cap J$ and $IJ$ is generated by $9,\ 6,\ 3(1+\sqrt{-5})$, and $3(1-\sqrt{-5})$.
Hence, $IJ = (3)$.

By **E.10.35**, $I$ and $J$ are projective $A$-modules such that $I \oplus J \simeq A^2$.
Since $I$ and $J$ are not principal, they cannot be isomorphic to $A$.

Finally, the relation

$$2 \cdot 3 - (1 + \sqrt{-5})(1 - \sqrt{-5}) = 0$$

between the generators of $I$ and $J$ respectively, shows that neither $I$ nor $J$ is free.

**Solution E. 10.41.** If $M$ is projective and finitely generated, then $M$ is a direct summand of $A^n$ for some $n \in \mathbb{N}$, and we write $A^n \simeq M \oplus N$ for some $A$-module $N$.
Since $A^n$ is finitely generated, its direct summand $N$ is finitely generated as well. Let $\{r_1, \dots, r_m\}$ be a set of generators of $N$.
We construct the diagram

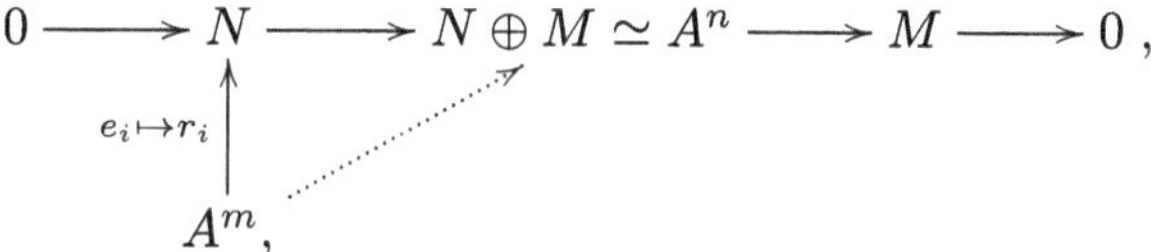

from which we obtain an exact sequence

$$A^m \longrightarrow A^n \longrightarrow M \longrightarrow 0,$$

as required.

Clearly, the reverse implication holds for any $A$-module.

**Solution E. 10.42.** $1 \Leftrightarrow 2$. The functor $\mathrm{Hom}_A(\bullet, E)$ is contravariant and left exact, see **T.4.16**.1. Hence, its exactness is equivalent to the property that, for any injective homomorphism $f\colon M \longrightarrow N$, the induced homomorphism

$$f^*\colon \mathrm{Hom}_A(N, E) \longrightarrow \mathrm{Hom}_A(M, E)$$

is surjective.
In other words, for any homomorphism $g\colon M \longrightarrow E$ there exists a homomorphism $\tilde{g}\colon N \longrightarrow E$ such that $f^*(\tilde{g}) = \tilde{g} \circ f = g$.
This means that $E$ is injective.

$3 \Leftrightarrow 4$. The implication $\Rightarrow$ is an immediate consequence of **T.4.18**.
Conversely, consider an exact sequence

$$0 \longrightarrow E \xrightarrow{f} M \longrightarrow N \longrightarrow 0,$$

where $E \simeq f(E)$ is a submodule $M$, and $N \simeq M/f(E)$.
By hypothesis, there exists a submodule $L$ of $M$ such that $M \simeq E \oplus L$.
Thus, we have

$$L \simeq M/E \simeq M/f(E) \simeq N \quad \text{and} \quad M \simeq E \oplus N,$$

which is one of the conditions ensuring the splitting of the sequence, see **T.4.18** again.

$1 \Rightarrow 3$. It suffices to consider the diagram

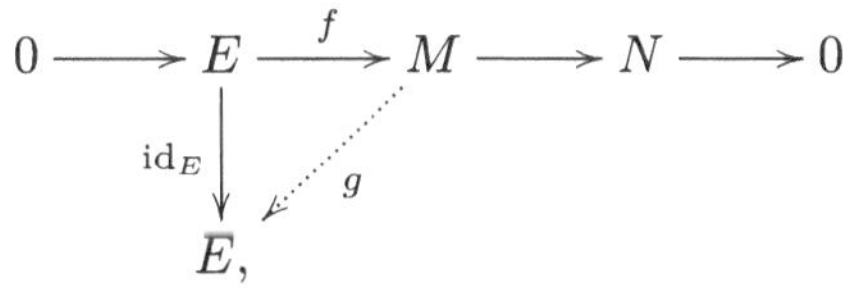

$$\begin{array}{ccccccccc} 0 & \longrightarrow & E & \xrightarrow{f} & M & \longrightarrow & N & \longrightarrow & 0 \\ & & {\scriptstyle \mathrm{id}_E}\downarrow & \swarrow{\scriptstyle g} & & & & & \\ & & E, & & & & & & \end{array}$$

where, by hypothesis, there exists $g$ such that $g \circ f = \mathrm{id}_E$.
Such a $g$ is a retraction of $f$, and its existence guarantees the splitting of the sequence.

$3 \Rightarrow 1$. Let $f\colon M \longrightarrow N$ and $g\colon M \longrightarrow E$ be homomorphisms, with $f$ injective. Define $U = (E \oplus N)/L$, where $L$ is the submodule of $E \oplus N$ generated by the pairs $(g(m), -f(m))$ as $m$ varies in $M$.
Consider the diagram

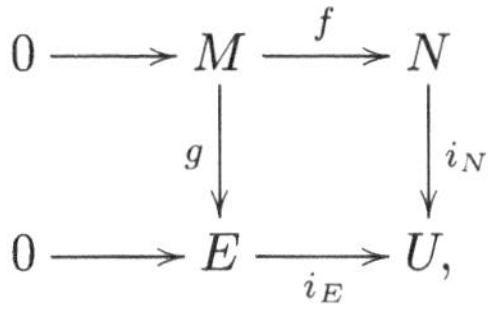

$$\begin{array}{ccccc} 0 & \longrightarrow & M & \xrightarrow{f} & N \\ & & {\scriptstyle g}\downarrow & & \downarrow{\scriptstyle i_N} \\ 0 & \longrightarrow & E & \xrightarrow[i_E]{} & U, \end{array}$$

where $i_E(e) = \overline{(e,0)}$ and $i_N(n) = \overline{(0,n)}$.
The square commutes by construction.
Moreover, $i_E$ is injective. Indeed, if $i_E(e) = 0$, then there exist elements $m_i \in M$ and $a_i \in A$ such that

$$(e,0) = \sum_i a_i(g(m_i), -f(m_i)) = (g(m), -f(m)) \quad \text{with} \quad m = \sum_i a_i m_i \in M.$$

Thus, $0 = f(m)$, and the injectivity of $f$ yields $m = 0$.
Accordingly, $e = g(m) = 0$.

By hypothesis, there exists $r\colon U \longrightarrow E$ such that $r \circ i_E = \mathrm{id}_E$.
We can define $\tilde{g}\colon N \longrightarrow E$ via composition as $\tilde{g} = r \circ i_N$, thus obtaining the diagram

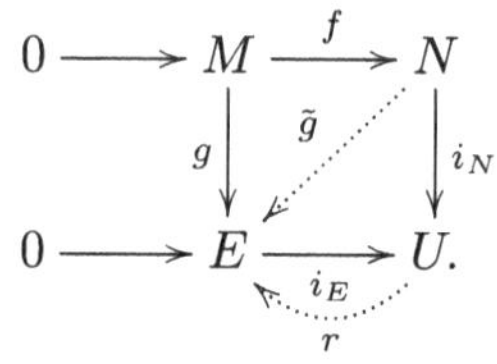

$$\begin{array}{ccccc} 0 & \longrightarrow & M & \xrightarrow{f} & N \\ & & \downarrow g & \swarrow \tilde{g} & \downarrow i_N \\ 0 & \longrightarrow & E & \underset{r \curvearrowleft}{\xrightarrow{i_E}} & U. \end{array}$$

Finally, to conclude that $E$ is injective, we need to prove that $\tilde{g} \circ f = g$. Indeed, for any $m \in M$, we have

$$(\tilde{g} \circ f)(m) = (r \circ (i_N \circ f))(m) = ((r \circ i_E) \circ g)(m) = g(m).$$

**Solution E. 10.43.** 1. If $A$ is a field, then $A$-modules are vector spaces and an exact sequence of vector spaces obviously verifies condition 3 of **E.10.42**.

Alternatively, assume that $f\colon M \longrightarrow N$ is an injective homomorphism. We can complete a basis of $f(M)$ to a basis of $N$, and write $N = f(M) \oplus L$, for some vector subspace $L$ of $N$.
It is easy to extend $g\colon M \longrightarrow F$ to a homomorphism $\tilde{g}\colon N \longrightarrow F$ such that $g = \tilde{g} \circ f$ by defining

$$\tilde{g}(f(m) + \ell) = g(m).$$

2. A free module is not injective in general.
Consider the exact sequence

$$0 \longrightarrow \mathbb{Z} \xrightarrow{n} \mathbb{Z} \longrightarrow \mathbb{Z}/(n) \longrightarrow 0,$$

and apply the functor $\mathrm{Hom}_{\mathbb{Z}}(\bullet, \mathbb{Z})$ to obtain the exact sequence

$$0 \longrightarrow \mathrm{Hom}_{\mathbb{Z}}(\mathbb{Z}/(n), \mathbb{Z}) \longrightarrow \mathrm{Hom}_{\mathbb{Z}}(\mathbb{Z}, \mathbb{Z}) \xrightarrow{n^*} \mathrm{Hom}_{\mathbb{Z}}(\mathbb{Z}, \mathbb{Z})$$

where $n^*$ is not surjective.
Indeed, it suffices to note that $\mathrm{Hom}_{\mathbb{Z}}(\mathbb{Z}, \mathbb{Z}) \simeq \mathbb{Z}$ and that $n^*$ still corresponds to the multiplication by $n$.
Thus, $\mathrm{Hom}_{\mathbb{Z}}(\bullet, \mathbb{Z})$ is not exact, and $\mathbb{Z}$ is not injective by **E.10.42**.

**Solution E. 10.44.** The implication $\Rightarrow$ immediately follows from the definition of injective module, because the inclusion homomorphism $I \longrightarrow A$ is injective.

Conversely, let $M$ and $N$ be $A$-modules, and consider homomorphisms $f\colon M \longrightarrow N$ and $g\colon M \longrightarrow E$, with $f$ injective.
Consider the set

$$S = \{(N', g')\colon N' \subseteq N \text{ and } g'\colon N' \longrightarrow E \text{ such that } g' \circ f = g\}$$

of pairs of submodules $N'$ of $N$ and homomorphisms $g'$ that extend $g$.
Note that $S$ is not empty, because it contains the pair $(f(M), \eta)$, where $\eta(f(m)) = g(m)$ for any $m \in M$.
It is easy to see that $S$, ordered by the relation

$$(N', g') < (N'', g'') \iff N' \subseteq N'' \text{ and } g''_{|N'} = g'$$

verifies the hypothesis of Zorn's Lemma.
Thus, $S$ has at least one maximal element $(\widehat{N}, \hat{g})$. If $\widehat{N} = N$ we are done.

Assume, by contradiction, that there exists $n \in N \setminus \widehat{N}$ and consider the ideal $I = \widehat{N} : n \subset A$.
We have a homomorphism

$$g_1 \colon I \xrightarrow{n} \widehat{N} \xrightarrow{\hat{g}} E,$$

which, by hypothesis, extends to $\widetilde{g_1} \colon A \longrightarrow E$, and we obtain a commutative diagram

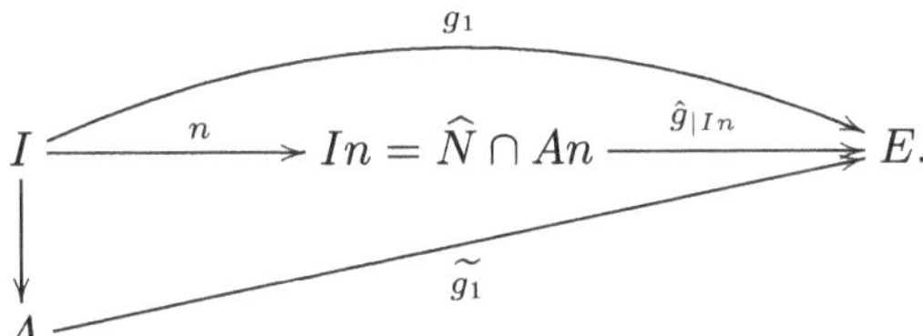


We can define $\tilde{g} \colon \widehat{N} + An \longrightarrow E$ by setting

$$\tilde{g}(\hat{n} + an) = \hat{g}(\hat{n}) + \widetilde{g_1}(a) \quad \text{for every } \hat{n} \in \widehat{N} \text{ and } a \in A.$$

The map $\tilde{g}$ is well-defined.
Indeed, if $\hat{n}_1 + a_1 n = \hat{n}_2 + a_2 n$, then $a_1 - a_2 \in \widehat{N} : n = I$, and

$$\begin{aligned}
\tilde{g}(\hat{n}_1 + a_1 n) - \tilde{g}(\hat{n}_2 + a_2 n) &= \hat{g}(\hat{n}_1) + \widetilde{g_1}(a_1) - \hat{g}(\hat{n}_2) - \widetilde{g_1}(a_2) \\
&= \hat{g}(\hat{n}_1 - \hat{n}_2) + \widetilde{g_1}(a_1 - a_2) \\
&= \hat{g}(a_2 n - a_1 n) + g_1(a_1 - a_2) \\
&= g_1(a_2 - a_1) + g_1(a_1 - a_2) = 0,
\end{aligned}$$

where the second-last equality immediately follows from the definition of $g_1$.
It is easy to verify that $\tilde{g}$ is a homomorphism and that $\tilde{g}_{|\widehat{N}} = \hat{g}$.
Therefore, we have a pair $(\widehat{N} + An, \tilde{g})$ in $S$ strictly bigger than $(\widehat{N}, \hat{g})$, contradicting the assumption that $(\widehat{N}, \hat{g})$ is maximal.

**Solution E. 10.45.** Since $A$ is a PID, projective $A$-modules are free, see **T.4.24**. A submodule $N$ of a free module $M$ is free by **T.4.23**, and hence, projective by **T.4.21**.

**Solution E. 10.46.** Let $m_1, m_2 \in T(M)$, and let $a_1, a_2 \in A \setminus \{0\}$ be such that $a_1 m_1 = a_2 m_2 = 0$. Then, $a_1 a_2 (m_1 + m_2) = 0$ with $a_1 a_2 \neq 0$.
For any $a \in A$, we also have $a_1(am_1) = a(a_1 m_1) = 0$.

**Solution E. 10.47.** Let $m_1$, $m_2 \in M_{[a]}$ and $b \in A$. Then, there exist $k_1$, $k_2 \in \mathbb{N}$ such that $a^{k_1} m_1 = a^{k_2} m_2 = 0$.
Therefore, $a^{\max\{k_1,k_2\}}(m_1+m_2) = 0$ and $a^{k_1}(bm_1) = 0$, that is, both $m_1+m_2$ and $bm_1$ are in $M_{[a]}$.

**Solution E. 10.48.** Since $M_{[a]}$, $M_{[b]}$ and $M_{[ab]}$ are finitely generated by **T.4.23**, there exists some $k \in \mathbb{N}$ such that

$$a^k M_{[a]} = b^k M_{[b]} = (ab)^k M_{[ab]} = 0.$$

Since $\gcd(a^k, b^k) = 1$, there exist $s$, $t \in A$ such that $sa^k + tb^k = 1$.
Now, set $c = tb^k$ and $d = sa^k$.
It is easy to see that:

i) $cM_{[ab]} \subseteq M_{[a]}$ and $dM_{[ab]} \subseteq M_{[b]}$;
ii) $cM_{[b]} = 0$ and $dM_{[a]} = 0$;
iii) $cm_a = (c+d)m_a = m_a$ for all $m_a \in M_{[a]}$, and $dm_b = (c+d)m_b = m_b$ for all $m_b \in M_{[b]}$.

1. The inclusion $M_{[a]} + M_{[b]} \subseteq M_{[ab]}$ is always true. Indeed, given $m_1 \in M_{[a]}$ and $m_2 \in M_{[b]}$, there exist integers $k_1$, $k_2$ such that $a^{k_1} m_1 = b^{k_2} m_2 = 0$. Consequently, $(ab)^{\max\{k_1,k_2\}}(m_1 + m_2) = 0$.
To show the opposite inclusion, note that if $m \in M_{[ab]}$, by i) we have

$$m = 1 \cdot m = (c+d)m \in M_{[a]} + M_{[b]}.$$

The sum is direct because if $m \in M_{[a]} \cap M_{[b]}$ then $m = (c+d)m = 0$ by ii).
2. From part 1, using ii) and iii), we obtain that multiplication by $c$, respectively by $d$, is the projection of $M_{[ab]}$ onto $M_{[a]}$, respectively onto $M_{[b]}$.
3. Let $M_{[ab]} = \langle m \rangle$. From part 2, it follows that $M_{[a]}$ is generated by $cm$ and $M_{[b]}$ by $dm$.
Conversely, if $M_{[a]} = \langle m_a \rangle$ and $M_{[b]} = \langle m_b \rangle$, let $m = m_a + m_b$. Then, $cm = m_a$ and $dm = m_b$. Thus, $\langle m \rangle \subseteq M_{[ab]}$.
For any $m' \in M_{[ab]}$, we have

$$m' = cm' + dm' = c'm_a + d'm_b = c'cm + d'dm = (c'c + d'd)m,$$

for some $c'$, $d' \in A$. Hence, $M_{[ab]} = \langle m \rangle$.

**Solution E. 10.49.** By hypothesis, $M \simeq A/(d)$, where $d \notin A^* \cup \{0\}$.
Note that $M$ is equal to its $d$-component $M_{[d]}$.
Let $d = \prod_{j=1}^h p_i^{e_i}$ be the factorization of $d$ into distinct irreducible factors, each counted with its multiplicity.

By **E.10.48**, we have

$$M = M_{[d]} = \bigoplus_{i=1}^{h} M_{[p_i^{e_i}]} = \bigoplus_{i=1}^{h} M_{[p_i]},$$

where $M_{[p_i]} \simeq A/(p_i^{e_i})$, as desired.

**Solution E. 10.50.** 1. Let $\{m_\alpha \colon \alpha \in \Lambda\}$ be a basis of $M$, and consider a non-zero element $m = \sum_{\alpha\in\Lambda} a_\alpha m_\alpha \in M$.
Suppose there exists $a \in A$ such that $am = 0$. Then, $\sum_{\alpha\in\Lambda} a a_\alpha m_\alpha = 0$ implies $a a_\alpha = 0$ for all $\alpha$. Since at least one $a_\alpha$ is not zero and $A$ is a domain, we obtain $a = 0$.
Therefore, we have shown that in $M$ there is no non-trivial torsion element.

2. It is a straightforward application of **T.4.33**.2.

3. Consider $A = K[x, y]$, which is a UFD but not a PID.
The ideal $I = (x,\ y)$ is finitely generated and torsion-free, but it is not free.

Now, consider $\mathbb{Q}$, which is a torsion-free, not finitely generated $\mathbb{Z}$-module. Clearly $\mathbb{Q}$ is not free, since each pair of elements of $\mathbb{Q}$ is linearly dependent over $\mathbb{Z}$.

**Solution E. 10.51.** Clearly, $M \simeq \operatorname{Coker} f$.
The matrix associated with $f$ with respect to the canonical bases is

$$\begin{pmatrix} 1 & 1 & 1 \\ -3 & 1 & 1 \\ 1 & -3 & -3 \\ 1 & 3 & 1 \end{pmatrix}$$

whose Smith form is

$$\begin{pmatrix} 1 & 0 & 0 \\ 0 & 2 & 0 \\ 0 & 0 & 4 \\ 0 & 0 & 0 \end{pmatrix}.$$

It follows that

$$M \simeq \mathbb{Z} \oplus \mathbb{Z}/(2) \oplus \mathbb{Z}/(4).$$

**Solution E. 10.52.** The Smith form of the matrix associated with $\varphi$ with respect to the canonical bases is

$$\begin{pmatrix} 1 & 0 & 0 & 0 \\ 0 & 1 & 0 & 0 \\ 0 & 0 & x & 0 \\ 0 & 0 & 0 & x^2(x-1) \end{pmatrix}.$$

Therefore,

$$\operatorname{Coker} \varphi \simeq 0 \oplus 0 \oplus \mathbb{Q}[x]/(x) \oplus \mathbb{Q}[x]/(x^2) \oplus \mathbb{Q}[x]/(x-1),$$

where this isomorphism is due to the fact that $(x^2)$ and $(x-1)$ are comaximal. Thus, $\dim_{\mathbb{Q}} \operatorname{Coker} \varphi = 4$.

Alternatively, if $d_1$, $d_2$, $d_3$ and $d_4$ denote the invariant factors of the matrix which represents $\varphi$, then $\operatorname{Coker} \varphi \simeq \bigoplus_{i=1}^{4} \mathbb{Q}[x]/(d_i)$.

It follows that $\dim_{\mathbb{Q}} \operatorname{Coker} \varphi = \sum \deg d_i = \deg(d_1 \cdots d_4) = 4$.

**Solution E. 10.53.** The matrix representing $\varphi$ is

$$\begin{pmatrix} 6 & 2 & 4 \\ 0 & a & 4 \\ 2 & 2 & 2 \end{pmatrix},$$

To compute the ideals $\Delta_i$, we first simplify it with the help of elementary operations, and obtain

$$\begin{pmatrix} 2 & 2 & 2 \\ 0 & 2 & 4 \\ 0 & 0 & a-8 \end{pmatrix}.$$

i) Let $a = 2k+1$. Then, $\Delta_1 = (1)$, $\Delta_2 = (2)$ and $\Delta_3 = (4(a-8))$, from which it follows $d_1 = 1$, $d_2 = 2$, $d_3 = 2(a-8)$ and

$$\operatorname{Coker} \varphi \simeq \mathbb{Z}/(2) \oplus \mathbb{Z}/(2a-16).$$

Therefore, for any odd $a$, $\operatorname{Coker} \varphi$ is finite.

ii) Let $a = 2k$. Then, $\Delta_1 = (2)$, $\Delta_2 = (4)$ and $\Delta_3 = (4(a-8))$. We obtain $d_1 = d_2 = 2$, $d_3 = a-8$, and

$$\operatorname{Coker} \varphi \simeq \mathbb{Z}/(2) \oplus \mathbb{Z}/(2) \oplus \mathbb{Z}/(a-8).$$

In this case, $\operatorname{Coker} \varphi$ is finite for all values $a \neq 8$.

In conclusion, $\operatorname{Coker} \varphi$ is infinite if and only if $a = 8$.

**Solution E. 10.54.** We have $\Delta_1 = (\gcd(a,b,c))$, $\Delta_2 = (\gcd(a^2, ab, b^2 - ac))$, and $\Delta_3 = (a^3)$.

1. $\operatorname{Coker} \varphi$ has at most two generators if and only if $d_1 = 1$.
This happens if and only if $\Delta_1 = (1)$.

2. $\operatorname{Coker} \varphi$ is cyclic if and only if $d_1 = d_2 = 1$.
If $\gcd(a,b) = 1$, then $\gcd(a,b,c) = 1$ and $\gcd(a^2, ab) = a$. Therefore, $\Delta_1 = (1)$ and $\Delta_2 = (a, b^2) = (1)$, and we obtain $d_1 = d_2 = 1$.

Conversely, let $d = \gcd(a,b)$. Then, $d \mid \gcd(a^2, ab, b^2 - ac) = 1$, and thus, $d = 1$.

**Solution E. 10.55.** With some elementary operations, we can simplify the matrix and obtain

$$\begin{pmatrix} a & 0 & 0 \\ -3 & 6 & 0 \\ 0 & 3 & 3 \end{pmatrix}.$$

As a result, $\Delta_1 = (a, 3)$, $\Delta_2 = (3a, 9)$ and $\Delta_3 = (18a)$.

1. In order for $\operatorname{Coker}\varphi$ to be finite, we must have $\Delta_3 \neq (0)$.
This condition is satisfied if and only if $a \neq 0$.

2. We have $d_1 = 1$ if $\gcd(a, 3) = 1$ and $d_1 = 3$, otherwise.
In the second case $\operatorname{Coker}\varphi$ is clearly not cyclic.
If $\gcd(a, 3) = 1$, we have $\Delta_2 = (3a, 9) = (3(a, 3)) = (3)$ and, again, $\operatorname{Coker}\varphi$ is not cyclic.

**Solution E. 10.56.** We compute the Smith form of the matrix

$$\begin{pmatrix} 0 & 3 & 3 \\ a & 3 & -1 \\ b & 0 & 0 \end{pmatrix},$$

whose columns are the vectors $m_1$, $m_2$, $m_3$
Since $\Delta_1 = (1)$, we obtain $d_1 = 1$, $\Delta_2 = (b, 3a, 12)$ and $\Delta_3 = (12b)$.
Thus, $M$ is finite if and only if $b \neq 0$.

Moreover, $M$ is cyclic if and only if $\Delta_2 = (1)$.
Since

$$\gcd(b, 3a, 12) = \gcd(b, 3\gcd(a, 4)) = \gcd(b, 3)\gcd(b, \gcd(a, 4)),$$

we have $b \not\equiv 0 \mod 3$ and $\gcd(b, \gcd(a, 4)) = 1$.
If $a$ is odd, the above condition is verified, otherwise either $\gcd(a, 4) = 2$ or $\gcd(a, 4) = 4$.

In conclusion, $M$ is cyclic if and only if

i) $b \not\equiv 0 \mod 3$ and $a$ is odd, or
ii) $b \not\equiv 0 \mod 3$, $a$ is even and $b$ is odd.

**Solution E. 10.57.** The Smith form of the matrix whose columns are $m_1$, $m_2$ and $m_3$ is

$$\begin{pmatrix} 2 & 0 & 0 \\ 0 & 4 & 0 \\ 0 & 0 & 8 \end{pmatrix}.$$

Thus,

$$M \simeq \mathbb{Z}/(2) \oplus \mathbb{Z}/(4) \oplus \mathbb{Z}/(8) \quad \text{and} \quad \operatorname{Ann}_{\mathbb{Z}} M = (8).$$

**Solution E. 10.58.** Since $\det A = 28$, the Smith forms of $A$ have $1, 1, 28$ or $1, 2, 14$ as diagonal values. From $\det B = 7$, we compute the Smith form of $B$, given by the invariant factors $1, 1, 7$.

Analogously, since $\det D = \det A \det B = 196$, the following are all the possible Smith forms of $D$, presented along with the corresponding matrices $A$ and $C$, and with $B = \operatorname{diag}(1, 1, 7)$:

$$\begin{aligned}
D_1 &= \operatorname{diag}(1,1,1,1,1,196), && \text{with } A = \operatorname{diag}(1,1,28),\ C = \operatorname{diag}(0,0,1);\\
D_2 &= \operatorname{diag}(1,1,1,1,7,28), && \text{with } A = \operatorname{diag}(1,1,28),\ C = 0;\\
D_3 &= \operatorname{diag}(1,1,1,1,2,98), && \text{with } A = \operatorname{diag}(1,2,14),\ C = \operatorname{diag}(0,0,2);\\
D_4 &= \operatorname{diag}(1,1,1,1,14,14), && \text{with } A = \operatorname{diag}(1,2,14),\ C = 0.
\end{aligned}$$

**Solution E. 10.59.** The matrix representing the relations on the elements of $M$ is

$$\begin{pmatrix} 3 & 2 & 1 \\ 0 & -2 & 4 \\ 1 & 1 & 2 \end{pmatrix},$$

whose Smith form is

$$\begin{pmatrix} 1 & 0 & 0 \\ 0 & 1 & 0 \\ 0 & 0 & 14 \end{pmatrix}.$$

Therefore, $M \simeq \mathbb{Z}/(14)$, and the possible orders for an element of $M$ are 1, 2, 7 and 14.

**Solution E. 10.60.** Since $M$ is a finitely generated $\mathbb{Z}$-module, by **T.4.33**.2 we obtain $M \simeq \mathbb{Z}^r \oplus T(M)$, where $r \geq 0$ and $T(M)$ is the torsion submodule of $M$. However, the hypothesis implies that $M$ is a torsion module, hence $r = 0$. By the structure theorem **T.4.31**, we have $M \simeq \bigoplus_{i=1}^{n} \mathbb{Z}/I_i$, where $I_i \neq 0$ for all $i$ and $I_i = 0 : m_i$ for some $m_i \in M$, see also the proof of **T.4.31**.
Assume that $I_1 = \ldots = I_s = \mathbb{Z}$ and $I_{s+1} \subsetneq \mathbb{Z}$.

If $s = n$, then $R\mathbb{Z}^n = \mathbb{Z}^n$. Consequently, $R$ is invertible with determinant $\pm 1$ and $M = 0$.

If $s < n$, then $(1) \neq I_i \supseteq (p_{m_i})$ for all $i = s+1, \ldots, n$.
Since the $p_{m_i}$ are prime, they generate maximal ideals. Therefore, $I_i = (p_{m_i})$ for all $i = s+1, \ldots, n$.
Moreover, $I_h \subseteq I_k$ for all $h \geq k$, and thus, we have $I_i = I_{s+1} = (p_{m_{s+1}})$ for all $i = s+1, \ldots, n$.

Let $p = p_{m_{s+1}}$. Then, $\det R$ is an associate of the determinant of the Smith form $D$ which is, for what we have shown above, $\det D = p^{n-s} \neq 0$, with $n - s \leq n$.

**Solution E. 10.61.** 1. Consider $\varphi \colon \mathbb{Z}^3 \longrightarrow M$ given by $\varphi(e_i) = m_i$, with $i = 1, 2, 3$. Then, $M \simeq \mathbb{Z}^3/\operatorname{Ker}\varphi$.
Let

$$\begin{pmatrix} 2 & 10 & 6 \\ -4 & -6 & -12 \\ -2 & 4 & a \end{pmatrix}$$

be the matrix of relations among the generators of $M$, and reduce it with some elementary row and column operations to the matrix

$$\begin{pmatrix} 2 & 0 & 0 \\ 0 & 14 & 0 \\ 0 & 0 & 6+a \end{pmatrix}.$$

From the latter, we recover the associated Smith form and the desired representation of $M$, with $a$ ranging over $\mathbb{Z}$:

| | |
|---|---|
| $\begin{pmatrix} 2 & 0 & 0 \\ 0 & 14 & 0 \\ 0 & 0 & 6+a \end{pmatrix}$ <br> $a \equiv 8 \mod 14$ <br> $M \simeq \mathbb{Z}/(2) \oplus \mathbb{Z}/(14) \oplus \mathbb{Z}/(a+6)$ | $\begin{pmatrix} 2 & 0 & 0 \\ 0 & 2 & 0 \\ 0 & 0 & 7(6+a) \end{pmatrix}$ <br> $a \equiv 0 \mod 2$ <br> $a \not\equiv 1 \mod 7$ <br> $M \simeq \mathbb{Z}/(2) \oplus \mathbb{Z}/(2) \oplus \mathbb{Z}/(7a+42)$ |
| $\begin{pmatrix} 1 & 0 & 0 \\ 0 & 14 & 0 \\ 0 & 0 & 2(6+a) \end{pmatrix}$ <br> $a \equiv 1 \mod 2$ <br> $a \equiv 1 \mod 7$ <br> $M \simeq \mathbb{Z}/(14) \oplus \mathbb{Z}/(2a+12)$ | $\begin{pmatrix} 1 & 0 & 0 \\ 0 & 2 & 0 \\ 0 & 0 & 14(6+a) \end{pmatrix}$ <br> $a \equiv 1 \mod 2$ <br> $a \not\equiv 1 \mod 7$ <br> $M \simeq \mathbb{Z}/(2) \oplus \mathbb{Z}/(14a+84)$. |

2. We have $\operatorname{Ann} M = 0$ only for $a = -6$.

**Solution E. 10.62.** Since $G_1$ and $G_2(\alpha)$ are finitely generated Abelian groups, they can be written as a direct sum of cyclic groups and they are isomorphic when they share the same invariant factors.
Consider the matrix

$$R = \begin{pmatrix} 2 & 0 & -4 & 6 & 12 \\ 2 & -2 & 4 & 4 & 4 \\ 1 & 1 & -3 & 1 & 1 \\ 3 & 3 & -15 & 9 & 21 \end{pmatrix}.$$

We have $G_1 \simeq \operatorname{Coker} \psi$, where $\psi\colon \mathbb{Z}^5 \longrightarrow \mathbb{Z}^4$ is the homomorphism associated with $R$ with respect to the canonical bases.
A computation of the Smith form of $R$ yields

$$\begin{pmatrix} 1 & 0 & 0 & 0 & 0 \\ 0 & 2 & 0 & 0 & 0 \\ 0 & 0 & 6 & 0 & 0 \\ 0 & 0 & 0 & 0 & 0 \end{pmatrix},$$

from which it follows $G_1 \simeq \mathbb{Z} \oplus \mathbb{Z}/(2) \oplus \mathbb{Z}/(6)$.

Now, we compute the Smith form associated with the matrix

$$S = \begin{pmatrix} 2 & 8 & -4 \\ \alpha & 6 & \alpha \\ -2 & -2 & 4 \end{pmatrix},$$

which represents $\varphi_\alpha$, and we obtain $G_2(\alpha)$.

In order to have an isomorphism, the invariant factors of $S$ must be 2, 6 and 0. Since $\Delta_3(S) = (\det S) = (36\alpha)$, the only possible value is $\alpha = 0$. For $\alpha = 0$ we also have $\Delta_1(S) = 2$ and $\Delta_2(S) = 12$. It follows that $d_1 = 2$, $d_2 = 6$. Thus, $G_1 \simeq G_2(0)$.

**Solution E. 10.63.** The Smith form of $R - xI$ is a diagonal matrix $D$, where

$$D = \mathrm{diag}(d_1, \ldots, d_6),\ d_1 \mid d_2 \mid \ldots \mid d_6 \text{ and } d_1 \cdots d_6 = (x-1)^\alpha (x-2)^\beta (x^2+1).$$

By the divisibility conditions and since $\deg(p_A(x)) = 6$, we have that $(x^2+1)$ divides only $d_6$, that $d_1 = d_2 = 1$, that $\alpha + \beta = 4$, and that the multiplicities $\gamma_i$ of $(x-1)$ as a divisor of $d_i$ must satisfy the following constraints:

$$\gamma_3 + \gamma_4 + \gamma_5 + \gamma_6 = \alpha \quad \text{and} \quad \gamma_3 \le \gamma_4 \le \gamma_5 \le \gamma_6.$$

Therefore, the possible 4-tuples $(\gamma_3, \gamma_4, \gamma_5, \gamma_6)_\alpha$ are

$$(0,0,0,0)_0,\ (0,0,0,1)_1,\ (0,0,0,2)_2,\ (0,0,1,1)_2,\ (0,0,0,3)_3,$$

$$(0,0,1,2)_3,\ (0,1,1,1)_3,\ (0,0,0,4)_4,\ (0,0,1,3)_4,\ (0,0,2,2)_4,$$
$$(0,1,1,2)_4 \text{ and } (1,1,1,1)_4.$$

In a similar way, we compute the possible 4-tuples for the multiplicities of $(x-2)$. To determine the Smith form, one has to consider all the couples of 4-tuples for which $\alpha + \beta = 4$ holds.

If $\alpha = 0$ and $\beta = 4$ or $\alpha = 4$ and $\beta = 0$, we have 5 possible Smith forms, as many as the 4-tuples whose coordinates sum to 4.

If $\alpha = 1$ and $\beta = 3$ or $\alpha = 3$ and $\beta = 1$, we have 3 possible Smith forms. Finally, if $\alpha = \beta = 2$, we have the 4 Smith forms:

$$\begin{aligned}
D_1 &= \mathrm{diag}(1,\ 1,\ 1,\ 1,\ 1,\ (x-1)^2(x-2)^2(x^2+1)); \\
D_2 &= \mathrm{diag}(1,\ 1,\ 1,\ 1,\ (x-1),\ (x-1)(x-2)^2(x^2+1)); \\
D_3 &= \mathrm{diag}(1,\ 1,\ 1,\ 1,\ (x-2),\ (x-1)^2(x-2)(x^2+1)); \\
D_4 &= \mathrm{diag}(1,\ 1,\ 1,\ 1,\ (x-1)(x-2),\ (x-1)(x-2)(x^2+1)).
\end{aligned}$$

**Solution E. 10.64.** We have $M \simeq \mathrm{Coker}\,\varphi$, where $\varphi\colon \mathbb{Z}^3 \longrightarrow \mathbb{Z}^4$ is the homomorphism represented by the matrix

$$\begin{pmatrix} 3 & a & 0 \\ 0 & 3 & b \\ 0 & 0 & 3 \\ 0 & 0 & 0 \end{pmatrix}.$$

To compute its Smith form, consider the ideals

$$\Delta_1 = (3, a, b) = (1), \quad \Delta_2 = (9, 3a, 3b, ab), \quad \text{and} \quad \Delta_3 = (27).$$

If $(3, ab) = 1$, then $\Delta_2 = (1)$ and $\Delta_3 = (27)$.
Therefore,

$$M \simeq \mathbb{Z} \oplus \mathbb{Z}/(27) \quad \text{and} \quad T(M) \simeq \mathbb{Z}/(27).$$

Otherwise, if $(3, ab) = 3$, then $\Delta_2 = (3)$ and $\Delta_3 = (27)$.
Therefore,

$$M \simeq \mathbb{Z} \oplus \mathbb{Z}/(3) \oplus \mathbb{Z}/(9) \quad \text{and} \quad T(M) \simeq \mathbb{Z}/(3) \oplus \mathbb{Z}/(9).$$

**Solution E. 10.65.** 1. Consider the matrix

$$\begin{pmatrix} 2 & 2 & 0 \\ 2 & a & 4 \\ a & 0 & 2 \end{pmatrix},$$

and compute

$$\Delta_1 = (2, a), \quad \Delta_2 = (4, 2a, a^2) \quad \text{and} \quad \Delta_3 = (4(2 - 3a)).$$

We have two cases:

i) if $\gcd(2, a) = 1$, then the Smith form is

$$\begin{pmatrix} 1 & 0 & 0 \\ 0 & 1 & 0 \\ 0 & 0 & 4(2 - 3a) \end{pmatrix}.$$

Therefore,

$$M \simeq \mathbb{Z}/(4(2 - 3a));$$

ii) if $\gcd(2, a) = 2$, then the Smith form is

$$\begin{pmatrix} 2 & 0 & 0 \\ 0 & 2 & 0 \\ 0 & 0 & 2 - 3a \end{pmatrix}.$$

Therefore,

$$M \simeq \mathbb{Z}/(2) \oplus \mathbb{Z}/(2) \oplus \mathbb{Z}/(2 - 3a).$$

2. By **E.10.17** and **E.10.19**.3, the module $\mathrm{Hom}_{\mathbb{Z}}(\mathbb{Z}/(7), M)$ is non-trivial if and only if $2 - 3a \equiv 0 \mod 7$, that is, $a \equiv 3 \mod 7$.

**Solution E. 10.66.** Let $M \simeq \mathrm{Coker}\, f$, where the homomorphism $f \colon \mathbb{Z}^3 \longrightarrow \mathbb{Z}^3$ is associated with the matrix

$$\begin{pmatrix} 2 & 1 & 1 \\ -1 & -3 & 1 \\ 0 & 0 & -a \end{pmatrix},$$

whose Smith form is

$$\begin{pmatrix} 1 & 0 & 0 \\ 0 & 1 & 0 \\ 0 & 0 & 5a \end{pmatrix}.$$

It immediately follows that $M \simeq \mathbb{Z}/(5a)$.

1. If $a = 3$, we may define $\varphi \colon \mathbb{Z}/(20) \longrightarrow \mathbb{Z}/(15)$ by letting $\varphi(n) = 3n$.

2. We have

$$\mathrm{Hom}_{\mathbb{Z}}(\mathbb{Z}/(20), M) \simeq \mathrm{Hom}_{\mathbb{Z}}(\mathbb{Z}/(20), \mathbb{Z}/(5a)) \simeq ((5a) : (20))/(5a),$$

see **E.10.19**.3.
If $a = 0$, then $\mathrm{Hom}_{\mathbb{Z}}(\mathbb{Z}/(20), \mathbb{Z}) = 0$.
Otherwise, $a \neq 0$ and

$$\mathrm{Hom}_{\mathbb{Z}}(\mathbb{Z}/(20), M) \simeq \begin{cases} \mathbb{Z}/(5) & \text{if } \gcd(a, 4) = 1; \\ \mathbb{Z}/(10) & \text{if } \gcd(a, 4) = 2; \\ \mathbb{Z}/(20) & \text{if } \gcd(a, 4) = 4. \end{cases}$$

**Solution E. 10.67.** Let $\varphi \colon \mathbb{Z}^3 \longrightarrow \mathbb{Z}^3$ be the homomorphism defined by $\varphi(e_i) = m_i$, with $i = 1, 2, 3$. Then, $M \simeq \mathbb{Z}^3 / \mathrm{Ker}\, \psi \simeq \mathrm{Coker}\, \varphi$.
The matrix representing $\varphi$ with respect to the canonical bases is

$$\begin{pmatrix} 2 & 0 & b \\ 4 & a & 4 \\ 6 & 2a & 6 \end{pmatrix}.$$

Therefore,

$$\Delta_1 = (2, a, b), \ \ \Delta_2 = (2a, ab, 4(b-2)) \ \text{and} \ \ \Delta_3 = (2a(2-b)).$$

By **E.10.15**, $M$ is simple if and only if $d_1 = d_2 = 1$ and $d_3$ is prime, that is, $\gcd(2, a, b) = \gcd(2a, ab, 4(b-2)) = 1$ and $2a(b-2) = \pm 2$.
Thus,

$$a = \pm 1 \ \text{and} \ \ b = 1 \ \text{or} \ \ b = 3.$$

Alternatively, reducing the matrix with elementary operations we obtain

$$\begin{pmatrix} 2 & 0 & 0 \\ 0 & a & 0 \\ 0 & 0 & b-2 \end{pmatrix}.$$

Thus, by **E.10.15**, $M$ is simple if and only if $M \simeq \mathbb{Z}/(2)$, *i.e.*, if and only if $a = \pm 1$ and $b - 2 = \pm 1$.

**Solution E. 10.68.** Let $B$ be the ring $K[x, y]/(x^2 - y^2)$.

1. The rings $A$ and $B$ are isomorphic and we claim that in both cases $\{\overline{1}, \overline{y}\}$ is a basis as a $K[x]$-module. Clearly, it is a generating set.
Moreover, there are no relations between $\overline{1}$ and $\overline{y}$.
Indeed, if $f(x)$, $g(x) \in K[x]$ are such that $f(x)\overline{1} + g(x)\overline{y} = \overline{0}$, then we have $f(x) + g(x)y \in (x^2 - y^2)$, that is,

$$f(x) + g(x)y = (x^2 - y^2)q(x, y)$$

for some $q(x, y) \in K[x, y]$.
The degree in $y$ of the polynomial on the left-hand side is at most 1, whereas it is at least 2 on the right-hand side. Therefore, equality is only possible when $q(x, y) = 0$. Hence, $f(x) + g(x)y = 0$, and finally, $f(x) = g(x) = 0$.
Thus, as a $K[x]$-module, $A$ is isomorphic to $K[x]^2$.

2. In this case

$$A \simeq K[x, y, y^{-1}]/(y^2 - y^{-1} + x^2).$$

In $A$, we have $\overline{y}^{-1} = \overline{y}^2 + \overline{x}^2$, that is, $\overline{1} = \overline{y}^3 + \overline{yx}^2$.
Thus, $\overline{1}$, $\overline{y}$, $\overline{y}^2$ generate $A$ as a $K[x]$-module.

We claim that this generating set is free.
If $f(x)$, $g(x)$, $h(x) \in K[x]$ are such that $f(x)\overline{1} + g(x)\overline{y} + h(x)\overline{y}^2 = \overline{0}$, then $f(x) + g(x)y + h(x)y^2 = (y^2 - y^{-1} + x^2)q(x, y, y^{-1})$ for some polynomial $q(x, y, z) \in K[x, y, z]$.
Therefore,

$$f(x)y + g(x)y^2 + h(x)y^3 = (y^3 - 1 + x^2y)q(x, y, y^{-1}).$$

Since $y^{-1}$ does not appear on the left-hand side, it cannot appear on the right-hand side as well. Thus, we can write $q(x, y, y^{-1}) = q(x, y)$.
Moreover, the degree in $y$ on the left-hand side is at most 3. If $\deg_y q(x, y) \geq 1$, the degree in $y$ on the right-hand side would be at least 4. Thus, $y$ does not appear in $q(x, y)$, and we can write $q(x, y) = q(x)$.
It follows that

$$f(x)y + g(x)y^2 + h(x)y^3 = (y^3 - 1 + x^2y)q(x).$$

Therefore, $h = q$, $g = 0$, $f = x^2q$ and $0 = -q$. Thus, $f = g = h = 0$, as we claimed.

We can conclude that, as a $K[x]$-module, $A$ is isomorphic to $K[x]^3$.

3. The ring $A$ is a quotient of the ring $B$ we defined above, modulo the principal ideal generated by $a = \overline{x}^4 - \overline{x}^3\,\overline{y} + \overline{y}$.
As we have seen in part 1, $B$ is a free $K[x]$-module generated by $\{\overline{1}, \overline{y}\}$. Therefore, any element of $(a)$ can be written as $ap(x, y) = ap_0(x) + ap_1(x)\overline{y}$. Thus, as a $K[x]$-module, $(a)$ is generated by

$$a = \overline{x}^4 - \overline{x}^3\,\overline{y} + \overline{y} = x^4\,\overline{1} + (1 - x^3)\overline{y} \quad \text{and} \quad a\overline{y} = (-x^5 + x^2)\overline{1} + x^4\,\overline{y}.$$

Therefore, if we let $\varphi\colon K[x]^2 \longrightarrow K[x]^2$ be the $K[x]$-module homomorphism defined by $\varphi(e_1) = x^4 e_1 + (1 - x^3)e_2$ and $\varphi(e_2) = (-x^5 + x^2)e_1 + x^4 e_2$, then $A \simeq \operatorname{Coker}\varphi$, and the associated matrix is

$$\begin{pmatrix} x^4 & -x^5 + x^2 \\ 1 - x^3 & x^4 \end{pmatrix}.$$

Since $\Delta_1 = (1)$ and $\Delta_2 = (2x^5 - x^2)$, we have that

$$A \simeq K[x]/(2x^5 - x^2)$$

is cyclic.

Using Gröbner bases may simplify the previous computations.
For instance, in part 3, a Gröbner basis of the ideal $(x^2 - y^2,\ x^4 - x^3y + y)$ with respect to the lex order with $y > x$ is

$$\{y + 2x^4,\ 2x^5 - x^2\},$$

and this fact immediately shows that $A \simeq K[x]/(2x^5 - x^2)$.

**Solution E. 10.69.** The reduced Gröbner basis of $I$ with respect to the lex order with $x > y > z$ is

$$\{x + y^3 - yz,\ y^4 - y^2z + 1\}.$$

1. We have

$$A/I \simeq K[z][y]/(y^4 - y^2z + 1),$$

which, as a $K[z]$-module, is generated by $\{\overline{1}, \overline{y}, \overline{y}^2, \overline{y}^3\}$.
2. Since $\overline{1}$, $\overline{y}$, $\overline{y}^2$ and $\overline{y}^3$ are independent modulo $I$, they form a basis of $A$, and thus, $A \simeq K[z]^4$.

**Solution E. 10.70.** 1. The reduced Gröbner basis of $I$ with respect to the lex order with $x > y > z$ is

$$\{x^2,\ xy,\ yz,\ z^2\}.$$

Therefore, $I$ is monomial, see **E.9.7**.
2. It is immediately verified that

$$M = \langle \overline{1}, \overline{x}, \overline{z}, \overline{xz} \rangle_{K[y]}.$$

3. Since $\overline{yx} = 0$ in $M$, we have that $0 \neq \overline{x} \in T(M)$, and thus, $M$ is not free.

4. From now on, we omit the bar which denotes equivalence classes.
Consider the homomorphism $\varphi\colon K[y]^4 \longrightarrow M$ given by

$$\varphi(e_1) = 1,\ \varphi(e_2) = x,\ \varphi(e_3) = z \ \text{ and } \ \varphi(e_4) = xz.$$

Assume $(a_1, a_2, a_3, a_4) \in \operatorname{Ker}\varphi$. Then,

$$0 = \varphi(a_1, a_2, a_3, a_4) = a_1(y) + a_2(y)x + a_3(y)z + a_4(y)xz,$$

*i.e.*, $a_1(y) + a_2(y)x + a_3(y)z + a_4(y)xz \in I$.
Since $I$ is monomial, we have $a_1 = 0,\ a_2 = ya_2',\ a_3 = ya_3'$ and $a_4 = ya_4'$.
Therefore,

$$\operatorname{Ker}\varphi = \{(0,\ yb_2,\ yb_3,\ yb_4)\colon b_2,\ b_3,\ b_4 \in K[y]\}$$

is free with basis

$$\{v_1 = (0, y, 0, 0),\ v_2 = (0, 0, y, 0),\ v_3 = (0, 0, 0, y)\}.$$

Now, call $f_1, f_2, f_3$ the vectors of the canonical basis of $K[y]^3$ and define $\psi\colon K[y]^3 \longrightarrow K[y]^4$ by letting $f_i \mapsto v_i$, with $i = 1, 2, 3$.
Then, $M \simeq \operatorname{Coker}\psi$ and the Smith form associated with the matrix that represents $\psi$ is

$$\begin{pmatrix} y & 0 & 0 \\ 0 & y & 0 \\ 0 & 0 & y \\ 0 & 0 & 0 \end{pmatrix}.$$

It immediately follows that

$$M \simeq (K[y]/(y))^3 \oplus K[y].$$

**Solution E. 10.71.** Recall that both $K[x, y]$, and thus $A$, as its quotient, are $K[x]$-modules by restriction of scalars via the inclusion homomorphism $K[x] \longrightarrow K[x, y]$.
Let $g = y^3 - xy^2 - y + x$.

1. We prove that $A$ as a $K[x]$-module is generated by $\overline{1}$, $\overline{y}$ and $\overline{y}^2$.
We write any element $f(x, y)$ of $K[x, y] = K[x][y]$ as

$$f(x, y) = p_0(x) + p_1(x)y + p_2(x)y^2 + p_3(x)y^3 + \dots.$$

Using the relation $\overline{g} = \overline{0}$, in $A$ we obtain

$$\overline{f} = \sum_{i=0}^{2} \overline{q_i(x)}\,\overline{y}^i,$$

for some $q_i \in K[x]$.

Alternatively, observe that, since $g$ is monic as a polynomial in $y$, the division of any polynomial $f$ by $g$ yields $f = qg + r$, with $\deg_y r < 3$. Therefore, in $A$ we have $\overline{f} = \overline{r} \in \langle 1, \overline{y}, \overline{y}^2 \rangle_{K[x]}$.

2. From now on, for simplicity, we omit the bar which denote equivalence classes in the quotient ring.
Let $B = K[x, y]/(g)$, and let $h = x^2 - xy + x - y$. Then, $A \simeq B/(h)$, and it is easy to see that $B$ is a free $K[x]$-module with basis $\{1, y, y^2\}$.

Let $f \colon K[x]^3 \longrightarrow B$ the $K[x]$-module homomorphism defined by

$$f(e_1) = h, \; f(e_2) = yh \;\text{ and }\; f(e_3) = y^2 h.$$

Then, $\operatorname{Im} f = \langle h, yh, y^2 h \rangle_B$, and therefore, $\operatorname{Im} f \subseteq (h)$.

To prove the opposite inclusion, let $s \in (h)$. Then, $s = p(x, y)h$ for some $p(x, y) \in B$. Therefore, $p(x, y) = a_0(x) + a_1(x)y + a_2(x)y^2$, and accordingly, $s \in \operatorname{Im} f$.
Thus, $\operatorname{Im} f = (h)$ and

$$A \simeq B/(h) = \operatorname{Coker} f.$$

The matrix representing $f$ with respect to the bases $\{e_1, e_2, e_3\}$ and $\{1, y, y^2\}$ is

$$\begin{pmatrix} x^2 + x & 0 & x^2 + x \\ -x - 1 & x^2 + x & -x - 1 \\ 0 & -x - 1 & 0 \end{pmatrix}.$$

Hence, $\Delta_1 = (x + 1)$, $\Delta_2 = (x + 1)^2$ and $\Delta_3 = (0)$.
It immediately follows that $d_1 = d_2 = x + 1, \; d_3 = 0$, and

$$A \simeq (K[x]/(x + 1))^2 \oplus K[x].$$

## 17.4 Chapter 11

**Solution E. 11.1.** It is immediate to see that $\mathrm{Bil}(M, N; P)$ is an Abelian group with respect to $+$, where the 0 map is the identity element.

As for scalar multiplication, we have $1_A b = b$.
For all $\alpha$, $\beta \in A$ and $b$, $b' \in \mathrm{Bil}(M, N; P)$, the equalities

$$(\alpha+\beta)b = \alpha b + \beta b, \quad \alpha(b+b') = \alpha b + \alpha b' \quad \text{and} \quad (\alpha\beta)b = \alpha(\beta b),$$

hold since $P$ is an $A$-module.
Indeed, for all $m \in M$ and $n \in N$, we have

$$\begin{aligned}
((\alpha+\beta)b)(m,n) &= (\alpha+\beta)b(m,n) = \alpha b(m,n) + \beta b(m,n) \\
&= (\alpha b + \beta b)(m,n); \\
(\alpha(b+b'))(m,n) &= \alpha((b+b')(m,n)) = \alpha(b(m,n) + b'(m,n)) \\
&= \alpha b(m,n) + \alpha b'(m,n) = (\alpha b + \alpha b')(m,n); \\
((\alpha\beta)b)(m,n) &= (\alpha\beta)b(m,n) = \alpha(\beta b(m,n)) \\
&= \alpha(\beta b)(m,n) = (\alpha(\beta b))(m,n).
\end{aligned}$$

**Solution E. 11.2.** By hypothesis, there exist $\alpha$, $\beta \in \mathbb{Z}$ such that $\alpha a + \beta b = 1$.
For any simple tensor $\overline{h} \otimes \overline{k}$, we have

$$\overline{h} \otimes \overline{k} = 1(\overline{h} \otimes \overline{k}) = (\alpha a + \beta b)(\overline{h} \otimes \overline{k}) = \alpha a h(\overline{1} \otimes \overline{k}) + \beta b k(\overline{h} \otimes \overline{1}).$$

Since $a(\overline{1} \otimes \overline{k}) = \overline{a} \otimes \overline{k} = 0$, and $b(\overline{h} \otimes \overline{1}) = 0$, every simple tensor is 0.
The conclusion follows from **T.5.3**.2.

**Solution E. 11.3.** Applying **T.5.4**.5, we have

$$A/I \otimes_A A/J \simeq (A/I)/J(A/I) \simeq (A/I)/(J + I/I),$$

and thus, it is isomorphic to $A/I + J$, by **T.4.3**.2.

Alternatively, we can use the universal property of tensor product **T.5.2**.
Construct the diagram

$$\begin{array}{ccc}
A/I \times A/J & \xrightarrow{\ f\ } & A/(I+J) \\
\Big\downarrow{\scriptstyle \tau} & \nearrow{\scriptstyle \tilde{f}} & \\
A/I \otimes A/J, & &
\end{array}$$

where $f(\overline{x}, \overline{y}) = \overline{xy}$. The map $f$ is well-defined.
Indeed, if $(\overline{x_1}, \overline{y_1}) = (\overline{x_2}, \overline{y_2})$, then $x_1 - x_2 \in I$, $y_1 - y_2 \in J$, and therefore,

$$x_1 y_1 - x_2 y_2 = x_1(y_1 - y_2) + y_2(x_1 - x_2) \in x_1 J + y_2 I \subseteq I + J,$$

that is, $\overline{x_1y_1} = \overline{x_2y_2}$ in $A/(I+J)$.

It is easy to verify that $f$ is bilinear. Next, we prove that $\tilde{f}$ is an isomorphism.

Since $\overline{a} = f(\overline{a}, \overline{1})$ for all $\overline{a} \in A/(I+J)$, we have that $f$ is surjective, and consequently, the induced map $\tilde{f}$ defined by $\tilde{f}(\overline{x} \otimes \overline{y}) = \overline{xy}$ is also surjective. To prove that $\tilde{f}$ is injective, observe that any element $\alpha$ of $A/I \otimes A/J$ can be written as a finite sum

$$\alpha = \sum_{i=1}^{k} \overline{x_i} \otimes \overline{y_i} = \Big(\sum_{i=1}^{k} \overline{x_i y_i}\Big) \otimes \overline{1} = \overline{\beta} \otimes \overline{1},$$

with $\overline{\beta} = \sum_{i=1}^{k} \overline{x_i y_i} \in A/I$.

Then, $\tilde{f}(\overline{\beta} \otimes \overline{1}) = \overline{\beta} = \overline{0}$ implies $\beta \in I + J$. Hence, $\beta = x + y$ for some $x \in I$, $y \in J$, and

$$\overline{\beta} \otimes \overline{1} = \overline{x+y} \otimes \overline{1} = \overline{y} \otimes \overline{1} = \overline{1} \otimes \overline{y} = \overline{1} \otimes \overline{0} = 0.$$

**Solution E. 11.4.** Let $\{e_i\}_{i \in I}$ and $\{e'_j\}_{j \in J}$ be bases of $M$ and $N$, respectively. We already know that $\{e_i \otimes e'_j : i \in I,\, j \in J\}$ is a generating set of $M \otimes N$, see **T.5.3**.3.

We now prove that this set is free.

Assume there is a finite linear combination

$$\sum_{i \in I_0,\, j \in J_0} c_{ij}(e_i \otimes e'_j) = 0.$$

To show that all $c_{ij}$ are zero, we fix indices $i_0$ and $j_0$, and prove that the corresponding coefficient $c_{i_0 j_0}$ is 0.

Let $f \colon M \times N \longrightarrow A$ be the bilinear map defined by $f(m,n) = a_{i_0} b_{j_0}$, for all $m = \sum_i a_i e_i$ and $n = \sum_j b_j e'_j$.

The diagram

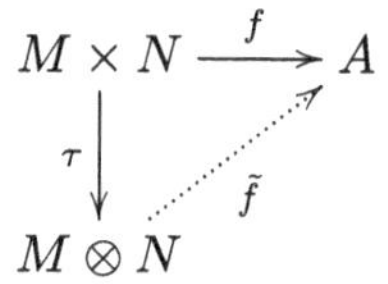

$$\begin{array}{ccc} M \times N & \xrightarrow{f} & A \\ \tau \big\downarrow & \nearrow_{\tilde{f}} & \\ M \otimes N & & \end{array}$$

is commutative, and thus, $\tilde{f}(m \otimes n) = a_{i_0} b_{j_0}$, for every simple tensor $m \otimes n$. In particular, $\tilde{f}(e_{i_0} \otimes e'_{j_0}) = 1$ and $\tilde{f}(e_i \otimes e'_j) = 0$ if $(i,j) \neq (i_0, j_0)$.

This implies

$$0 = \tilde{f}(0) = \tilde{f}\Big(\sum_{i,j} c_{ij}(e_i \otimes e'_j)\Big) = c_{i_0 j_0}.$$

**Solution E. 11.5.** 1. Consider the diagram

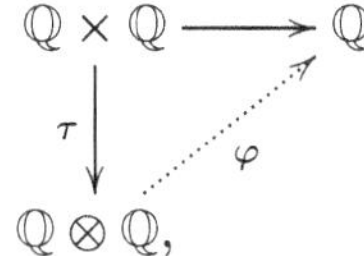


where the horizontal map is the usual product in $\mathbb{Q}$, which is $\mathbb{Z}$-bilinear. Hence, there exists a unique homomorphism $\varphi$ that makes the diagram commute, *i.e.*, such that $\varphi(x \otimes y) = xy$.

We show next that $\varphi$ is an isomorphism.
Since $\varphi(x \otimes 1) = x$ for all $x \in \mathbb{Q}$, we have that $\varphi$ is surjective.
To prove that $\varphi$ is injective, note that

$$L = \{x \otimes 1 \colon x \in \mathbb{Q}\}$$

is a generating set of $\mathbb{Q} \otimes \mathbb{Q}$.
In fact, for any simple tensor $m \otimes n$, where $n = a/b$,

$$m \otimes \frac{a}{b} = \frac{mb}{b} \otimes \frac{a}{b} = \frac{m}{b} \otimes a = \frac{am}{b} \otimes 1 \text{ with } \frac{am}{b} \in \mathbb{Q}.$$

Every element of $\mathbb{Q} \otimes \mathbb{Q}$ is a finite linear combination with coefficients in $\mathbb{Z}$ of simple tensors, and therefore, it can be written as $x \otimes 1$, for some $x \in \mathbb{Q}$. Thus, if $\varphi(x \otimes 1) = 0$, then $x = 0$, which implies $x \otimes 1 = 0$.

2. Note that, by hypothesis, $\mathbb{C}$ is equipped with the usual $\mathbb{R}$-vector space structure, and it is a free $\mathbb{R}$-module with basis $\{1, i\}$.
A simple tensor $x \otimes y$ in $\mathbb{C} \otimes_{\mathbb{R}} \mathbb{C}$ can be expressed as

$$\begin{aligned} x \otimes y &= (a + ib) \otimes (c + id) \\ &= a \otimes c + a \otimes id + ib \otimes c + ib \otimes id \\ &= ac(1 \otimes 1) + ad(1 \otimes i) + bc(i \otimes 1) + bd(i \otimes i), \end{aligned}$$

for some $a, b, c, d \in \mathbb{R}$.
The elements $1 \otimes 1$, $1 \otimes i$, $i \otimes 1$ and $i \otimes i$ form a basis of $\mathbb{C} \otimes_{\mathbb{R}} \mathbb{C}$, see the solution of **E.11.4**.
Now, consider the element

$$1 \otimes 1 + i \otimes i \in \mathbb{C} \otimes_{\mathbb{R}} \mathbb{C}.$$

For it to be a simple tensor, say $(a + ib) \otimes (c + id)$, it should satisfy the equations $ac = 1$, $ad = 0$, $bc = 0$ and $bd = 1$, but this system has no solution in $\mathbb{R}$.

3. It is not possible.
Indeed, consider the diagram, similar to the one in part 1,

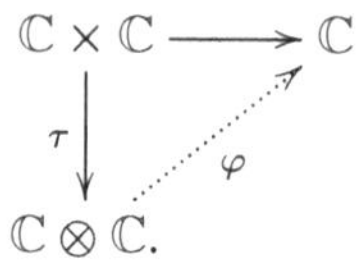

$$\begin{array}{ccc} \mathbb{C}\times\mathbb{C} & \longrightarrow & \mathbb{C} \\ {\scriptstyle\tau}\downarrow & \nearrow_{\varphi} & \\ \mathbb{C}\otimes\mathbb{C}. & & \end{array}$$

For any simple tensor $x \otimes_{\mathbb{R}} y$, if $0 = \varphi(x \otimes y) = xy$, then either $x = 0$ or $y = 0$. Thus, $\varphi$ is injective on simple tensors.

However, $\varphi$ is not injective. For instance, we have

$$\varphi(1 \otimes i - i \otimes 1) = i - i = 0.$$

The correct isomorphism is $\mathbb{C} \otimes_{\mathbb{R}} \mathbb{C} \simeq \mathbb{C}^2$, that follows from **T.5.4**.6.

**Solution E. 11.6.** Construct the diagram

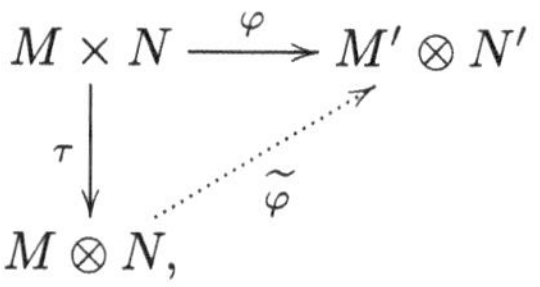

$$\begin{array}{ccc} M\times N & \xrightarrow{\varphi} & M'\otimes N' \\ {\scriptstyle\tau}\downarrow & \nearrow_{\widetilde{\varphi}} & \\ M\otimes N, & & \end{array}$$

where $\varphi(m,n) = f(m) \otimes g(n)$ is $A$-bilinear.
Then, by the universal property, $\widetilde{\varphi}$ is well-defined, and $\widetilde{\varphi} = f \otimes g$, since the diagram commutes.

**Solution E. 11.7.** For any simple tensor $m \otimes n$, we have

$$\begin{aligned} \big((f' \circ f) \otimes (g' \circ g)\big)(m \otimes n) &= (f' \circ f)(m) \otimes (g' \circ g)(n) \\ &= f'(f(m)) \otimes g'(g(n)) \\ &= (f' \otimes g')(f(m) \otimes g(m)) \\ &= (f' \otimes g')((f \otimes g)(m \otimes n)) \\ &= \big((f' \otimes g') \circ (f \otimes g)\big)(m \otimes n). \end{aligned}$$

**Solution E. 11.8.** It is enough to verify that any simple tensor $q \otimes_{\mathbb{Z}} \overline{a}$ is trivial:

$$q \otimes_{\mathbb{Z}} \overline{a} = \frac{nq}{n} \otimes_{\mathbb{Z}} \overline{a} = \frac{q}{n} \otimes_{\mathbb{Z}} n\overline{a} = \frac{q}{n} \otimes_{\mathbb{Z}} \overline{0} = 0.$$

**Solution E. 11.9.** 1. If $N_1$ and $N_2$ are projective, there exist $A$-modules $M_1$ and $M_2$, and free $A$-modules $F_1$ and $F_2$ such that $F_1 \simeq M_1 \oplus N_1$ and $F_2 \simeq M_2 \oplus N_2$, see **T.4.22**.
Hence,

$$(M_1 \oplus M_2) \oplus (N_1 \oplus N_2) \simeq (M_1 \oplus N_1) \oplus (M_2 \oplus N_2) \simeq F_1 \oplus F_2,$$

and $N_1 \oplus N_2$ is a direct summand of a free module.

Conversely, if $N_1 \oplus N_2$ is projective, there exist a free module $F$ and a module $M$ such that

$$F \simeq (N_1 \oplus N_2) \oplus M \simeq (M \oplus N_1) \oplus N_2 \simeq (M \oplus N_2) \oplus N_1.$$

Thus, $N_1$ and $N_2$ are projective.

2. Let $N_1$ and $N_2$ be projective. Then, with the same notation used in part 1, by **T.5.4**.4,

$$\begin{aligned} F_1 \otimes F_2 &\simeq (N_1 \oplus M_1) \otimes (N_2 \oplus M_2) \\ &\simeq (N_1 \otimes N_2) \oplus (N_1 \otimes M_2) \oplus (M_1 \otimes N_2) \oplus (M_1 \otimes M_2), \end{aligned}$$

where $F_1 \otimes F_2$ is free according to **E.11.4**.

3. Viewing $\mathbb{Z}/(n)$ as a $\mathbb{Z}$-module, we have, for example, $\mathbb{Z}/(2) \otimes \mathbb{Z}/(3) = 0$, see **E.11.3**.

Clearly, 0 is a projective module.

However, neither $\mathbb{Z}/(2)$ nor $\mathbb{Z}/(3)$ are projective modules, since neither of them can be a direct summand of any $\mathbb{Z}^n$.

4. Since tensor product is right-exact, see **T.5.6**, it is enough to examine injective maps.

Let $f\colon M \longrightarrow N$ be an injective homomorphism and consider the following diagram

$$\begin{array}{ccc} M \otimes (N_1 \oplus N_2) & \xrightarrow{f\otimes(\mathrm{id}_{N_1},\mathrm{id}_{N_2})} & N \otimes (N_1 \oplus N_2) \\ \Big\downarrow{\scriptstyle \alpha} & & \Big\downarrow{\scriptstyle \beta} \\ (M \otimes N_1) \oplus (M \otimes N_2) & \xrightarrow{(f\otimes\mathrm{id}_{N_1},f\otimes\mathrm{id}_{N_2})} & (N \otimes N_1) \oplus (N \otimes N_2), \end{array}$$

where $\alpha$ and $\beta$ are the isomorphisms provided by **T.5.4**.4.

The homomorphism $f \otimes (\mathrm{id}_{N_1}, \mathrm{id}_{N_2})$ is injective if and only if $f \otimes \mathrm{id}_{N_1}$ and $f \otimes \mathrm{id}_{N_2}$ are injective.

5. Let $f\colon M \longrightarrow N$ be an injective homomorphism. Then,

$$M \otimes N_1 \xrightarrow{f\otimes\mathrm{id}_{N_1}} N \otimes N_1$$

is injective, and accordingly, also

$$(M \otimes N_1) \otimes N_2 \xrightarrow{(f\otimes\mathrm{id}_{N_1})\otimes\mathrm{id}_{N_2}} (N \otimes N_1) \otimes N_2$$

is injective.

Clearly, $(f \otimes \mathrm{id}_{N_1}) \otimes \mathrm{id}_{N_2} = f \otimes (\mathrm{id}_{N_1} \otimes \mathrm{id}_{N_2})$.

6. Consider $\mathbb{Z}/(2) \otimes \mathbb{Z}/(3) = 0$, where 0 is flat but neither $\mathbb{Z}/(2)$ nor $\mathbb{Z}/(3)$ are flat.

**Solution E. 11.10.** Note that the residue field $K = A/\mathfrak{m}$ is an $(A, K)$-bimodule. By definition,

$$\mu(M) = \dim_K M/\mathfrak{m}M = \dim_K (M \otimes_A K),$$

see **T.4.13** and **T.5.4**.5.
To compute $\mu(M \otimes_A N)$ we use the properties of tensor product and **T.5.8**, and we obtain

$$\begin{aligned}(M \otimes_A N) \otimes_A K &\simeq M \otimes_A (N \otimes_A K) \simeq M \otimes_A (K \otimes_A N) \\ &\simeq M \otimes_A ((K \otimes_K K) \otimes_A N) \simeq M \otimes_A (K \otimes_K (K \otimes_A N)) \\ &\simeq (M \otimes_A K) \otimes_K (N \otimes_A K) \simeq K^{\mu(M)} \otimes_K K^{\mu(N)} \\ &\simeq K^{\mu(M)\mu(N)}.\end{aligned}$$

**Solution E. 11.11.** A direct application of **E.11.10** yields $\mu(M) = 0$ or $\mu(N) = 0$. Hence, either $M = \mathfrak{m}M$ or $N = \mathfrak{m}N$.
The conclusion follows by Nakayama's Lemma.

**Solution E. 11.12.** To verify that $M$ is a $\mathbb{Z}$-module is easy and left to the reader.

As for the second statement, it is sufficient to prove that each simple tensor is zero.
Let $\alpha = \frac{a}{p^m}$ and $\beta = \overline{\frac{b}{p^n}}$, with $a, b \in \mathbb{Z}$ and $m, n \in \mathbb{N}$. Then,

$$\begin{aligned}\alpha \otimes \beta &= \frac{ap^n}{p^{n+m}} \otimes \beta = p^n \frac{a}{p^{n+m}} \otimes \beta = \frac{a}{p^{n+m}} \otimes p^n \beta \\ &= \frac{a}{p^{n+m}} \otimes \overline{\frac{bp^n}{p^n}} = \frac{a}{p^{n+m}} \otimes \overline{\frac{b}{1}} = \frac{a}{p^{n+m}} \otimes \overline{0} = 0.\end{aligned}$$

**Solution E. 11.13.** 1. By **E.11.3**,

$$\mathbb{Q}[x]/(x) \otimes_{\mathbb{Q}[x]} \mathbb{Q}[x]/(x^2+1) \simeq \mathbb{Q}[x]/(x, x^2+1) = 0.$$

Thus, the dimension of the given vector space is zero.

2. Since $\mathbb{Q}[\alpha] \simeq \mathbb{Q}[x]/(x^5 - 3) \simeq \mathbb{Q}^5$ as a $\mathbb{Q}$-vector space, by **T.5.4**.1 and 4 we obtain $\mathbb{C} \otimes \mathbb{Q}[\alpha] \simeq \mathbb{C}^5$.

**Solution E. 11.14.** Let $M$ be a finitely generated projective $A$-module with minimal generating set $\{m_1, \ldots, m_n\}$.
Consider the homomorphism $\varphi \colon A^n \longrightarrow M$ defined by $e_i \mapsto m_i$, where $\{e_1, \ldots, e_n\}$ is the canonical basis of $A^n$.

We have a short exact sequence

$$0 \longrightarrow N = \operatorname{Ker} \varphi \longrightarrow A^n \longrightarrow M \longrightarrow 0.$$

Since, by hypothesis, $M$ is projective, the sequence splits and $A^n \simeq M \oplus N$. Tensoring by $A/\mathfrak{m}$, we get

$$K^n \simeq M/\mathfrak{m}M \oplus N/\mathfrak{m}N$$

as $K$-vector spaces.

Since $\dim_K M/\mathfrak{m}M = \dim_K K^n = n$, by **T.4.13**, we obtain $N/\mathfrak{m}N = 0$, that is, $N = \mathfrak{m}N$. The module $N$ is finitely generated, as a direct summand of a finitely generated module. Thus, an application of Nakayama's Lemma yields $N = 0$.
Therefore $M \simeq A^n$ is free.

**Solution E. 11.15.** Let $H$ be any ideal of $A$.
Tensoring the exact sequence

$$0 \longrightarrow H \longrightarrow A \longrightarrow A/H \longrightarrow 0$$

with $B$, which is $A$-flat, we obtain the commutative diagram

$$\begin{array}{ccccccccc} 0 & \longrightarrow & H \otimes_A B & \longrightarrow & A \otimes_A B & \longrightarrow & A/H \otimes_A B & \longrightarrow & 0 \\ & & \downarrow & & \downarrow \simeq & & \downarrow \simeq & & \\ 0 & \longrightarrow & HB & \longrightarrow & B & \longrightarrow & B/HB & \longrightarrow & 0. \end{array}$$

Thus, the Snake Lemma implies

$$H \otimes_A B \simeq HB \quad \text{for any ideal } H \text{ of } A.$$

Consider the short exact sequence of $A$-modules

$$0 \longrightarrow I \cap J \xrightarrow{f} I \oplus J \xrightarrow{g} I + J \longrightarrow 0,$$

where $f(a) = (a, -a)$ and $g(a, b) = a + b$.
Tensoring with $B$, the exactness is preserved, thus we have

$$0 \longrightarrow (I \cap J)B \longrightarrow IB \oplus JB \longrightarrow IB + JB \longrightarrow 0.$$

Comparing this with the sequence

$$0 \longrightarrow IB \cap JB \longrightarrow IB \oplus JB \longrightarrow IB + JB \longrightarrow 0,$$

the conclusion follows easily.

**Solution E. 11.16.** If $a = 0$ or $a$ is invertible, all statements are clearly true, hence, we assume $a \neq 0$ and $a \notin A^*$.

$1 \Rightarrow 2$. We will prove that the sequence

$$0 \longrightarrow (a) \xrightarrow{i} A \xrightarrow{\pi} A/(a) \longrightarrow 0$$

splits, by showing that there exists $r\colon A \longrightarrow (a)$ such that $r \circ i = \mathrm{id}_{(a)}$, see **T.4.18**.
Since $a \in (a^2)$, there exists $c \in A$ such that $a = ca^2$. Let $r(b) = ca \cdot b$ for all $b \in A$. Then, $r \circ i(a) = ca^2 = a$, that implies $r \circ i = \mathrm{id}_{(a)}$.

$2 \Rightarrow 3$. Since $A$ is free, and therefore, flat, also its direct summands are flat by **E.11.9**.4. By hypothesis, there exists an $A$-module $M$ such that $A \simeq (a) \oplus M$. Consequently, $M \simeq A/(a)$ is flat.

$3 \Rightarrow 1$. Since the inclusion homomorphism $i$ of $(a)$ into $A$ is injective, and $A/(a)$ is flat, the homomorphism

$$i \otimes \mathrm{id}_{A/(a)} : (a) \otimes A/(a) \longrightarrow A \otimes A/(a)$$

is injective.
Therefore, the induced homomorphism $(a)/(a^2) \simeq (a) \otimes A/(a) \longrightarrow A/(a)$ is injective, and it is also the zero homomorphism.
This implies $(a)/(a^2) = 0$.

**Solution E. 11.17.** 1. Let $I \neq A$ be an ideal. Then, there exists a maximal ideal $\mathfrak{m}$ such that $I \subseteq \mathfrak{m}$.
Hence, $IM \subseteq \mathfrak{m}M \subsetneq M$, and the conclusion immediately follows.

2. Let $n \in N \setminus \{0\}$ and $N_1 = \langle n \rangle \simeq A/\operatorname{Ann} n$.
Consider the inclusion homomorphism $f : N_1 \longrightarrow N$ and tensor with $M$. Since $M$ is flat, we obtain an injective homomorphism

$$f \otimes \mathrm{id}_M : N_1 \otimes M \longrightarrow N \otimes M.$$

We have

$$N_1 \otimes M \simeq A/\operatorname{Ann} n \otimes M \simeq M/(\operatorname{Ann} n)M,$$

and it is enough to prove $M/(\operatorname{Ann} n)M \neq 0$.
Since $n \neq 0$, the ideal $\operatorname{Ann} n$ is proper. The conclusion follows from part 1.

**Solution E. 11.18.** Since $a$ is not a zero-divisor, the multiplication map $\varphi : A \xrightarrow{a} A$ is injective. Tensoring with $N$, we obtain an injective homomorphism $\varphi \otimes \mathrm{id}_N : A \otimes_A N \longrightarrow A \otimes_A N$.
Therefore, the induced homomorphism $N \xrightarrow{a} N$ is injective.

**Solution E. 11.19.** 1. Proving that $I$, respectively $J$, and $I \cap J = IJ$ are free is equivalent to showing that there is no $a \neq 0$ such that $ax$, respectively $ay$, and $axy$ is zero.
This is trivially true since $A$ is a domain.

2. Clearly, also $I + J = (x, y)$ has no torsion.
Consider the homomorphism $\varphi : A/I \xrightarrow{y} A/I$, which is obviously injective. Tensoring with $I + J$, we obtain

$$\varphi \otimes \mathrm{id}_{I+J}(\overline{1} \otimes x) = \overline{y} \otimes x = \overline{1} \otimes xy = \overline{x} \otimes y = 0.$$

Since $\overline{1} \otimes x$ is non-zero in $A/I \otimes (I + J) \simeq (x, y)/(x^2, xy)$, see **T.5.4**.5, $\varphi \otimes \mathrm{id}_{I+J}$ is not injective, and consequently, $I + J$ is not flat.

**Solution E. 11.20.** Recall that, by **T.4.2**, $\operatorname{Hom}_A(A, M) \simeq M$.
Additionally, by **E.10.17** and **T.5.4**.6,

$$\begin{aligned}\mathrm{Hom}(M,M)\otimes\mathrm{Hom}(N,N) &\simeq \mathrm{Hom}(A^r,A^r)\otimes\mathrm{Hom}(A^s,A^s)\\ &\simeq \mathrm{Hom}(A,A^r)^r\otimes\mathrm{Hom}(A,A^s)^s\\ &\simeq A^{r^2}\otimes A^{s^2}\simeq A^{r^2s^2}\end{aligned}$$

for some $r,s\in\mathbb{N}_+$. The latter is isomorphic to

$$\begin{aligned}\mathrm{Hom}(A^r\otimes A^s,A^r\otimes A^s) &\simeq \mathrm{Hom}(A^{rs},A^{rs})\\ &\simeq \mathrm{Hom}(A,A^{rs})^{rs}\simeq A^{(rs)^2}.\end{aligned}$$

**Solution E. 11.21.** The matrix associated with $\varphi$ is

$$\begin{pmatrix}4 & 8\\ 4 & -4\\ 16 & 20\end{pmatrix},$$

whose Smith form is

$$\begin{pmatrix}4 & 0\\ 0 & 12\\ 0 & 0\end{pmatrix}.$$

Hence, $\mathrm{Coker}\,\varphi\simeq\mathbb{Z}\oplus\mathbb{Z}/(4)\oplus\mathbb{Z}/(12)$.
By **T.5.4**.4 and **E.11.3**, we have

$$\mathbb{Z}/(15)\otimes_{\mathbb{Z}}(\mathbb{Z}/(4)\oplus\mathbb{Z}/(12))\simeq\mathbb{Z}/(15,4)\oplus\mathbb{Z}/(15,12)\simeq\mathbb{Z}/(3).$$

**Solution E. 11.22.** The matrix of relations among the elements of $M_a$ is

$$\begin{pmatrix}2 & 1 & 1\\ -1 & 1 & a\\ 0 & 1 & 0\end{pmatrix},$$

and its Smith form is

$$\begin{pmatrix}1 & 0 & 0\\ 0 & 1 & 0\\ 0 & 0 & 2a+1\end{pmatrix}.$$

Therefore, $M_a\simeq\mathbb{Z}/(2a+1)$, and

$$M_a\otimes\mathbb{Z}/(n)\neq 0 \quad\text{if and only if}\quad \gcd(2a+1,n)\neq 1.$$

## 17.5 Chapter 12

**Solution E. 12.1.** Consider the localization $A_a$. We claim that $A_a \neq 0$. Indeed, if $\frac{1}{1} = \frac{0}{1}$, then there exists $n \geq 1$ such that $a^n = 1a^n = 0$, contradicting the hypothesis $a \notin \mathcal{N}(A)$.

Thus, in $A_a$ we find at least one maximal ideal. By the one-to-one correspondence between prime ideals of $A$ with empty intersection with $\{a^n : n \in \mathbb{N}\}$ and primes of $A_a$, its contraction with respect to the localization homomorphism $A \longrightarrow A_a$ is a prime ideal of $A$ not containing $a^n$ for any $n \in \mathbb{N}$.

Note that this is an alternative proof of the fact that

$$\bigcap_{\mathfrak{p} \in \operatorname{Spec} A} \mathfrak{p} \subseteq \mathcal{N}(A),$$

see **T.1.14**.1.

**Solution E. 12.2.** The inclusion $\sigma_S(A^*) \subseteq (S^{-1}A)^*$ always holds.
If $\sigma_S$ is an isomorphism, then

$$\sigma_S(s) = \frac{s}{1} \in (S^{-1}A)^* = \sigma_S(A^*) \text{ for all } s \in S.$$

Hence, $s \in \sigma_S^{-1}((S^{-1}A)^*) = A^*$, and therefore, $S \subseteq A^*$.

Conversely, if $S \subseteq A^*$, then, by the universal property of ring of fractions **T.6.4**, the identity homomorphism $\mathrm{id}_A$ factorizes through a homomorphism $\varphi\colon S^{-1}A \longrightarrow A$, which is obviously surjective.
Since $\varphi\left(\frac{a}{s}\right) = as^{-1} = 0$ implies $a = 0$, the homomorphism $\varphi$ is also injective.
Finally, note that $\varphi^{-1} = \sigma_S$.

**Solution E. 12.3.** We recall that $A = A^* \sqcup \mathcal{D}(A)$, since $A$ is finite, see **T.1.1**.
Since $\sigma_S$ is injective, by **T.6.3**.1 we have $S \cap \mathcal{D}(A) = \emptyset$. Hence $S \subseteq A^*$.
The conclusion follows from **E.12.2**.

**Solution E. 12.4.** 1. Let $A_{(p)} = \left\{\frac{a}{b} \in \mathbb{Q} \colon b \not\equiv 0 \mod p\right\}$.
It is easy to verify that this is a subring of $\mathbb{Q}$, see **E.8.41**.
Let $f\colon \mathbb{Z} \longrightarrow A_{(p)}$ be defined as $f(a) = \frac{a}{1}$. Then, $f(S) \subseteq (A_{(p)})^*$, and by the universal property, we obtain a homomorphism

$$\tilde{f}\colon S^{-1}\mathbb{Z} \longrightarrow A_{(p)}$$

which is obviously surjective.
It is also injective, since $\tilde{f}\left(\frac{a}{s}\right) = \frac{a}{s} = 0$ if and only if $a = 0$.
Thus, $S^{-1}\mathbb{Z} \simeq A_{(p)}$.

2. Let $B = \left\{\frac{a}{b} \in \mathbb{Q} \colon b \not\equiv 0 \mod p_i \text{ for all } i\right\}$.
From the homomorphism $f\colon \mathbb{Z} \longrightarrow B$ defined by $f(a) = \frac{a}{1}$, we obtain an isomorphism $S^{-1}\mathbb{Z} \simeq B$ as in part 1.

3. To simplify notation, the elements of $A$ will be denoted by $0, \ldots, 11$.
We have $S = \{1, 2, 4, 8\}$, and for any $s \in S$, the elements $\frac{2}{s}$, $\frac{4}{s}$, $\frac{8}{s}$ are invertible in $S^{-1}A$ .
Moreover,

$$\frac{0}{s} = \frac{3}{s} = \frac{6}{s} = \frac{9}{s}, \quad \frac{1}{s} = \frac{4}{s} = \frac{7}{s} = \frac{10}{s}, \quad \text{and} \quad \frac{2}{s} = \frac{5}{s} = \frac{8}{s} = \frac{11}{s} \quad \text{for any } s \in S.$$

Finally, note that $\frac{1}{2^k} = \frac{2^k}{1}$ for any $k = 1, 2, 3$.
As a result, every element of $S^{-1}A$ can be written as $\frac{n}{1}$ for some $n = 0, 1, 2$.
Consequently, $S^{-1}A$ is a field with 3 elements, *i.e.*, isomorphic to $\mathbb{Z}/(3)$.

4. In this case $S = \{1, 3, 5, 7, 9, 11\}$.
It is straightforward to verify that $\frac{a}{s} = \frac{b}{s}$, respectively $\frac{a}{s} = \frac{a}{t}$, if and only if $4 \mid a - b$, respectively $4 \mid s - t$.
Hence, it suffices to consider numerators 0, 1, 2, 3 and denominators 1, 3.
Moreover, $\frac{1}{3} = \frac{3}{1}$ and $\frac{2}{3} = \frac{2}{1}$.
Thus, $|S^{-1}A| = 4$, $(S^{-1}A)^* = \{\frac{1}{1}, \frac{3}{1}\}$, and $(\frac{2}{1})$ is the unique prime and maximal ideal.
We can conclude that

$$S^{-1}A \simeq \mathbb{Z}/(4).$$

5. In this case $S = \{1, 2, 4, 5, 7, 8, 10, 11\}$.
Similarly to the previous case, we verify that it is enough to consider numerators 0, 1, 2 and denominators 1, 2 to describe all elements of $S^{-1}A$.
Moreover, $\frac{1}{2} = \frac{2}{1}$, and therefore, $S^{-1}A$ is a field with 3 elements, *i.e.*, isomorphic to $\mathbb{Z}/(3)$.

**Solution E. 12.5.** 1. Let $\frac{a}{s} \in S^{-1}A$. We have to find an element $b \in A$ such that $\frac{a}{s} = \frac{b}{1} = \sigma_S(b)$.
Since $S$ is a finite set, for any $s \in S$, there exist $r, k \in \mathbb{N}$ such that $k > r$ and $s^k = s^r$. Hence, $s^r(a - as^{k-r}) = 0$, and we have

$$\frac{a}{s} = \frac{as^{k-r-1}}{1} = \sigma_S(as^{k-r-1}).$$

Observe that, in this way, $S^{-1}A \simeq A/\operatorname{Ker}\sigma_S$, and this isomorphism can be used for instance to obtain simpler proofs of **E.12.4**.3, 4, and 5.

2. We have $\operatorname{Ker}\sigma_S = \{a \in A\colon\ as = 0 \ \text{ for some } \ s \in S\} = (\overline{3})$.
Thus, by part 1,

$$S^{-1}A \simeq A/\operatorname{Ker}\sigma_S \simeq \mathbb{Z}/(3).$$

**Solution E. 12.6.** 1. Since $I$ is an ideal, $0 \in I$, and therefore, $1 \in S$.
If $1 + i, 1 + j \in S$, then $(1 + i)(1 + j) \in 1 + I + I^2 \subseteq 1 + I = S$.
Therefore, $S$ is a multiplicative subset.

By the characterization of the Jacobson radical **T.1.15**, it is sufficient to prove that every element $\frac{x}{s} \in S^{-1}I$ is such that $1 + \frac{x}{s}\frac{y}{t} \in (S^{-1}A)^*$ for all $\frac{y}{t} \in S^{-1}A$.

Since

$$1+\frac{x}{s}\frac{y}{t}=\frac{st+xy}{st} \quad \text{and} \quad st+xy\in 1+I=S,$$

the conclusion follows.

2. To simplify notation, the elements of $A$ will be denoted by $0,\dots,59$.
The distinct non-trivial proper ideals of $A$ are (2), (3), (4), (5), (6), (10), (12), (15), (20), and (30).
We have $-3=1-4\in S$ and $5=1+4\in S$. Thus, $\frac{3}{1}$, $\frac{5}{1}$, and $\frac{15}{1}$ are invertible in $S^{-1}A$. It follows that

$$S^{-1}(2)=S^{-1}(6)=S^{-1}(10)=S^{-1}(30),$$

$$S^{-1}(3)=S^{-1}(5)=S^{-1}(15)=S^{-1}A.$$

Moreover, $\frac{4}{1}=0$ implies that

$$S^{-1}(4)=S^{-1}(12)=S^{-1}(20)=S^{-1}(0).$$

Therefore, $S^{-1}A$ has only two proper ideals, namely (0) and $S^{-1}(2)$.
Finally, $S^{-1}A$ is a local ring with maximal ideal generated by $\frac{2}{1}$.

Note that $\operatorname{Ker}\sigma_S=(4)$ and $S^{-1}A\simeq\mathbb{Z}/(4)$ by **E.12.5**.1.

3. We recall that primes of $T^{-1}A$ are in one-to-one correspondence with primes of $A$ which have empty intersection with $T$, see **T.6.8**.
We have $(p)\cap T\neq\emptyset$ if and only if there exists $a\in(p)$ such that $a\equiv 1 \mod m$, that is, if and only if $(m,p)=1$.
Choosing an $m$ which is a product of at least two prime divisors of 60, for example $m=6$ or 15, we will to obtain a ring $T^{-1}A$ which is not local.

**Solution E. 12.7.** 1. Consider the homomorphism $f\colon A\times B\longrightarrow A$ defined by $f(a,b)=a$. Note that:

i) for any $s\in S$, $f(s)=1$ is invertible;
ii) if $f(a,b)=a=0$, then $(1,0)(a,b)=(0,0)$ with $(1,0)\in S$;
iii) for any $a\in A$, we have $a=f(a,b)f(1,0)^{-1}$.

Then, by **T.6.5**, $f$ can be extended to an isomorphism $\tilde{f}\colon S^{-1}C\longrightarrow A$.

2. The proof of part 1 only used the element $(1,0)\in S$.
Since $(1,0)\in T$, the same proof works.

3. For any $b\in B$ we have

$$((1,b),(1,1))\sim((1,0),(1,0)) \quad \text{and} \quad ((1,0),(1,0))\sim((1,0),(1,1)).$$

However, if $b\neq 0$, then $((1,b),(1,1))\not\sim((1,0),(1,1))$, and hence, $\sim$ is not transitive.

**Solution E. 12.8.** We define $B=A[x]/(1-fx)$.
If $f$ is nilpotent, then there exists $n\in\mathbb{N}$ such that $f^n=0$. Hence, $A_f=0$.

Since

$$1 = (1 - (fx)^n) = (1 - fx)(1 + fx + \ldots + (fx)^{n-1}) \in (1 - fx),$$

we have that $B = 0$ also.

Now, assume $f^n \neq 0$ for all $n$.
We define $\varphi\colon A \longrightarrow A[x] \xrightarrow{\pi} B$ by $\varphi(a) = \overline{a}$.
Since $\overline{fx} = \overline{1}$ in $B$, the elements $\overline{f^k}$ are invertible in $B$ for all $k$. Therefore, $\varphi$ factorizes through $A_f$ yielding a homomorphism $\psi\colon A_f \longrightarrow B$ defined by

$$\psi\left(\frac{a}{f^k}\right) = \varphi(a)\varphi(f^k)^{-1}.$$

To prove that $\psi$ is an isomorphism, we apply **T.6.5**. It is sufficient to prove that if $\overline{a} = \varphi(a) = \overline{0}$, there exists $s$ such that $f^s a = 0$ in $A$, and for any $\overline{b} \in B$, there exist $a \in A$ and an integer $k$ such that $\overline{b} = \varphi(a)\varphi(f^k)^{-1}$.

Let $a \in A$ be such that $\overline{a} = \overline{0}$. Then, there exists $p(x) = \sum_{i=0}^{h} b_i x^i \in A[x]$ such that $a = p(x)(1 - fx)$.
Hence, $b_0 = a$, $b_1 = fb_0 = fa$, …, $b_h = fb_{h-1} = f^h a$, and $fb_h = f^{h+1}a = 0$.

Finally, each $\overline{b} \in B$ can be written as

$$\overline{b} = \sum_{i=0}^{k} \overline{b_i}\overline{x}^i = \overline{f}^{-k} \sum_{i=0}^{k} \overline{b_i\ f^{k-i}}.$$

Therefore, $\overline{b}$ is of the form $\varphi(a)\varphi(f^k)^{-1}$.

**Solution E. 12.9.** 1. By the previous exercise, we have

$$\mathbb{Z}\left[\tfrac{2}{3}\right] \subseteq \mathbb{Z}\left[\tfrac{1}{3}\right] \simeq \mathbb{Z}[x]/(1 - 3x) \simeq \mathbb{Z}_3.$$

To prove the opposite inclusion, since $1 - \frac{2}{3} = \frac{1}{3}$, we can write each element $\frac{1}{3^n}$ as a finite sum of powers of $\frac{2}{3}$ with integer coefficients.
2. We will assume that all rational numbers $\frac{a}{b}$ are reduced, *i.e.*, such that $\gcd(a, b) = 1$.
Let $A \subsetneq \mathbb{Q}$ be a subring. Denote by $P$ the set of all prime divisors of denominators of elements of $A$, and let $S$ be the smallest multiplicative set containing $P$.
Note that if $p \in P$, then there exists an element $a = \frac{m}{ps} \in A$. Hence, $\frac{m}{p} = sa \in A$, and there exist $b$, $c \in \mathbb{Z}$ such that $bm + cp = 1$.
Consequently, $\frac{1}{p} = bsa + c \in A$, and this implies $S^{-1}\mathbb{Z} \subseteq A$.

The opposite inclusion immediately follows from the definition of $S$.

**Solution E. 12.10.** We know that $I = (xz + 2z,\ y - z,\ z^2 - z)$.
Since the elements $x + 2$ and $z - 1$ are invertible in $\mathbb{Q}[x, y, z]_{(x,y,z)}$, we have that $J$ is the extension of the ideal $(y,\ z)$.
Hence, the image of $f = x^3 z - y^2$ is an element of $J$.

**Solution E. 12.11.** We know that $I = (x^2 + 2y^2 - 3,\ xy - y^2,\ y^3 - y)$.
Since $(y)$, $(y+1)$, and $(y-1)$ are pairwise comaximal, we have

$$I = (x^2 - 3,\ y) \cap (x - 1,\ y - 1) \cap (x + 1,\ y + 1).$$

Therefore, $I \subset \mathfrak{p}_1$ and $I_{\mathfrak{p}_1} \cap \mathbb{C}[x, y] \neq (1)$.
Since $y^3 - y = (y^2 + y)(y - 1)$ and $y^2 + y$ is invertible in $\mathbb{C}[x, y]_{\mathfrak{p}_1}$, the ideal $I_{\mathfrak{p}_1} \cap \mathbb{C}[x, y]$ contains $y - 1$. The element $y$ is also invertible, hence $xy - y^2 \in I$ implies that $x - 1 \in I_{\mathfrak{p}_1}$.
Thus,

$$I_{\mathfrak{p}_1} \cap \mathbb{C}[x, y] = \mathfrak{p}_1.$$

For $\mathfrak{p}_2$, note that $I \not\subset (x,\ y)$, hence $I_{\mathfrak{p}_2} \cap \mathbb{C}[x, y] = (1)$.

**Solution E. 12.12.** In the solution of **E.9.15** we saw that

$$\sqrt{I} = (x,\ y) \cap (x^4 + 1,\ x^3 + y).$$

Therefore, by **T.6.9**.4,

$$\sqrt{I_{(x,y)}} = (\sqrt{I})_{(x,y)} = (x,\ y)_{(x,y)}.$$

Since $f = x^2 + 5x$, we have $\frac{f}{1} \in \sqrt{I_{(x,y)}}$.

**Solution E. 12.13.** From the solution of **E.9.18**.2, it follows that the minimal primes of $I$ are $(x,\ z)$, $(x,\ y)$ and $(y,\ z)$.
Since $y$ is invertible in $\mathbb{R}[x, y, z]_{(x,z)}$, from **E.9.18**.3 we obtain that

$$I_{(x,z)} = (x^2,\ z)_{(x,z)}.$$

Thus, by **T.6.13**.3,

$$\begin{aligned} A_{\mathfrak{p}} = (\mathbb{R}[x, y, z]/I)_{\mathfrak{p}} &\simeq \mathbb{R}[x, y, z]_{(x,z)}/I_{(x,z)} \\ &\simeq \mathbb{R}[x, y, z]_{(x,z)}/(x^2,\ z)_{(x,z)} \simeq (\mathbb{R}[x, y]/(x^2))_{(x)}. \end{aligned}$$

**Solution E. 12.14.** We know that $I = (x - yz,\ yz^2 - y)$.
Since $z^2 - 1$ is invertible in $S^{-1}A$, we have

$$S^{-1}(A/I) \simeq K[x, y, z]_{(x,y)}/I_{(x,y)} \simeq K[x, y, z]_{(x,y)}/(x,\ y)_{(x,y)} \simeq K(z).$$

**Solution E. 12.15.** We know that $I = (x^2 - yz,\ xz - yz,\ y^2 - yz)$.
Since $\overline{z}$ and $\overline{y - z}$ are invertible in $S^{-1}A$, the equalities

$$(\overline{y - z})\overline{y} = \overline{z}(\overline{x - y}) = \overline{0}$$

in $A$ yield $\frac{\overline{y}}{1} = \frac{\overline{x}}{1} = 0$.
Thus,

$$S^{-1}A \simeq \mathbb{Q}(z).$$

**Solution E. 12.16.** It is sufficient to prove that each simple tensor $\frac{a}{p^n} \otimes \overline{\frac{b}{p^m}}$ is zero in $\mathbb{Z}_p \otimes_{\mathbb{Z}} (\mathbb{Z}_p/\mathbb{Z})$.
As $b \in \mathbb{Z}$, we have

$$\frac{a}{p^n} \otimes \overline{\frac{b}{p^m}} = \frac{ap^m}{p^{n+m}} \otimes \overline{\frac{b}{p^m}} = \frac{a}{p^{n+m}} \otimes \overline{\frac{p^m b}{p^m}} = \frac{a}{p^{n+m}} \otimes \bar{b} = 0.$$

**Solution E. 12.17.** 1. The homomorphism $\varphi\colon S^{-1}A \longrightarrow T^{-1}A$ defined by

$$\varphi\left(\frac{a}{s}\right) = \frac{a}{s}$$

is well-defined and verifies $\varphi(T_1) \subseteq (T^{-1}A)^*$ by hypothesis.
By the universal property, $\varphi$ induces a homomorphism

$$\widetilde{\varphi}\colon T_1^{-1}(S^{-1}A) \longrightarrow T^{-1}A$$

and a commutative diagram

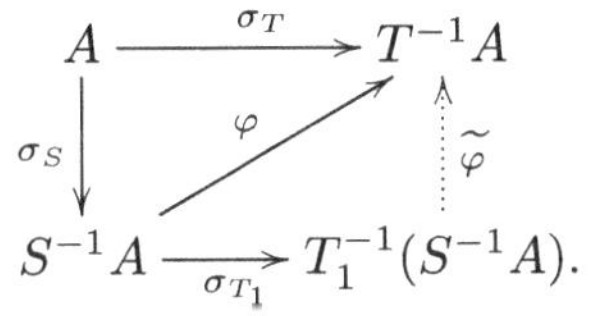

$$\begin{array}{ccc} A & \xrightarrow{\sigma_T} & T^{-1}A \\ {\scriptstyle \sigma_S}\downarrow & \overset{\varphi}{\nearrow} & \uparrow{\scriptstyle \widetilde{\varphi}} \\ S^{-1}A & \xrightarrow[\sigma_{T_1}]{} & T_1^{-1}(S^{-1}A). \end{array}$$

Note that $\varphi\left(\frac{a}{s}\right) = 0$ if and only if there exists $t \in T$ such that $ta = 0$. Consequently,

$$\sigma_S(t)\frac{a}{s} = \frac{ta}{s} = 0, \quad \text{with} \quad \sigma_S(t) \in T_1.$$

Moreover, for any $\frac{a}{t} \in T^{-1}A$ we can write $\frac{a}{t} = \varphi\left(\frac{a}{1}\right)\varphi(\sigma_S(t))^{-1}$.
As a result, $\widetilde{\varphi}$ is an isomorphism by **T.6.5**.
2. Let $U = \{st\colon s \in S \text{ and } t \in T\}$.
It is easy to verify that $U$ is a multiplicative subset containing both $S$ and $T$.
By part 1, we obtain

$$\sigma_S(U)^{-1}(S^{-1}A) \simeq U^{-1}A \simeq \sigma_T(U)^{-1}(T^{-1}A).$$

Now, we prove the isomorphism $\sigma_S(U)^{-1}(S^{-1}A) \simeq \sigma_S(T)^{-1}(S^{-1}A)$.
Consider the homomorphism

$$\psi\colon S^{-1}A \longrightarrow \sigma_S(T)^{-1}(S^{-1}A), \quad \text{defined by} \quad \psi\left(\frac{a}{s}\right) = \frac{\frac{a}{s}}{1}.$$

For any $\frac{u}{1} = \frac{st}{1} \in \sigma_S(U)$, we have

$$\psi\left(\frac{st}{1}\right) \cdot \frac{\frac{1}{s}}{\frac{t}{1}} = \frac{\frac{st}{1}}{1} \cdot \frac{\frac{1}{s}}{\frac{t}{1}} = \frac{\frac{st}{s}}{\frac{t}{1}} = \frac{\frac{t}{1}}{\frac{t}{1}} = 1.$$

It follows that $\sigma_S(U) \subset (\sigma_S(T)^{-1}(S^{-1}A))^*$.
By the universal property, we obtain an induced homomorphism

$$\widetilde{\psi}\colon \sigma_S(U)^{-1}(S^{-1}A) \longrightarrow \sigma_S(T)^{-1}(S^{-1}A),$$

which is obviously surjective.
To prove that $\widetilde{\psi}$ is injective, note that if $\psi\left(\frac{a}{s}\right) = 0$, then there exists $t \in T$ such that $\frac{t}{1} \cdot \frac{a}{s} = \frac{ta}{s} = 0$ in $S^{-1}A$. Thus, there exists $s_1 \in S$ such that $s_1 ta = 0$ in $A$. Hence, $\sigma_S(s_1 t)\frac{a}{s} = 0$ with $s_1 t \in U$.
Finally, the desired isomorphism is a consequence of **T.6.5**.

A similar argument provides the isomorphism

$$\sigma_T(U)^{-1}(T^{-1}A) \simeq \sigma_T(S)^{-1}(T^{-1}A).$$

Therefore, we can conclude that

$$\sigma_S(T)^{-1}(S^{-1}A) \simeq U^{-1}A \simeq \sigma_T(S)^{-1}(T^{-1}A).$$

**Solution E. 12.18.** By definition, $\mathbb{Z}_2 = \{\frac{a}{2^n} \in \mathbb{Q}\colon\ a \in \mathbb{Z},\ n \in \mathbb{N}\} = S^{-1}\mathbb{Z}$, where $S$ is the multiplicative subset consisting of all powers of 2.
If $p \neq 2$, then $(p) \cap S = \emptyset$. Thus, $S^{-1}(p) = (p)\mathbb{Z}_2$ is a prime ideal of $\mathbb{Z}_2$.
Since $\frac{2}{1} \in \mathbb{Z}_2 \setminus (p)\mathbb{Z}_2$, we can define

$$f\colon \mathbb{Z}_2 \times \mathbb{Z}_{(p)} \longrightarrow (\mathbb{Z}_2)_{(p)\mathbb{Z}_2}\,, \qquad f\left(\frac{a}{2^n}, \frac{c}{d}\right) = \frac{ac}{2^n d}\,,$$

for any $a, c \in \mathbb{Z}$, $n \in \mathbb{N}$, and $d \notin (p)$, see **E.12.17**.
It is easy to verify that $f$ is a $\mathbb{Z}$-bilinear map.
Hence, we obtain a commutative diagram

$$\begin{array}{ccc} \mathbb{Z}_2 \times \mathbb{Z}_{(p)} & \xrightarrow{\ f\ } & (\mathbb{Z}_2)_{(p)\mathbb{Z}_2} \\ \downarrow & \nearrow_{\tilde{f}} & \\ \mathbb{Z}_2 \otimes_{\mathbb{Z}} \mathbb{Z}_{(p)}. & & \end{array}$$

Since

$$\frac{a}{2^n} \otimes \frac{c}{d} = \frac{1}{2^n} \otimes \frac{ac2^n}{d2^n} = 1 \otimes \frac{ac}{d2^n}$$

and $\tilde{f}$ is obviously surjective, it is not hard to prove that $\tilde{f}$ is an isomorphism.

If $p = 2$, we consider the following commutative diagram

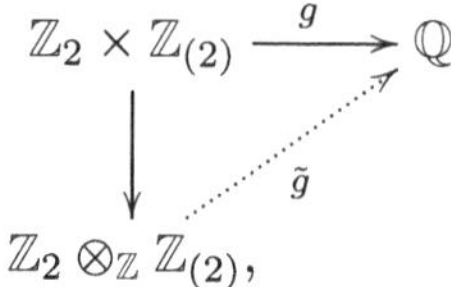

which is induced by $g$, where $g$ is defined by $g\left(\frac{a}{2^n}, \frac{c}{d}\right) = \frac{ac}{2^n d}$.
The conclusion follows from an argument similar to the one used in the previous case.

**Solution E. 12.19.** 1. If $a \in A$ is such that $\sigma_S(a) = 0$, then there exists $s \in S$ such that $as = 0$. Hence, $a = 0$, since $s \notin \mathcal{D}(A)$.
Moreover, if $T \supsetneq S$, then $T$ contains at least one zero-divisor $b$ such that $bc = 0$ for some $c \neq 0$.
Thus, $\sigma_T(c) = \frac{c}{1} = \frac{0}{1}$ and $\sigma_T$ is not injective.

2. Let $\frac{a}{s} \in Q(A)$, and assume it is not a zero-divisor. Then, $\frac{a}{1}$ is not a zero-divisor.

We claim that $a \notin \mathcal{D}(A)$. Indeed, if $ab = 0$ for some $b \in A$, then $\frac{a}{1} \cdot \frac{b}{1} = 0$, and therefore, $\sigma_S(b) = \frac{b}{1} = 0$. The injectivity of $\sigma_S$ yields $b = 0$.
Thus, $a \in S$ and $\frac{a}{s}$ is invertible.

3. Assume $A = A^* \sqcup \mathcal{D}(A)$. Then, $S = A^*$, and $\sigma_S(as^{-1}) = \frac{as^{-1}}{1} = \frac{a}{s}$ for any $\frac{a}{s} \in S^{-1}A$.
Thus, $\sigma_S$ is surjective and the conclusion follows from part 1.

4. When $\mathcal{D}(A) = \{0\}$ and $S = A \setminus \{0\}$, every non-zero element of $Q(A)$ is invertible by part 2, hence $Q(A)$ is a field containing $A$ by part 1.

Now, let $K$ be a field containing $A$. Then, each element of $S$ is invertible in $K$, and, by the universal property, the inclusion of $A$ in $K$ induces an inclusion of $Q(A)$ in $K$.

**Solution E. 12.20.** Let $S = A \setminus \mathfrak{p}$, and consider the homomorphism defined by

$$f\colon A \xrightarrow{\sigma_S} A_{\mathfrak{p}} \xrightarrow{\pi} A_{\mathfrak{p}}/\mathfrak{p}A_{\mathfrak{p}}.$$

It is easy to verify that $\operatorname{Ker} f = \mathfrak{p}$, hence $f$ induces an injective homomorphism

$$g\colon A/\mathfrak{p} \longrightarrow A_{\mathfrak{p}}/\mathfrak{p}A_{\mathfrak{p}}\,, \quad \bar{a} \mapsto \frac{a}{1} + \mathfrak{p}A_{\mathfrak{p}}.$$

Now, let $\overline{S} = A/\mathfrak{p} \setminus \{\overline{0}\}$. Then, for any $\overline{s} \in \overline{S}$, we have $g(\overline{s}) \in (A_{\mathfrak{p}}/\mathfrak{p}A_{\mathfrak{p}})^*$, and $g$ induces a homomorphism

$$\tilde{g}\colon Q(A/\mathfrak{p}) = \overline{S}^{-1}(A/\mathfrak{p}) \longrightarrow A_{\mathfrak{p}}/\mathfrak{p}A_{\mathfrak{p}}$$

such that the diagram

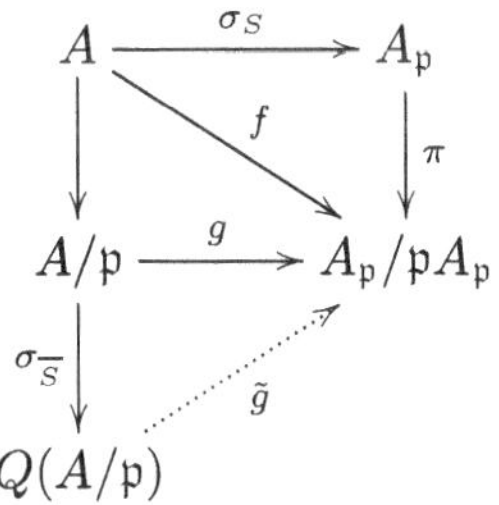

is commutative.
Each element $A_\mathfrak{p}/\mathfrak{p}A_\mathfrak{p}$ can be written as $\frac{a}{s}+\mathfrak{p}A_\mathfrak{p}=g(\overline{a})g(\overline{s})^{-1}$. Moreover, $g$ is injective.
The conclusion follows from **T.6.5**.

**Solution E. 12.21.** Since $A$ is a domain, the inclusion $A_a\subseteq Q(A)$ holds for any $a\neq 0$.
Let $(0)=\mathfrak{p}_0,\ \mathfrak{p}_1,\dots,\mathfrak{p}_k$ be the distinct prime ideals of $A$.
If $k=0$, then $A$ is a field and $A_a=A=Q(A)$ for all $a\neq 0$.

Otherwise, if $k>0$, consider the set $\bigcap_{i=1}^k\mathfrak{p}_i$. If this intersection is zero, then, by **T.1.12**.2, $\bigcap_{i=1}^k\mathfrak{p}_i=\mathfrak{p}_0$ yields $\mathfrak{p}_i=\mathfrak{p}_0$ for some $i>0$, which contradicts the hypothesis.
Therefore, we can take $0\neq a\in\bigcap_{i=1}^k\mathfrak{p}_i$ and consider the ring $A_a$. Since $\mathfrak{p}_iA_a=A_a$ for any $i>0$, the only prime ideal of $A_a$ is the zero ideal. Hence, $A_a$ is a field containing $A$.
Since $Q(A)$ is the smallest field containing $A$, we obtain $Q(A)\subseteq A_a$ also.

Alternatively, to prove the non-trivial inclusion take $\frac{b}{c}\in Q(A)$ and consider the ideal $\sqrt{(c)}$.
Since

$$a\in\bigcap_{i=1}^k\mathfrak{p}_i\subseteq\bigcap_{i:\ c\in\mathfrak{p}_i}\mathfrak{p}_i=\sqrt{(c)},$$

there exists $t\in\mathbb{N}$ such that $a^t=dc$ for some $d\in A$.
This immediately yields $\frac{b}{c}=\frac{bd}{a^t}$, and accordingly, $Q(A)\subseteq A_a$.

**Solution E. 12.22.** 1. At first, we prove that

$$\mathcal{D}(A)=\bigcup_{i=1}^n\overline{\mathfrak{p}_i}.$$

By **T.1.12**.2, we have $\mathcal{D}(A)\supseteq\overline{\mathfrak{p}_i}$ for all $i$. Therefore, $\mathcal{D}(A)\supseteq\bigcup_{i=1}^n\overline{\mathfrak{p}_i}$.

To prove the opposite inclusion, note that $\overline{a}\in\mathcal{D}(A)$ if and only if there exists $b\in B$ such that $\overline{b}\neq\overline{0}$ and $ab\in\mathfrak{p}_i$ for all $i$. Since $\overline{b}\neq\overline{0}$, there exists $i$ such that $b\notin\mathfrak{p}_i$. Thus, $a\in\mathfrak{p}_i$.

Now, let $S=A\setminus\bigcup_{i=1}^n\overline{\mathfrak{p}_i}$. Then, $Q(A)=S^{-1}A$, and the maximal ideals of $Q(A)$ are exactly the ideals $S^{-1}\overline{\mathfrak{p}_i}$, see the remark after **T.6.8**.
From **T.6.13**.3 it follows that

$$S^{-1}(A/\overline{\mathfrak{p}_i})\simeq S^{-1}A/S^{-1}\overline{\mathfrak{p}_i}$$

is a field containing $A/\overline{\mathfrak{p}_i}$, since $A/\overline{\mathfrak{p}_i}$ is a domain, for all $i=1,\dots,n$.

Moreover, since $S\subseteq A\setminus\overline{\mathfrak{p}_i}$, we can consider $S^{-1}(A/\overline{\mathfrak{p}_i})$ as a subring of $Q(A/\overline{\mathfrak{p}_i})$, see the proof of **T.6.19**.6.
By **E.12.19**.4,

$$S^{-1}(A/\overline{\mathfrak{p}_i})=Q(A/\overline{\mathfrak{p}_i}).$$

By **T.6.13**.2, we know that $\bigcap_{i=1}^n S^{-1}\overline{\mathfrak{p}_i}=S^{-1}\left(\bigcap_{i=1}^n\overline{\mathfrak{p}_i}\right)=0$ in $S^{-1}A$.

Hence, by the Chinese Remainder Theorem,

$$Q(A) = S^{-1}A \simeq \bigoplus_{i=1}^{n} S^{-1}A/S^{-1}\overline{\mathfrak{p}_i} \simeq \bigoplus_{i=1}^{n} S^{-1}(A/\overline{\mathfrak{p}_i})$$
$$\simeq \bigoplus_{i=1}^{n} Q(A/\overline{\mathfrak{p}_i}) \simeq \bigoplus_{i=1}^{n} Q(B/\mathfrak{p}_i).$$

2. We have $A \simeq \mathbb{C}[x,y]/(x) \cap (y)$ and $\mathcal{D}(A) = (\overline{x}) \cup (\overline{y})$.
By part 1,

$$S^{-1}A \simeq \mathbb{C}(x) \oplus \mathbb{C}(y).$$

3. The ideal $(x^2 - y^3)$ is prime, and therefore, $\mathcal{D}(A) = (0)$.
Thus, $S^{-1}A = Q(A)$ is the quotient field of $A$.

**Solution E. 12.23.** The statement is true when $B$ is a local ring, see **E.10.22**.
Consider the localization of $B$ at one of its maximal ideals $\mathfrak{m}$, and assume that $K$ is finitely generated as a $B$-module.
Since $K_\mathfrak{m} = K$ is a finitely generated $B_\mathfrak{m}$-module, this contradicts **E.10.22**.

**Solution E. 12.24.** The ring $A$ is a domain, and the $A$-module structure of $Q(A)$ is defined by restriction of scalars via the localization homomorphism $A \longrightarrow Q(A)$.

1. Observe that, by **E.12.17**, we have $Q(A) = K(x)$.
Since $m = x\frac{m}{x}$ for any $m \in Q(A)$, we have $Q(A) = (x)Q(A)$, and therefore, $Q(A)/(x)Q(A) = 0$, which is finitely generated.

2. Assume that $Q(A)$ is a finitely generated $A$-module.
Since $A$ is local with maximal ideal $(x)$ and $Q(A) = (x)Q(A)$, by Nakayama's Lemma, we have $Q(A) = 0$, but this is a contradiction.

**Solution E. 12.25.** 1. Let $s \in \operatorname{Ann} M \cap S$. Then, for any $\frac{m}{t} \in S^{-1}M$ we have $\frac{m}{t} = \frac{0}{s}$, that is, $S^{-1}M = 0$.

2. Let $M = \langle m_1, \ldots, m_r \rangle$.
Since $S^{-1}M = 0$, for any $i = 1, \ldots, r$ there exists $s_i \in S$ such that $s_i m_i = 0$.
Letting $s = \prod_{i=1}^{r} s_i$, we have $s \in \operatorname{Ann} M \cap S$.

3. Take $A = \mathbb{Z}$ and $S = \{2^n : n \in \mathbb{N}\}$. Consider the $A$-module

$$M = \bigoplus_{n \in \mathbb{N}} \mathbb{Z}/(2^{n+1}).$$

It is easy to verify that $\operatorname{Ann} M = 0$, hence $S \cap \operatorname{Ann} M = \emptyset$.
For each element $\alpha \in S^{-1}M$ there exists $k \in \mathbb{N}$ such that $\alpha$ can be written as

$$\alpha = \left(\frac{\overline{a_0}}{s_0}, \ldots, \frac{\overline{a_k}}{s_k}, 0, \ldots, 0, \ldots\right), \quad \text{with } \overline{a_i} \in \mathbb{Z}/(2^{i+1}) \text{ and } s_i \in S \text{ for all } i.$$

Obviously, $2^{k+1}\alpha = 0$, hence $S^{-1}M = 0$.

**Solution E. 12.26.** Note that

$$(1) = (f_h \colon h \in H) \subseteq \sqrt{(f_h^{n_h} : h \in H)},$$

for any choice of integers $n_h$.
It follows that $(f_h^{n_h} : h \in H) = A$.

Since $m = 0$ in $M_{f_h}$ for all $h$, there exist integers $n_h$ such that $f_h^{n_h} m = 0$ in $M$. By the previous remark, we can write 1 as a finite sum of elements $a_h f_h^{n_h}$ with $a_h \in A$.
Therefore, $m = 1m = \sum a_h f_h^{n_h} m = 0$.

**Solution E. 12.27.** 1. This immediately follows from **E.12.25**.1 and 2.

2. By **T.6.12**, the sequence

$$0 \longrightarrow M'_{\mathfrak{p}} \longrightarrow M_{\mathfrak{p}} \longrightarrow M''_{\mathfrak{p}} \longrightarrow 0$$

is exact for any $\mathfrak{p} \in \operatorname{Spec} A$.
Hence, $M_{\mathfrak{p}} \neq 0$ if and only if either $M'_{\mathfrak{p}} \neq 0$ or $M''_{\mathfrak{p}} \neq 0$.

3. Let $\mathfrak{p} \in \operatorname{Spec} A$. By **T.6.14**.2, we have

$$(M \otimes_A N)_{\mathfrak{p}} = M_{\mathfrak{p}} \otimes_{A_{\mathfrak{p}}} N_{\mathfrak{p}}.$$

Since $A_{\mathfrak{p}}$ is a local ring, the module $M_{\mathfrak{p}} \otimes_{A_{\mathfrak{p}}} N_{\mathfrak{p}}$ is non-zero if and only if $M_{\mathfrak{p}} \neq 0$ and $N_{\mathfrak{p}} \neq 0$, see **E.11.11**. The conclusion easily follows.

**Solution E. 12.28.** 1. By **E.12.27**.1, since $M$ is finitely generated, we have that $M_{(p)} = 0$ if and only if there exists $s \in (\mathbb{Z} \setminus (p)) \cap \operatorname{Ann}_{\mathbb{Z}} M$, *i.e.*, if and only if

$$(60) = \operatorname{Ann}_{\mathbb{Z}} M \not\subset (p).$$

Therefore, $M_{(p)} \neq 0$ only for $p = 2, 3, 5$.

2. Since $10 \notin (3)$, we have

$$M_{(3)} \simeq (\mathbb{Z}/(10))_{(3)} \oplus (\mathbb{Z}/(12))_{(3)} \simeq (\mathbb{Z}/(12))_{(3)},$$

and $(\mathbb{Z}/(12))_{(3)} \simeq \mathbb{Z}/(3)$ as already seen in **E.12.4**.5.

**Solution E. 12.29.** 1. By the properties of tensor product

$$M \simeq A/(xyz - z^2,\, xy^2 - 4,\, yz,\, x - y^2),$$

which is isomorphic to

$$\begin{aligned} A/(xy^2 - 4,\, x - y^2,\, yz,\, z^2) &\simeq A/(x - y^2,\, y^4 - 4,\, yz,\, z^2) \\ &\simeq A/(x - y^2,\, y^4 - 4,\, z), \end{aligned}$$

because the $S$-polynomial of $y^4 - 4$ and $yz$ is $-4z$.
Thus, $M$ has dimension 4 over $\mathbb{Q}$.

2. Recall that, since $M$ is finitely generated, $M_{\mathfrak{p}} \neq 0$ if and only if $\mathfrak{p} \supseteq \operatorname{Ann} M$, see **E.12.27**.1.
Since $\operatorname{Ann} M = (x - y^2,\, y^4 - 4,\, z)$ and

$$\sqrt{\operatorname{Ann} M} = \sqrt{(x-2,\, y^2-2,\, z)} \cap \sqrt{(x+2,\, y^2+2,\, z)} = \mathfrak{p}_1 \cap \mathfrak{p}_2$$

is the intersection of two maximal ideals, the only primes containing $\operatorname{Ann} M$ are $\mathfrak{p}_1$ and $\mathfrak{p}_2$. Thus, $\operatorname{Supp} M = \{\mathfrak{p}_1,\, \mathfrak{p}_2\}$.

**Solution E. 12.30.** Let $\mathfrak{m} \subset A$ be a maximal ideal.
If $I_{\mathfrak{m}} = 0$, then $\operatorname{Ann} I \not\subset \mathfrak{m}$ by **E.12.27**.1.
On the other hand, if $I_{\mathfrak{m}} = A_{\mathfrak{m}}$, then $I \not\subset \mathfrak{m}$, and therefore, $I + \operatorname{Ann} I \not\subset \mathfrak{m}$ for every maximal ideal $\mathfrak{m}$ of $A$. Hence,

$$I + \operatorname{Ann} I = A,$$

and there exist $i \in I$ and $j \in \operatorname{Ann} I$ such that $i + j = 1$.
Note that $i \neq 0$, otherwise $j = 1$ and $I = 0$, which contradicts the hypothesis.
Moreover, $i = i(i+j) = i^2$ is idempotent.
Finally, for any $a \in I$, we have $a = a(i+j) = ai$, that is, $I = (i)$.

**Solution E. 12.31.** 1. Obviously, $1 \in S$. Moreover, if $(q_1, p) = (q_2, p) = 1$, then $(q_1 q_2, p) = 1$. It follows that $S$ is closed under multiplication. We also note that $0 \notin S$.

Alternatively, we can observe that $S = \mathbb{Q}[x] \setminus ((x-1) \cup (x^2+2))$, where $(x-1)$ and $(x^2+2)$ are prime ideals.

2. Reasoning as in **E.12.4**.1 and 2, we find

$$A = \left\{ \frac{f}{g} \in \mathbb{Q}(x) \colon (g, p) = 1 \right\}.$$

The prime ideals of $A$ are in one-to-one correspondence with the prime ideals $\mathfrak{p}$ of $\mathbb{Q}[x]$ such that $\mathfrak{p} \cap S = \emptyset$. Since all non-zero primes of $\mathbb{Q}[x]$ are principal and generated by irreducible elements, the only prime ideals of $A$ are

$$\mathfrak{p}_1 = (0), \;\; \mathfrak{p}_2 = (x-1), \;\; \text{and} \;\; \mathfrak{p}_3 = (x^2+2).$$

Obviously, the maximal ones are $\mathfrak{p}_2$ and $\mathfrak{p}_3$.

3. Clearly $A/(0) \simeq A$, and note that

$$A/IA \simeq S^{-1}\left(\mathbb{Q}[x]/I\right)$$

for any ideal $I$ of $\mathbb{Q}[x]$.
Hence, $A/(x-1)A \simeq S^{-1}\mathbb{Q} \simeq \mathbb{Q}$ and

$$A/(x^2+2)A \simeq S^{-1}\left(\mathbb{Q}[x]/(x^2+2)\right) \simeq \mathbb{Q}(\sqrt{-2}).$$

4. Let $\mathfrak{p}_1$, $\mathfrak{p}_2$, and $\mathfrak{p}_3$ be the ideals computed in part 2.
It is clear that $A/\mathfrak{p}_1 \otimes_A A/\mathfrak{p}_i \simeq A/\mathfrak{p}_i$, and we described these rings in part 3.

In the remaining case, since $(x-1)$ and $(x^2+2)$ are comaximal, we have

$$A/\mathfrak{p}_2 \otimes_A A/\mathfrak{p}_3 \simeq A/(\mathfrak{p}_2 + \mathfrak{p}_3) = 0.$$

**Solution E. 12.32.** Let $\overline{A} = A/I$, $\overline{M} = M/IM$, and $\overline{\mathfrak{m}} = \mathfrak{m}/I$ for any $\mathfrak{m} \in \operatorname{Max} A$ containing $I$.
It is sufficient to prove that the $\overline{A}$-module $\overline{M}$ is trivial.

Since the maximal ideals of $\overline{A}$ are exactly the maximal ideals of $A$ containing $I$, it is enough to prove that $\overline{M}_{\overline{\mathfrak{m}}} = 0$ for all maximal ideals $\mathfrak{m} \supseteq I$, see **T.6.15**.
This holds, because

$$\overline{M}_{\overline{\mathfrak{m}}} \simeq M_{\mathfrak{m}}/IM_{\mathfrak{m}} = 0.$$

Alternatively, we can assume, by contradiction, that $IM \subsetneq M$ and take $m \in M \setminus IM$. Then,

$$I \subseteq IM : m \neq A.$$

Let $\mathfrak{m}$ be a maximal ideal such that $IM : m \subseteq \mathfrak{m}$. Since, by hypothesis, $M_{\mathfrak{m}} = 0$, we have $\frac{m}{1} = 0$, and hence, there exists $a \in A \setminus \mathfrak{m}$ such that $am = 0$. This contradicts the choice of $\mathfrak{m}$.

**Solution E. 12.33.** 1. Since prime ideals of $A_{\mathfrak{p}}$ are in one-to-one correspondence with prime ideals of $A$ contained in $\mathfrak{p}$, the minimality hypothesis implies that $\mathfrak{p}A_{\mathfrak{p}}$ is the unique prime of $A_{\mathfrak{p}}$.
Therefore, it is the nilradical of $A_{\mathfrak{p}}$.

2. Let $a \in \mathfrak{p}$. Then, by the previous part, $\frac{a}{1} \in A_{\mathfrak{p}}$ is nilpotent. Hence, there exists $s \in A \setminus \mathfrak{p}$ such that $sa^n = 0$ for some $n \in \mathbb{N}_+$. Let $n$ be minimal with that property, then $sa^{n-1} \neq 0$.
It follows that $a$ is a zero-divisor.

3. By **T.6.17**.1, being reduced is a local property, hence $A_{\mathfrak{p}}$ is reduced. By part 1, this yields $\mathfrak{p}A_{\mathfrak{p}} = \mathcal{N}(A_{\mathfrak{p}}) = 0$.
Therefore, the zero ideal is the only maximal ideal of $A_{\mathfrak{p}}$, which is then a field.

**Solution E. 12.34.** Since $1 \cdot 1 \in S$, we have $1 \in \overline{S}$.

Let $s, t \in \overline{S}$. Then, there exist $a, b \in A$ such that $as$, $bt \in S$. Since $S$ is multiplicative, we have $asbt \in S$.
By definition, this implies that $st \in \overline{S}$, and this proves that $\overline{S}$ is a multiplicative subset.

Now, assume $st \in \overline{S}$. Then, $ast \in S$ for some $a \in A$.
Therefore, $(at)s = (as)t \in S$, which implies $s, t \in \overline{S}$.

**Solution E. 12.35.** 1. Obviously, $1 \in f^{-1}(T)$.
Moreover, $a$, $b \in f^{-1}(T)$ yields $f(ab) = f(a)f(b) \in T$, that is, $ab \in f^{-1}(T)$.

2. Let $a \in \overline{f^{-1}(T)}$. Then, there exists $b \in A$ such that $ba \in f^{-1}(T)$, that is, $f(b)f(a) = f(ba) \in T$.
Since $T$ is saturated, we have $f(a) \in T$, that is, $a \in f^{-1}(T)$.

3. Assume that $f^{-1}(T)$ is multiplicative. Then, $1 \in T$.
Since $f$ is a surjective homomorphism, for any $s$, $t \in T$, there exist $a$, $b$ in $f^{-1}(T)$ such that $f(a) = s$ and $f(b) = t$.
Since $f^{-1}(T)$ is multiplicative, we have $ab \in f^{-1}(T)$, and thus, $st = f(ab) \in T$.

Now, assume that $f^{-1}(T)$ is also saturated.
Let $s \in \overline{T}$, and let $t \in B$ be such that $st \in T$.
By hypothesis, there exist $a$, $b \in A$ such that $f(a) = s$ and $f(b) = t$. Then, $ab \in f^{-1}(T)$ implies $a$, $b \in \overline{f^{-1}(T)} = f^{-1}(T)$.
Hence, $s = f(a) \in T$.

**Solution E. 12.36.** 1. The first statement immediately follows from the fact that $1 \in S$.

If $a$, $b \in I^S$, then there exist $s$, $t \in S$ such that $as$, $bt \in I$, and hence $(a+b)st \in I$. Moreover, $cas \in I$ for any $c \in A$.
Thus, $a + b$, $ac \in I^S$.

Alternatively, we can observe that, for any ideal $I$ of $A$, **T.6.7**.1c immediately implies $I^S = I^{ec}$ with respect to the homomorphism $\sigma_S$.

2. We have $\sigma_S(a) = \frac{a}{1} = 0$ if and only if there exists $u \in S$ such that $au = 0$, that is, if and only if $a \in (0)^S$.

3. Let $a \in I^S$. Then, there exists $s \in S$ such that $as \in I \subset J$, which implies $a \in J^S$.

4. By part 1, we have $I^S \subseteq (I^S)^S$.

To prove the opposite inclusion, let $a \in (I^S)^S$. Then, there exists $s \in S$ such that $as \in I^S$. Hence, there exists $t \in S$ such that $ast \in I$.
Since $st \in S$, we obtain $a \in I^S$.

5. By part 1, we have $IJ \subseteq I^S J^S$, and hence, $(IJ)^S \subseteq (I^S J^S)^S$ by part 3.

To prove the opposite inclusion, let $a = \sum_i a_i b_i \in I^S J^S$ with $a_i \in I^S$ and $b_i \in J^S$ for all $i$. Then, for all $i$, there exist $s_i$, $t_i \in S$ such that $s_i a_i \in I$ and $t_i b_i \in J$.
Let $u = \prod_i s_i t_i$. Then, we have $ua \in IJ$, which yields $a \in (IJ)^S$, and thus, $I^S J^S \subseteq (IJ)^S$.
It immediately follows that

$$(I^S J^S)^S \subseteq ((IJ)^S)^S = (IJ)^S$$

by parts 3 and 4.

**Solution E. 12.37.** The proofs of parts 1 and 2 are analogous to those of **E.12.36**.1 and 3, respectively.

3. Part 1 yields $\sigma_S^{-1}(Q) \subseteq \sigma_S^{-1}(Q)^S$.

To prove the opposite inclusion, take $q \in \sigma_S^{-1}(Q)^S$. Then, there exists $s \in S$ such that $\frac{sq}{1} = \sigma_S(sq) \in Q$.
Hence, $\frac{q}{1} \in Q$ and $q \in \sigma_S^{-1}(Q)$.

4. Since any submodule of $S^{-1}M$ is equal to $S^{-1}N$ for some submodule $N$ of $M$, the statements immediately follow from part 3.

5. The proof is analogous to the one of **E.12.36**.4.
6. The statement immediately follows from part 4 and **T.6.13**.2.
7. This follows from part 4 and from

$$\sigma_S^{-1}(S^{-1}N) + \sigma_S^{-1}(S^{-1}P) \subseteq \sigma_S^{-1}(S^{-1}N + S^{-1}P).$$

**Solution E. 12.38.** Consider the multiplicative subsets

$$S_A = \{\overline{18}^n \in \mathbb{Z}/(200) \colon n \in \mathbb{N}\} \quad \text{and} \quad S_B = \{\overline{6}^n \in \mathbb{Z}/(200) \colon n \in \mathbb{N}\}.$$

By **T.6.19**.3, we have

$$\overline{S}_B = (\mathbb{Z}/(200)) \setminus \bigcup_{\mathfrak{p} \cap S_B = \emptyset} \mathfrak{p} = (\mathbb{Z}/(200)) \setminus \bigcup_{\overline{2}, \overline{3} \notin \mathfrak{p}} \mathfrak{p} = \overline{S}_A .$$

Hence, by **T.6.19**.6,

$$\begin{aligned}(\mathbb{Z}/(200))_{\overline{18}} &= S_A^{-1}(\mathbb{Z}/(200)) = \overline{S}_A^{-1}(\mathbb{Z}/(200)) \\ &= \overline{S}_B^{-1}(\mathbb{Z}/(200)) = S_B^{-1}(\mathbb{Z}/(200)) = (\mathbb{Z}/(200))_{\overline{6}} .\end{aligned}$$

Moreover, $\overline{S}_B^{-1}(\mathbb{Z}/(200)) \simeq \overline{S}_B^{-1}(\mathbb{Z}/(8)) \times \overline{S}_B^{-1}(\mathbb{Z}/(25))$, and since $\overline{2} \in \overline{S}_B$, we have $\overline{S}_B^{-1}(\mathbb{Z}/(8)) = 0$.
Finally, the homomorphism $\sigma_{\overline{S}_B} : \mathbb{Z}/(25) \longrightarrow \overline{S}_B^{-1}(\mathbb{Z}/(25))$ is surjective, see **E.12.5**, and injective because $5 \notin \overline{S}_B$, see **E.12.36**.2.
Therefore,

$$S_A^{-1}(\mathbb{Z}/(200)) \simeq S_B^{-1}(\mathbb{Z}/(200)) \simeq \mathbb{Z}/(25).$$

From **E.11.3** we obtain $C \simeq \mathbb{Z}/(25, 40) \simeq \mathbb{Z}/(5)$.
As a result, $A \simeq B \not\simeq C$.

Now, consider the homomorphism

$$\varphi \colon \mathbb{Z}_{(3)}[x] \longrightarrow (\mathbb{Z}_{(3)})_6 \quad \text{defined by} \quad \varphi(x) = \tfrac{1}{6}.$$

The kernel of $\varphi$ contains $(6x - 1)$. Since $6x - 1$ has degree 1 and is primitive, it is irreducible in $\mathbb{Z}_{(3)}[x]$. Hence, $\operatorname{Ker}\varphi = (6x - 1)$.
Moreover, $\varphi$ is surjective, because every element of $(\mathbb{Z}_{(3)})_6$ can be written as

$$\frac{a}{6^k} = a\varphi(x^k) = \varphi(ax^k).$$

Thus, $D \simeq (\mathbb{Z}_{(3)})_6$.
Alternatively, we obtain $D \simeq (\mathbb{Z}_{(3)})_6$ as a special case of **E.12.8**.

Finally, $\mathbb{Z}_{(3)}$ is a local ring with unique maximal ideal generated by the image of 3. In $(\mathbb{Z}_{(3)})_6$ we have $\frac{1}{3} = \frac{2}{6}$, that is, 3 is invertible. Consequently, $(\mathbb{Z}_{(3)})_6 \subseteq \mathbb{Q}$ is a field containing $\mathbb{Z}$. Therefore, $D \simeq \mathbb{Q} \not\simeq A,\ B,\ C$.

**Solution E. 12.39.** Note that, by **T.6.3**.2, $U^{-1}A \neq 0$ implies that there exists $\mathfrak{q} \in \operatorname{Spec} A$ such that $U \cap \mathfrak{q} = \emptyset$ (why?).
By **T.6.19**.1, 3 and 6, we can assume without loss of generality that

$$S = \overline{S} = A \setminus \mathfrak{p},$$

for some $\mathfrak{p} \in \operatorname{Spec} A$.
If $\mathfrak{p}$ is not minimal, then $\mathfrak{p}$ properly contains at least one $\mathfrak{q} \in \operatorname{Min} A$. Hence, $S = A \setminus \mathfrak{p} \subsetneq A \setminus \mathfrak{q}$ is not maximal.

Conversely, let $\mathfrak{p}$ be a minimal prime of $A$. Define $S = A \setminus \mathfrak{p}$. Suppose, by contradiction, that there is a multiplicative subset $T$ such that $T \supsetneq S$ and $T^{-1}A \neq 0$. By **T.6.19**.1, 3 and 6, we can assume, without loss of generality, that $T$ is saturated. Hence,

$$T = A \setminus \bigcup_{\substack{\mathfrak{q} \in \operatorname{Spec} A \\ \mathfrak{q} \cap T = \emptyset}} \mathfrak{q}.$$

If $\mathfrak{q} \cap T = \emptyset$, then $\mathfrak{q} \subseteq A \setminus T \subsetneq A \setminus S = \mathfrak{p}$, contradicting the minimality of $\mathfrak{p}$.

**Solution E. 12.40.** By **T.6.19**.3, we have that $V = \{2^n 3^m\}_{n,m \in \mathbb{N}}$ is the saturation of both $S$ and $T$. Therefore, by **T.6.19**.6,

$$S^{-1}\mathbb{Z} = V^{-1}\mathbb{Z} = T^{-1}\mathbb{Z}.$$

**Solution E. 12.41.** 1. By **T.6.19**.3, it is sufficient to prove that

$$A \setminus \overline{S} = \bigcup_{\mathfrak{q} \cap S = \emptyset} \mathfrak{q} = \bigcup_{\mathfrak{p} \in \mathcal{V}(I)} \mathfrak{p}.$$

Obviously, if $\mathfrak{p} \in \mathcal{V}(I)$, then $\mathfrak{p} \cap S = \emptyset$, otherwise $1 \in \mathfrak{p}$. This implies the inclusion $\supseteq$.

To prove the opposite inclusion, let $\mathfrak{q}$ be a prime that does not intersect $S$. In this case, $(I, \mathfrak{q}) \neq 1$, otherwise there would be some $a \in I$ and $b \in \mathfrak{q}$ such that $a + b = 1$, that is, $b = 1 - a \in \mathfrak{q} \cap S$, which is a contradiction.
Therefore, there exists a prime $\mathfrak{q}' \supseteq (I, \mathfrak{q})$, which implies $\mathfrak{q} \subseteq \mathfrak{q}' \in \mathcal{V}(I)$.

2. For any prime $\mathfrak{p}$ of $A$ we have $S_f \cap \mathfrak{p} = \emptyset$ if and only if $f \notin \mathfrak{p}$.
Hence, we can write $\overline{S_f} = A \setminus \bigcup_{f \notin \mathfrak{p}} \mathfrak{p}$.
Therefore,

$$\overline{S_f} = A \setminus \bigcup_{f \notin \mathfrak{p}} \mathfrak{p} \subseteq A \setminus \bigcup_{g \notin \mathfrak{q}} \mathfrak{q} = \overline{S_g}$$

if and only if $\bigcup_{f \notin \mathfrak{p}} \mathfrak{p} \supseteq \bigcup_{g \notin \mathfrak{q}} \mathfrak{q}$, *i.e.*, if and only if, for any prime $\mathfrak{q}$ which does not contain $g$, we have $\mathfrak{q} \subseteq \bigcup_{f \notin \mathfrak{p}} \mathfrak{p}$.
By **T.1.12**.1, this is equivalent to saying that $\mathfrak{q} \subseteq \mathfrak{p}$ for some prime $\mathfrak{p}$ not containing $f$.

We can conclude that $\overline{S_f} \subseteq \overline{S_g}$ if and only if, for any prime ideal $\mathfrak{q}$, $g \notin \mathfrak{q}$ implies $f \notin \mathfrak{q}$. This is equivalent to saying that, for every prime $\mathfrak{p}$, if $f \in \mathfrak{p}$, then $g \in \mathfrak{p}$, *i.e.*,

$$\sqrt{(f)} = \bigcap_{f \in \mathfrak{p}} \mathfrak{p} \supseteq \bigcap_{g \in \mathfrak{q}} \mathfrak{q} = \sqrt{(g)}.$$

**Solution E. 12.42.** 1. The primes of $A_{\mathfrak{q}}$ are in one-to-one correspondence with the primes of $A$ contained in $\mathfrak{q}$. These primes correspond to the primes $I$ of $K[x, y]$ such that $(x, y) \supseteq I \supseteq (x^2 - y^2)$, and thus, $I \supseteq (x + y)$ or $I \supseteq (x - y)$.

Note that if $(x, y) \supset I \supsetneq (x \pm y)$, then $I = (x, y)$.
Indeed, there is an isomorphism $K[x, y]/(x \pm y) \simeq K[x]$, which sends $(\overline{x}, \overline{y})$ to $(x)$, and, in $K[x]$, there are no prime ideals between $(0)$ and $(x)$.
Therefore, the only primes of $A_{\mathfrak{q}}$ are $(x - y)A_{\mathfrak{q}}$, $\mathfrak{p}A_{\mathfrak{q}}$, and $\mathfrak{q}A_{\mathfrak{q}}$, which is the unique maximal ideal.

2. Let $S = A_{\mathfrak{q}} \setminus \mathfrak{p}A_{\mathfrak{q}}$.
In the localization $S^{-1}A_{\mathfrak{q}}$, there is only one maximal ideal, namely $S^{-1}(\mathfrak{p}A_{\mathfrak{q}})$.

Moreover, since $\operatorname{char} K \neq 2$, we have $x - y \notin (x + y)$ and $\frac{x-y}{1} \in S$. Thus,

$$S^{-1}(\mathfrak{p}A_{\mathfrak{q}}) = \left(\frac{x+y}{1}\right) = \left(\frac{0}{x-y}\right)$$

is the zero ideal, and therefore, $S^{-1}A_{\mathfrak{q}}$ is a field.
To describe this field, note that $(A_{\mathfrak{q}})_{\mathfrak{p}A_{\mathfrak{q}}} \simeq A_{\mathfrak{p}}$ by **E.12.17**.
Furthermore,

$$S^{-1}((x^2 - y^2)_{\mathfrak{q}}) = S^{-1}((x + y)_{\mathfrak{q}}).$$

Hence,

$$\begin{aligned}(A_{\mathfrak{q}})_{\mathfrak{p}A_{\mathfrak{q}}} &\simeq A_{\mathfrak{p}} \simeq S^{-1}(K[x, y]/(x^2 - y^2)) \\ &\simeq S^{-1}K[x, y]/S^{-1}(x^2 - y^2) \simeq S^{-1}K[x, y]/S^{-1}(x + y) \\ &\simeq S^{-1}(K[x, y]/(x + y)) \simeq K(x).\end{aligned}$$

## 17.6 Chapter 13

**Solution E. 13.1.** 1. Let $(a)$, $(b) \subseteq \mathbb{Z}$. Then, we have $(a) \subseteq (b)$ if and only if $b \mid a$. Therefore, any strictly ascending chain starting with $(a)$ has at most as many elements as the proper divisors of $a$.

For the same reason, given an element $a \neq 0, \pm 1$ of $\mathbb{Z}$, there exists an infinite descending chain $(a) \supsetneq (a^2) \supsetneq \ldots \supsetneq (a^k) \supsetneq \ldots$.

2. It is sufficient to observe that $A$ is a field isomorphic to $\mathbb{R}(\sqrt{-2}) \simeq \mathbb{C}$.

**Solution E. 13.2.** Since $K$ is Noetherian and Artinian, and $V \simeq K^n$, we have $1 \Rightarrow 2$ and $1 \Rightarrow 3$ by **T.7.2**.3 and **T.7.3**, respectively.
We show that $2 \Rightarrow 1$ and $3 \Rightarrow 1$. Assume that $V$ has infinite dimension. Then, there exists a set $\{v_i\}_{n \in \mathbb{N}}$ of linearly independent vectors of $V$. As $n$ varies in $\mathbb{N}$, the subspaces $V_n = \langle v_1, \ldots, v_n \rangle_K$, respectively $W_n = \langle v_n, v_{n+1}, \ldots \rangle_K$, form an ascending, respectively a descending, chain of subspaces of $V$ which is not stationary.

**Solution E. 13.3.** We want to show that if $M$, $N$, and $P$ are $A$-modules, and

$$0 \longrightarrow N \xrightarrow{f} M \xrightarrow{g} P \longrightarrow 0$$

is a short exact sequence, then $M$ is Noetherian if and only if both $N$ and $P$ are Noetherian.

Assume $M$ is Noetherian, and let $N'$ a submodule of $N$. Then, $N' \simeq f(N')$ is a submodule of $M$, and thus, is finitely generated.
Analogously, given a submodule $P'$ of $P$, the submodule $g^{-1}(P')$ of $M$ is finitely generated, and thus, $P' = g(g^{-1}(P'))$ is finitely generated.

Conversely, assume $N$ and $P$ to be Noetherian, and consider a submodule $M'$ of $M$. Then, $g(M')$ and $f^{-1}(M' \cap f(N))$ are submodules of $P$ and $N$, respectively, and therefore, are finitely generated.
Hence,

$$M' \cap f(N) = f(f^{-1}(M' \cap f(N))) \simeq f^{-1}(M' \cap f(N))$$

is finitely generated. Applying **E.10.29** to the exact sequence

$$0 \longrightarrow M' \cap f(N) \xrightarrow{i} M' \xrightarrow{g_{|M'}} g(M') \longrightarrow 0$$

the conclusion follows.

**Solution E. 13.4.** 1. Consider the ideals $H_i = \operatorname{Ker}(\varphi^i)$. Since $A$ is Noetherian, the chain $H_1 \subseteq H_2 \subseteq \ldots$ is stationary, and there exists a positive integer $n$ such that $H_n = H_{n+1}$.
Let $a \in H_1$. We need to show that $a = 0$. Since $\varphi$ is surjective, also $\varphi^n$ is surjective, and there exists $b \in A$ such that $\varphi^n(b) = a$.
Hence, $\varphi^{n+1}(b) = \varphi(a) = 0$, that is, $b \in H_{n+1} = H_n$, and thus, $a = 0$.

2. Since $\varphi$ is surjective, $\varphi(I)$ is an ideal of $A$ for every ideal $I \subset A$.
As $\varphi(I \cap J) \subseteq \varphi(I) \cap \varphi(J)$ always holds, it is sufficient to prove the opposite inclusion. Let $a \in \varphi(I) \cap \varphi(J)$. Then, $a = \varphi(i) = \varphi(j)$ for some $i \in I$ and $j \in J$. Hence, $\varphi(i - j) = 0$. Since $\varphi$ is injective by part 1, $i = j \in I \cap J$.

3. In general, when $A$ is not Noetherian, neither of these statements are true. As an example, consider the ring $A = K[x_i : i \in \mathbb{N}_+]$, and let $\varphi$ be the endomorphism defined by $\varphi(x_1) = 0$ and $\varphi(x_i) = x_{i-1}$ for $i > 1$. Then, $\varphi$ is surjective but not injective.

Now, consider the ideals $I = (x_1 + x_2)$ and $J = (x_2)$ of $A$.
Since $I$ and $J$ are principal and generated by coprime elements, $I \cap J = IJ$. Therefore, $\varphi(I \cap J) = (\varphi(x_1 + x_2)\varphi(x_2)) = (x_1^2)$, while $\varphi(I) \cap \varphi(J) = (x_1)$.

**Solution E. 13.5.** Since $A$ is Noetherian, $I$ is a finitely generated $A$-module. By Nakayama's Lemma, there exists an element $b \in A$ such that $b \equiv 1 \mod I$ and $bI = 0$.
Let $a = 1 - b \in I$. Then,

$$(a) \subseteq I = 1 \cdot I = (a + b)I = aI \subseteq (a),$$

which shows $I = (a)$.
Moreover, $a$ is an idempotent, since $0 = ab = a(1 - a)$.

**Solution E. 13.6.** It is easy to verify that the sequence

$$0 \longrightarrow N_1 \cap N_2 \longrightarrow M \xrightarrow{f} M/N_1 \oplus M/N_2,$$

where the first homomorphism is the inclusion, and $f$ is given by the natural projections, is exact.
Therefore, $M/(N_1 \cap N_2) \simeq \operatorname{Im} f \subseteq M/N_1 \oplus M/N_2$, and all these modules are Noetherian, see **T.7.2**.2 and 3.

**Solution E. 13.7.** If $M$ is Noetherian, then any quotient of $M$ is Noetherian, see **T.7.2**.2.

Conversely, suppose that $M/\overline{f}M$ and $M/\overline{g}^2M$ are Noetherian $A$-modules. By **E.13.6**, it is enough to prove that $\overline{f}M \cap \overline{g}^2M = \overline{fg}^2M = 0$.
The containment $\overline{f}M \cap \overline{g}^2M \supseteq \overline{fg}^2M$ is obvious.

To prove the opposite inclusion, note that, by hypothesis, there exist polynomials $r, s \in K[x]$ such that $rf + sg^2 = 1$.
Let $m \in \overline{f}M \cap \overline{g}^2M$. Then, $m = \overline{a}\overline{f}m_1 = \overline{b}\overline{g}^2n_1$, with $a, b \in K[x]$ and $m_1, n_1 \in M$.
Therefore,

$$m = (\overline{rf + sg^2})m = \overline{r}\overline{f}(\overline{b}\overline{g}^2n_1) + \overline{s}\,\overline{g}^2(\overline{a}\overline{f}m_1) = (\overline{rb}\,\overline{fg^2})n_1 + (\overline{sa}\overline{fg^2})m_1$$

belongs to $\overline{fg}^2M$.

**Solution E. 13.8.** Note that, by hypothesis, $d \neq 0$.

1. Since $(d) = IJ$, we have $d = \sum_{i=1}^{k} f_i g_i$ for some $f_i \in I$, $g_i \in J$, and some integer $k$. Let $\widetilde{J} = (g_1, \dots, g_k)$. Then, $(d) \subseteq I\widetilde{J} \subseteq IJ = (d)$.

2. We will show that every ideal $0 \neq I$ of $A$ is finitely generated.
Let $\widetilde{J}$ be the finitely generated ideal from part 1. For every $f \in I$ and every $i$, we have $fg_i \in IJ = (d)$, and thus, $fg_i = h_i d$ for some $h_i \in A$. Moreover, $fd \in I\widetilde{J}$, and therefore,

$$fd = f\sum_{i=1}^{k} f_i g_i = \sum_{i=1}^{k} f_i(fg_i) = d\sum_{i=1}^{k} f_i h_i.$$

Since $A$ is a domain, $f = \sum_{i=1}^{k} f_i h_i$ and $I = (f_1, \dots, f_k)$.

**Solution E. 13.9.** 1. It easily follows from the linearity of $f$ and $g$.

2. Let $\pi_A \colon A \times_C B \longrightarrow A$ and $\pi_B \colon A \times_C B \longrightarrow B$ be the natural projections. We first prove that they are surjective. Since $f$ and $g$ are surjective, if $a \in A$, then $f(a) = c$ for some $c \in C$, and there exists $b \in B$ such that $g(b) = c = f(a)$. Therefore, $(a, b) \in A \times_C B$ and $\pi_A(a, b) = a$.
In a similar way we can prove that $\pi_B$ is surjective.

By **E.10.3**, the ideals of $A$ and $B$ are the $(A \times_C B)$-submodules of $A$ and $B$, respectively. Then, since by hypothesis $A$ and $B$ are Noetherian rings, $A$ and $B$ are Noetherian as $(A \times_C B)$-modules. Hence, $A \times B$ is Noetherian as an $(A \times_C B)$-module, and $A \times_C B$, being a submodule of a Noetherian $(A \times_C B)$-module, is also a Noetherian $(A \times_C B)$-module by **T.7.2**.2, *i.e.*, $(A \times_C B)$ is a Noetherian ring.

**Solution E. 13.10.** Note that, if $m = 0$, then $A$ is a field and all statements are obvious. Thus, we can assume $\mathfrak{m} = (m) \neq 0$.

1. Assume that every non-zero element of $\mathfrak{m}$ admits such a factorization. Suppose, by contradiction, that there exists $0 \neq a = um^k$, with $u \in A^*$ and $a \in \bigcap_{n \in \mathbb{N}} \mathfrak{m}^n$. Then, $a \in \mathfrak{m}^n$ for all $n$, and we have

$$\mathfrak{m}^k \subseteq (a) \subseteq \mathfrak{m}^n \subseteq \mathfrak{m}^k \quad \text{for every} \quad n > k.$$

Therefore, for $n > k$, we have $\mathfrak{m}^n = \mathfrak{m}^k$, which implies $m^k = bm^n$ and $m^k(1 - bm^{n-k}) = 0$. Since $1 - bm^{n-k}$ is invertible by **T.1.15**, we have $m^k = 0$, and then, $a = 0$, and we have found a contradiction.

Conversely, assume $\bigcap_{n \in \mathbb{N}} \mathfrak{m}^n = 0$. Let $0 \neq a \in \mathfrak{m}$, and let $k$ be the largest exponent such that $a \in \mathfrak{m}^k$. Then, $a = um^k$ for some $u \notin \mathfrak{m}$. Since $A$ is local, $u$ is invertible.

2. Let $I \subset A$ be a proper ideal. Then, $I \subset \mathfrak{m}$ and, by part 1, every $a \in I$ can be written as $a = um^k$, where $u \in A^*$.

Hence, $a \in I$ implies $m^k \in I$ for some integer $k$. Let $h$ be the smallest exponent such that $m^h \in I$. Then, $I \supseteq \mathfrak{m}^h$, and every $a \in I$ can be expressed as $a = um^k = um^{k-h}m^h \in \mathfrak{m}^h$.

3. We prove that, if $A$ is Noetherian, then every $0 \neq a \in \mathfrak{m}$ admits a factorization $a = um^k$. The conclusion will follow from parts 1 and 2.
Let $0 \neq a = b_1 m \in \mathfrak{m}$. If $b_1$ is invertible, we are done. Otherwise, there exists $b_2 \in A$ such that $b_1 = b_2 m \in \mathfrak{m}$, and thus, $a = b_2 m^2$ and $(b_1) \subsetneq (b_2)$.
Proceeding in this way we construct an ascending chain

$$(b_1) \subsetneq (b_2) \subsetneq \ldots \subsetneq (b_n) \subsetneq \ldots$$

of ideals of $A$, whose generators are such that $b_i m^i = a$ for all $i$.
If all $b_i$ are non-invertible, the chain is infinite, which is not possible since $A$ is Noetherian.
Thus, there exists a $b_k$ which is invertible and such that $a = b_k m^k$.

**Solution E. 13.11.** Every ideal of an ascending chain $\mathcal{C}$ of prime ideals of $A$ is contained in a maximal ideal $\mathfrak{m}_{\mathcal{C}}$ which is principal by hypothesis. Localizing $A$ at $\mathfrak{m}_{\mathcal{C}}$, by **T.7.2**.7 and applying **E.13.10**, we obtain that every non-zero ideal of $A_{\mathfrak{m}_{\mathcal{C}}}$ has the form $\mathfrak{m}_{\mathcal{C}}^k A_{\mathfrak{m}_{\mathcal{C}}}$.
Thus, the length of the chain $\mathcal{C}_{\mathfrak{m}_{\mathcal{C}}}$ of the localized primes of $\mathcal{C}$ is equal to 1 if $A_{\mathfrak{m}_{\mathcal{C}}}$ is a domain, 0 otherwise.

From the one-to-one correspondence between prime ideals of $A$ and prime ideals of its localizations, it immediately follows

$$\dim A \leq \sup\{\dim A_{\mathfrak{m}} : \mathfrak{m} \in \operatorname{Max} A\} \leq 1.$$

**Solution E. 13.12.** Since, by hypothesis, $M$ is Noetherian, we have that $M = \langle m_1, \ldots, m_n \rangle$ is finitely generated.
We consider the homomorphism

$$\varphi\colon A \longrightarrow M^n \quad \text{defined by} \quad \varphi(a) = (am_1, \ldots, am_n),$$

and we prove that $\operatorname{Ker} \varphi = \operatorname{Ann}_A M$.
The containment $\operatorname{Ann}_A M \subseteq \operatorname{Ker} \varphi$ is obvious.

To prove the opposite inclusion, we observe that if $a \in \operatorname{Ker} \varphi$, then $am_i = 0$ for all $i = 1, \ldots, n$, and thus, $a \in \operatorname{Ann}_A M$.
Therefore,

$$A / \operatorname{Ann}_A M \simeq \operatorname{Im} \varphi \subseteq M^n.$$

Since $M$ is a Noetherian $A$-module, by **T.7.2**.2 we have that $\operatorname{Im} \varphi \subseteq M^n$ is a Noetherian $A$-module. Hence, $A / \operatorname{Ann}_A M$ is Noetherian as an $A$-module and as an $A / \operatorname{Ann}_A M$-module by **T.7.2**.6.

**Solution E. 13.13.** 1. Assume $I + J \subsetneq A$. Then, there exists a maximal ideal $\mathfrak{m}$ such that $I, J \subseteq I + J \subseteq \mathfrak{m}$, and this contradicts the hypothesis.

2. The ideal $IJ$ is contained in both $I$ and $J$, hence, by hypothesis, is contained in every prime ideal of $A$, and thus, in the nilradical $\mathcal{N}(A)$. Since $A$ is Noetherian, the ideal $\mathcal{N}(A)$ is nilpotent by **T.7.5**.
Therefore, there exists $n \in \mathbb{N}_+$ such that $(IJ)^n \subseteq \mathcal{N}(A)^n = 0$.

**Solution E. 13.14.** 1. Let $\{g_1, \ldots, g_n\}$ be a minimal generating set of $I$, where $n = \mu(I) > 1$ by hypothesis.
Then, $I^2$ is generated by all distinct products $g_i g_j$, which are $\frac{n(n+1)}{2} < n^2$. Therefore, $\mu(I^2) < n^2$.

2. Since every ideal $J$ of $A$ is flat, by tensoring with $J$ the exact sequence

$$0 \longrightarrow I \longrightarrow A \longrightarrow A/I \longrightarrow 0,$$

we have that the induced sequence

$$0 \longrightarrow I \otimes J \longrightarrow J \longrightarrow J/IJ \longrightarrow 0$$

is exact.
We obtain a commutative diagram

$$\begin{array}{ccccccccc} 0 & \longrightarrow & I \otimes J & \longrightarrow & J & \longrightarrow & J/IJ & \longrightarrow & 0 \\ & & \downarrow & & \downarrow & & \downarrow & & \\ 0 & \longrightarrow & IJ & \longrightarrow & J & \longrightarrow & J/IJ & \longrightarrow & 0, \end{array}$$

where the last two vertical arrows are the identity homomorphisms.
From the Snake Lemma, or directly from **T.4.20**, it follows $I \otimes J \simeq IJ$.

Now, let $J = I$. By **E.11.10**, we have $\mu(I)^2 = \mu(I \otimes I) = \mu(I^2)$, and therefore, part 1 implies $\mu(I) \leq 1$ for all $I$, *i.e.*, $A$ is a PIR.
To prove that $A$ is a domain, let $a, b \in A$ be such that $ab = 0$. Then, $(a) \otimes (b) \simeq (a)(b) = 0$.
Since $A$ is local, from **E.11.11** it follows that either $(a) = 0$ or $(b) = 0$.

**Solution E. 13.15.** 1. We have

$$\sqrt{I} = \bigcap_{\mathfrak{p}_i \in \operatorname{Min} I} \mathfrak{p}_i \quad \text{and} \quad \sqrt{J} = \bigcap_{\mathfrak{q}_j \in \operatorname{Min} J} \mathfrak{q}_j.$$

Hence, if $I$ and $J$ have the same minimal primes, they also have the same radical.

Conversely, let $\mathfrak{p}$ be a minimal prime of $I$. Then,

$$\sqrt{J} = \bigcap_{\mathfrak{q}_j \in \operatorname{Min} J} \mathfrak{q}_j = \sqrt{I} \subseteq \mathfrak{p}.$$

Since $A$ is Noetherian, $\operatorname{Min} J$ is finite by **T.7.10**.2. Thus, there exists an ideal $\mathfrak{q}_j \in \operatorname{Min} J$ such that $\mathfrak{q}_j \subseteq \mathfrak{p}$, see **T.1.12**.2. For the same reason, there exists

$\mathfrak{p}_i \in \operatorname{Min} I$ such that $\mathfrak{p}_i \subseteq \mathfrak{q}_j \subseteq \mathfrak{p}$. Thus, by the minimality of $\mathfrak{p}$, we have $\mathfrak{p} = \mathfrak{q}_j$.
We have proven that the minimal primes of $I$ are minimal primes of $J$. The conclusion follows exchanging the roles of $I$ and $J$.

2. This follows from part 1 and the definitions of height and Krull dimension. Observe that, for any ideal $I$, we have

$$\operatorname{ht} I = \min\{\operatorname{ht} \mathfrak{p} \colon \mathfrak{p} \in \operatorname{Min} I\},$$

and that the chains of prime ideals of $A/I$ involved in determining the dimension of $A/I$ correspond to the chains of prime ideals of $A$ that start with one of the minimal primes of $I$.

**Solution E. 13.16.** Let $\bigcap_{i=1}^{n} \mathfrak{q}_i$ be a minimal primary decomposition of $\operatorname{Ann} M$.
Since $\operatorname{Ann} M$ is 0-dimensional, the ideals $\mathfrak{m}_i = \sqrt{\mathfrak{q}_i}$ are maximal, for all $i$.
By the first finiteness theorem **T.7.5**, there exist $s_i$ such that $\mathfrak{m}_i^{s_i} \subseteq \mathfrak{q}_i$. For each $i$ choose $s_i$ minimal with respect to this property, and let

$$J = \mathfrak{m}_1^{s_1-1} \cap \bigcap_{i=2}^{n} \mathfrak{m}_i^{s_i}.$$

By construction, $J \not\subseteq \operatorname{Ann} M$, and thus, $0 \neq JM \subseteq M$ and $\operatorname{Ann}(JM) = \mathfrak{m}_1$.
Now, we prove that, if $0 \neq m \in JM$, then the submodule $\langle m \rangle \subseteq M$ is simple. The homomorphism from $A$ to $\langle m \rangle$ defined by $1 \mapsto m$ is surjective, and its kernel is the proper ideal $\operatorname{Ann} m \supseteq \operatorname{Ann}(JM) = \mathfrak{m}_1$.
The conclusion follows from **E.10.14**.1.

**Solution E. 13.17.** Let $\mathfrak{p} \in \operatorname{Min} A$ be a minimal prime.
Consider the decomposition of the nilradical

$$\mathcal{N}(A) = \sqrt{(0)} = \mathfrak{p} \cap \bigcap_i \mathfrak{p}_i$$

as a finite intersection of minimal primes, see **T.7.10**.3.
By the minimality of the ideals in the decomposition, we can choose $b \in \bigcap_i \mathfrak{p}_i$ such that $b \notin \mathfrak{p}$. Then, $b$ is not nilpotent, and $b\mathfrak{p} \subset \mathcal{N}(A)$.
Therefore, there exists $n \in \mathbb{N}$ such that $(b\mathfrak{p})^n = (0)$, and we take $a = b^n$.

Conversely, let $\mathfrak{p} \in \operatorname{Spec} A$ and $\mathcal{N}(A) = \bigcap_i \mathfrak{p}_i$, where the ideals $\mathfrak{p}_i$ are the minimal primes of $A$. If there exists $a \in A \setminus \mathcal{N}(A)$ such that $a\mathfrak{p}^n = 0$, then $a^n\mathfrak{p}^n = 0$ and $a\mathfrak{p} \subseteq \mathcal{N}(A)$. Then, since $a \notin \mathcal{N}(A)$, there exists $\mathfrak{p}_j$ such that $a \notin \mathfrak{p}_j$. Hence, $\mathfrak{p} \subset \mathfrak{p}_j$, because $\mathfrak{p}_j$ is prime.
From this and the minimality of $\mathfrak{p}_j$, it follows that $\mathfrak{p} = \mathfrak{p}_j$ is a minimal prime.

**Solution E. 13.18.** Let $I$ be a 0-dimensional radical ideal contained in $J$. Then, we can write $I = \sqrt{I} = \bigcap_i \mathfrak{m}_i$ as a finite intersection of distinct maximal ideals. Hence,

$$J = \left(\bigcap_i \mathfrak{m}_i\right) + J = \left(\prod_i \mathfrak{m}_i\right) + J = \prod_i (\mathfrak{m}_i + J) = \bigcap_i (\mathfrak{m}_i + J),$$

where $\mathfrak{m}_i + J$ is equal to $\mathfrak{m}_i$ or to $(1)$, depending on whether $J$ is contained in $\mathfrak{m}_i$ or not. From the previous equality, we obtain that if $\mathfrak{m}_i + J = (1)$ for all $i$, then $J = (1)$, otherwise $J = \bigcap_{i\colon J \subseteq \mathfrak{m}_i} \mathfrak{m}_i$. Thus, $J$ is radical.

We conclude by observing that $A/J$ is isomorphic to a quotient of $A/I$, and therefore, it is 0-dimensional whenever $J \neq (1)$.

**Solution E. 13.19.** It is sufficient to find a decomposition of

$$\sqrt{I} = (xzt,\, yt,\, xyzt,\, xz) = (yt,\, xz).$$

We have $\sqrt{I} = (x,\, y) \cap (y,\, z) \cap (x,\, t) \cap (z,\, t)$.
Hence,

$$\operatorname{Min} I = \{(x,\, y),\, (y,\, z),\, (x,\, t),\, (z,\, t)\}.$$

**Solution E. 13.20.** 1. First, we observe that $I = (x^2z,\, x^2y^4t + x^2y^3,\, xt^2)$. Then, we compute its reduced Gröbner basis with respect to the lex order with $x > y > z > t$, and obtain

$$\{x^2z,\, x^2y^3,\, xt^2\}.$$

Thus, $I$ is monomial.

2. Repeatedly using **T.2.6**.1, we obtain

$$I = (x) \cap (x^2,\, t^2) \cap (x^2,\, z,\, t^2) \cap (y^3,\, z,\, t^2).$$

It is easy to see that $(x) \cap (x^2,\, t^2) \cap (y^3,\, z,\, t^2)$ is a minimal primary decomposition of $I$, with

$$\operatorname{Ass} I = \{(x),\, (x,\, t),\, (y,\, z,\, t)\} \quad \text{and} \quad \operatorname{Min} I = \{(x),\, (y,\, z,\, t)\}.$$

3. Since

$$\mathcal{D}(A/I) = \bigcup_{\mathfrak{p}_i \in \operatorname{Ass}(\overline{0})} \mathfrak{p}_i \quad \text{and} \quad \mathcal{N}(A/I) = \bigcap_{\mathfrak{p}_i \in \operatorname{Min}(\overline{0})} \mathfrak{p}_i,$$

see **T.7.10**.3, from part 2 we have

$$\mathcal{D}(A/I) = (\overline{x},\, \overline{t}) \cup (\overline{y},\, \overline{z},\, \overline{t}) \quad \text{and} \quad \mathcal{N}(A/I) = (\overline{y},\, \overline{z},\, \overline{t}) \cap (\overline{x}) = (\overline{xy},\, \overline{xz},\, \overline{xt}).$$

**Solution E. 13.21.** 1. We have $(x^5 - 3x^2) = (x^2) \cap (x^3 - 3)$. Hence, in $\mathbb{Q}[x]/(x^5 - 3x^2)$, we have

$$\operatorname{Ass}(\overline{0}) = \left\{(\overline{x}),\, \left(\overline{x^3 - 3}\right)\right\},$$

and $\operatorname{Min}(\overline{0}) = \operatorname{Ass}(\overline{0})$.

In the ring $A$, the prime ideals are

$$\mathfrak{p}_1 = (\overline{x}) \oplus (\overline{1}),\ \mathfrak{p}_2 = \left(\overline{x^3 - 3}\right) \oplus (\overline{1}),\ \mathfrak{p}_3 = (\overline{1}) \oplus (\overline{2}), \text{ and } \mathfrak{p}_4 = (\overline{1}) \oplus (\overline{3}),$$

and they are all associated primes of $(0_A)$ (why?).
Hence, by **T.7.10**.3,

$$\mathcal{N}(A) = \bigcap_i \mathfrak{p}_i = \left(\overline{x^4 - 3x}\right) \oplus (\overline{6}) \text{ and } \mathcal{D}(A) = \bigcup_{i=1}^{4} \mathfrak{p}_i.$$

2. For $i = 1, \ldots, 4$, write $\mathfrak{p}_i = \mathfrak{q}_i \oplus \mathfrak{t}_i$, and let $S_i = A \setminus \mathfrak{p}_i$. We will examine the 4 cases.
Every element in $S_1^{-1}A$ of the form $\sigma_{S_1}((\overline{1}, \overline{\beta}))$ is invertible. In particular, $(\overline{1}, \overline{0}) \in S_1$ implies $\sigma_{S_1}((\overline{a}, \overline{b})) = \sigma_{S_1}((\overline{a}, \overline{0}))$ in $S_1^{-1}A$, and thus, it is sufficient to consider pairs with second coordinate equal to $\overline{0}$.
Hence,

$$\begin{aligned} S_1^{-1}A &\simeq (\mathbb{Q}[x]/(x^5 - 3x^2))_{\mathfrak{q}_1} \oplus \overline{0} \\ &\simeq (\mathbb{Q}[x]/(x^2))_{\mathfrak{q}_1} \oplus (\mathbb{Q}[x]/(x^3 - 3))_{\mathfrak{q}_1} \simeq (\mathbb{Q}[x]/(x^2))_{(x)} \simeq \mathbb{Q}[x]/(x^2). \end{aligned}$$

In particular, $S_1^{-1}A$ contains a non-trivial nilpotent, and therefore, is not a domain.
Analogously,

$$S_2^{-1}A \simeq (\mathbb{Q}[x]/(x^5 - 3x^2))_{\mathfrak{q}_2} \oplus \overline{0} \simeq \left(\mathbb{Q}[x]/(x^3 - 3)\right)_{(x^3-3)} \simeq \mathbb{Q}(\sqrt[3]{3})$$

is a field.
Finally, for $i = 3, 4$, we have

$$S_i^{-1}A \simeq \overline{0} \oplus (\mathbb{Z}/(12))_{\mathfrak{t}_i} \simeq (\mathbb{Z}/(4))_{\mathfrak{t}_i} \oplus (\mathbb{Z}/(3))_{\mathfrak{t}_i}.$$

Thus,

$$S_3^{-1}A \simeq (\mathbb{Z}/(4))_{(2)} \simeq \mathbb{Z}/(4),\quad S_4^{-1}A \simeq (\mathbb{Z}/(3))_{(3)} \simeq \mathbb{Z}/(3),$$

and only the latter is a field.

**Solution E. 13.22.** 1. Let $\mathfrak{p}$ be a prime ideal containing $I$. Then, either $\mathfrak{p} \supseteq (I, 3)$ or $\mathfrak{p} \supseteq (I, 5)$.
It is easy to see that $(I, 3) = (y - 2,\ (x + y)^3,\ 3) = (y - 2,\ (x + 2)^3,\ 3)$ and $(I, 5) = (x^2 - 4,\ (x + y)^3,\ 5)$. Let

$$\mathfrak{m}_1 = (x + 2,\ y - 2,\ 3),\ \mathfrak{m}_2 = (x + 2,\ y - 2,\ 5) \text{ and } \mathfrak{m}_3 = (x - 2,\ y + 2,\ 5).$$

These ideals are maximal, and thus, pairwise comaximal. Observe that $\sqrt{(I, 3)} = \mathfrak{m}_1$.

Moreover,

$$\begin{aligned}(I,\,5) &= (x+2,\,(x+y)^3,\,5)\cap(x-2,\,(x+y)^3,\,5)\\ &= (x+2,\,(y-2)^3,\,5)\cap(x-2,\,(y+2)^3,\,5).\end{aligned}$$

The ideals in this intersection are primary, since their radicals are $\mathfrak{m}_2$ and $\mathfrak{m}_3$, respectively.
By **T.7.10**.3,

$$\mathcal{D}(A)=\mathfrak{m}_1\cup\mathfrak{m}_2\cup\mathfrak{m}_3.$$

2. Since (9) and (5) are comaximal, also $\mathfrak{q}_1=(I,\,9)$, $\mathfrak{q}_2=(x+2,\,(y-2)^3,\,5)$ and $\mathfrak{q}_3=(x-2,\,(y+2)^3,\,5)$ are pairwise comaximal.
Therefore, $I=\mathfrak{q}_1\cap\mathfrak{q}_2\cap\mathfrak{q}_3$ and

$$\begin{aligned}A &\simeq A/\mathfrak{q}_1\oplus A/\mathfrak{q}_2\oplus A/\mathfrak{q}_3\\ &\simeq \mathbb{Z}[x,y]/\mathfrak{q}_1\oplus\mathbb{Z}[x,y]/\mathfrak{q}_2\oplus\mathbb{Z}[x,y]/\mathfrak{q}_3\\ &\simeq \mathbb{Z}[x,y]/(I,9)\oplus\mathbb{Z}/(5)[y]/(y-2)^3\oplus\mathbb{Z}/(5)[y]/(y+2)^3.\end{aligned}$$

Note that $A_{\mathfrak{m}_i}\simeq\bigoplus_{j=1}^{3}(A/\mathfrak{q}_j)_{\mathfrak{m}_i}$ and $(A/\mathfrak{q}_j)_{\mathfrak{m}_i}=0$ for all $i\neq j$, because $\mathfrak{q}_j\not\subseteq\mathfrak{m}_i$.
Therefore, for all $i=1,2,3$ we have $(A/\mathfrak{q}_i)_{\mathfrak{m}_i}\simeq A_{\mathfrak{m}_i}$, and none of these rings is a domain because they all contain non-zero nilpotents.
In conclusion, $A_{\mathfrak{p}}$ is not a domain for all $\mathfrak{p}\in\operatorname{Spec}A$.

**Solution E. 13.23.** $1\Rightarrow 2$. Since $I$ is 0-dimensional, we can write $\sqrt{I}=\bigcap_i\mathfrak{m}_i$ as a finite intersection of distinct maximal ideals. Therefore, $\mathbb{V}(I)$ is a finite union of points, and, by **T.3.17**, $d=\dim_K A/I<\infty$.
Thus, for all $i$ there exists a non-trivial $K$-linear combination $\sum_{j=0}^{d}a_{ij}\overline{x_i}^j=\overline{0}$, and the polynomials $h_i(x_i)=\sum_{j=0}^{d}a_{ij}x_i^j$ are non-zero elements of $I\cap K[x_i]$.
$2\Rightarrow 1$. Let $0\neq h_i(x_i)\in I\cap K[x_i]$, and let $\deg(h_i)=m_i$. Then,

$$\{x_1^{t_i}\cdots x_n^{t_n}\,:\,t_i<m_i\ \text{ for all }\ i\}$$

is a generating set of the $K$-vector space $A/I$.
By **T.3.17**, the associated variety $\mathbb{V}(I)$ is finite, and the conclusion follows from **T.3.19**.

**Solution E. 13.24.** Observe that, since the ideal $I$ is 0-dimensional, by **E.13.23**, the polynomials $h_i$ exist. Moreover, they can be computed using **T.2.25**.
The ideal $(\sqrt{h_1},\ldots,\sqrt{h_n})$ is 0-dimensional, again by **E.13.23**, and radical by **E.8.60**.
Since for all $i$, we have $\sqrt{h_i}\in\sqrt{I}$, we obtain

$$(\sqrt{h_1},\ldots,\sqrt{h_n})\subseteq(I,\sqrt{h_1},\ldots,\sqrt{h_n})\subseteq\sqrt{I}\subsetneq K[x_1,\ldots,x_n].$$

The conclusion easily follows from **E.13.18** with $J=(I,\sqrt{h_1},\ldots,\sqrt{h_n})$.

**Solution E. 13.25.** 1. Let $0 \neq m \in M$ be such that $\operatorname{Ann} m$ is maximal in $\Sigma$, and let $a$, $b$ in $A$ be such that $ab \in \operatorname{Ann} m$.
If $bm = 0$, then $b \in \operatorname{Ann} m$. Otherwise, $(1) \supsetneq \operatorname{Ann}(bm) \supseteq \operatorname{Ann} m$.
Then, since $\operatorname{Ann} m$ is maximal by hypothesis, we have $\operatorname{Ann}(bm) = \operatorname{Ann} m$, and therefore, $a \in \operatorname{Ann} m$.

2. Let $A$ be Noetherian, and let $M \neq 0$. Then, the non-empty family

$$\Sigma = \{\operatorname{Ann} m\colon\ 0 \neq m \in M\}$$

has maximal elements, which, by part 1, are elements of $\operatorname{Ass} M$.

**Solution E. 13.26.** Let $\mathfrak{p} = \operatorname{Ann} m$, and let $f\colon A \longrightarrow M$ be the homomorphism defined by $f(1) = m$. Then, $\operatorname{Ker} f = \operatorname{Ann} m = \mathfrak{p}$, and the conclusion follows from homomorphism theorem **T.1.9**.1.

Conversely, suppose there exists an injective homomorphism $j\colon A/\mathfrak{p} \longrightarrow M$, with $m = j(\overline{1})$. If $a \in \mathfrak{p}$, then $am = aj(\overline{1}) = j(\overline{a}) = 0$, which yields $\mathfrak{p} \subseteq \operatorname{Ann} m$.

To prove the opposite inclusion, let $a \in \operatorname{Ann} m$. Then, $0 = am = j(\overline{a})$, and since $j$ is injective, we have $a \in \mathfrak{p}$. Therefore, $\operatorname{Ann} m \subseteq \mathfrak{p}$.

**Solution E. 13.27.** 1. Let $\mathfrak{p} \in \operatorname{Ass} N$. Then, there exists $j\colon A/\mathfrak{p} \longrightarrow N$, which is injective by **E.13.26**. Since $f \circ j\colon A/\mathfrak{p} \longrightarrow M$ is also injective, $\mathfrak{p} \in \operatorname{Ass} M$, again by **E.13.26**.

Let $\mathfrak{p} = \operatorname{Ann} m \in \operatorname{Ass} M$. Then, there exists $j\colon A/\mathfrak{p} \longrightarrow M$, which is injective and with $\overline{1} \mapsto m$. Consider $j(A/\mathfrak{p}) \cap \operatorname{Ker} g$.
There are two cases: if $j(A/\mathfrak{p}) \cap \operatorname{Ker} g = 0$, then $g \circ j\colon A/\mathfrak{p} \longrightarrow P$ is also injective, and $\mathfrak{p} \in \operatorname{Ass} P$.
Otherwise, $j(A/\mathfrak{p}) \cap \operatorname{Ker} g \neq 0$, and there exists $0 \neq m_1 \in j(A/\mathfrak{p}) \cap \operatorname{Im} f$. Let $m_1 = f(n) = j(\overline{a}) = am$ for some $0 \neq n \in N$ and $a \notin \mathfrak{p}$. Then, $\mathfrak{p} = \operatorname{Ann} m = \operatorname{Ann}(am)$. Indeed, clearly, $\operatorname{Ann} m \subseteq \operatorname{Ann}(am)$.

To prove the opposite inclusion, let $b$ be an element such that $bam = 0$. Then, $ba \in \mathfrak{p}$, and we obtain $b \in \mathfrak{p}$.
Since $f$ is injective, we can conclude that

$$\operatorname{Ann} n = \operatorname{Ann} f(n) = \operatorname{Ann}(am) = \mathfrak{p},$$

and that $\mathfrak{p} \in \operatorname{Ass} N$.

2. Since $M \neq 0$, we have $\operatorname{Ass} M \neq \emptyset$ by **E.13.25**.2. Let $\mathfrak{p}_1 = \operatorname{Ann} m_1 \in \operatorname{Ass} M$, for some $m_1 \in M$. Then, by **E.13.26**, there exists an injective homomorphism $j_1\colon A/\mathfrak{p}_1 \longrightarrow M$. Define $M_0 = 0$, and $M_1 = j_1(A/\mathfrak{p}_1) = \langle m_1 \rangle \subset M$. Then,

$$M_1 \simeq M_1/M_0 \simeq A/\mathfrak{p}_1.$$

If $M_1 = M$, the proof is complete.
Otherwise, $\operatorname{Ass}(M/M_1) \neq \emptyset$. Let $\mathfrak{p}_2 = \operatorname{Ann}(\overline{m_2}) \in \operatorname{Ass}(M/M_1)$, for some $m_2 \in M$. Let also $M_2 = \langle m_1, m_2 \rangle \supsetneq M_1$. By **E.13.26**, we know that there exists an injective homomorphism $j_2\colon A/\mathfrak{p}_2 \longrightarrow M/M_1$. If $j_2$ is also surjective,

then $M_2 = M$. If this is not the case, repeating these arguments, we construct an ascending chain of submodules that satisfies the required properties. Since $M$ is Noetherian, this chain is stationary, and there exists $t \geq 1$ such that $M_t = M$.

3. Let $0 = M_0 \subsetneq M_1 \subsetneq \ldots \subsetneq M_t = M$ be as in part 2. We proceed by induction on $t$. If $t = 1$, we have $M \simeq A/\mathfrak{p}$, with $\mathfrak{p} \in \operatorname{Spec} A$, and thus, $\operatorname{Ass} M = \operatorname{Ass}(A/\mathfrak{p}) = \{\mathfrak{p}\}$.

Now, by induction, assume that $\operatorname{Ass} M_{t-1}$ is finite.
Since $M_t/M_{t-1} \simeq A/\mathfrak{p}_t$, from the exact sequence

$$0 \longrightarrow M_{t-1} \longrightarrow M_t \longrightarrow M_t/M_{t-1} \longrightarrow 0$$

and part 1, we obtain $\operatorname{Ass} M_t \subseteq \operatorname{Ass} M_{t-1} \cup \{\mathfrak{p}_t\}$.
Thus, $\operatorname{Ass} M$ is finite.

**Solution E. 13.28.** Observe that $N \subseteq M$ implies $I \subseteq \operatorname{Ann} M \subseteq \operatorname{Ann} N$, and that $N$ is an $A$-submodule if and only if $N$ is an $A/I$-submodule.

**Solution E. 13.29.** Consider the descending chain of submodules

$$\operatorname{Im} \varphi \supseteq \operatorname{Im} \varphi^2 \supseteq \ldots \supseteq \operatorname{Im} \varphi^n \supseteq \ldots.$$

Since $M$ is Artinian, the chain is stationary, and there exists a positive integer $k$ such that $\operatorname{Im} \varphi^k = \operatorname{Im} \varphi^{k+1}$. Hence, for any $m \in M$, there exists $m_1 \in M$ such that $\varphi^k(m) = \varphi^{k+1}(m_1)$, that is, $\varphi^k(m - \varphi(m_1)) = 0$.
The conclusion follows, since $\varphi$, and thus $\varphi^k$, are injective.

**Solution E. 13.30.** 1. Let $a \in A \setminus A^*$. We will prove that $a$ is a zero-divisor. Consider the chain $(a) \supseteq (a^2) \supseteq \ldots$. Since d.c.c. holds, there exists an integer $k$ such that $(a^k) = (a^{k+1})$. Thus, there exists $b \in A$ such that $a^k(ab - 1) = 0$. Since $a$ is not invertible, $ab - 1 \neq 0$.
Hence, $a^k$, and consequently $a$, are zero-divisors.

2. Let $\mathfrak{m}$ be the maximal ideal of $A$.
If $S$ contains an element of $\mathfrak{m} = \mathcal{N}(A)$, see **T.7.13**.1, then $0 \in S$, and $S^{-1}A = 0$. Otherwise, by part 1, every element of $S$ is invertible, and therefore, $\sigma_S$ is an isomorphism by **E.12.2**.
In both cases, $\sigma_S$ is evidently surjective.

**Solution E. 13.31.** We prove the statement by induction on $n$.
If $n = 1$ and $\mathfrak{m}_1 M = 0$, then $\mathfrak{m}_1 \subseteq \operatorname{Ann} M$ and $M$ is an $A/\mathfrak{m}_1$-module, thus a vector space. Since, by **E.13.2**, $M$ is Noetherian if and only if it is Artinian, the conclusion follows from **T.7.2**.6 and **E.13.28**.

Now, let $n > 1$, and consider the exact sequence

$$0 \longrightarrow \mathfrak{m}_n M \longrightarrow M \longrightarrow M/\mathfrak{m}_n M \longrightarrow 0.$$

Arguing as above, $M/\mathfrak{m}_n M$ is an $A/\mathfrak{m}_n$-vector space, and thus, it is Artinian if and only if is Noetherian. Since $(\mathfrak{m}_1 \cdots \mathfrak{m}_{n-1})\mathfrak{m}_n M = 0$, by the inductive hypothesis, we obtain that $\mathfrak{m}_n M$ is Noetherian if and only if it is Artinian. Since the sequence is exact, we can conclude using **T.7.2**.2 and **T.7.3**.

## 17.7 Chapter 14

**Solution ToF. 14.1.** True. An element $a \in A$ is not a zero-divisor in $A/I$ if and only if for any $b \in A$ such that $ab \in I$, we have $b \in I$.
Therefore, $b \in I : (a)$ implies $b \in I$, and since $I$ is always contained in $I : (a)$, the conclusion follows.

**Solution ToF. 14.2.** True. The product $(p, q) \mapsto pq$ in $Q[x]$ is $A[x]$-bilinear, hence it induces a homomorphism defined by $p \otimes q \mapsto pq$ for any $p, q \in Q[x]$, which is obviously surjective. It is also injective, since if $p \in Q[x]$, then there exists $0 \neq d \in A$ such that $dp = p' \in A[x]$.
Therefore, if $p, q \in Q[x]$, then

$$p \otimes q = \frac{p'}{d} \otimes q = \frac{1}{d} \otimes p'q = 1 \otimes pq.$$

**Solution ToF. 14.3.** True. Since $\varphi$ is a surjective homomorphism, by homomorphism theorem **T.1.9**.1, $B \simeq A/\operatorname{Ker}\varphi$. Hence, $\operatorname{Ker}\varphi$ is a prime ideal. If $\operatorname{Ker}\varphi = 0$, then we have an isomorphism.

Otherwise, $\operatorname{Ker}\varphi$ is a non-trivial prime in a PID, thus it is maximal, see **T.1.23** and **E.8.57**, and in this case, $B$ is a field.

**Solution ToF. 14.4.** Statements 1 and 2 are true, see **T.1.6**.7 and **E.8.27**.2. Statement 3 is false, see **E.8.28**.

**Solution ToF. 14.5.** False. See the discussion following **T.1.5**.

**Solution ToF. 14.6.** False. For example, $\mathcal{N}(\mathbb{Z}/(6)) = (\bar{0})$, but $(6)$ is not prime in $\mathbb{Z}$.

**Solution ToF. 14.7.** True. By **E.11.3**, we have

$$\mathbb{Q}[x]/(x^2-1) \otimes_{\mathbb{Q}[x]} \mathbb{Q}[x]/(x^2+1) \simeq \mathbb{Q}[x]/(x^2-1, x^2+1),$$

which is clearly zero.

**Solution ToF. 14.8.** True. The ring $A$ is a quotient of $A[x]$, hence it is Noetherian by **T.7.2**.2.

**Solution ToF. 14.9.** True. We recall that, by **T.2.7**.1, if $\mathrm{Lt}_>(I)$ is prime, then its minimal set of monomial generators is a subset of the variables. Reordering the variables, if necessary, we can assume that $\mathrm{Lt}_>(I) = (x_1, \dots, x_k)$, and hence a reduced Gröbner basis of $I$ is of the form

$$G = \{x_1 - f_1(x_{k+1}, \dots, x_n), \dots, x_k - f_k(x_{k+1}, \dots, x_n)\}$$

for some $f_1, \dots, f_k$.
Therefore, $K[X]/I \simeq K[x_{k+1}, \dots, x_n]$ is a domain, *i.e.*, $I$ is prime.

**Solution ToF. 14.10.** True. Let $b \in \sqrt{I : J}$. Then, there exists $n$ such that $b^n J \subseteq I$. We need to prove that if $c \in \sqrt{J}$, then $bc \in \sqrt{I}$. Since $c^m \in J$ for some $m$, we have $b^n c^m \in I$. If $n \geq m$, then, multiplying by $c^{n-m}$, we obtain $(bc)^n \in I$, thus $bc \in \sqrt{I}$.

If $n < m$ the proof is similar.

**Solution ToF. 14.11.** True. If $I = J$, then $(0)$ is maximal and $A$ is a field. Otherwise, since $I$ and $J$ are comaximal and $I \cap J = (0)$, the Chinese Remainder Theorem yields $A \simeq A/I \oplus A/J$, *i.e.*, $A$ is a direct sum of fields. Hence, $A$ is Artinian because d.c.c holds in $A$.

**Solution ToF. 14.12.** False. Consider the $\mathbb{Z}$-modules $M = \mathbb{Z}/(4)$ and $N = \mathbb{Z}$. Then, $M$ is a torsion module and $N$ is torsion-free.
Thus, $T(M \otimes_{\mathbb{Z}} N) \simeq T(M) = M = \mathbb{Z}/(4)$, while $T(M) \otimes T(N) = \mathbb{Z}/(4) \otimes 0 = 0$.

**Solution ToF. 14.13.** True. By **T.3.9**.3, if $\gcd(f, g) = 1$, then the polynomials $p(x) = \operatorname{Res}_y(f, g)$ and $q(y) = \operatorname{Res}_x(f, g)$ are non-zero, and belong to $I$ by **T.3.7**. Thus, for any monomial ordering, $\operatorname{Lt}(I)$ contains powers of both the variables and the conclusion follows from **T.3.17**.

Conversely, let $h = \gcd(f, g) \neq 1$, and write $f = f_1 h$ and $g = g_1 h$ for some $f_1, g_1 \in \mathbb{C}[x, y]$. Then, by **T.3.1**.6, we have $\mathbb{V}_{\mathbb{C}}(I) = \mathbb{V}_{\mathbb{C}}(h) \cup \mathbb{V}_{\mathbb{C}}(f_1, g_1)$.
Since $h$ obviously cannot satisfy condition 2 of **T.3.17**, $\mathbb{V}_{\mathbb{C}}(h)$ is infinite, and the conclusion follows.

**Solution ToF. 14.14.** True. Consider the homomorphism $\varphi\colon A^n \longrightarrow (A/I)^n$ defined by $\varphi(a_1, \ldots, a_n) = (\overline{a_1}, \ldots, \overline{a_n})$.
It is easy to verify that $\varphi$ is surjective and that its kernel is $IA^n$.

**Solution ToF. 14.15.** False. Take $K = \mathbb{Q}$ and $p(x) = x^2 + 1$. Then, $p(x)$ is irreducible in $K[x]$, but the ideal

$$(x^2 + 1, y^2 + 1) = (x^2 - y^2, y^2 + 1) = (x - y, y^2 + 1) \cap (x + y, y^2 + 1)$$

is a non-trivial intersection of distinct primes.

**Solution ToF. 14.16.** False. Let $A = \mathbb{Z}/(4)$, $M = A$, and $N = (2)A \simeq \mathbb{Z}/(2)$. Then, $M$ is free, hence projective. However, $N$ is not projective, since, for example, the sequence $0 \longrightarrow \mathbb{Z}/(2) \xrightarrow{2} \mathbb{Z}/(4) \longrightarrow N \longrightarrow 0$ does not split, because $\mathbb{Z}/(4) \not\simeq \mathbb{Z}/(2) \oplus \mathbb{Z}/(2)$.

**Solution ToF. 14.17.** True. If $I$ is maximal and $J \not\subseteq I$, then there exists $0 \neq a \in J \setminus I$. Therefore, $A = I + (a) \subseteq I + J$.

Conversely, we consider $A/I$ and claim that every non-zero element is invertible. Let $\overline{0} \neq \overline{a} \in A/I$. Since, by hypothesis, $(a) \not\subseteq I$, we have $I + (a) = A$. It follows that there exist $i \in I$ and $b \in A$ such that $i + ba = 1$, that is, $\overline{a}\overline{b} = \overline{1}$ and $\overline{a}$ is invertible.

**Solution ToF. 14.18.** True. Since $A/\mathfrak{p}$ is a finite domain, it is a field, see **T.1.1**.

**Solution ToF. 14.19.** 1. True. It suffices to prove that if $I + J = A$, then $I^n + J^n = A$ for every $n > 1$, because the reverse implication is trivial.
Let $i \in I$ and $j \in J$ be such that $i + j = 1$. Then, for any $n > 1$,

$$1=(i+j)^{2n-1}=i^n\sum_{k=0}^{n-1}\binom{2n-1}{k}i^{n-1-k}j^k+j^n\sum_{k=n}^{2n-1}\binom{2n-1}{k}i^{2n-1-k}j^{k-n}$$

belongs to $I^n+J^n$.

2. True. Take $a \in \sqrt{I:J}$. Then, there exists $m$ such that $a^m j \in I$ for all $j \in J$. Hence, $(aj)^m \in I$ yields $aj \in \sqrt{I}$ for any $j$, that is, $a \in \sqrt{I}:J$.

3. False. Take the ideals $I=(8)$ and $J=(6)$ in $\mathbb{Z}$. Then, $\sqrt{I}:J=(2):(6)=\mathbb{Z}$, while $\sqrt{I:J}=\sqrt{(4)}=(2)$, see **E.8.20**.4.

4. False. Consider the ideals $I=(12)$ and $J=(8)$ in $\mathbb{Z}$. Then, $I:J=(3)$, while $I:\sqrt{J}=(12):(2)=(6)$.

**Solution ToF. 14.20.** False. Fix the lex order with $x>y$. The given generators of $I$ are a Gröbner basis, while a Gröbner basis of $J$ is

$$\{x,\, y^2+1\}.$$

Hence, the rings $A_1=\mathbb{Q}[x,y]/I$ and $A_2=\mathbb{Q}[x,y]/J$ are $\mathbb{Q}$-vector spaces with bases $\{1,\, x,\, y,\, xy\}$ and $\{1,\, y\}$, respectively.

We show that they cannot be isomorphic as rings. Assume, by contradiction, that there exists a ring isomorphism $\varphi\colon A_1 \longrightarrow A_2$. Then, $\varphi_{|\mathbb{Q}}=\mathrm{id}_{\mathbb{Q}}$ and the elements $\varphi(1)$, $\varphi(x)$, $\varphi(y)$, $\varphi(xy)$ are linearly dependent over $\mathbb{Q}$, *i.e.*, there exist $a_1,\ldots,a_4 \in \mathbb{Q}$ not all zero such that

$$0=a_1\varphi(1)+a_2\varphi(x)+a_3\varphi(y)+a_4\varphi(xy)=\varphi(a_1+a_2x+a_3y+a_4xy).$$

This implies that $1$, $x$, $y$, $xy$ are linearly dependent in $A_1$, which is a contradiction.

**Solution ToF. 14.21.** True. If $I=I^{ec}$ and $I^e$ is prime, then $I$ is prime by **T.1.17**.7.

Conversely, note that $K(x)[y]=S^{-1}(K[x,y])$, with $S=K[x]\setminus\{0\}$, and there exists a one-to-one correspondence between primes $I$ of $K[x,y]$ such that $I\cap S=\emptyset$ and primes of $K(x)[y]$.
This correspondence is defined by $I \longrightarrow I^e$ and satisfies the property $I^{ec}=I$, see **T.6.8**.

**Solution ToF. 14.22.** True. It suffices to prove that $\sqrt{(f,g)} \subseteq \sqrt{(f^2,g^3)}$, because the opposite inclusion is trivial.
Since $(f,g)\subseteq \sqrt{(f^2,g^3)}$, the conclusion follows by taking radicals.

**Solution ToF. 14.23.** True. Since $A$ is local, $A^*=A\setminus\mathfrak{m}$ and $A/\mathfrak{m}$ is a field. Hence, $\pi(a)\in(A/\mathfrak{m})^*$ if and only if $a\notin\mathfrak{m}$, *i.e.*, if and only if $a\in A^*$.

**Solution ToF. 14.24.** True. It is a special case of **T.4.12**.

**Solution ToF. 14.25.** False. Consider $A=K[x_n\colon n\in\mathbb{N}]$. Then, $A$ is a non-Noetherian domain, but $A$ is contained in its quotient field, which is Noetherian.

**Solution ToF. 14.26.** 1. True. It is easy to verify that $A = (\overline{2}) \oplus (\overline{9})$.
Thus, $M_1$ is projective because it is a direct summand of the free module $A$.

The statements 2, 3 and 4 are false. The exact sequence

$$0 \longrightarrow (\overline{6}) \longrightarrow A \xrightarrow{3} (\overline{3}) \longrightarrow 0$$

does not split, because $\operatorname{Ann}_A\big((\overline{3}) \oplus (\overline{6})\big) = (\overline{6}) \neq (\overline{0}) = \operatorname{Ann}_A A$.
Hence, the module $M_2$ is not projective and cannot be free.
Finally, $\operatorname{Ann}_A M_1 = (\overline{9})$, hence $M_1$ is not free.

**Solution ToF. 14.27.** False. Since $\operatorname{Ann}_{K[x]} K \neq (0)$, the field $K$ is a torsion $K[x]$-module. Hence, it cannot be a direct summand of the free module $K[x]$.

**Solution ToF. 14.28.** True. Since $M$ is finitely generated, $\mathfrak{p} \in \operatorname{Supp} M$ if and only if $\operatorname{Ann} M \subseteq \mathfrak{p}$, see **E.12.27**.1. Moreover, $M = \mathbb{Z}/(12) \otimes_{\mathbb{Z}} \mathbb{Z}/(30) \simeq \mathbb{Z}/(6)$, hence its annihilator is $(6)$ and $\operatorname{Supp} M = \{(2), (3)\}$.

**Solution ToF. 14.29.** False. The ring $A$ is Noetherian, hence $\mathcal{D}(A)$ is the union of the associated primes of $(0)$, see **T.7.10**.3. Thus, $\mathcal{D}(A) = (\overline{x}) \cup (\overline{y})$, but the element $a = \overline{x} + \overline{y}$ belongs to neither $\mathcal{D}(A)$ nor $K$.

**Solution ToF. 14.30.** True. Since $f$ is irreducible and $K[X]$ is a UFD, the ideal $(f)$ is prime by **T.1.25** and **T.1.22**.5.
By the strong form of the Nullstellensatz and **T.3.1**.4, we have

$$\sqrt{(g)} = \mathbb{I}(\mathbb{V}(g)) \subseteq \mathbb{I}(\mathbb{V}(f)) = (f),$$

and the conclusion follows.

**Solution ToF. 14.31.** True. Since $Q$ is projective, the exact sequence

$$0 \longrightarrow \operatorname{Ker}\varphi \longrightarrow P \longrightarrow Q \longrightarrow 0$$

splits and $P \simeq \operatorname{Ker}\varphi \oplus Q$. Thus, $\operatorname{Ker}\varphi$ is a direct summand of a projective module.
The conclusion follows from **E.11.9**.1.

**Solution ToF. 14.32.** False. We have $\mathbb{C} \otimes_{\mathbb{Q}} \mathbb{Q}[x]/(f) \simeq \mathbb{C}[x]/(f)$, see the proof of **E.11.13**.2. Since $f$ is squarefree, we have a factorization

$$f(x) = a \prod_{i=1}^{k} (x - \alpha_i)$$

in $\mathbb{C}[x]$, with $a \in \mathbb{Q}$ and $\alpha_i \neq \alpha_j$ when $i \neq j$.
By the Chinese Remainder Theorem,

$$\mathbb{C}[x]/(f) \simeq \prod_{i=1}^{k} \mathbb{C}[x]/(x - \alpha_i) \simeq \mathbb{C}^k,$$

and hence, there are no non-zero nilpotents.

**Solution ToF. 14.33.** False. Take $A = \mathbb{Z} = M$ and $S = A \setminus \{0\}$. Then, $S^{-1}M = \mathbb{Q}$ is not a finitely generated $\mathbb{Z}$-module.

**Solution ToF. 14.34.** True. By **T.5.4**.2, 3 and 5 we have

$$M/\mathfrak{m}M \otimes M/\mathfrak{n}M \simeq (A/\mathfrak{m} \otimes M) \otimes (A/\mathfrak{n} \otimes M) \simeq M \otimes (A/\mathfrak{m} \otimes A/\mathfrak{n}) \otimes M$$
$$\simeq (A/(\mathfrak{m}+\mathfrak{n})) \otimes M \otimes M,$$

which is 0 since $\mathfrak{m} + \mathfrak{n} = A$.

**Solution ToF. 14.35.** False. Consider the ideal $I = (x^2 - 1) = (x+1) \cap (x-1)$ of $\mathbb{Q}[x]$. Clearly, $\mathrm{Lt}(I) = (x^2)$ is primary.
Nevertheless, $I$ is not primary, because it is a non-trivial intersection of two distinct primes.

**Solution ToF. 14.36.** True. The matrix is already in diagonal form, hence

$$\operatorname{Coker} \varphi \simeq \mathbb{Q}[x]/(x-1) \oplus \mathbb{Q}[x]/(x) \oplus \mathbb{Q}[x]/(x(x-1)^3),$$

which has dimension 6 over $\mathbb{Q}$.

**Solution ToF. 14.37.** True. If $A$ is a field, then $M$ is a vector space, and therefore it is free.

Conversely, let $I \subset A$ be an ideal. Then, by hypothesis, the $A$-module $A/I$ is free, hence $I = \mathrm{Ann}(A/I) = 0$.
Thus, in $A$ there are no non-trivial ideals and $A$ is a field by **T.1.2**.3.

**Solution ToF. 14.38.** True. Assume, by contradiction, that there is $m \neq 0$ in $T(M)$. Then, $A \supsetneq \operatorname{Ann} m \neq 0$. Hence, there exists $0 \neq a \in A$ such that $am = 0$, and there exists $\mathfrak{m} \in \operatorname{Max} A$ such that $\operatorname{Ann} m \subseteq \mathfrak{m}$. For any $b \in A \setminus \mathfrak{m}$, we have $bm \neq 0$, which yields $\frac{m}{1} \neq 0$ in $M_{\mathfrak{m}}$.
Finally, $\frac{a}{1} \neq 0$ and $\frac{a}{1}\frac{m}{1} = \frac{am}{1} = 0$, that is, $\frac{m}{1}$ is a non-trivial torsion element in $M_{\mathfrak{m}}$, which is a contradiction.

**Solution ToF. 14.39.** True. If $M$ is free of rank $k$, then $M \simeq A^k$.
Therefore, $M \otimes_A B \simeq A^k \otimes_A B \simeq B^k$ by **T.5.4**.1 and 4.

**Solution ToF. 14.40.** True. If $f \in I$, then $IA_f \cap A = A$ and the statement obviously holds.
Now, assume that $f \notin I$. Since $I \subsetneq (I, f)$ and $I \subseteq IA_f \cap A$, by **T.1.17**.3, the inclusion $\subseteq$ holds.

To prove the opposite inclusion, take $a \in \sqrt{IA_f \cap A} \cap \sqrt{(I, f)}$. Then, there exist $n \in \mathbb{N}$, $i \in I$, $b \in A$, and $j \in IA_f \cap A$ such that $a^n = j = i + bf$. Since $j \in IA_f \cap A$, there exists $m$ such that $f^m j \in I$.
Hence,

$$bf^{m+1} = (j - i) f^m = f^m j - f^m i \in I.$$

Therefore, $a^{n(m+1)} = (i + bf)^{m+1} \in I$ and $a \in \sqrt{I}$.

**Solution ToF. 14.41.** True. Take $0 \neq a \in A$. By hypothesis, the descending chain $(a) \supseteq (a^2) \supseteq \ldots \supseteq (a^n) \supseteq \ldots$ is stationary, and hence, there exists $k$ such that $(a^k) = (a^{k+1})$. Thus, there exists $b \in A$ such that $a^k = ba^{k+1}$.

Since $A$ is a domain, this yields $1 = ba$, that is, $a$ is invertible.

Alternatively, since $A$ is a domain, the ideal $(0)$ is prime, hence maximal by **T.7.13**.1, *i.e.*, $A$ is a field.

**Solution ToF. 14.42.** False. Consider $A = \mathbb{Z}_{(p)}$ with $p$ prime.
Since $\mathbb{Z}$ is a domain, $A$ is a domain. Since in $A$ every ideal is the extension of an ideal of $\mathbb{Z}$ by **T.6.7**.2, $A$ is a PID.
However, since $A$ is local by **T.6.6**, $\mathcal{J}(A) = (p)A \neq 0$.

**Solution ToF. 14.43.** True. Consider the map $\varphi\colon \mathbb{Q} \times \mathbb{R} \longrightarrow \mathbb{R}$ defined by $\varphi(x, y) = xy$. It is obviously $\mathbb{Z}$-bilinear, hence it induces a unique $\mathbb{Z}$-module homomorphism

$$\widetilde{\varphi}\colon \mathbb{Q} \otimes_{\mathbb{Z}} \mathbb{R} \longrightarrow \mathbb{R}, \quad \widetilde{\varphi}(x \otimes y) = xy,$$

which is clearly surjective.
Since

$$\frac{a}{b} \otimes_{\mathbb{Z}} y = \frac{a}{b} \otimes_{\mathbb{Z}} \frac{by}{b} = a \otimes_{\mathbb{Z}} \frac{y}{b} = 1 \otimes_{\mathbb{Z}} \frac{ay}{b} \quad \text{for all} \quad \frac{a}{b} \in \mathbb{Q} \quad \text{and} \quad y \in \mathbb{R},$$

every element of $\mathbb{Q} \otimes_{\mathbb{Z}} \mathbb{R}$ can be written as $1 \otimes y$ for some $y \in \mathbb{R}$.
Thus, $0 = \widetilde{\varphi}(1 \otimes y) = y$ yields $y = 0$, that is, $\widetilde{\varphi}$ is injective as well.

**Solution ToF. 14.44.** True. Consider the $A$-module $N'/(N' \cap N)$, and localize it at an ideal $\mathfrak{m} \in \operatorname{Max} A$.
Using **T.6.13**.2 and 3, we obtain

$$(N'/(N' \cap N))_{\mathfrak{m}} \simeq N'_{\mathfrak{m}}/(N' \cap N)_{\mathfrak{m}} \simeq N'_{\mathfrak{m}}/(N'_{\mathfrak{m}} \cap N_{\mathfrak{m}}).$$

By hypothesis, $N'_{\mathfrak{m}} \subseteq N_{\mathfrak{m}}$, hence $(N'/(N' \cap N))_{\mathfrak{m}} = 0$ for all $\mathfrak{m} \in \operatorname{Max} A$.
The conclusion immediately follows from **T.6.15**.

**Solution ToF. 14.45.** False. Consider the ring $\mathbb{Z}/(6)$ whose only prime ideals are $(\overline{2})$ and $(\overline{3})$.
It is easy to verify that both $(\mathbb{Z}/(6))_{(\overline{2})} \simeq \mathbb{Z}/(2)$ and $(\mathbb{Z}/(6))_{(\overline{3})} \simeq \mathbb{Z}/(3)$ are domains, but $\mathbb{Z}/(6)$ is not.

**Solution ToF. 14.46.** True. The module $M$ is free with basis $\{\overline{1}, \overline{x}\}$, hence it is flat.

**Solution ToF. 14.47.** False. Consider $\mathbb{Q}$, which is not finitely generated as a $\mathbb{Z}$-module. Clearly, $\mathbb{Q}$ is torsion-free, but, for any pair of elements $\frac{a}{b} \neq \frac{c}{d}$ of $\mathbb{Q}$, we have a non-trivial relation

$$bc\frac{a}{b} - ad\frac{c}{d} = 0.$$

Hence, $\mathbb{Q}$ is not free over $\mathbb{Z}$, see **E.10.50**.

**Solution ToF. 14.48.** True. By **T.6.19**.3, we have that $S$ is the saturation of $T$, and the conclusion follows from **T.6.19**.6.

**Solution ToF. 14.49.** True. The statement obviously holds if one of the two modules is trivial, hence we assume $M \neq 0$ and $N \neq 0$. We recall that the statement is true when $A$ is local, by **E.11.11**.
Assume, by contradiction, that $J = \operatorname{Ann} M + \operatorname{Ann} N \subsetneq A$. Then, there exists a maximal ideal $\mathfrak{m}$ which contains $J$.
Localizing at $\mathfrak{m}$, we obtain

$$0 = (M \otimes_A N)_{\mathfrak{m}} \simeq M_{\mathfrak{m}} \otimes_{A_{\mathfrak{m}}} N_{\mathfrak{m}}.$$

Therefore, either $M_{\mathfrak{m}} = 0$ or $N_{\mathfrak{m}} = 0$. Assume, for instance, that $M_{\mathfrak{m}} = 0$. Since $M$ is finitely generated, there exists $s \notin \mathfrak{m}$ such that $sM = 0$, see **E.12.25**.1. This contradicts the fact that $\operatorname{Ann} M \subseteq J \subseteq \mathfrak{m}$.

**Solution ToF. 14.50.** True. Clearly, $\mathfrak{m} \subseteq \operatorname{Ann}(M/\mathfrak{m}M)$.

To prove the opposite inclusion, note that $\operatorname{Ann}(M/\mathfrak{m}M) \neq A$, because if $M = \mathfrak{m}M$, Nakayama's Lemma yields $M = 0$, which is not possible.
Then, $\operatorname{Ann}(M/\mathfrak{m}M) = \mathfrak{m}$, and obviously, $\mathcal{V}(\mathfrak{m}) = \{\mathfrak{m}\}$.

**Solution ToF. 14.51.** True. The proof is similar to the first part of the proof of **E.11.15**.

**Solution ToF. 14.52.** True. See the proof of **E.13.4**.1.

**Solution ToF. 14.53.** False. Let $A = \prod_{n \in \mathbb{N}} K$ be a direct product of fields. The elements of $A$ are sequences of elements of $K$, and the operations are defined componentwise. Let $S \subset A$ be the set

$$\{s_0 = (0, 1, \ldots, 1, \ldots), (1, \ldots, 1, \ldots) = 1_A\}.$$

The set $S$ is clearly multiplicative, and the ring homomorphism $\tau\colon A \longrightarrow A$ defined by $(a_0, a_1, a_2, \ldots) \mapsto (a_1, a_2, \ldots)$ satisfies:

i) $\tau(S) = \{1_A\} \subset A^*$;

ii) $\tau(a) = 0$ yields $a = (a_0, 0, \ldots, 0, \ldots)$, hence $s_0 a = 0$;

iii) every $a = (a_0, a_1, \ldots) \in A$ can be written as $\tau((0, a_0, a_1, \ldots,))\tau(1_A)^{-1}$.

From **T.6.5** we obtain that $A \simeq S^{-1}A$, but $S \not\subseteq A^*$ because $s_0 \notin A^*$.

**Solution ToF. 14.54.** True. Since $A$ is free over itself, every ideal $I \subset A$ is free by hypothesis. Since every principal ideal $(a)$ is free, the ring $A$ is a domain. Assume, by contradiction, that there exists an ideal $I$ which is not principal, and take $a_1 \in I \setminus 0$ and $a_2 \in I \setminus (a_1)$. Since $a_1 a_2 = a_2 a_1$, we immediately obtain a non-trivial relation in $I$, namely, $a_2 a_1 - a_1 a_2 = 0$.
This a contradiction, since $I$ is free.

**Solution ToF. 14.55.** True. Take $B = A$ and $\mathfrak{p}$ the maximal ideal of $A$.

**Solution ToF. 14.56.** True. Let $A = \mathbb{Z}/(540)$. Then, by **E.8.5**.1, it suffices to observe that $\overline{7} \in A^*$ and $\overline{30}$, $\overline{60}$, $\overline{90} \in \mathcal{N}(A)$.

**Solution ToF. 14.57.** False. The ideal $(2x + 1)$ is principal and maximal because $\mathbb{Z}_{(2)}[x]/(2x + 1) \simeq \mathbb{Q}$.

To prove the latter isomorphism we simply note that

$$\mathbb{Z}_{(2)}[x]/(2x+1) \simeq (\mathbb{Z}_{(2)})_2 \simeq \mathbb{Z}_2 \otimes_{\mathbb{Z}} \mathbb{Z}_{(2)} \simeq \mathbb{Q},$$

by **E.12.8**, **T.6.14**.1, and **E.12.18**.

**Solution ToF. 14.58.** False. $(\mathbb{Z}/(15) \oplus \mathbb{Z}/(18))_{(3)} \simeq \mathbb{Z}/(3) \oplus \mathbb{Z}/(9)$.

**Solution ToF. 14.59.** True. The matrix of relations is

$$\begin{pmatrix} 2 & 1 & 1 \\ -1 & 1 & a \\ 0 & 1 & 0 \end{pmatrix},$$

which has Smith canonical form

$$\begin{pmatrix} 1 & 0 & 0 \\ 0 & 1 & 0 \\ 0 & 0 & 2a+1 \end{pmatrix}.$$

Therefore, $M \simeq \mathbb{Z}/(2a+1)$.
Since $a > 0$, then $2a+1 \neq \pm 1$, and hence, $M \otimes_{\mathbb{Z}} \mathbb{Z}/(n) \neq 0$ for any $n$ such that $\gcd(2a+1, n) \neq 1$.

**Solution ToF. 14.60.** False. Consider $I = (12) = (3) \cap (4) \subset \mathfrak{p} = (2) \subset \mathbb{Z}$. Then, $I$ is not primary, $\mathfrak{p}$ is prime, and $I_{(2)} = (4)_{(2)}$ is primary.

**Solution ToF. 14.61.** True. Let $ax + by + c = 0$ be the equation of $\ell$. By **T.3.1**.5, we have $\mathcal{C} \cap \ell = \mathbb{V}(f, ax + by + c)$.
Therefore, if $b \neq 0$, then $\mathcal{C} \cap \ell = \mathbb{V}(f(x, -\frac{ax+c}{b}))$, otherwise $\mathcal{C} \cap \ell = \mathbb{V}(f(-\frac{c}{a}, y))$. Since, by hypothesis, $\ell \not\subseteq \mathcal{C}$, in both cases $\mathcal{C} \cap \ell$ is a variety defined by a non-zero polynomial in one variable of degree $\leq n$, which has at most $n$ roots.

**Solution ToF. 14.62.** True. Let $m \in T(M)$, and let $a \in A \setminus \{0\}$ be such that $am = 0$. Consider the exact sequence

$$0 \longrightarrow A \xrightarrow{a} A \longrightarrow A/(a) \longrightarrow 0.$$

Tensoring with $M$, we obtain

$$M \xrightarrow{a} M \longrightarrow M/aM \longrightarrow 0,$$

where the first homomorphism is not injective.
This contradicts the flatness of $M$.

**Solution ToF. 14.63.** True. For any $a \in A \setminus \{0\}$, we have $Q(A) \otimes_A A/(a) = 0$, because

$$\frac{b}{c} \otimes_A \overline{d} = \frac{ab}{ac} \otimes_A \overline{d} = \frac{b}{ac} \otimes_A \overline{ad} = 0.$$

Using **T.5.4**.1 and 4, we obtain

$$Q(A) \otimes_A M \simeq Q(A) \otimes_A A^n \simeq Q(A)^n.$$

**Solution ToF. 14.64.** False. Take $A = \mathbb{Q}[x]_{(x)}$ and $F = A^{\mathbb{N}}$ with basis $\{e_i\}_{i\in\mathbb{N}}$. Let $N \subset F$ be the submodule $N = \langle xe_0, xe_1 - e_0, xe_2 - e_1, \dots, xe_{i+1} - e_i, \dots\rangle$. Clearly, $\mathcal{J}(A) = (x)$ is finitely generated.
Consider the quotient $M = F/N$. It is not hard to prove that $M \neq 0$ and

$$(x)M = (\langle xe_i : i \in \mathbb{N}\rangle + N)/N = M.$$

Compare this statement with the one of **E.10.24**.

**Solution ToF. 14.65.** False. Consider the ideal $I = (x^2, 2x) = (x^2, 2) \cap (x)$. We have $\sqrt{I} = (x)$, but $I$ is not primary.

**Solution ToF. 14.66.** True. Let $M = N \oplus P$, where we assume $M$ is finitely generated. If $M = \langle (n_1, p_1), \dots, (n_h, p_h)\rangle$, then $N = \langle n_1, \dots, n_h\rangle$ is finitely generated.

**Solution ToF. 14.67.** False. To compute $\sqrt{I}$ we use the decomposition

$$\begin{aligned}\sqrt{I} &= \sqrt{(y-z,\, x^3y^3 - y^2)} \cap \sqrt{(y+z,\, x^3y^3 + y^2)}\\ &= \sqrt{(y-z,\, y^2)} \cap \sqrt{(y-z,\, x^3y-1)} \cap \sqrt{(y+z,\, y^2)} \cap \sqrt{(y+z,\, x^3y+1)}\\ &= (y,\, z) \cap (y-z,\, x^3y-1) \cap (y+z,\, x^3y+1),\end{aligned}$$

where the last equality depends of the fact that $x^3y \pm 1$ are irreducible in $\mathbb{Q}[x, y]$, and hence, the ideals $(y \pm z,\, x^3y \pm 1)$ are prime.

Note that the generators of the ideals found in the above decomposition are already reduced Gröbner bases with respect to the deglex order with $x > z > y$, hence

$$f = (x^3y + 1)(y + z) \notin (y - z,\, x^3y - 1).$$

**Solution ToF. 14.68.** True. We recall that $I \otimes A/I \simeq I/I^2$.
Consider the inclusion $I \longrightarrow A$ and tensor with $A/I$. By the flatness hypothesis, we obtain an injective homomorphism $I \otimes A/I \longrightarrow A \otimes A/I$. Each element $i \otimes \overline{a} \in I \otimes A/I$ has image $i \otimes \overline{a} = 1 \otimes \overline{ia} = 0$ in $A \otimes A/I$, hence the injectivity yields $I \otimes A/I = 0$, that is, $I = I^2$ as desired.

**Solution ToF. 14.69.** 1. All the inclusions are true.
Let $a \in \operatorname{Ann} N$. Then, $0 = af(m) = f(am)$, and the injectivity of $f$ yields $am = 0$ for any $m \in M$, that is, $a \in \operatorname{Ann} M$.
Moreover, for any $\ell \in L$, there exists $n \in N$ such that $g(n) = \ell$, hence $a\ell = ag(n) = g(an) = g(0) = 0$, that is, $a \in \operatorname{Ann} L$.
Finally, take $b \in \operatorname{Ann} M$ and $c \in \operatorname{Ann} L$. Since $g$ induces an isomorphism $L \simeq \operatorname{Coker} f = N/\operatorname{Im} f$, we have $cn \in \operatorname{Im} f$ for any $n \in N$. Hence, $cn = f(m)$, for some $m \in M$, and $bcn = bf(m) = f(bm) = f(0) = 0$, that is, $bc \in \operatorname{Ann} N$. Since $\operatorname{Ann} M \cdot \operatorname{Ann} L$ is generated by the products $bc$, the conclusion follows.

2. True. In general, if $\varphi\colon M_1 \longrightarrow M_2$ is an $A$-module homomorphism, then $\varphi(T(M_1)) \subseteq T(M_2)$. Indeed, if $m \in T(M_1)$, then there exists $a \in A \setminus \{0\}$ such that $am = 0$. Therefore, $a\varphi(m) = \varphi(am) = 0$ and $\varphi(m) \in T(M_2)$.

3. False. For a counterexample, consider the sequence of $\mathbb{Z}$-modules

$$0 \longrightarrow \mathbb{Z} \xrightarrow{2} \mathbb{Z} \xrightarrow{\pi} \mathbb{Z}/(2) \longrightarrow 0,$$

where $T(M) = T(N) = 0$ and $T(L) = \mathbb{Z}/(2)$.

4. True. Obviously, $f$ injective yields $f_{|T(M)}$ injective, and $g_{|T(N)} \circ f_{|T(M)} = 0$ immediately follows from $g \circ f = 0$.
Therefore, $\operatorname{Im} f_{|T(M)} \subseteq \operatorname{Ker} g_{|T(N)}$.

To prove the opposite inclusion, take $n \in \operatorname{Ker} g_{|T(N)}$. Then, $g(n) = 0$ yields $n = f(m)$ for some $m \in M$. We have to prove that $m \in T(M)$.
Since $n$ is a torsion element, there exists $a \in A \setminus \{0\}$ such that $an = 0$.
Therefore, $f(am) = af(m) = 0$ and the injectivity of $f$ yields $am = 0$, that is, $m \in T(M)$.

**Solution ToF. 14.70.** True. Let $\varphi_5 \colon \mathbb{Z}[x] \longrightarrow \mathbb{Z}$ be the evaluation homomorphism defined by $p(x) \mapsto p(5)$. Then,

$$I = (p(x) \in \mathbb{Z}[x] \colon \varphi_5(p(x)) \in (2)) = (2)^c \supseteq (2, x-5).$$

Since $1 \notin (2)^c$ and $(2, x-5)$ is maximal, we obtain that $I = (2, x-5)$ and $\mathbb{Z}[x]/(2, x-5) \simeq \mathbb{Z}/(2)$.

**Solution ToF. 14.71.** False. Note that every ring homomorphism $\varphi \colon A \longrightarrow B$ verifies $\varphi(-1_A) = -1_B$.
Let $\varphi \colon \mathbb{C} \longrightarrow \mathbb{R}$ be a ring homomorphism. Then $\varphi(i)^2 = \varphi(i^2) = \varphi(-1) = -1$, which is not satisfied by any $\varphi(i) \in \mathbb{R}$.
Hence, there is no ring homomorphism from $\mathbb{C}$ to $\mathbb{R}$.

**Solution ToF. 14.72.** True. Since $M = \mathfrak{p}M$, by Nakayama's Lemma, there exists $a \equiv 1 \mod \mathfrak{p}$ such that $aM = 0$.
Hence, $\operatorname{Ann} M \not\subset \mathfrak{p}$, and we can conclude that $M_{\mathfrak{p}} = 0$, see **E.12.27**.1.

**Solution ToF. 14.73.** True. Assume, by contradiction, that there exist elements $a, b \in A \setminus I$ such that $ab \in I$. Then,

$$(I, a), (I, b) \supsetneq I \quad \text{and} \quad (I, a) \cap (I, b) \supseteq I.$$

Since

$$(I, a) \cap (I, b) \subseteq \sqrt{(I, a) \cap (I, b)} = \sqrt{(I, ab)} \subseteq \sqrt{I} = I,$$

we obtain that $(I, a) \cap (I, b) = I$, and this contradicts the irreducibility of $I$.

**Solution ToF. 14.74.** True. Take $a, b \in A$ such that $ab \in I$. Then, for all $j$, we have either $a \in I_j$ or $b \in I_j$. If $a \in I_j$ for all $j$, then $a \in I$. Otherwise, let $h$ be the smallest index such that $a \notin I_h$.
Since $I_h \supset I_k$ and $ab \in I_k$ for all $k \geq h$, we have $b \in I_k$ for all $k$. Thus, $b \in I$.

**Solution ToF. 14.75.** True. The statement is obvious for $n = 0$. For $n = 1$, it follows from **E.8.73**.4.
We use induction on $n$, assuming that $(K[[x_1, \dots, x_n]], (x_1, \dots, x_n))$ is local.
An element $f \in K[[x_1, \dots, x_{n+1}]] = K[[x_1, \dots, x_n]][[x_{n+1}]]$ can be written as

$$f = \sum_{i \in \mathbb{N}} f_i(x_1, \ldots, x_n) x_{n+1}^i,$$

and, by **E.8.73**.1, it is invertible if and only if $f_0(x_1, \ldots, x_n) \in K[[x_1, \ldots, x_n]]^*$, *i.e.*, by the inductive hypothesis, if only if $f_0 \notin (x_1, \ldots, x_n)$.
Since $f \notin (x_1, \ldots, x_{n+1})$ if and only if

$$f_0 \notin (x_1, \ldots, x_{n+1}) \cap K[[x_1, \ldots, x_n]] = (x_1, \ldots, x_n),$$

the invertible elements of $K[[x_1, \ldots, x_{n+1}]]$ are precisely those not belonging to $(x_1, \ldots, x_{n+1})$.
Hence, $K[[x_1, \ldots, x_{n+1}]]$ is local with maximal ideal $(x_1, \ldots, x_{n+1})$.

**Solution ToF. 14.76.** True. Since all $\mathfrak{p}_i$ are minimal, we have $\dim A_{\mathfrak{p}_i} = 0$ for all $i$. Moreover, $\dim A \geq \dim A/\mathfrak{p}_i$ for all $i$.

To prove the opposite inequality, note that any chain of prime ideals

$$\mathfrak{q}_1 \subsetneq \mathfrak{q}_2 \subsetneq \cdots \subsetneq \mathfrak{q}_n \quad \text{such that} \quad \mathfrak{q}_1 \neq \mathfrak{p}_i \quad \text{for all} \quad i,$$

can be extended with a minimal prime $\mathfrak{p}_j \subsetneq \mathfrak{q}_1$.

**Solution ToF. 14.77.** True. Take $0 \neq a \in A$.
Since $(a^2) = (a)(a) = (a) \cap (a) = (a)$, we have $a = ba^2$ for some $b \in A$.
Thus, $a(1 - ba) = 0$, which yields $ab = 1$.

**Solution ToF. 14.78.** True. Any free module is flat.

Conversely, assume that $A$ is a PID, and let $M$ be a finitely generated flat $A$-module. By **T.4.33**.2, we can write $M = A^k \oplus T(M)$, where $T(M) \simeq \bigoplus_i A/(d_i)$ is the torsion part and $d_i \neq 0$ for all $i$.
Since $M$ is flat, the exact sequence

$$0 \longrightarrow A \xrightarrow{d_i} A \longrightarrow A/(d_i) \longrightarrow 0$$

tensored with $M$ remains exact.
This is possible only if $T(M) = 0$, *i.e.*, only if $M$ is free.

**Solution ToF. 14.79.** False. The ideals $\mathfrak{p} \subset \mathfrak{q} \subset A$ are prime, hence $A \setminus \mathfrak{q} \subset A \setminus \mathfrak{p}$ is a containment between multiplicative subsets of $A$. Nevertheless, it is not possible to find an injective homomorphism between their localizations.
Indeed, since $xy = 0$ in $A$, we have $\frac{x}{1} = \frac{0}{1}$ in $A_{\mathfrak{p}}$. Therefore, the maximal ideal of $A_{\mathfrak{p}}$ is 0 and $A_{\mathfrak{p}}$ is a field.
On the other hand, in $A_{\mathfrak{q}}$ we have $\frac{x}{1} \neq 0$, and $\frac{y}{1} \neq 0$, but $\frac{x}{1}\frac{y}{1} = \frac{0}{1}$.
Hence, $A_{\mathfrak{q}}$ contains zero-divisors, and there is no injective homomorphism from $A_{\mathfrak{q}}$ into the field $A_{\mathfrak{p}}$.

**Solution ToF. 14.80.** True. By the characterization of projective modules, $M$ is projective if and only if there exist $A$-modules $N$ and $F$ such that $F \simeq M \oplus N$ is free. Tensoring over $A$ with the flat $A$-module $S^{-1}A$, we obtain $S^{-1}F \simeq S^{-1}M \oplus S^{-1}N$.

Recalling that if $F \simeq A^H$ for some $H$, then $S^{-1}F \simeq (S^{-1}A)^H$ is a free $S^{-1}A$-module, the conclusion follows from the same characterization.

**Solution ToF. 14.81.** False. Take $f = x^2+1$ and $g = y^2+1$. Then,

$$\mathbb{Q}[x,y]/(f) \otimes_{\mathbb{Q}[x,y]} \mathbb{Q}[x,y]/(g) \simeq \mathbb{Q}[x,y]/(x^2+1,\, y^2+1),$$

which has dimension 4 as a $\mathbb{Q}$-vector space.

**Solution ToF. 14.82.** False. By **E.12.8**, the ring $K[x,y]/(xy-1)$ is isomorphic to $K[x]_x = S^{-1}K[x]$, where $S = \{x^k\colon\ k \in \mathbb{N}\}$.
Assume $\mathfrak{p}$ is a prime such that $\mathfrak{p}^c = (x)$. Then, $K[x]_x = S^{-1}(x) = \mathfrak{p}^{ce} = \mathfrak{p}$, which contradicts the fact that a prime ideal is proper.

**Solution ToF. 14.83.** True. The non-trivial implication is a special case of **ToF.14.44**.

**Solution ToF. 14.84.** True. Any free module is projective.
The reverse implication is in **E.11.14**.

**Solution ToF. 14.85.** True. Any homomorphism $\varphi\colon \mathbb{Z}/(n) \longrightarrow \mathbb{Q}/\mathbb{Z}$ is determined by $\varphi(\overline{1})$, which has to satisfy $n\varphi(\overline{1}) = \varphi(\overline{n}) = 0$.
Hence, $\varphi(\overline{1}) \in \left\{\frac{a}{n} + \mathbb{Z}\colon\ a = 0,\dots,n-1\right\}$.
The map

$$\mathbb{Z}/(n) \longrightarrow \operatorname{Hom}_{\mathbb{Z}}(\mathbb{Z}/(n), \mathbb{Q}/\mathbb{Z}), \quad \overline{a} \mapsto \varphi_a \quad \text{where} \quad \varphi_a(\overline{1}) = \frac{a}{n} + \mathbb{Z},$$

is well-defined and provides the required isomorphism.

**Solution ToF. 14.86.** False. It is easy to verify that

$$\begin{aligned} M &\simeq A/(x-1) \oplus A/(x^3-1) \oplus A/(x^2-1) \\ &\simeq (A/(x-1))^3 \oplus A/(x+1) \oplus A/(x^2+x+1). \end{aligned}$$

Hence, $\operatorname{Ann}_A M = ((x-1)(x+1)(x^2+x+1))$, while

$$\operatorname{Ann}_A \mathbb{Q}^6 = \operatorname{Ann}_A \mathbb{Q} = (x-a)$$

for some $a \in \mathbb{Q}$ (why?).

**Solution ToF. 14.87.** False. Let $n = 1$, $K = \mathbb{Q}$, and $S = \mathbb{Z} \subset \mathbb{Q}$. Since every polynomial $0 \neq p \in \mathbb{Q}[x]$ has finitely many zeroes, we have $\mathbb{I}(S) = (0)$ and $\mathbb{V}(\mathbb{I}(S)) = \mathbb{Q} \neq S$.

**Solution ToF. 14.88.** True. Let $P_0 \subsetneq \dots \subsetneq P_m$ be a chain of primes in $S^{-1}A$. By the one-to-one correspondence between prime ideals of $S^{-1}A$ and primes of $A$ which do not intersect $S$, there exist primes $\mathfrak{p}_0,\dots,\mathfrak{p}_m$ of $A$ such that

$$S^{-1}\mathfrak{p}_i = P_i \quad \text{and} \quad \mathfrak{p}_0 \subsetneq \dots \subsetneq \mathfrak{p}_m.$$

By definition, $\dim A = n \geq m$. Since the inequality holds for any chain in $S^{-1}A$, we have $\dim A \geq \dim S^{-1}A$.

Now, let $\mathfrak{p}_0 \subsetneq \ldots \subsetneq \mathfrak{p}_n$ be a chain of primes of $A$.
Taking $T = A \setminus \mathfrak{p}_n$, we obtain a chain of distinct primes $T^{-1}\mathfrak{p}_0 \subsetneq \ldots \subsetneq T^{-1}\mathfrak{p}_n$ in $T^{-1}A$. Thus, $\dim T^{-1}A \geq n$, and the equality follows.

**Solution ToF. 14.89.** True. If $\varphi$ is a surjective homomorphism, then $\operatorname{Ker}\varphi$ is a maximal ideal of $\mathbb{Z}[x]$. Hence, $\operatorname{Ker}\varphi = (p, f)$ with $p \in \mathbb{Z}$ prime and $f \in \mathbb{Z}[x]$ irreducible modulo $p$, see **E.8.78**.
Thus,

$$K \simeq \mathbb{Z}[x]/(p, f) \simeq (\mathbb{Z}/(p))[x]/(\overline{f})$$

is a finite field with $p^{\deg \overline{f}}$ elements.

**Solution ToF. 14.90.** True. Since $P$ is finitely generated and projective, there exist $n \in \mathbb{N}_+$ and an $A$-module $M$ such that $A^n \simeq P \oplus M$. Hence,

$$\begin{aligned} A^n &\simeq \operatorname{Hom}_A(A, A)^n \simeq \operatorname{Hom}_A(A^n, A) \simeq \operatorname{Hom}_A(P \oplus M, A) \\ &\simeq \operatorname{Hom}_A(P, A) \oplus \operatorname{Hom}_A(M, A). \end{aligned}$$

It follows that $\operatorname{Hom}_A(P, A)$ is projective, because it is a direct summand of a free $A$-module.

**Solution ToF. 14.92.** False. Since $nm \equiv 0 \mod (n)$ for any $m \in \mathbb{Z}/(n)$, we have $S^{-1}(\mathbb{Z}/(n)) = 0$ for all $n \in \mathbb{N}_+$. On the other hand, the element $\dfrac{(1, 1, \ldots, 1, \ldots)}{1}$ is non-zero in $S^{-1}\left(\prod_{n \in \mathbb{N}_+} \mathbb{Z}/(n)\right)$. Indeed, for any $n > 0$, the $k$ th component of $n(1, 1, \ldots, 1, \ldots)$ is non-zero for $k > n$.

**Solution ToF. 14.91.** False. Consider the $\mathbb{Z}$-modules $\mathbb{Z}/(n)$ for $n \in \mathbb{N}_+$, and let $N = \mathbb{Q}$. Then, by **E.11.8**, $\mathbb{Z}/(n) \otimes_{\mathbb{Z}} \mathbb{Q} = 0$ for any $n \in \mathbb{N}_+$, and hence,

$$\prod_{n \in \mathbb{N}_+} (\mathbb{Z}/(n) \otimes_{\mathbb{Z}} \mathbb{Q}) = 0.$$

On the other hand, $\left(\prod\limits_{n \in \mathbb{N}_+} \mathbb{Z}/(n)\right) \otimes_{\mathbb{Z}} \mathbb{Q} \neq 0$. Indeed, consider the cyclic submodule generated by $m = (\overline{1}_{\mathbb{Z}/(n)})_{n \in \mathbb{N}_+} \in \prod\limits_{n \in \mathbb{N}_+} \mathbb{Z}/(n)$.

We have the inclusion

$$0 \longrightarrow \langle m \rangle_{\mathbb{Z}} \longrightarrow \prod_{n \in \mathbb{N}_+} \mathbb{Z}/(n),$$

and, tensoring with $\mathbb{Q}$, which is a flat $\mathbb{Z}$-module by **T.6.12** and **T.6.14**, we obtain the inclusion

$$\begin{array}{ccccc} 0 & \longrightarrow & \langle m \rangle_{\mathbb{Z}} \otimes_{\mathbb{Z}} \mathbb{Q} & \longrightarrow & \left(\prod\limits_{n \in \mathbb{N}_+} \mathbb{Z}/(n)\right) \otimes_{\mathbb{Z}} \mathbb{Q} \\ & & \Big\downarrow \simeq & & \\ & & \mathbb{Z} \otimes_{\mathbb{Z}} \mathbb{Q} \neq 0. & & \end{array}$$

## 17.8 Chapter 15

**Solution E. 15.1.** Let $m_1 = x_{i_1}^{a_1} \cdots x_{i_r}^{a_r}$, with $a_1 > 0$, and let $m_2 = x_{j_1}^{b_1} \cdots x_{j_s}^{b_s}$. Assume, by contradiction, that for every $i \neq j$ there exists a positive integer $k = k(i,j)$ such that $x_i^k > x_j$. Therefore, for all $h = 1, \ldots, s$, we can choose $k_h \geq 1$ such that $x_{i_1}^{k_h} > x_{j_h}$. Thus, $x_{i_1}^{k_h b_h} > x_{j_h}^{b_h}$.
Setting $k_0 = \sum\limits_{h=1}^{s} k_h b_h$, we obtain

$$m_1^{k_0} \geq x_{i_1}^{a_1 k_0} \geq x_{i_1}^{k_0} = x_{i_1}^{k_1 b_1} \cdots x_{i_1}^{k_s b_s} > m_2,$$

which is the desired contradiction.

**Solution E. 15.2.** 1. Define $\varphi$ by letting

$$\varphi(x^a y^b z^c) = t^{a+2b+3c}, \quad \text{for all } \; x^a y^b z^c \in \operatorname{Mon} A$$

and extending by linearity.
Then, $\varphi$ is the evaluation homomorphism given by $p(x,y,z) \mapsto p(t,t^2,t^3)$ for all $p(x,y,z) \in A$. For every $q(t) \in B$ we have $\varphi(q(x)) = q(t)$, hence the homomorphism is surjective.
2. Clearly, $f_1 = y - x^2$, $f_2 = z - x^3 \in \operatorname{Ker} \varphi$.
Now, fix the lex order with $z > y > x$, and let $r = r(x,y,z) \in A$ be the reduction of a given element $f = f(x,y,z) \in A$ modulo $F = \{f_1, f_2\}$. Since none of the monomials of $r$ belongs to $(\operatorname{lt}(f_1), \operatorname{lt}(f_2)) = (y,\, z)$, we can write $r = r(x)$.
Let $f = g_1 f_1 + g_2 f_2 + r \in \operatorname{Ker} \varphi$. Then, $0 = \varphi(f) = 0 + r(t)$. Hence, $r = 0$ and $f \in (f_1, f_2)$.
Therefore,

$$\operatorname{Ker} \varphi = (y - x^2,\, z - x^3).$$

Alternatively, we compute $\operatorname{Ker} \varphi$ using a technique that can be generalized and applied to the problem of elimination of parameters from a system of polynomial equations.
Let $\varphi \colon K[x,y,z] \longrightarrow K[t]$ be the composition

$$K[x,y,z] \longrightarrow K[x,y,z,t] \xrightarrow{\psi} K[t],$$

of the inclusion homomorphism and $\psi$, where $\psi$ is defined by

$$\psi(x) = p_1(t), \quad \psi(y) = p_2(t), \quad \psi(z) = p_3(t) \;\text{ and }\; \psi_{|K[t]} = \mathrm{id}_{K[t]},$$

and

$$p_1(t) = t, \quad p_2(t) = t^2, \quad p_3(t) = t^3.$$

Clearly, $\psi_{|K[x,y,z]} = \varphi$ and $\operatorname{Ker} \varphi = \operatorname{Ker} \psi \cap K[x,y,z]$.

Moreover,

$$\operatorname{Ker}\psi = (x - p_1(t),\, y - p_2(t),\, z - p_3(t)).$$

Indeed, the containment $\supseteq$ is obvious.

The other one can be easily obtained dividing an element $f(x,y,z,t)$ of $\operatorname{Ker}\psi$ by the set $\{x-p_1(t),\, y-p_2(t),\, z-p_3(t)\}$, using any monomial ordering on $K[x,y,z,t]$ such that

$$t^k < x, \quad t^k < y, \quad t^k < z \quad \text{for all} \quad k \in \mathbb{N},$$

*e.g.*, the lex order with $x > y > z > t$.

Now, by the Elimination Theorem, we can find $\operatorname{Ker}\varphi$ by computing a Gröbner basis $G$ of $\operatorname{Ker}\psi$ with respect to the lex order with $t > x > y > z$. We obtain

$$G = \{t - x,\, x^2 - y,\, xy - z,\, xz - y^2,\, y^3 - z^2\}.$$

Thus,

$$\operatorname{Ker}\varphi = (x^2 - y,\, xy - z,\, xz - y^2,\, y^3 - z^2) = (x^2 - y,\, x^3 - z).$$

**Solution E. 15.3.** 1. We proceed by induction on $n$. The statement is obvious for $n = 1$.

For $n > 1$, by the inductive hypothesis and the distributivity of direct sum over Hom and tensor product, we have

$$\begin{aligned}
\operatorname{Hom}(A^n, M) \otimes N &\simeq \big(\operatorname{Hom}(A^{n-1}, M) \oplus \operatorname{Hom}(A, M)\big) \otimes N \\
&\simeq \big(\operatorname{Hom}(A^{n-1}, M) \otimes N\big) \oplus (\operatorname{Hom}(A, M) \otimes N) \\
&\simeq \operatorname{Hom}(A^{n-1}, M \otimes N) \oplus \operatorname{Hom}(A, M \otimes N) \\
&\simeq \operatorname{Hom}(A^n, M \otimes N).
\end{aligned}$$

2. Let $Q$ be an $A$-module. Observe that the map

$$\varphi_Q\colon \operatorname{Hom}_A(Q, M) \times N \longrightarrow \operatorname{Hom}_A(Q, M \otimes N)$$

defined by

$$(\varphi_Q(f,p))(q) = f(q) \otimes p \quad \text{for all } q \in Q,$$

is bilinear, and therefore, induces a unique homomorphism

$$\widetilde{\varphi}_Q\colon \operatorname{Hom}(Q, M) \otimes N \longrightarrow \operatorname{Hom}_A(Q, M \otimes N),$$

which, by part 1, is an isomorphism when $Q = A^n$.

Since $L$ is finitely generated, there exists an exact sequence

$$0 \longrightarrow L' \longrightarrow A^n \longrightarrow L \longrightarrow 0$$

for some $n \in \mathbb{N}$. Applying the contravariant functor $\operatorname{Hom}(\bullet, M)$, we obtain the exact sequence

$$0 \longrightarrow \operatorname{Hom}(L, M) \longrightarrow \operatorname{Hom}(A^n, M) \longrightarrow \operatorname{Hom}(L', M),$$

which is still exact after tensoring with $N$, since $N$ is flat.

On the other hand, applying $\operatorname{Hom}(\bullet, M \otimes N)$, we have the exact sequence

$$0 \longrightarrow \operatorname{Hom}(L, M \otimes N) \longrightarrow \operatorname{Hom}(A^n, M \otimes N) \longrightarrow \operatorname{Hom}(L', M \otimes N).$$

Thus, we can form a commutative diagram

$$\begin{array}{ccccccc}
0 \longrightarrow & \operatorname{Hom}(L, M) \otimes N & \longrightarrow & \operatorname{Hom}(A^n, M) \otimes N & \longrightarrow & \operatorname{Hom}(L', M) \otimes N \\
& \downarrow \widetilde{\varphi}_L & & \downarrow \widetilde{\varphi}_{A^n} & & \downarrow \widetilde{\varphi}_{L'} \\
0 \longrightarrow & \operatorname{Hom}(L, M \otimes N) & \longrightarrow & \operatorname{Hom}(A^n, M \otimes N) & \longrightarrow & \operatorname{Hom}(L', M \otimes N).
\end{array}$$

Since $\widetilde{\varphi}_{A^n}$ is an isomorphism, $\operatorname{Ker} \widetilde{\varphi}_L = 0$.

**Solution E. 15.4.** Consider the family of ideals

$$\Sigma = \{I \subset A\colon\ I \text{ is not finitely generated}\}$$

and assume, by contradiction, $\Sigma \neq \emptyset$.
For every ascending chain of elements of $\Sigma$, the union of the elements in the chain is an ideal which is not finitely generated. Therefore, Zorn's Lemma yields that $\Sigma$ has a maximal element $P$.

Since $P$ is not finitely generated, $P$ is a proper ideal which is not prime. Hence, there exist $a$, $b \notin P$ with $ab \in P$. By the maximality of $P$, the ideal $(P, a)$ is finitely generated, and thus, there exist a positive integer $k$ and elements $d_1 = p_1 + s_1 a, \ldots, d_k = p_k + s_k a$, with $p_i \in P$ and $s_i \in A$ for all $i$, such that

$$(P, a) = (d_1, \ldots, d_k).$$

Now, consider the ideal $J = P : (a)$, that properly contains $P$, because $b \in J \setminus P$. Once again, due to maximality, we have that both $J$ and, accordingly, $aJ$ are finitely generated.

We will show that

$$P = (p_1, \ldots, p_k) + aJ.$$

As a consequence, $P$ is finitely generated, which is a contradiction.
Clearly, $\supseteq$ holds.

To prove the opposite inclusion, let $c \in P \subseteq (P, a)$, where

$$c = \sum_i c_i d_i = \sum_i c_i (p_i + s_i a) = \sum_i c_i p_i + ja \quad \text{for some} \quad c_i, j \in A.$$

Since $c - \sum_i c_i p_i \in P$, we have

$$j \in J \quad \text{and} \quad c \in (p_1, \dots, p_k) + aJ.$$

**Solution E. 15.5.** 1. It is easy to see that

$$I \subsetneq (x) \subsetneq (x, z) \subsetneq (x, y, z).$$

By the one-to-one correspondence between ideals of $A$ containing $I$ and ideals of $A/I$, the chain of prime ideals $(\overline{x}) \subsetneq (\overline{x}, \overline{z}) \subsetneq (\overline{x}, \overline{y}, \overline{z})$ has length 2, and therefore, $\dim A/I \geq 2$.

2. The reduced Gröbner basis with respect to both given orderings is

$$G = \{x^2z - x^2,\ xyz + xz,\ x^2y + x^2\}.$$

Thus, the escaliers are the same.

3. From part 2 and the Elimination Theorem, it immediately follows that $I \cap \mathbb{Q}[y, z] = 0$.

Now, consider the lex order with $z > x > y$ and compute the reduced Gröbner basis of $I$. Since we obtain the same basis, we can conclude that

$$I \cap \mathbb{Q}[x, y] = (x^2y + x^2).$$

4. By repeatedly applying $\sqrt{(J,\, fg)} = \sqrt{(J,\, f)} \cap \sqrt{(J,\, g)}$, we obtain

$$\begin{aligned}
\sqrt{I} &= \sqrt{(x^2,\, xyz + xz)} \cap \sqrt{(z - 1,\, xy + x)} \\
&= \sqrt{(x)} \cap \sqrt{(y + 1,\, x^2)} \cap \sqrt{(z,\, x^2)} \cap \sqrt{(x,\, z - 1)} \cap \sqrt{(y + 1,\, z - 1)} \\
&= (x) \cap (y + 1,\, x) \cap (z,\, x) \cap (x,\, z - 1) \cap (y + 1,\, z - 1).
\end{aligned}$$

Finally, $\mathbb{V}(I) = \mathbb{V}(\sqrt{I})$ implies

$$\mathbb{V}(I) = \mathbb{V}(x) \cup \mathbb{V}(y + 1,\, x) \cup \mathbb{V}(z,\, x) \cup \mathbb{V}(x,\, z - 1) \cup \mathbb{V}(y + 1,\, z - 1).$$

Therefore,

$$\mathbb{V}(I) = \mathbb{V}(x) \cup \mathbb{V}(y + 1,\, z - 1)$$

is a decomposition of $\mathbb{V}(I)$ as a union of irreducible components.

5. Since $(A/I)_{\mathfrak{p}} \simeq A_{\mathfrak{p}}/I_{\mathfrak{p}}$, we have that $(A/I)_{\mathfrak{p}} \neq 0$ if and only if $I_{\mathfrak{p}} \neq A_{\mathfrak{p}}$, *i.e.*, if and only if $I \subseteq \mathfrak{p}$.

Therefore, the localization of $A/I$ at $\mathfrak{p}$ is not trivial for all associated primes and, *a fortiori*, for all minimal primes.

The converse does not hold. For instance,

$$\mathfrak{m} = (x,\, y,\, z^2 + 1) \supset I \quad \text{and} \quad \mathfrak{m} \notin \operatorname{Ass}(I).$$

6. The minimal primes of $I$ are $\mathfrak{p}_1 = (x)$ and $\mathfrak{p}_2 = (y+1,\, z-1)$, and the corresponding primary components are $(I_{\mathfrak{p}_i})^c$ for $i = 1, 2$.
Therefore,

$$(I_{\mathfrak{p}_1})^c = ((x^2z - x^2,\, xyz + xz)_{\mathfrak{p}_1})^c = ((x^2,\, x)_{\mathfrak{p}_1})^c = (x),$$

$$(I_{\mathfrak{p}_2})^c = ((x^2z - x^2,\, xyz + xz)_{\mathfrak{p}_2})^c = ((z-1,\, y+1)_{\mathfrak{p}_2})^c = (y+1,\, z-1).$$

**Solution E. 15.6.** Let $I_1I_2 = \bigcap_{i=1}^n \mathfrak{q}_i$ be a primary decomposition of $I_1I_2$. After reordering indices, if necessary, we may assume there exists an integer $0 \le r \le n$ such that

$$I_1 \subseteq \sqrt{\mathfrak{q}_i} \;\text{ for }\; i = 1, \dots, r \quad\text{and}\quad I_1 \nsubseteq \sqrt{\mathfrak{q}_i} \;\text{ for }\; i = r+1, \dots, n.$$

Set $J = \bigcap_{i=1}^r \mathfrak{q}_i$ and $J' = \bigcap_{i=r+1}^n \mathfrak{q}_i$.
For every $i = 1, \dots, r$, there exists $t_i \in \mathbb{N}_+$ such that $I_1^{t_i} \subseteq \mathfrak{q}_i$, thus there exists $t$ such that $I_1^t \subseteq J$.
For every $i = r+1, \dots, n$, there exists $a_i \in I_1 \setminus \sqrt{\mathfrak{q}_i}$. Since $a_ib \in I_1I_2 \subseteq \mathfrak{q}_i$ for all $b \in I_2$, we obtain $b \in \mathfrak{q}_i$, that is, $I_2 \subseteq J'$.
To conclude, it is now sufficient to observe that

$$I_1I_2 = I_1I_2 \cap I_2 = (J \cap J') \cap I_2 = J \cap I_2.$$

**Solution E. 15.7.** Recall that, for every $A$-module $N$, the sequence

$$N \otimes M_1 \xrightarrow{\mathrm{id}_N \otimes \varphi} N \otimes M_2 \xrightarrow{\mathrm{id}_N \otimes \psi} N \otimes M_3 \longrightarrow 0$$

is exact.

We need to show that, if $M_3$ is flat, then $\mathrm{id}_N \otimes \varphi$ is injective.
Consider a short exact sequence of $A$-modules

$$0 \longrightarrow L \xrightarrow{g} F \xrightarrow{f} N \longrightarrow 0,$$

where $F$ is free, and hence, flat.
Tensoring with $M_2$ and $M_3$, we obtain a commutative diagram

$$\begin{array}{ccccccccc}
 & & L \otimes M_2 & \xrightarrow{g \otimes \mathrm{id}_{M_2}} & F \otimes M_2 & \xrightarrow{f \otimes \mathrm{id}_{M_2}} & N \otimes M_2 & \longrightarrow & 0 \\
 & & \downarrow{\scriptstyle \mathrm{id}_L \otimes \psi} & & \downarrow{\scriptstyle \mathrm{id}_F \otimes \psi} & & \downarrow{\scriptstyle \mathrm{id}_N \otimes \psi} & & \\
0 & \longrightarrow & L \otimes M_3 & \xrightarrow[g \otimes \mathrm{id}_{M_3}]{} & F \otimes M_3 & \xrightarrow[f \otimes \mathrm{id}_{M_3}]{} & N \otimes M_3 & \longrightarrow & 0,
\end{array}$$

where the vertical arrows are surjective, the first row is exact by the properties of tensor product, and the second row is exact because $M_3$ is flat by hypothesis.
The Snake Lemma yields an exact sequence

$$\mathrm{Ker}(\mathrm{id}_L \otimes \psi) \longrightarrow \mathrm{Ker}(\mathrm{id}_F \otimes \psi) \longrightarrow \mathrm{Ker}(\mathrm{id}_N \otimes \psi) \longrightarrow 0,$$

where the homomorphisms are obtained by properly restricting $g \otimes \mathrm{id}_{M_2}$ and $f \otimes \mathrm{id}_{M_2}$.
Since $F$ is flat,

$$\mathrm{Ker}(\mathrm{id}_F \otimes \psi) = \mathrm{Im}(\mathrm{id}_F \otimes \varphi) \simeq F \otimes M_1.$$

Tensoring the exact sequence $0 \longrightarrow L \longrightarrow F \longrightarrow N \longrightarrow 0$ with $M_1$, we obtain a commutative diagram with exact rows

$$\begin{array}{ccccccc}
L \otimes M_1 & \xrightarrow{g\otimes \mathrm{id}_{M_1}} & F \otimes M_1 & \xrightarrow{f\otimes \mathrm{id}_{M_1}} & N \otimes M_1 & \longrightarrow & 0 \\
\downarrow \mathrm{id}_L \otimes\varphi & & \downarrow \mathrm{id}_F \otimes\varphi \ \simeq & & \downarrow \mathrm{id}_N \otimes\varphi & & \\
\mathrm{Ker}(\mathrm{id}_L \otimes\psi) & \longrightarrow & \mathrm{Ker}(\mathrm{id}_F \otimes\psi) & \longrightarrow & \mathrm{Ker}(\mathrm{id}_N \otimes\psi) & \longrightarrow & 0 \\
\downarrow & & \downarrow & & \downarrow & & \\
0 & & 0 & & 0, & &
\end{array}$$

where the vertical arrows are surjective because $\mathrm{Im}(\mathrm{id}_\bullet \otimes\varphi) = \mathrm{Ker}(\mathrm{id}_\bullet \otimes\psi)$.
The fact that $\mathrm{id}_N \otimes\varphi$ is injective follows from diagram chasing.

**Solution E. 15.8.** 1. It is clear from the definition that $\mathfrak{p}^{(n)} = (\mathfrak{p}^n)^{ec}$, with respect to the localization homomorphism $A \longrightarrow A_\mathfrak{p}$. Thus, $\mathfrak{p}^n \subseteq \mathfrak{p}^{(n)}$.

To prove the opposite inclusion, let $a \in \mathfrak{p}^{(n)}$. Then, $\frac{a}{1} = \frac{b}{s} \in \mathfrak{p}^n A_\mathfrak{p}$, with $b \in \mathfrak{p}^n$ and $s \notin \mathfrak{p}$. Hence, there exists $u \notin \mathfrak{p}$ such that $usa = ub \in \mathfrak{p}^n$, where $\mathfrak{p}^n$ is primary. Since $us \notin \mathfrak{p}$, we obtain $a \in \mathfrak{p}^n$.

2. Let $A = K[x, y, z]/(xy - z^2)$ and $\overline{\mathfrak{p}} = (\overline{x}, \overline{z}) \subset A$. Then, $\overline{\mathfrak{p}}$ is prime and $\overline{\mathfrak{p}}^2 = (\overline{x}^2, \overline{z}^2, \overline{xz})$ is not primary.
Since $\frac{1}{\overline{y}} \in A_{\overline{\mathfrak{p}}}$ and $\overline{xy} \in \overline{\mathfrak{p}}^2$, we have

$$\frac{\overline{x}}{1} = \overline{xy} \cdot \frac{1}{\overline{y}} \in \overline{\mathfrak{p}}^2 A_{\overline{\mathfrak{p}}}.$$

Therefore, $\overline{x} \in \overline{\mathfrak{p}}^{(2)} \setminus \overline{\mathfrak{p}}^2$, and the equality does not hold in general.

**Solution E. 15.9.** Denote by $\varphi$ the endomorphism of $A^4$ defined by the matrix

$$X = \begin{pmatrix} x^2-1 & 0 & x^2-1 & x^2-1 \\ 3x+3 & x^2+x & x+1 & 2x+2 \\ 3x+3 & x+1 & x+1 & 2x+2 \\ 3x+3 & x+1 & x+1 & 2x+2 \end{pmatrix}.$$

In this way $M \simeq \mathrm{Coker}\,\varphi$.
The Smith normal form of $X$ is the diagonal matrix

$$\mathrm{diag}(x+1,\, x^2-1,\, x^2-1,\, 0).$$

1. From the computation of the Smith normal form, we obtain $M \simeq A \oplus T(M)$ where
$$T(M) \simeq A/(x+1) \oplus (A/(x^2-1))^2.$$
Hence, $\operatorname{Ann}_A M = 0$. Also observe that $\operatorname{Ann}_A T(M) = (x^2-1)$.

2. We have
$$M \otimes_A A/(x-1) \simeq A/(x+1, x-1) \oplus (A/(x^2-1, x-1))^2 \oplus A/(x-1),$$
which is isomorphic to $(A/(x-1))^3$, whereas
$$M \otimes_A A/(x-i) \simeq A/(x+1, x-i) \oplus (A/(x^2-1, x-i))^2 \oplus A/(x-i),$$
is isomorphic to $A/(x-i)$.

3. We have
$$\operatorname{Hom}_A(M,N) \simeq \operatorname{Hom}_A(A,N) \oplus \operatorname{Hom}_A(T(M),N) \simeq N \oplus \operatorname{Hom}_A(T(M),N).$$
Then, it is enough to consider any $A$-module $N$ such that $\operatorname{Hom}_A(T(M),N) = 0$. For instance, take a cyclic module of the form $N = A/(f)$ such that
$$((f):(x+1))/(f) = ((f):(x^2-1))/(f) = 0.$$
Any $f$ coprime with $x^2-1$ satisfies the required condition.

**Solution E. 15.10.** 1. Let $\mathfrak{p}_1 = (a)$, $\mathfrak{p}_2 = (ab)$, and $\mathfrak{q} = (ac)$.
By hypothesis, $b$ is not invertible and $a \notin \mathfrak{p}_2$. Since $ab \in \mathfrak{p}_2$ and $\mathfrak{p}_2$ is prime, we have $b \in \mathfrak{p}_2$. Therefore, $b = abb_1$ for some $b_1 \in A$. This implies $b(1-ab_1) = 0 \in \mathfrak{q}$. Since $\mathfrak{q}$ is primary and $1 - ab_1 \notin \sqrt{\mathfrak{q}}$, because it does not belong to $\mathfrak{p}_1$, we obtain $b \in \mathfrak{q}$.
Consequently, $\mathfrak{p}_2 \subseteq \mathfrak{q}$, as desired.

2. This is a straightforward application of part 1.

3. Let $I$ be the intersection of all primary ideals contained in $\mathfrak{p}_1$. Then, $\mathfrak{p}_2 \subseteq I$ by part 1.

To prove the opposite inclusion, observe that $\mathfrak{p}_2$ is primary, and thus, it belongs to the set of ideals whose intersection is $I$.

4. By hypothesis, there exists a maximal ideal $\mathfrak{m}$ containing both $\mathfrak{p}_1$ and $\mathfrak{p}_2$. If $\mathfrak{p}_1 \subsetneq \mathfrak{m}$, then, by part 3, $\mathfrak{p}_1$ is the intersection of all primary ideals contained in $\mathfrak{m}$, and the same holds for $\mathfrak{p}_2$. Hence, either $\mathfrak{p}_1 = \mathfrak{p}_2$ or one of them is $\mathfrak{m}$.

**Solution E. 15.11.** 1. If $x > y$, we have $\operatorname{lt}(g_1) = x^2y$ and $\operatorname{lt}(g_2) = x^3$.
Otherwise, if $y > x$, we have $\operatorname{lt}(g_1) = -y^2$ and $\operatorname{lt}(g_2) = -yx$.
Hence, in both cases $S(g_1, g_2) = xg_1 - yg_2 = 0$ and $G = \{g_1, G_2\}$ is a Gröbner basis with respect to all lexicographic orders.
Note that $G$ is reduced with respect to the lex order with $x > y$. However, $G$ is not reduced with respect to the lex order with $y > x$, since $yx \mid yx^2$ and $\operatorname{lc}(g_1) = \operatorname{lc}(g_2) = -1$.

In this case, the reduced basis is

$$G' = \{y^2 - x^4,\, yx - x^3\}.$$

2. We have

$$\sqrt{I} = \sqrt{(x^2 - y)} \cap \sqrt{(y,\, x^3)} = (x^2 - y) \cap (x,\, y) = (x^2 - y).$$

It immediately follows that $I \neq \sqrt{I}$. Indeed, $x^2 - y \notin I$ because, in all cases, $x^2$ does not belong to $\mathrm{Lt}(I)$.

3. Since $\mathbb{V}_{\mathbb{R}}(I) = \mathbb{V}_{\mathbb{R}}(\sqrt{I})$, by part 2 we have

$$\mathbb{V}_{\mathbb{R}}(I) = \{(a, a^2) \colon a \in \mathbb{R}\}.$$

Since $(a, a^2, 0) \in \mathbb{V}_{\mathbb{R}}(J)$ for all $a \in \mathbb{R}$, the statement is true.

**Solution E. 15.12.** The reduced Gröbner basis of $I$ with respect to the lex order with $x > y$ is

$$G = \left\{x^2 - y^2,\, xy + y^2,\, y^4 + \tfrac{1}{2}y\right\}.$$

1. We have

$$\begin{aligned}\sqrt{I} &= \sqrt{(I,\, y)} \cap \sqrt{(I,\, y^3 + \tfrac{1}{2})}\\ &= \sqrt{(x^2,\, y)} \cap \sqrt{(x + y,\, y^3 + \tfrac{1}{2})} \cap \sqrt{(x - y,\, 2y^2,\, y^3 + \tfrac{1}{2})}\\ &= (x,\, y) \cap (x + y,\, y^3 + \tfrac{1}{2})\end{aligned}$$

and both these ideals are maximal.

2. Since $(I,\, y) + (I,\, y^3 + \frac{1}{2}) = 1$, it follows that

$$I = (I,\, y) \cap (I,\, y^3 + \tfrac{1}{2}) = (x^2,\, y) \cap (x + y,\, y^3 + \tfrac{1}{2}) = \mathfrak{q}_1 \cap \mathfrak{q}_2.$$

Since $\sqrt{\mathfrak{q}_1} = (x,\, y)$ is maximal, $(x^2,\, y)$ is primary. Moreover, $(x + y,\, y^3 + \frac{1}{2})$ is maximal, and thus, primary.

3. The intersection $\mathfrak{q}_1 \cap \mathfrak{q}_2$ is a minimal primary decomposition of $I$. Hence,

$$\mathcal{D}(A/I) = (\overline{x},\, \overline{y}) \cup \left(\overline{x + y},\, \overline{y^3 + \tfrac{1}{2}}\right).$$

Moreover, since the ideals we found in part 1 are comaximal,

$$\sqrt{I} = (x,\, y)(x + y,\, y^3 + \tfrac{1}{2}) = (x^2 + xy,\, xy^3 + \tfrac{1}{2}x,\, xy + y^2,\, y^4 + \tfrac{1}{2}y).$$

Finally, $\mathcal{N}(A/I) = (\overline{x + y})$, since the only generator that does not reduce to zero modulo $G$ is

$$xy^3 + \tfrac{1}{2}x \xrightarrow{G}_{*} \tfrac{1}{2}(x + y).$$

4. Since $\mathfrak{p} = \sqrt{\mathfrak{q}_1}$ and $(A \setminus \mathfrak{p}) \cap (x+y,\, y^3+\frac{1}{2}) \neq \emptyset$, we have

$$I_{\mathfrak{p}} \cap A = \big((x^2,\, y)_{\mathfrak{p}} \cap A\big) \cap \big((x+y,\, y^3+\tfrac{1}{2})_{\mathfrak{p}} \cap A\big) = (x^2,\, y).$$

Moreover, since both radicals $\sqrt{\mathfrak{q}_1}$ and $\sqrt{\mathfrak{q}_2}$ intersect $A \setminus \mathfrak{q}$, we have

$$I_{\mathfrak{q}} \cap A = A_{\mathfrak{q}} \cap A = A.$$

5. From the previous parts, it follows that

$$(A/I)_{\mathfrak{p}} \simeq (A/\mathfrak{q}_1 \oplus A/\mathfrak{q}_2)_{\mathfrak{p}} \simeq (A/\mathfrak{q}_1)_{\mathfrak{p}} \simeq A/\mathfrak{q}_1 = \mathbb{Q}[x]/(x^2),$$

whereas $(A/I)_{\mathfrak{q}} \simeq A_{\mathfrak{q}}/I_{\mathfrak{q}} = 0$.

**Solution E. 15.13.** 1. The module $\operatorname{Coker} \varphi_{ab}$ is cyclic if and only if in the Smith normal form of $B_{ab}$ we have $d_1 = d_2 = 1$.
Since

$$(d_1) = \Delta_1 = (x-a,\, 1-x,\, b,\, a-x^2),$$

we proceed by analyzing two cases.

i) $b \neq 0$. In this case, clearly $d_1 = 1$, and we need to compute $d_2$. Since $\Delta_2 = (x-a, 1-x)$, we have $\Delta_2 = (1)$, and hence $d_2 = 1$, if and only if $a \neq 1$.
ii) $b = 0$. In this case, $\Delta_1 = (x-a,\, 1-x,\, a-x^2) = (1)$ if and only if $a \neq 1$. Moreover, $\Delta_2 = (x-a)$, and thus, $\operatorname{Coker} \varphi_{ab}$ cannot be cyclic.

Therefore, $\operatorname{Coker} \varphi_{ab}$ is cyclic for $b \neq 0$ and $a \neq 1$.

2. Let $b \neq 0$ and $a = 1$. Then, we have $d_1 = 1$, $d_2 = x-1$, and

$$\Delta_3 = ((x-1)^2(1-x^2)) = ((x-1)^3(x+1)).$$

Therefore, $d_3 = (x-1)^2(x+1)$ and

$$\operatorname{Coker} \varphi_{1b} \simeq \mathbb{Q}[x]/(x-1) \oplus \mathbb{Q}[x]/(x-1)^2 \oplus \mathbb{Q}[x]/(x+1).$$

For $b = 0$ and $a \neq 1$, we have $d_1 = 1,\ d_2 = x-a$, and

$$\Delta_3 = ((x-a)^2(a-x^2)).$$

Therefore,

$$d_3 = \begin{cases} (x-a)(x+c)(x-c)) & \text{if } a = c^2 \text{ for some } c \in \mathbb{Q}, \\ (x-a)(x^2-a) & \text{otherwise}, \end{cases}$$

and

$$\begin{aligned}\operatorname{Coker}\varphi_{a0} &\simeq \mathbb{Q}[x]/(x-a) \oplus \mathbb{Q}[x]/(x-a) \oplus \mathbb{Q}[x]/(x^2-a)\\ &\simeq \begin{cases} (\mathbb{Q}[x]/(x-a))^2 \oplus \mathbb{Q}[x]/(x-c) \oplus \mathbb{Q}[x]/(x+c) & \text{if } a = c^2,\\ (\mathbb{Q}[x]/(x-a))^2 \oplus \mathbb{Q}[x]/(x^2-a) & \text{otherwise.}\end{cases}\end{aligned}$$

Finally, for $b = 0$ and $a = 1$, we have

$$B_{10} = \begin{pmatrix} x-1 & 0 & 0\\ 0 & 1-x & x-1\\ 0 & 1-x^2 & 0 \end{pmatrix}.$$

Hence, $\Delta_1 = (x-1)$, $\Delta_2 = (x-1)^2$, $\Delta_3 = ((x-1)^3(x+1))$, and

$$\operatorname{Coker}\varphi_{10} \simeq (\mathbb{Q}[x]/(x-1))^3 \oplus \mathbb{Q}[x]/(x+1).$$

**Solution E. 15.14.** The reduced Gröbner basis of $I$ with respect to the lex order with $x > y > z$ is

$$\{x - yz,\, y^2 z\}.$$

We compute the intersection $\mathfrak{q}_1 \cap \mathfrak{q}_2$ using the identity

$$\mathfrak{q}_1 \cap \mathfrak{q}_2 = ((t-1)\mathfrak{q}_1, t\mathfrak{q}_2) \cap K[x,y,z].$$

We have

$$((t-1)\mathfrak{q}_1, t\mathfrak{q}_2) = (f_1 = tx - x,\ f_2 = tz - z,\ f_3 = ty^2,\ f_4 = tx - tyz)$$

and

$$\begin{array}{lll} S(f_1,f_2) = 0, & S(f_2,f_3) = -y^2z = -f_7, & S(f_4,f_5) \xrightarrow{f_7} 0,\\ S(f_1,f_3) = -x^2z = -f_5, & S(f_2,f_4) \xrightarrow{f_2,f_6} 0, & S(f_4,f_6) = 0,\\ S(f_1,f_4) \xrightarrow{f_2} -x+yz = -f_6, & S(f_2,f_7) \xrightarrow{f_7} 0, & S(f_5,f_6) \xrightarrow{f_7} 0,\\ S(f_1,f_5) \xrightarrow{f_5} 0, & S(f_3,f_4) \xrightarrow{f_7} 0, & S(f_5,f_7) = 0,\\ S(f_1,f_6) \xrightarrow{f_2,f_6} 0, & S(f_3,f_5) = S(f_3,f_7) = 0, & S(f_6,f_7) \xrightarrow{f_7} 0. \end{array}$$

Since all the remaining pairs have pairwise coprime leading monomials, the corresponding $S$-polynomials reduce to zero. Therefore, $\{f_1,\ldots,f_7\}$ is a Gröbner basis of $((t-1)\mathfrak{q}_1, t\mathfrak{q}_2)$ and

$$\mathfrak{q}_1 \cap \mathfrak{q}_2 = (x^2z,\, x - yz,\, y^2z) = (x - yz,\, y^2z) = I.$$

Evidently, $\mathfrak{q}_1$ is prime, and hence, primary. Moreover, the zero-divisors of $K[x,y,z]/\mathfrak{q}_2 \simeq K[y,z]/(y^2)$ are nilpotent, and therefore, also $\mathfrak{q}_2$ is primary.

Finally, $\sqrt{\mathfrak{q}_1} = (x,z) \neq \sqrt{\mathfrak{q}_2} = (x,y)$, and the decomposition is clearly minimal.

**Solution E. 15.15.** Let $I$ be an ideal of $A$. If $I = 0$, clearly $I$ is finitely generated.
Otherwise, there exists $0 \neq x \in I$. If $\mathfrak{m} \in \operatorname{Max} A$ is such that $\mathfrak{m} \supseteq I$, then $\mathfrak{m} \in \mathcal{M}_x$. Therefore, every $I \neq 0$ is contained in a finite number of maximal ideals.

Now, consider a non-trivial ascending chain $\{I_\alpha\}_{\alpha \in \Lambda}$ of ideals of $A$. Let $I_{\overline{\alpha}}$ be the first non-zero element in the chain, and let $\mathfrak{m}_1, \dots, \mathfrak{m}_s$ be the maximal ideals of $A$ which contain $I_{\overline{\alpha}}$. Since $A_{\mathfrak{m}_i}$ is Noetherian, for all $i = 1, \dots, s$ there exists an index $\beta_i$ such that the chain $\{I_\alpha A_{\mathfrak{m}_i}\}_{\alpha \in \Lambda}$ is stationary starting from $\beta_i$.
Moreover, if $\mathfrak{m} \in \operatorname{Max} A$ and $I_{\overline{\alpha}} \not\subset \mathfrak{m}$, then $I_{\overline{\alpha}} A_{\mathfrak{m}} = A_{\mathfrak{m}}$.
Thus, there exists $\beta = \max_i\{\beta_i\} \geq \overline{\alpha}$ such that, for all $\gamma \geq \beta$, $I_\gamma A_{\mathfrak{m}} = I_\beta A_{\mathfrak{m}}$, *i.e.*,

$$(I_\gamma / I_\beta)_{\mathfrak{m}} \simeq I_\gamma A_{\mathfrak{m}} / I_\beta A_{\mathfrak{m}} = 0,$$

for all $\mathfrak{m} \in \operatorname{Max} A$. This yields $I_\gamma / I_\beta = 0$.
Hence, the chain is stationary and $A$ is Noetherian.

**Solution E. 15.16.** By definition, $I^{[n]} \subseteq I^n$.
1. The ideal $I^2$ is generated by all elements $fg$, with $f, g \in I$. Since

$$2fg = (f+g)^2 - f^2 - g^2 \in I^{[2]} \text{ for all } f,\, g \in I,$$

we have $I^2 \subseteq I^{[2]}$.
2. Let $B = \mathbb{Z}/(2)[x, y]$ and $I = (x,\, y)$. We have

$$I^2 = (x^2,\, xy,\, y^2) \supsetneq (x^2,\, y^2),$$

and we prove that $I^{[2]} = (x^2,\, y^2)$.
Since the characteristic is 2, $(ax+by)^2 = a^2x^2 + b^2y^2$ for all $a,\, b \in B$. Thus, $I^{[2]} \subseteq (x^2, y^2)$.

The opposite inclusion is obvious.
3. The ideal $I^n$ is generated by all monomials $x^n$, $x^{n-1}y, \dots, y^n$ of degree $n$. If we consider $(x + a_i y)^n = \sum_j \binom{n}{j} a_i^j y^j x^{n-j}$ for distinct values $a_1, \dots, a_{n+1}$ in $\mathbb{Q}$, we obtain $n+1$ relations, which can be written as

$$\begin{pmatrix} (x+a_1y)^n \\ (x+a_2y)^n \\ \vdots \\ (x+a_{n+1}y)^n \end{pmatrix} = M \begin{pmatrix} \binom{n}{0} x^n \\ \binom{n}{1} x^{n-1}y \\ \vdots \\ \binom{n}{n} y^n \end{pmatrix},$$

where

$$M = \begin{pmatrix} 1 & a_1 & a_1^2 & \dots & a_1^n \\ \vdots & \vdots & \vdots & & \vdots \\ 1 & a_{n+1} & a_{n+1}^2 & \dots & a_{n+1}^n \end{pmatrix}$$

is a Vandermonde matrix.
Since $\det M = \prod_{i \neq j} (a_i - a_j) \neq 0$, this matrix is invertible and, accordingly, $I^n \subseteq I^{[n]}$.

**Solution E. 15.17.** 1. From the chain $J \supseteq J^2 \supseteq \ldots \supseteq J^k \supseteq \ldots$, we obtain the ascending chain

$$I : J \subseteq I : J^2 \subseteq \ldots \subseteq I : J^k \subseteq \ldots.$$

Hence, $I : J^\infty = \bigcup_{n=1}^{\infty} I : J^n$ is an ideal of $A$.

2. At first we compute $I : J^\infty$ when $I$ is $\mathfrak{p}$-primary for some prime $\mathfrak{p}$.

If $J$ is an ideal not contained in $\mathfrak{p}$, then for every element $j \in J \setminus \mathfrak{p}$ we have $I : (j) = I$. Therefore,

$$I : J = I : (J, j) = (I : J) \cap (I : (j)) = (I : J) \cap I = I$$

and $I : J^\infty = I$.

If $J \subseteq \mathfrak{p}$, since $A$ is Noetherian, there exists $k$ such that $J^k \subseteq I$. Hence,

$$A = I : J^k \subseteq I : J^\infty \subseteq A.$$

Now, take $a, b \in A$, and consider a minimal primary decomposition

$$I = \bigcap_{\alpha \in \Lambda} \mathfrak{q}_\alpha.$$

Let $\mathfrak{p}_\alpha = \sqrt{\mathfrak{q}_\alpha}$, $\Lambda_b = \{\alpha \in \Lambda \colon b \notin \mathfrak{p}_\alpha\}$ and $\Lambda_{a,b} = \{\alpha \in \Lambda \colon a \notin \mathfrak{p}_\alpha \text{ or } b \notin \mathfrak{p}_\alpha\}$. Then, from the previous discussion we deduce

$$I : (b)^\infty = \bigcap_{\alpha \in \Lambda} \mathfrak{q}_\alpha : (b)^\infty = \bigcap_{\alpha \in \Lambda_b} \mathfrak{q}_\alpha \quad \text{and} \quad I : (a,b)^\infty = \bigcap_{\alpha \in \Lambda_{a,b}} \mathfrak{q}_\alpha.$$

Since the decomposition is minimal, it follows that $I : (b)^\infty = I : (a,b)^\infty$ if and only if $\Lambda_b = \Lambda_{a,b}$, *i.e.*, if and only if every associated prime of $I$ containing $b$ also contains $a$.

**Solution E. 15.18.** 1. We have $y^3 - 1 = (y-1)(y-\alpha)(y-\alpha^2)$, where $\alpha$ is a primitive third root of unity.
The only possible value of $y$ for a point $(x,y) \in \mathbb{V}_{\mathbb{Q}}(I)$ is $y = 1$. However, for $y = 1$ the corresponding value of $x$ must satisfy the relation $x^2 + x + 1 = 0$, which has no rational roots. Therefore, $\mathbb{V}_{\mathbb{Q}}(I) = \emptyset$.

We decompose $I \subseteq \mathbb{Q}[x,y]$ as follows:

$$\begin{aligned}
I &= (x^2 + xy + y^2,\, y - 1) \cap (x^2 + xy + y^2,\, y^2 + y + 1) \\
&= (x^2 + x + 1,\, y - 1) \cap (x^2 + xy + (-y - 1),\, y^2 + y + 1) \\
&= (x^2 + x + 1,\, y - 1) \cap (x - 1,\, y^2 + y + 1) \cap (x + y + 1,\, y^2 + y + 1) \\
&= \mathfrak{p}_1 \cap \mathfrak{p}_2 \cap \mathfrak{p}_3,
\end{aligned}$$

where $\mathfrak{p}_i \in \operatorname{Spec} \mathbb{Q}[x,y]$ for $i = 1,2,3$.
We can further decompose $I \subseteq \mathbb{C}[x,y]$ as the intersection of maximal ideals

$$\begin{aligned} I =& (x-\alpha,\, y-1) \cap (x-\alpha^2,\, y-1) \cap (x-1,\, y-\alpha) \\ & \cap (x-1,\, y-\alpha^2) \cap (x+\alpha+1,\, y-\alpha) \cap (x+\alpha^2+1,\, y-\alpha^2). \end{aligned}$$

From this decomposition we obtain

$$\mathbb{V}_{\mathbb{C}}(I) = \{\alpha,1),\, (\alpha^2,1),\, (1,\alpha),\, (1,\alpha^2),\, (\alpha^2,\alpha),\, (\alpha,\alpha^2)\}.$$

2. In the decomposition $I = \mathfrak{p}_1 \cap \mathfrak{p}_2 \cap \mathfrak{p}_3$, all ideals are prime.
Therefore,

$$I = \sqrt{I},\ \ \mathcal{N}(\mathbb{Q}[x,y]/I) = (\overline{0}), \ \ \text{and}\ \ \mathcal{D}(\mathbb{Q}[x,y]/I) = \overline{\mathfrak{p}}_1 \cup \overline{\mathfrak{p}}_2 \cup \overline{\mathfrak{p}}_3.$$

3. From the previous parts, the associated primes of $I$ are all maximal.
Hence, $\overline{\mathfrak{p}}_1$, $\overline{\mathfrak{p}}_2$, and $\overline{\mathfrak{p}}_3$ are the only primes of $\mathbb{Q}[x,y]/I$.
4. Localizing at any of the primes $\overline{\mathfrak{p}}_i$, we obtain $(\overline{\mathfrak{p}}_i)_{\overline{\mathfrak{p}}_i} = 0$, because for any element $\overline{a} \in \overline{\mathfrak{p}}_i$ there exists $\overline{s} \in (\mathbb{Q}[x,y]/I) \setminus \overline{\mathfrak{p}}_i$ such that $\overline{as} = \overline{0}$.
Thus, for all three primes, $(\mathbb{Q}[x,y]/I)_{\overline{\mathfrak{p}}_i}$ is a field.

**Solution E. 15.19.** 1. Since, by hypothesis, $I\,I^{-1} = A$, there exist $i_j \in I$ and $m_j \in I^{-1}$ such that

$$1 = \sum_{j=1}^{n} i_j m_j$$

for some positive integer $n$.
Therefore, for all $i \in I$, we have $i = \sum_{j=1}^{n} (im_j) i_j$, where $im_j \in A$ for all $j$.
Hence, $I$ is generated by the elements $i_j$.
2. By part 1, the homomorphism $\pi\colon A^n \longrightarrow I$, defined on the elements of the canonical basis of $A^n$ by $\pi(e_j) = i_j$, with $j = 1, \dots, n$, is surjective. Hence, the sequence of $A$-modules

$$0 \longrightarrow \operatorname{Ker}\pi \longrightarrow A^n \xrightarrow{\ \pi\ } I \longrightarrow 0,$$

is exact, and it is sufficient to show that the sequence splits.
Consider the map $s\colon I \longrightarrow A^n$ defined by $s(i) = \sum_{j=1}^{n} im_j e_j$. It is easy to see that $s$ is an $A$-module homomorphism such that $\pi \circ s = \mathrm{id}_I$, that is, $s$ is a section of $\pi$.

**Solution E. 15.20.** Since $I = \sqrt{I}$, we have

$$m = |\,\mathbb{V}(I)| = \dim_{\mathbb{C}} \mathbb{C}[x_1, \dots, x_n]/I.$$

By hypothesis, the monomials $\overline{1}, \overline{x_n} \dots, \overline{x_n}^{m-1}$ are a basis of $\mathbb{C}[x_1, \dots, x_n]/I$.
Therefore, for every $i < n$ there exist $b_{ij} \in \mathbb{C}$ such that $\overline{x_i} = \sum_{j=0}^{m-1} b_{ij} \overline{x_n}^j$.
Hence, the polynomials

$$x_i - p_i(x_n) = x_i - \sum_{j=0}^{m-1} b_{ij} x_n^j$$

belong to $I$ and $(x_1 - p_1, \ldots, x_{n-1} - p_{n-1}, p_n) \subseteq I$.

To prove the opposite inclusion, note that by using the Division Algorithm, every $f \in \mathbb{C}[x_1, \ldots, x_n]$ can be written as

$$f = \sum_{i=1}^{n-1} a_i(x_i - p_i) + r(x_n) \quad \text{for some } a_i \in \mathbb{C}[x_1, \ldots, x_n].$$

If $f \in I$, then $r(x_n) \in I \cap \mathbb{C}[x_n]$, and consequently,

$$I \subseteq (x_1 - p_1, \ldots, x_{n-1} - p_{n-1}, p_n).$$

Finally, since $p_i \in \mathbb{C}[x_n]$ for all $i$, and $\deg p_i < \deg p_n$ for $i \neq n$, the set

$$\{x_1 - p_1, \ldots, x_{n-1} - p_{n-1}, p_n\}$$

is the reduced Gröbner basis of $I$.

**Solution E. 15.21.** 1. Consider the commutative diagram

$$\begin{array}{ccccc} P & \xrightarrow{\psi} & M & & \\ {\scriptstyle f}\downarrow & & \downarrow{\scriptstyle \mathrm{id}_M} & & \\ P' & \xrightarrow[\psi']{} & M & \longrightarrow & 0, \end{array}$$

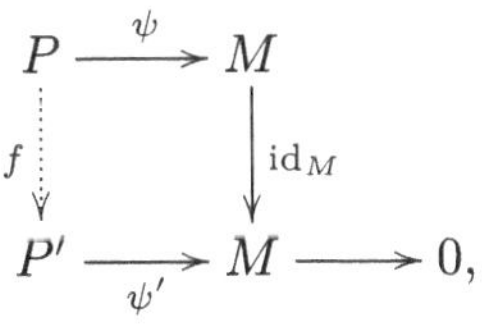

where $f$ exists because $\psi'$ is surjective and $P$ is projective.
If $p = \varphi(n) \in \operatorname{Im}\varphi = \operatorname{Ker}\psi$, then

$$0 = (\mathrm{id}_M \circ \psi)(p) = (\psi' \circ f)(p), \quad \textit{i.e.}, \quad f(p) \in \operatorname{Ker}\psi' = \operatorname{Im}\varphi'.$$

Hence, $f_{|\varphi(N)}$ has values in $\varphi'(N')$, and we can define the homomorphism $g\colon N \longrightarrow N'$ as the map that associates to every $n \in N$ the unique element $g(n) \in N'$ such that $f(\varphi(n)) = \varphi'(g(n))$. Since $\varphi'$ is injective, $g$ is well-defined. The conclusion follows from a straightforward application of the Snake Lemma to the commutative diagram

$$\begin{array}{ccccccccc} 0 & \longrightarrow & N & \xrightarrow{\varphi} & P & \xrightarrow{\psi} & M & \longrightarrow & 0 \\ & & \downarrow{\scriptstyle g} & & \downarrow{\scriptstyle f} & & \downarrow{\scriptstyle \mathrm{id}_M} & & \\ 0 & \longrightarrow & N' & \xrightarrow{\varphi'} & P' & \xrightarrow{\psi'} & M & \longrightarrow & 0. \end{array}$$

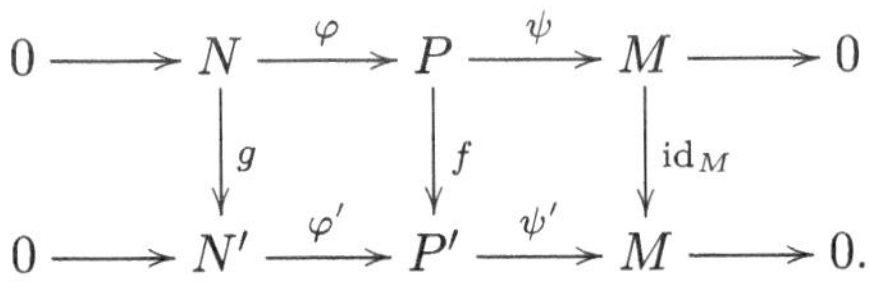

2. Define a sequence

$$0 \longrightarrow N \xrightarrow{\lambda} P \oplus N' \xrightarrow{\eta} P' \longrightarrow 0,$$

where $\lambda(n) = (\varphi(n), g(n))$ and $\eta(p, n') = f(p) - \varphi'(n')$.
Clearly, $\lambda$ and $\eta$ are homomorphisms and $\lambda$ is injective because $\varphi$ is injective. Moreover,

$$(\eta \circ \lambda)(n) = \eta(\varphi(n), g(n)) = f(\varphi(n)) - \varphi'(g(n)) = 0.$$

Therefore, $\operatorname{Im} \lambda \subseteq \operatorname{Ker} \eta$.

To prove the opposite inclusion, let $(p, n') \in \operatorname{Ker} \eta$. Then,

$$f(p) = \varphi'(n') \quad \text{and} \quad 0 = \psi'(\varphi'(n')) = \psi'(f(p)) = \mathrm{id}_M(\psi(p)).$$

Hence, $p \in \operatorname{Ker} \psi = \operatorname{Im} \varphi$ and, since $\varphi$ is injective, there exists a unique $n \in N$ such that $\varphi(n) = p$. Thus,

$$\varphi'(n') = f(p) = f(\varphi(n)) = \varphi'(g(n)) \quad \text{and} \quad n' = g(n)$$

since $\varphi'$ is injective.
Therefore,

$$(p, n') = (\varphi(n), g(n)) = \lambda(n) \in \operatorname{Im} \lambda.$$

Finally, let $p' \in P'$. Then, $\psi'(p') \in M$ and there exists $p \in P$ such that $\psi(p) = \psi'(p')$.
Therefore,

$$\psi'(f(p)) = \psi(p) = \psi'(p') \quad \text{and} \quad f(p) - p' \in \operatorname{Ker} \psi' = \operatorname{Im} \varphi'.$$

Hence, there exists $n' \in N'$ such that $f(p) - p' = \varphi'(n')$, *i.e.*,

$$p' = f(p) - \varphi'(n') = \eta(p, n')$$

and $\eta$ is surjective.
Since $P'$ is projective and we have proven that the sequence is exact, there is an isomorphism $P \oplus N' \simeq N \oplus P'$, as desired.

# References

[1] W.W. Adams and P. Loustaunau: An introduction to Gröbner bases. Graduate Studies in Mathematics. American Mathematical Society, Providence, RI (1994).

[2] A. Altman and S. Kleiman: A term of commutative algebra. Worldwide Center of Mathematics (2017).

[3] M.F. Atiyah and I.G. Macdonald: Introduction to commutative algebra. Addison-Wesley Series in Mathematics. Westview Press, Boulder, CO (1969).

[4] H. Cartan and S. Eilenberg. Homological Algebra. Princeton Landmarks in Mathematics. Princeton University Press, Princeton, NJ (1956).

[5] D.A. Cox, J. Little, D. O'Shea: Ideals, Varieties, and Algorithms. Undergraduate Texts in Mathematics. Springer (2015).

[6] S. Lang: Algebra. Graduate Texts in Mathematics, vol. 211. Springer-Verlag, New York (2002).

[7] R. Mines and F. Richman and W. Ruitenberg: A course in constructive algebra. Universitext. Springer New York, NY (1988).

[8] M. Reid: Undergraduate Commutative Algebra. London Mathematical Society Student Texts, vol. 29. Cambridge University Press, Cambridge (1995).



A. Bandini et al., *Commutative Algebra through Exercises*, UNITEXT 159,
https://doi.org/10.1007/978-3-031-56910-4

# Index

## Symbols





A. Bandini et al., *Commutative Algebra through Exercises*, UNITEXT 159,
https://doi.org/10.1007/978-3-031-56910-4

## G



## H



## I



## J



## K



## L

## M



## N



## O



## P

## R



## S



## T